Arsenic in Plants

Arsenic in Plants

Uptake, Consequences and Remediation Techniques

Edited by

Prabhat Kumar Srivastava
Department of Botany
KS Saket PG College
Ayodhya, Uttar Pradesh, India

Rachana Singh
Department of Botany
University of Allahabad
Prayagraj, Uttar Pradesh, India

Parul Parihar
Department of Botany
University of Allahabad
Prayagraj, Uttar Pradesh, India
and
Department of Bioscience and Biotechnology
Banasthali Vidyapith
Rajasthan, India

Sheo Mohan Prasad
Department of Botany
University of Allahabad
Prayagraj, Uttar Pradesh, India

WILEY

This edition first published 2023

Registered Offices
John Wiley & Sons, Inc., 111 River Street, Hoboken, NJ 07030, USA
John Wiley & Sons Ltd, The Atrium, Southern Gate, Chichester, West Sussex, PO19 8SQ, UK

Editorial Office
The Atrium, Southern Gate, Chichester, West Sussex, PO19 8SQ, UK

For details of our global editorial offices, customer services, and more information about Wiley products visit us at www.wiley.com.

Wiley also publishes its books in a variety of electronic formats and by print-on-demand. Some content that appears in standard print versions of this book may not be available in other formats.

Library of Congress Cataloging-in-Publication Data applied for
Hardback ISBN: 9781119791423

Cover Design: Wiley
Cover Image: Courtesy of Sarvesh Kumar Singh

Set in 9.5/12.5pt STIXTwoText by Straive, Pondicherry, India

Contents

List of Contributors

Sajid Rashid Ahmad
College of Earth and Environmental Sciences
University of the Punjab
Lahore, Pakistan

Duraid K.A. Al-Taey
Department of Horticulture
Faculty of Agriculture
University of AL-Qasim Green
AL-Qasim, Babylon, Iraq

Samina Aslam
Department of Chemistry
Women University Multan
Multan, Punjab, Pakistan

Ummey Aymen
Department of Botany
Lovely Professional University
Phagwara, Punjab, India

Irshad Bibi
Institute of Soil and Environmental Sciences
University of Agriculture Faisalabad
Faisalabad, Pakistan

Asok K. Biswas
Plant Physiology and Biochemistry Laboratory
Centre for Advanced Study
Department of Botany
University of Calcutta
Kolkata, West Bengal, India

Isabel Caçador
MARE – Marine and Environmental Sciences Centre
Faculty of Sciences
University of Lisbon
Lisbon, Portugal

and

Department of Plant Biology
Faculty of Sciences
University of Lisbon
Lisbon, Portugal

Debasis Chakrabarty
Biotechnology and Molecular Biology Division
CSIR-National Botanical Research Institute
Lucknow, Uttar Pradesh, India

and

Academy of Scientific and Innovative Research (AcSIR)
Ghaziabad, Uttar Pradesh, India

Moumita Chatterjee
Department of Botany
V. Sivaram Research Foundation
Bengaluru, Karnataka, India

Charu Chaturvedi
Plant Molecular Biology Laboratory
Department of Botany
Dayanand Anglo-Vedic (PG) College
Chhatrapati Shahu Ji Maharaj University
Kanpur, Uttar Pradesh, India

Nilanjana Roy Chowdhury
School of Environmental Studies
Jadavpur University
Kolkata, West Bengal, India

Antara Das
School of Environmental Studies
Jadavpur University
Kolkata, West Bengal, India

Ayan De
School of Environmental Studies
Jadavpur University
Kolkata, West Bengal, India

Letúzia Maria de Oliveira
Center for Water Resources
College of Agriculture and Food Sciences
Florida A&M University
Tallahassee, FL, USA

Daljeet Singh Dhanjal
Department of Biotechnology
School of Bioengineering and Biosciences
Lovely Professional University
Phagwara, Punjab, India

Bernardo Duarte
MARE – Marine and Environmental Sciences Centre
Faculty of Sciences
University of Lisbon
Lisbon, Portugal

Neeraj Kumar Dubey
Department of Botany
Rashtriya Snatkottar Mahavidyalaya
Jaunpur, Uttar Pradesh, India

Mujahid Farid
Department of Environmental Sciences
University of Gujrat
Gujrat, Pakistan

Sumaya Farooq
School of Bioengineering and Biosciences
Lovely Professional University
Phagwara, Punjab, India

Sharmistha Ganguly
Department of Botany
Ranchi University
Ranchi, Jharkhand, India

Neelam Gautam
Biotechnology and Molecular Biology Division
CSIR-National Botanical Research Institute
Lucknow, Uttar Pradesh, India

and

Academy of Scientific and Innovative Research (AcSIR)
Ghaziabad, Uttar Pradesh, India

Kavita Ghosal
Department of Botany
P.D. Women's College
Jalpaiguri, West Bengal, India

Aysegul Yagmur Goren
Department of Environmental Science and Engineering
Izmir Institute of Technology
Izmir, Turkey

Govind Gupta
Environmental Science Discipline
Department of Chemistry
Manipal University Jaipur
Rajasthan, India

Sunil Kumar Gupta
CAS Key Laboratory of Tropical Forest Ecology
Xishuangbanna Tropical Botanical Garden
Chinese Academy of Sciences
Mengla, Yunnan, People's Republic of China

Israr Masood ul Hasan
State Key Laboratory for Modification of Chemical Fibers and Polymer Materials
College of Environmental Science and Engineering
Donghua University
Shanghai, China

Emad Jafarzadeh
Department of Toxicology and Pharmacology
Faculty of Pharmacy
Tehran University of Medical Sciences (TUMS)
Tehran, Iran

Asim Jilani
Center of Nanotechnology
King Abdul Aziz University
Jeddah, Saudi Arabia

Madhurima Joardar
School of Environmental Studies
Jadavpur University
Kolkata, West Bengal, India

Anuja Joseph
School of Environmental Studies
Jadavpur University
Kolkata, West Bengal, India

Dhriti Kapoor
Department of Botany
School of Bioengineering and Biosciences
Lovely Professional University
Phagwara, Punjab, India

Harry Kaur
Department of Biosciences and Bioengineering
Indian Institute of Technology
Roorkee, Uttarakhand, India

Jabbar Khan
Environmental Science Discipline
Department of Chemistry
Manipal University Jaipur
Rajasthan, India

***Marya Khan*†**
Department of Botany
Lovely Professional University
Phagwara, Punjab, India

Maria Kidwai
Biotechnology and Molecular Biology Division
CSIR-National Botanical Research Institute
Lucknow, Uttar Pradesh, India

Mehmet Kobya
Department of Environmental Engineering
Gebze Technical University
Kocaeli, Turkey

and

Department of Environmental Engineering
Kyrgyz-Turkish Manas University
Bishkek, Kyrgyzstan

Tanzeela Kokab
College of Earth and Environmental Sciences
University of the Punjab
Lahore, Pakistan

Arun Kumar
Center of Advanced Study in Botany
Institute of Science
Banaras Hindu University
Varanasi, Uttar Pradesh, India

Sanjay Kumar
Institute of Multidisciplinary Research for Advanced Materials (IMRAM)
Tohoku University
Sendai, Japan

***Vijay Kumar*†**
Department of Chemistry Central
Ayurveda Research Institute Jhansi
Uttar Pradesh, India

Marta Irene Litter
IIIA (CONICET-UNSAM)
Instituto de Investigación e Ingeniería Ambiental
Escuela de Hábitat y Sostenibilidad
Universidad Nacional de General San Martín
Buenos Aires Province, Argentina

Muzaffar Majid
College of Earth and Environmental Sciences
University of the Punjab
Lahore, Pakistan

Naina Marwa
Plant Ecology and Climate Change Science Division
CSIR-National Botanical Research Institute
Lucknow, Uttar Pradesh, India

and

Department of Botany
University of Lucknow
Lucknow, Uttar Pradesh, India

Mohammad Mehdizadeh
Department of Agronomy and Plant Breeding
University of Mohaghegh Ardabili
Ardabil, Iran

Shraddha Mishra
Department of Biological Sciences
Birla Institute of Technology and Science
Pilani, Rajasthan, India

Deepanjan Mridha
School of Environmental Studies
Jadavpur University
Kolkata, West Bengal, India

Waseem Mushtaq
Allelopathy Laboratory
Department of Botany
Aligarh Muslim University
Aligarh, Uttar Pradesh, India

Lucy Ngatia
Center for Water Resources
College of Agriculture and Food Sciences
Florida A&M University
Tallahassee, FL, USA

Nabeel Khan Niazi
Institute of Soil and Environmental Sciences
University of Agriculture Faisalabad
Faisalabad, Pakistan

Subhamita Sen Niyogi
Anushka Soham Purified Water Manufacturing Co. Private Limited
Hooghly, West Bengal, India

Anahita Omidi
Department of GIS and Remote Sensing
Faculty of Geography
University of Tehran
Tehran, Iran

Neha Pandey
Department of Botany
CMP PG College
Prayagraj, Uttar Pradesh, India

Vivek Pandey
Plant Ecology and Climate Change Science Division
CSIR-National Botanical Research Institute
Lucknow, Uttar Pradesh, India

Parul Parihar
Department of Botany
University of Allahabad
Prayagraj, Uttar Pradesh, India

and

Department of Bioscience and Biotechnology
Banasthali Vidyapith
Rajasthan, India

Sílvia Pedro
MARE – Marine and Environmental Sciences Centre
Faculty of Sciences
University of Lisbon
Lisbon, Portugal

Mamta Pujari
Department of Botany
School of Bioengineering and Biosciences
Lovely Professional University
Phagwara, Punjab, India

Amit Prakash Raghuvanshi
Plant Molecular Biology Laboratory
Department of Botany
Dayanand Anglo-Vedic (PG) College
Chhatrapati Shahu Ji Maharaj University
Kanpur, Uttar Pradesh, India

Praveen C. Ramamurthy
Interdisciplinary Centre for Water Research (ICWaR)
Indian Institute of Science
Bengaluru, Karnataka, India

Varsha Rani
Department of Pharmaceutical Engineering and Technology
Indian Institute of Technology
Banaras Hindu University
Varanasi, Uttar Pradesh, India

Iravati Ray
School of Environmental Studies
Jadavpur University
Kolkata, West Bengal, India

Tarit Roychowdhury
School of Environmental Studies
Jadavpur University
Kolkata, West Bengal, India

Gauri Saxena
Department of Botany
University of Lucknow
Lucknow, Uttar Pradesh, India

Muhammad Bilal Shakoor
College of Earth and Environmental Sciences
University of the Punjab
Lahore, Pakistan

Pooja Sharma
Department of Environmental Microbiology
School for Environmental Sciences
Babasaheb Bhimrao Ambedkar Central University
Lucknow, Uttar Pradesh, India

Hamidreza Sharifan
Department of Natural Science
Albany State University
Albany, GA, USA

Riddhi Shrivastava
Environmental Science Discipline
Department of Chemistry
Manipal University Jaipur
Rajasthan, India

and

Department of Chemistry
Poornima College of Engineering
Jaipur, Rajasthan, India

Dhouha Belhaj Sghaier
UR MaNE
Faculté des Sciences de Bizerte
Université de Carthage
Tunis, Tunisia

Shahida Anusha Siddiqui
Department of Biotechnology and Sustainability
Technical University of Munich (TUM)
Straubing, Bavaria, Germany

and

DIL e.V.–German Institute of Food Technologies
D-Quakenbrück, Lower Saxony, Germany

Palin Sil
Plant Physiology and Biochemistry Laboratory
Centre for Advanced Study
Department of Botany
University of Calcutta
Kolkata, West Bengal, India

Anita Singh
Center of Advanced Study in Botany
Institute of Science
Banaras Hindu University
Varanasi, Uttar Pradesh, India

Joginder Singh
Department of Biotechnology
School of Bioengineering and Biosciences
Lovely Professional University
Phagwara, Punjab, India

Nandita Singh
Plant Ecology and Climate Change Science Division
CSIR-National Botanical Research Institute
Lucknow, Uttar Pradesh, India

Naveen Kumar Singh
Environmental Science Discipline
Department of Chemistry
Manipal University Jaipur
Rajasthan, India

Rachana Singh
Department of Botany
University of Allahabad
Prayagraj, Uttar Pradesh, India

Simranjeet Singh[†]
Interdisciplinary Centre for Water Research (ICWaR)
Indian Institute of Science
Bengaluru, Karnataka, India

Surendra Pratap Singh
Plant Molecular Biology Laboratory
Department of Botany
Dayanand Anglo-Vedic (PG) College
Chhatrapati Shahu Ji Maharaj University
Kanpur, Uttar Pradesh, India

Debapriya Sinha
School of Environmental Studies
Jadavpur University
Kolkata, West Bengal, India

Dwaipayan Sinha
Department of Botany
Government General Degree College
West Bengal, India

Noomene Sleimi
UR MaNE
Faculté des Sciences de Bizerte
Université de Carthage
Tunis, Tunisia

Zahra Souri
Department of Biology
Faculty of Science
Razi University
Kermanshah, Iran

Koko Tampubolon
Program Study of Agrotechnology
Faculty of Agriculture and Animal Husbandry
Universitas Tjut Nyak Dhien
Medan, Sumatera Utara, Indonesia

Ankita Thakur
Department of Botany
School of Bioengineering and Biosciences
Lovely Professional University
Phagwara, Punjab, India

Anuj Kumar Tiwari
Department of Botany
Bhavan's Mehta Mahavidyalaya
Kaushambi, Uttar Pradesh, India

Madhu Tiwari
Biotechnology and Molecular Biology Division
CSIR-National Botanical Research Institute
Lucknow, Uttar Pradesh, India

Sanjay Kumar Verma
Department of Biological Sciences
Birla Institute of Technology and Science
Pilani, Rajasthan, India

Pradeep Kumar Yadav
Center of Advanced Study in Botany
Institute of Science
Banaras Hindu University
Varanasi, Uttar Pradesh, India

Gerald Zvobgo
Department of Crop Productivity and Molecular Technology
Tobacco Research Board
Harare, Zimbabwe

Preface

Arsenic contamination in agricultural lands has become a global problem extending from Middle-East countries, South Asian countries including Afghanistan and Pakistan, India, Bangladesh, South-East Asian countries, China, Japan, Canada, USA, Mexico, Brazil, Argentina, Chile, New Zealand to European and African countries. This metalloid severely affects the plant as well as the human system by interrupting important physiological and molecular processes. The food chain is infiltrated by arsenic through arsenic-loaded groundwater and industrial and municipal wastewater contaminated with arsenic used for irrigation purposes. Arsenic also penetrates the food chain through the usage of fertilizers and herbicides (arsenicals) in agricultural fields. Arsenic severely intoxicates plants via various physiological and biochemical anomalies and reduces their growth and development. Toxicity symptoms range from biomass reduction to morphological impairments leading to the loss in fruit and grain yield that culminates into the complete death of the plants. Severe toxic effects of arsenic change the concentration, accumulation, and translocation of nutrient elements in plants, inhibit seed germination, and increase arsenic levels in the edible parts of vegetables.

This book gives an overview of arsenic, prominently covers the occurrence of arsenic in our environment, usage of arsenicals in crop fields, its chemistry, speciation, its transportation and metabolism in plants, phytotoxicity, i.e. impact on plant metabolism, alteration in different plant groups, from plants' overall structure, their physiology up to the changes at their ultrastructural level; and mechanisms involved therein and interaction/interruption with phytohormones and metabolic processes and future perspectives. The book covers the morphological, anatomical, and other quantitative and qualitative traits of plants including their physiological, biochemical, and molecular responses under arsenic stress. The impact of exogenous phytohormones and growth-regulating substances and mineral nutrients has been covered. It discusses *-omics* approaches, i.e. regulation at genomic, transcriptomic, proteomic, ionomic, and metabolomic levels adapted by plants to combat this stress condition and the models used to explain these adaptations.

This book brings forth ideas being explored by scientists and environmentalists to overcome this menace. This book emphasizes the differences in the mechanism of tolerance in hyper-accumulator and non-accumulator plants. It discusses the management of arsenic contamination in the soil-plant continuum, major arsenic remediation techniques including the removal of arsenic from soil and water through physical and biological methods.

Thus, this book is a comprehensive compilation of studies to date and is an endeavor to bridge the gap between the research from the past to the current time. This book will serve as a reference book for environmentalists, toxicologists, and risk assessors. The compilation of various studies in the form of an edited book enriches the existing knowledge about arsenic pollution and opens newer avenues to be exercised. The students and scholars would find many studies, researches, reviews of literature, views, and opinions in one book.

This book is the result of an arduous effort of many scholars working in different parts of the world along with all four editors. All the editors thankfully acknowledge their contributions. All editors also gratefully acknowledge the team at JohnWiley & Sons Ltd. that made possible the proposed book in its present form. We hope that this book will remain relevant for the upcoming many years for the students of environmental sciences, stress physiology, agronomy, life sciences, and crop sciences at the university level.

Prabhat Kumar Srivastava
Rachana Singh
Parul Parihar
Sheo Mohan Prasad

1

An Introduction to Arsenic: Sources, Occurrence, and Speciation

Jabbar Khan[1], *Govind Gupta*[1], *Riddhi Shrivastava*[1,2], *and Naveen Kumar Singh*[1]

[1] *Environmental Science Discipline, Department of Chemistry, Manipal University Jaipur, Rajasthan, India*
[2] *Department of Chemistry, Poornima College of Engineering, Jaipur, Rajasthan, India*

1.1 Introduction

Naturally, arsenic is present in rocks and water in the environment, and its concentration depends on geological and anthropogenic activities. Generally, the concentrations of arsenic in noncontaminated soils are usually less than 10 mg kg^{-1}. Arsenic (As) contamination has become a worldwide problem due to its toxicity and increasing contamination of soil, water, and crops around the world. It occurs as a result of geological processes and anthropogenic activities. Arsenic is a toxic metal that occurs by a natural and anthropogenic process such as the burning of fossil fuels, mining, and uses of agrochemicals (Mandal and Suzuki 2002; Bissen and Frimmel 2003). Excess arsenic in water and soil accumulates in plants and leads to food chain contamination. Arsenic causes toxic effects in plants and carcinogenic effects in human beings through water, soil, and food contamination (Zhao et al. 2010; Naujokas et al. 2013). Litter et al. (2010) reported that regular arsenic consumption through food and water causes arsenicosis, affects the central nervous system detrimentally, and causes hyperkeratosis, hepatic damage, skin cancer, hair fall, etc. Chakraborty et al. (2018) investigated the contamination of arsenic in groundwater and food materials in different regions of the Ganga River Basin (GRB), which includes Nepal, Bangladesh, and Tibet, where arsenic concentration was above the permissible limit of the World Health Organization's (WHO) standards. Anderson and Bruand (1991) reported the position of arsenic in the Group 15 of the periodic table, and it exists in the environment with the combination of oxygen, chlorine, and sulfur. Saeki et al. (2000) reported that arsenic has long been toxic and teratogenic (risk for a birth defect in a baby). In soil, dust, rocks, and air, arsenic is present in small quantities. In many industrial goods and processes, arsenic is used. Therefore, through waste and environmental pollution, arsenic becomes a major contaminant (Berg et al. 2001; Reboredo et al. 2019). The mobilization and occurrence of

Arsenic in Plants: Uptake, Consequences and Remediation Techniques, First Edition.
Edited by Prabhat Kumar Srivastava, Rachana Singh, Parul Parihar, and Sheo Mohan Prasad.

heavy metals in the environment include various procedures such as soil weathering, rock and coal, biological processes, volcanic processes, etc. Similarly, high amount of urban waste, burning of fossil fuels, mining, use of fertilizers, biocides, sewage sludge, seed desiccants, alloys, and anthropogenic activities account for the widespread dispersion of arsenic (Smedley and Kinniburgh 2002). However, this causes adverse effects such as atmospheric accumulation, especially in crops, such as increased cancer risk, teratogenicity, and mutagenicity (Farid et al. 2003). Due to high-arsenic concentrations, a cereal yield decrease of about 20% was observed. The potential to accumulate arsenic and its relative toxicity increase the threat to the ecosystem. Reducing As and other heavy metal contamination in crops, with particular emphasis on horticultural products, has been one of the key objectives of the research over the past decades (Wilson et al. 2014; Mancinelli et al. 2019). Due to overexploitation of water, more arsenic may be released into the aquifers as arsenopyrite minerals oxidized by exposure to oxygen-rich water. Therefore, due to oxidation of arsenic-containing rocks and release, more arsenic concentration is reported in the groundwater from Bangladesh (Mandal et al. 1996; Nickson et al. 1998). Chen et al. (1992) reported that arsenic is known as one of the most significant environmental contaminants due to its toxic effects on human health. Arsenic toxicity is due to the replacement of phosphate by arsenic (+5), the protein thiol groups' affinity of arsenic (+3), and the cross-linking of protein–DNA and DNA–DNA. Arsenic contamination is a regular occurrence in many countries due to its pervasiveness in the environment, and millions of people have been continuously exposed to arsenic through geological contamination of potable water (International Agency for Research on Cancer 2004). Arsenic contamination in marine habitats is primarily due to the indiscriminate disposal of effluents containing high arsenic from household and industrial discharge (Huysmans and Frankenberger 1990; Filali et al. 2000). Aquatic plants growing in contaminated water may accumulate arsenic, causing a health risk to animals and humans through the food chain. The concentration of arsenic in seafood and fish can be high due to accumulation and biomagnification (International Agency for Research on Cancer 2004). Arsenic is a notorious neurotoxin that affects the nervous system in the exposed species. Arsenic can be tolerated to a certain extent in humans because it is eliminated from the body through urine, stool, skin, hair, nails, and breathing. Arsenic is accumulated in tissues as a result of excessive exposure affecting cellular functions and metabolism (Mukhopadhyay et al. 2002). Aside from toxicity, arsenic's inhibitory effects are influenced by background concentrations and the type of organism (Birnboim and Doly 1979). In plant and animal tissues, it can be actively sequestered. Arsenicals have been used medicinally for a long time and were among the first chemotherapeutic agents to be used in the treatment of infectious diseases such as syphilis and trypanosomiasis. Salvarsan, an arsenic-based drug, was introduced by Paul Ehrlich as a "magic bullet" in syphilis treatment (Waxman and Anderson 2001).

1.2 Status of Arsenic Contamination Around the World

The geogenic and anthropogenic degradation of persistent toxic substances poses significant threats to the environment (Nordstrom 2002; Hoang et al. 2010). Arsenic contamination in groundwater is reported in various countries, including Argentina, Bangladesh, Chile,

India, Mexico, Hungary, and Taiwan (Nikolaidis et al. 2004; Yin et al. 2011a). Contamination of arsenic in water and soil is a worldwide issue due to its carcinogenic effects on human health. Contamination of arsenic in groundwater is prevalent in different areas of West Bengal, India (Tripathi et al. 2007; Chakraborti et al. 2009; Singh et al. 2016). The Ganga-Brahmaputra-Meghna plain is presently the world's most acutely contaminated area with the arsenic concentration >4000 $\mu g l^{-1}$ (Rahman et al. 2006). Similarly, arsenic concentration above the WHO standard of 10 $\mu g l^{-1}$ has been reported in other regions of India, including Uttar Pradesh, Bihar, Jharkhand, and Haryana along with West Bengal (Chakraborti et al. 2004; Mishra et al. 2016; Gupta et al. 2017; Satyapal et al. 2018). There is a high incidence of poverty, in addition to the problems of handling environmental pollution, global warming, etc. (Usfar et al. 2010; McIntyre et al. 2013). Almost every country in the world is affected by environmental pollution, and exposure to environmental contaminants is implicated in the pathogenesis of cardiovascular disease, diabetes, and obesity (Carpenter 2006). The latest WHO data has shown that China and India are the two most affected countries in terms of air pollution, with about 6.5 million related deaths per year (WHO 2016). These high rates of death can be expected to increase further. Non-airborne contamination (e.g. polluted food and water) is more difficult to assess. Despite our awareness about the adverse health effects due to exposure to pollution, levels of environmental contaminants, particularly in developing countries, have continued to increase over the past few years. Over the past decade, great strides have been made to significantly reduce the amount of emission generated, while a complete elimination is desirable, so it is not a practical or feasible objective (Fulekar 2010).

1.3 Arsenic in the Aquatic and Terrestrial Environment

Arsenic uptake, translocation, accumulation, and its toxicity are largely influenced by its oxidation state. Arsenite (As(III)) is the main form of arsenic (Abedin et al. 2002a). It is more mobile and more toxic under flooded rice field conditions. Like anaerobic conditions in flooded rice fields, asphyxiated conditions exist in the estuarine ecosystem of Sundarbans (India) owing to high salinity, tidal flushing, and waterlogging. Arsenate (As(V)) and arsenite (As(III)) are the most common As species in soils that are controlled by chemical and microbial transformations (Chen et al. 2017). Arsenic mobility and availability to plants can also be affected by soil microbes and algae (Neubauer et al. 2007; Meng et al. 2011; Yin et al. 2011b; Wang et al. 2013; Kohfahl et al. 2016).

Plant growth-promoting bacteria (PGPB) may improve plant competitiveness and responses to metal stress during phytoremediation (Zhang et al., 2008). Recent advances in manufacturing and technology have contributed to the degradation of the natural environment in response to increasing human populations, representing a significant threat to living species (Deng et al. 2007). Aquatic water bodies like rivers and lakes are frequently polluted with metals and metalloids, along with their surrounding terrestrial areas (Fu et al. 2014). Nonessential metals and metalloids are harmful to living organisms and ecosystems (Ackerman et al. 2016). Especially, birds are very susceptible to higher concentrations of metals. According to Burger and Gochfeld (2005), birds are always not high on the food chain but often cover large food and breeding distances and are thus exposed more

(Shahbaz et al. 2013). High levels of toxic trace metals and metalloids in the environment and the associated health effects have become a global concern (Wu et al. 2016). These metals are essentially released from mining, smelting, photovoltaic cells, pigments, batteries, synthetic plastics, insecticides, and leather industries into the environment (Lucia et al. 2010; Yang et al. 2018). They avoid photolytic, biological, and chemical degradation and thus survive for a longer period (Ali and Khan 2018). The trend in bioaccumulation through the food chain is rising with metals and As contamination in water and soil. Organisms at higher trophic levels are more susceptible since the toxicity increases with the increase in trophic level (Cai et al. 2009: Green et al. 2010).

1.4 Absolute Bioavailability and Bioaccessibility of As in Plants and Agronomic Systems

Humans get exposed to arsenic through food from plant and animal sources or with ingestion of arsenic-contaminated drinking water (Smith et al. 2002a; Kana et al. 2018). Rice and fish are considered to be the most important sources of food contamination with arsenic. Arsenic contamination of rice reported with more than hundreds $ng\,g^{-1}$ causes chronic poisoning in the regular consuming human population (Williams et al. 2007; Chen and Chen 2014; Shigehiro et al. 2015). According to Cascio et al. (2011), inorganic arsenic (iAs) exposure has been linked to rice intake in the human population of Bangladesh. Wheat plants were grown in England in a field experiment where the soil contained 12 $mg\,As\,kg^{-1}$ and the wheat grain contained 2–17 $\mu g\,As\,kg^{-1}$ (Zhao et al. 2007). Inorganic arsenic (iAs) can enter into higher trophic levels through bioaccumulation and biomagnification. In the fragile ecosystems where the biota is adapted to persist in particular abiotic stress, As contamination can result in changes in ecosystem function and services.

1.5 Factors Determining Arsenic Speciation and Bioavailability in Soil

1.5.1 Effect of Redox Potential (Eh) and pH

The interface of the root and chemical composition of soil depend on the flow of water, ion diffusion and convention methods, plant absorption, pH, root exudation, etc. and can contribute to the phenomenon of precipitation, favoring immobilization as well. Arsenic occurs in two oxidation states in nature, i.e. As^{3+} and As^{5+} (Sharma et al. 2018). In an aerobic environment, it is present as arsenate (As^{5+}), and under anaerobic conditions, it is present as arsenite (As^{3+}) (Ronci et al. 2017) where As^{3+} is more toxic as compared to As^{5+}. Due to its oxidation state, its mobility can be controlled by redox potential (Kumar et al. 2019). Arsenic oxidation states, alterations in redox potential, and pH have direct effects on its solubility in soil. At higher redox levels in soil, arsenic solubility decreases, and it is present as As^{5+}. The E_h value decreases with the decrease in pH value, as on highly oxidizing conditions at pH 3, the E_h value will be +250 to +450 mV, and at pH 11, the value will be −100 to +75 mV. As the pH value directly affects the functional groups of different metal ions, it plays a vital role in the adsorption of As^{3+} and As^{5+} (Tang et al. 2020). Adsorption of arsenic by metal oxyhydroxides depends on pH value (Smedley et al. 2002). pH value affects metal ions to react with the OH^- groups to complete their electron shell. Due to this, OH^- group can bind or release H^+ by developing a charge on the surface. At the pH range of 6.5–8.5, arsenic shows high sensitivity to mobilize its ions. Due to this, it is a very peculiar element among all the metalloids (Kim et al. 2002). At pH > 6.9, under oxidizing conditions, As(V) predominates over As(III), in the form of $HAsO_4^{2-}$ where predominating species will be H_3AsO_4, whereas in extremely basic conditions, the species will be AsO_4^{3-} (Al-Abed et al. 2007). On the other hand, at pH < 9.2, under reducing conditions, the neutral species of As(III) will be predominant. In aquatic systems, the mineral surface is already covered with OH^- group that releases H^+ with the change of pH value. The adsorption capacity of these aquatic minerals depends on the presence of OH, OH^{-2}, and O^- functional groups. The adsorption of As(V) decreases as the pH rises on the adsorbent's basic isoelectric point, but the adsorption of As(III) is less pH-dependent and often achieves its highest value at pH 8–9. At the pH of about 7–10, As(III) preferably sorbed to hydrous oxide with an optimal pH range of 7, whereas at the pH of about 4–7, As(V) is preferably sorbed with an optimal pH range of about 4 (Wolthers et al. 2005). It has also been found that arsenic sorption on clay minerals depends on the pH. On the other hand, adsorption of As on clean quartz is very less at pH of above 3. As the pH rises, the desorption of arsenic occurs from mineral surfaces to groundwater (Hatje et al. 2003). Adsorption of As^{3+} and As^{5+} to humic acid also depends on the rate of pH. At pH 5.5 the adsorption rate of As^{5+} on humic substance is maximum, whereas for As^{3+} adsorption rate is high at a pH of about 8.5.

1.5.2 Interactions with Al, Fe, and Mn Oxides and Oxyhydroxides

Uptake, translocation, toxicity, and accumulation of arsenic are largely influenced by its oxidation state. However, the oxidative environment in the rhizosphere alters the speciation of arsenic on the root surface by Fe plaque formation, interaction with Mn, and transformation to As(V). As a result of effective As fixation and detoxification, Fe plaque has formed in the rhizosphere of salt marsh plants due to O_2 leaking through the aerenchyma. Arsenic, which is found in sulfide minerals, is directly released to the environment by the processes of mobilization, dissolution, redox reaction, and adsorption–desorption process (Zhang et al. 2018). The arsenic mobility is determined by parent mineral form, mobilization, and oxidation state.

Zecchin et al. (2019) reported As(III) and As(V) as the two most common and prevalent arsenic species in the environment. As(III) is found at the neutral pH value and under reduced conditions, whereas As(V) is found under well oxidation conditions. This form is more toxic and less soluble as compared to reduced form of As(III); however, both forms are present in the environment. Arsenic mobility mainly depends on the interaction with heavy metals (Vithanage et al. 2017). As(V) easily connects with metal oxides due to the creation of thermodynamically stable inner-sphere complexes with metal ions (Vithanage et al. 2017; Zecchin et al. 2019). Whereas As(III) depends directly on the pH and redox potential of the environment, As(V) is more reactive with oxides of Fe and Mn (Lin et al. 2017). The adsorption affinity of arsenic species depends on the pH value. For instance, adsorption affinity of As(V) at a low pH value is high, whereas for As(III), it is low at a high pH value (Zhou et al. 2018). As the pH decreases, the arsenic mobility increases due to the mineral dissolution and increased level of surface potential. On the other hand, as the pH increases, desorption of arsenic occurs due to the low stability of the metal oxide-arsenic complex (Lin et al. 2017; Zhou et al. 2018). Due to the reductive environment, arsenic dissolution occurs from the iron oxides and oxyhydroxides. Because of the low cost and simplicity of installation and maintenance of metal oxides, it is used for the effective removal of arsenic by the adsorption process (Dixit and Heringet al. 2003; Benjwal et al. 2015). Commonly used metal oxide adsorbents are manganese, ferric hydroxide, and alumina, while iron or alumina-based adsorbents are effectively used for the removal of As. The utilization of iron-based adsorbents for arsenic removal can also be used for water purification. Adsorption of arsenic by iron hydroxide mainly depends on the pH value (Giles et al. 2011; Prathna et al. 2018). It has also been found that the adsorption of arsenic by iron mainly depends on the concentration of As(V) and Fe(III) as well as pH value (4–10). The high concentration of sulfate decreases the efficiency of As(V) removal, and at low pH, the concentration of sulfate remains high. On the other hand, arsenic can also be effectively removed using alumina as an adsorbent (Hlavay and Polyák 2005), but as compared to alumina, hydrated ferric oxide (HFO) has the better efficiency for the removal of As(V) from water due to its desorption behavior. The presence of sulfate and phosphate also affects efficiency. It has also been found that HFO activated alumina (Al_2O_3) and acidified or HFO-coated activated alumina can be used as an adsorbent for arsenic, but due to the desorption behavior, iron was more effective (Hlavay and Polyák 2005; Chiavola et al. 2016). Ferrihydrite, a kind of iron oxide, also plays an important role to control arsenic mobility and its concentration, whereas manganese oxide with ferrihydrite plays an important role in the adsorption of As. Activated alumina is also used as an adsorbent for the removal of As(V) due to its large surface area. At the level of pH from 5 to 7, activated alumina can easily remove more than 90% arsenic. The adsorption of As(V) by an adsorbent based on iron and aluminum mainly depends on the pH value. Arsenate adsorption increases the concentration of adsorbent and the amount of adsorbent at pH 7. The following reactions should clarify that:

$$\mathrm{Fe{-}OH + H_2AsO_4^- + H^+ \rightarrow Fe{-}H_2AsO_4 + H_2O;\ pH\,2-7}$$

$$\mathrm{Fe{-}OH + H_3AsO_3 \rightarrow Fe{-}H_2AsO_3 + H_2O;\ pH \leq 2}$$

1.5.3 Interactions with P, Si, and Other Elements' Concentration in the Soil

Arsenic (As) and phosphorus (P) belong to a similar chemical group with the same dissociation constant and solubility resulting in similar geochemical activity in the soil. Absorption of P, Fe, and As may be related to each other and may therefore be influenced in the rhizosphere by phosphate solubilization and iron plaque formation (Adriano 2001). The fate of arsenic in the rhizosphere that has not yet been studied can therefore be influenced by root-microbe-induced transformation processes. Due to the existence of iron plaques that interfere with Si—As interaction, it has been reported that both internal and exterior Si inhibit the uptake of As and P (Guo et al. 2007). Arsenic plays a significant role in soil and water contamination. Soil, especially that is under the flooded location, has a high percentage of arsenic with its solubility and bioavailability. The presence of arsenic in soil mainly depends upon the oxidation–reduction and adsorption–desorption processes. In well-oxidized conditions, As(V) normally predominates in the soil as $H_2AsO_4^-$ and $H_2AsO_4^{2-}$, whereas in the reduced condition, it predominates as H_3AsO_3 and $H_3AsO_3^-$ (Bissen and Frimmel 2003). As compared to As(V), As(III) is weaker to adsorb the soil minerals, so it is comparatively more toxic and mobile (Dias et al. 2009; Campbell and Nordstrom 2014). As(V)-reducing bacteria immediately convert As(V) to As(III) in reduced conditions, which results in releasing As(III) into the aqueous phase (Yamamura et al. 2008). Iron also plays a significant role because it serves as an electron acceptor. As Fe $(OH)_2$ is worked as the adsorbent for arsenic, it also reduces the adsorption capacity of arsenic for soil (Kumarathilaka et al. 2018). So arsenic solubility and bioavailability mainly depend upon the oxidation and reduction processes. In soil, as a fertilizer, phosphate is used where the presence of PO_4^{-3} directly suppresses the adsorption capacity of arsenic to both oxides and soil. Phosphates, which are an important constituent of soil, and their geochemical behavior are same as arsenate (AsO_4^{-3}), and both can be easily adsorbed by Fe-and Al-hydroxides (Violante and Pigna 2002). The presence of phosphates also stimulates microbial activity; thus, it also promotes the redox reaction, so it affects the affinity of arsenic (Deng et al. 2018; Smith et al. 2002b). As(V) has a higher affinity for Fe and phyllosilicates than $PO4^{-3}$, but Al-hydroxides, allophane, and kaolinite have a higher affinity for phosphates. The presence of phosphates in soil affects the As solubility under both suboxic and anoxic conditions. The adsorption of PO_4^{-3} leads to a decreased level of arsenic reduction, which also leads to a decreased level of As(III) but increases the level of As(V). In suboxic conditions under the dominance of As(III), the phosphate concentration will decrease under low pH, while it increases as the pH increases. On the other hand, under anoxic conditions due to the phosphates, fertilization directly stimulates the reduction of Fe-hydroxide and As(V) (Deng et al. 2018). In suboxic conditions under low pH, addition of phosphates reduces the As(V) reduction, hence decreasing the As solubility.

1.5.4 Interactions with Organic Matter

Organic matter is a complex mixture of polyfunctional organic acids produced mostly by the decomposition of a wide range of marine and terrestrial plants and animals (Redman et al. 2002). Its reactivity is mostly determined by the environment; for example, in the aquatic environment, it is highly reactive for soluble metal and metal hydroxide, which is significant for bioavailability, solubility, and mobility of organic and inorganic contaminants (McArthur et al. 2004). Organic matter reacts with As that affects its speciation and mobility especially in an aquatic environment (Wang and Mulligan 2006; Wu et al. 2019). The organic matter with a high number of sediments shows a high concentration of As (Anawar et al. 2003). This organic matter shows different association levels and deprotonation at different pH levels. At low pH values, they remain highly protonated and unchanged with cross-linked confirmation, but at high pH values, they dissociate and become more negatively charged (Wei et al. 2016). Because of this negatively charged and cross-linked conformation, the organic matter has the properties with sorption of other metal ions. Because of the ligand exchange-surface complex mechanism, organic matter has the sorption property for As(III) and As (V) (Rahman et al. 2013). Inorganic materials, such as humic and fulvic acids, are both negatively charged and have a significant tendency to adsorb to the solid surfaces of As(III) and As(V) (Fakour and Lin 2014). Organic matter strongly binds to As(III) and As(V) due to the formation of organic matter-metal complexes. Humic acids and fulvic acids can easily make the aqueous complex with As(III) and As(V). Arsenic bonds to these organic materials, forming a coating on the material's surface. Thus, the surface of the metal is directly dominated by organic matter. The nucleophilic property of amino and sulfhydryl groups strongly reacts with As(III) due to its electrophile property (Wang and Mulligan 2006).

1.5.5 Clay Minerals and Other Factors

Arsenic contamination in soil and groundwater affects water and soil quality. Arsenic is usually found in mineral shales, such as magmatic sulfides, iron ores, etc., as arsenopyrite (Fe AsS), realgar (AsS), and orpiment (As_2S_3) (Garcia-Sanchez et al. 2002). As(III) and As(V) can be easily adsorbed on sediment particles as a result of the oxidation and dissolving process from As-containing minerals (Sarkar and Paul 2016). Some anthropogenic sources like pesticides, fertilizers, and agricultural wastes increased the levels of As in water and soil too. As compared to As(V), As(III) is more toxic for soil minerals because it is less adsorbed as compared to As(V) (Hu et al. 2020). Clay minerals, such as iron, have the highest affinity for adsorbing As(V), forming high-affinity Fe—As(V) surface complexes on the inner sphere. On the goethite surfaces, As(III) has the same affinity as As(V) (Garcia-Sanchez et al. 2002). Activated alumina at pH 7 has a high affinity for As(V) as compared to As(III) (Uddin and Jeong 2020). In soil, phyllosilicate minerals have a low affinity for arsenic as compared to iron and aluminum oxides. Due to the different behaviors of As(III) and As(V), heterogeneous oxidation is the common factor to control the mobility of arsenic. As(III) adsorption to clay minerals is low at low pH, while at high pH 7.5–9.5, it is maximum. The adsorption rate of clay minerals by $Al(OH)_3$, especially for As(III), is maximum at pH 8, whereas phyllosilicates' maximum adsorption rate is at pH 9. Phyllosilicates component SiO_4 tetrahedra are less reactive than $Al(OH)_3$ functional groups. The Si/Al ratio of kaolinitic minerals like KGa-1 is 1, while impurities (IMt-2) and smectites (SWy-1) have a Si/Al ratio of 2 (Iriel et al. 2020). Thus, kaolinitic minerals have a high affinity to adsorb arsenic comparatively to Illite and smectites. For As(V), the extraction efficiency by KGa-1, SWy-1, and IMt-2 was 60, 80, and 33% at the level of pH 7, but at pH 8, the extractability for KGa-1 and SWy-1 was 100%, while for IMt-2, it was less than 50%. Oxidation of As(III) to As(V) is affected by the presence of MnO_2 in clay minerals (Zhang et al. 2020). As(III) is normally stable in the clay minerals within the range of pH 4 and 9; however, oxidation of As(III) to As(V) occurs in the range of 9.2 or above. This oxidation process can be enhanced by heterogeneous reactions with minerals in the presence of KGa-1 and IMt-2.

1.6 Arsenic Speciation in Plants

1.6.1 Methods of Determination of As and As Species in Plants

Marcinkowska and Barałkiewicz (2016) reported that the elemental speciation analysis has been a very helpful technique in environmental control, assessment of human health risk, and quality of food management. In recent years, the most prevalent equipment for element speciation has been high-performance liquid chromatography (HPLC) combined with inductively coupled plasma mass spectrometry (ICP-MS). Fiamegkos et al. (2016) reported that the hydride generation atomic absorption spectrometry (HG-AAS)-based methods are easier to implement and bear lower expenses in terms of the analytical instrumentation. There is a possibility for nonchromatographic speciation of the HG method coupled to inductively coupled plasma optical emission spectrometry (ICP-OES) (Welna and Pohl 2017). HG-AAS or hydride generation atomic fluorescence spectrometry (HG-AFS), on the other hand, methods are ineffective for determining non-hydride-forming species such as arsenic (Braeuer and Goessler 2019). X-ray absorption spectroscopy (XAS) and electrospray ionization mass spectrometry (ESI-MS) are used free of charge to substantially increase the data obtained from the HPLC-ICP-MS study (Nearing et al. 2014). For solid samples, laser ablation with inductively coupled plasma mass spectrometry (LA-ICP-MS) is used as an innovative approach for analyzing the content of As (Sajnóg et al., 2018). Recently, LA-ICP-MS was used to estimate arsenic in farmed market mushrooms (Li et al. 2018). Because a hydride does not generate at high acidity di-methyl arsenate (DMA) conditions, selective arsine generation with atomic absorption can be used to determine total inorganic arsenic as an alternative for liquid chromatography. Arsenic-contaminated irrigation also influences rice cultivation (Abedin et al. 2002a; Dos Santos et al. 2017). Mustard (Jedynak and Kowalska, 2011), bean (Sadee et al. 2016), carrot (Mayorga et al. 2013; Bergqvist et al. 2014), spinach (Shahid et al. 2017), and lettuce were other vegetables examined with arsenic toxicity (Ma et al. 2017). De las Torres et al. (2020) used HG-AFS to evaluate the total arsenic concentration in strawberry plants. Oven-dried plant samples were ground and allowed to stand overnight in concentrated HNO_3. Plant samples of 0.5 g digest with concentrated HCl and concentrated HNO_3 (2 : 1 v/v) at 80 °C for one hour and then are controlled to 120–130 °C for 20 hours to avoid As volatilization. To confirm the precision and accuracy of the digestion process and subsequent analysis with a blank reagent and a standard reference, the solution was purified after digestion for arsenic analysis by ICP-MS/AAS.

1.6.2 Uptake and Efflux Mechanism of Arsenate and Arsenite Species

Arsenic is not needed for life; therefore, there are no arsenic-specific uptake systems for any known organism. Via many forms of transporters, all arsenic absorption pathways are adventitious. Their physiological functions include the transport of phosphate, glucose, and glycerol minerals and nutrients, but arsenic can imitate certain natural substrates well enough to share their absorption systems (Garbinski et al. 2019). When the climate was anoxic, life emerged, and arsenic was mainly in the form of As(III). Uptake and accumulation of As(III) in living cells have more toxic effects than As(V). The first As(III) uptake pathway was discovered in 1997 (Rosen and Liu 2009). The primary species of arsenic is arsenate. It has an almost high affinity with soil iron oxides/hydroxides; thus, arsenate concentrations in soil are typically low. Wenzel et al. (2002) discovered arsenate concentrations 53 μM in soil solutions up to 2.3 μM in contaminated soil. Fitz and Wenzel (2002) reported high concentrations of arsenate using hydroponics studies in soil solution and questioned their environmental significance. Arsenate and phosphate share the same transport channel in plants, according to electrophysiological and physiological studies. Phosphate has a higher affinity for transporters than arsenate (Asher and Reay, 1979;

Ullrich-Eberius et al. 1989; Meharg et al.1994). The mechanism of uptake requires cotransport with stoichiometry for each of at least $2H^+$ or H_2PO^{4-} (Ullrich-Eberius et al. 1989). The presence of phosphate transporters in plants has been reported in a variety of ways (Rausch and Bucher 2002; Bucher 2006). Reduced arsenate uptake is a well-known arsenate resistance mechanism used by many plant organisms, which is accomplished in resistant plants by suppressing the high-affinity phosphate/arsenate uptake method (Meharg and Hartley-Whitaker 2002). Arsenite (As(III)) is particularly effective in reducing situations like flooded paddy fields (Marin et al. 1993; Takahashi et al. 2004; Xu et al. 2008). As bioavailability of arsenic to the rice plant is more in flooded paddy soil which leads to increase mobilization of arsenite into the soil (Xu et al. 2008). Arsenite concentrations in soil solutions vary from 0.01 to 3 μM, as compared to flooded paddy soils, which are higher than nonpolluted aerobic soil. There is a pK_a of 9.22 for arsenous acid ($As(OH)_3$). Arsenite in soil solution was found as an undissociated neutral molecule at $pH < 8$. Arsenite is transported by certain aquaglyceroporins in *Escherichia coli*, yeast, and human beings (Bhattacharjee and Rosen 2007). The *E. coli* contains GlpF facilitator of glycerol, Fps1p protein of the yeast glycerol channel, and AQP7 and AQP9 mammalian aqua glycerol porins. Alternative mechanisms have also been established in yeast other than those involving aqua glycerin porin. According to Xu et al. (2007), arsenate given to an aerated nutrient solution is easily converted to arsenite. Reduction of arsenate to arsenite was little contributed by microbes or root exudate solution. This is odd, as, in the aerobic setting, arsenate is supposed to be stable. Phosphate prevents the absorption of arsenate and the development of arsenite in the soil solution as after arsenate reduction within the cells, the arsenite is squeezed out by root cells. Arsenite efflux has been observed by the roots of various plants like wheat (*Triticum aestivum*), barley (*Hordeum vulgare*), and maize (*Zea mays*); however, not in the *Pteris vittata*, an arsenic hyperaccumulator plant. Zhao et al. (2009) reported that within 24 hours of arsenate exposure, arsenite outflow was nearly three times the quantity of As accumulated in plants, indicating fast As uptake and translocation by roots from the medium. Arsenite accumulation in the sunflower and maize roots is shown by speciation in the rhizosphere (Ultra et al. 2007; Vetterlein et al. 2007), indicating that arsenite efflux occurs in a plant grown in the contaminated soil. It is essential to elucidate arsenite efflux processes from plant roots. A typical and essential detoxification process for microbes is the reduction of arsenate followed by arsenite efflux (Bhattacharjee and Rosen 2007). Bienert et al. (2008) reported that the channel-mediated arsenite extrusion of aqua porins/aqua glyceroporin is an ancient detoxification mechanism for As. This opinion does not rule out the concept that arsenite outflow is a constituent rather than an adaptive form of detoxification. While roots can reabsorb arsenite, the existence of rapid efflux machinery will logically lead to a decrease in the root cells as a burden. Notwithstanding, determining and differentiating the speciation of arsenite and arsenate in aqueous solutions are difficult as it lacks a reliable method for its measurements. Speciation of arsenite and arsenate can be done using the electrochemical methods more accurately. These studies will be carried out using different electrodes, such as boron-doped diamond electrodes, metal-modified gold electrodes, carbon nanotubes, mercury, and bismuth films (Ma et al. 2014). The oxidative determination of As(III) was performed by Mays and Hussam (2009) using platinum nanotube array electrodes that showed much better As(III) performance and better limit of detection (LOD) and reproducibility than the comparison electrodes stated. The studies on the determination of arsenic contamination, speciation, and its bio-concentration with selective and high sensitivity electrodes are scant and need to be explored. Through the phosphate transporters, plants take up arsenate, and through the nodulin 26-like intrinsic (NIP) aquaporin channels, plants take up arsenite and undissociated methylated As species. In some plants arsenic uptake and accumulation lead to phytotoxicity in terms of protein degradation, peroxidation of lipids, cell organelle, and damage of DNA (Rascio et al.and Navari-Izzo 2011). Rahman et al. (2008) and Singh et al. (2016) reported the accumulation of As by various aquatic plants like *Lemna minor* and *Spirodela polyrhiza*. According to Tripathi et al. (2007), aquatic macrophytes absorb As(V) mostly through phosphate uptake transporters, but physicochemical adsorption on root surfaces could potentially be a mechanism for As species uptake (Rahman et al. 2008). Arsenic is regarded as an element that is toxic and carcinogenic (IARC 2004). Human health issues may be caused by high arsenic levels in soil, drinking water, crops, animals, vegetables, and food chain (Nagajyoti et al. 2010; Zhao et al. 2010; Abdul et al. 2015). According to WHO (2011), arsenic amounts of up to 1.27 and 2.20 $mg\,kg^{-1}$ were found in fruits and vegetables. Recently, Upadhyay et al. (2019) reported the presence of arsenic in crops, fruits, food products, and animals. Arsenic is present in some regions of the aquifer near agricultural areas, and its presence in strawberry plants motivates analysis. The rate of absorption of arsenic information is not available in strawberry plants.

1.6.3 Uptake and Efflux Mechanism of Methylated Arsenic Species

In plant samples, methylated arsenic species (MMA, DMA) and trimethylarsine oxide (TMAO) have been discovered (Francesconi and Kuehnelt 2002; Meharg and Hartley-Whitaker 2002). These methylated organisms may be obtained from the soil in field-collected samples. Methylated As species are reported to be present in medium and low concentrations in hydroponic cultures, where DMA or MMA is identified in plant tissues, basically greater than 1% of the total arsenic content (Quaghebeur et al. 2003; Raab et al. 2007; Xu et al. 2007), implying that de novo methylation of As persists in plants. DMA can only account for a small portion of rice grains of the total arsenic (Williams et al. 2005; Zavala et al. 2008). Methylated As percentage varied between rice genotypes in total As (Liu et al. 2006), and addition of various As in water and soil management regimes has been shown to increase with total arsenic grain concentration (Xu et al. 2008). Role of arsenic is not clear in rice methylation. Mechanism and enzymology of As methylation in plants have not been fully understood. *S*-adenosyl-L-methionine (SAM) was used as the methyl donor in an in vitro assay. Because the plants were pre-exposed to arsenate, the movement was observed in leaf extracts but not in root extracts. MMA is the primary result of methylation, but DMA accumulated over a long assay duration. Methylation in plants is likely to follow the Challenger path, which has been well known in bacteria and fungi (Bentley and Chasteen 2002). In bacteria, using the methyl donor SAM, arsenite is methylated by *S*-adenosyl methyltransferase. MMA(V) pentavalent is the product of this reaction, which is reduced with thiols through a reductase to MMA(III) (e.g. glutathione). The total As content in sunflower remained stable for the next 32 days without As exposure to roots following 1 day of arsenate exposure, indicating that any conceivable As volatilization is unlikely (Raab et al. 2007).

1.6.4 Arsenic and Rhizosphere Interaction (Mycorrhizal Fungi, Rhizofiltration)

Hiltner (1904) described the word "rhizosphere effect." The number of multiple microorganisms increases near the plant roots because of the nutrients exuded by the plant roots, which provide an environment rich in carbon. Root and microorganisms initiate colonization by producing and exchanging chemical signaling. Several authors have identified this process in detail (Lugtenberg et al. 2001). Plant–microbe interactions promote plant growth through various mechanisms, including atmospheric nitrogen fixation, tolerance against biotic and abiotic stress, and plant growth promoting rhizobacteria (PGPRR) (Moulin et al. 2002; Schardl et al. 2004; Gray and Smith 2005). Glick (1995) reported that the plant roots play an essential role in associating large numbers of microorganisms with heavy metal degradation, and these interactions lead to their significant role in phytoremediation. Chaudhry et al. (1998) characterized as the plants use lower concentrations of contaminants in their roots to absorb contaminants from contaminated sources. Mycorrhizal fungi form mutualistic connections with plants and contribute to plant growth by enhancing soil mineral uptake and also protecting the plant from heavy metal stress in polluted soil (Leyval et al. 1997). Metals allocated inside the roots by the fungal symbiont and numerous mechanisms have been proposed for this process, including the metal binding on the cell wall and vacuole deposition, siderophore-mediated absorption, and transporter involvement (Galli et al. 1994; Schutzendubel and Polle 2002). The rhizofiltration system has been used to filtrate polluted water using aquatic macrophytes and macroalgae, including groundwater, stormwater, and other effluents (Mazej and Germ 2009). Plants transpire volatile arsenic from their leaves, roots, and stems in the phytovolatilization process. Plants accumulate the contaminants in the transpiration stream from soil and water and volatilize them into the environment in an altered form at comparatively low concentrations in this process. This process includes methylated and volatile arsenic molecules (Frankenberger and Arshad 2002; Adriano et al. 2004). However, the literature on phytovolatilized arsenic is sparse. The presence of sulfate and salinity has been shown to inhibit the volatilization process in soil (Terry and Zayed 1994). There is a lack of information on the kinetics of arsenic release and absorption from the soil in plants (Abedin et al. 2002b; Irtelli and Navari-Izzo 2008). Dissimilar reduction and detoxifying activities of microorganisms are known to cause As(V) to As(III) microbial reduction. The detoxification of As by bacteria is required by an AsV reductase and AsIII extrusion by an AsIII-efflux pump. Although there is evidence of arsenite in the rhizosphere, arsenate is the main species of plants that grow in aerobic soil (Vetterlein et al. 2007; Zhao et al. 2009). Arsenite occurrence is most likely due to arsenite efflux from the roots, which is transported from the rhizosphere to other plant components by arsenate and arsenite absorption (Logoteta et al. 2009). A synchronized interaction of the microalgae *Chlorella vulgaris* and the bacterium *Pseudomonas putida* on rice plant growth and arsenic accumulation under excess arsenic and phosphate reported reduced As uptake through enhanced P: As and reduced As biomass ratio by modulating P trafficking and indicated microbial interactions mitigate As stress during P-enriched conditions in rice (Srivastava et al. 2018).

1.7 Thiolated Arsenic and Bioavailability of Thiolated As Species in Plants and Terrestrial Environments

Arsenic and sulfur coexist frequently in the biogeochemical cycle due to their strong coordination (Härtig and Planer-Friedrich 2012). Thioarsenic compounds increase the complexity of As species and their biogeochemical and environmental processes (Maher et al. 2013). Planer-Friedrich et al. (2017) observed that thioarsenic compounds may arise in sulfide waters because thioarsenates and thioarsenites play a significant role in influencing the biogeochemistry of As under reduced conditions. Härtig and Planer-Friedrich (2012) and Planer-Friedrich and Wilson (2012) reported thioarsenates in geothermal fluids, slags, wetland, and flooded rice paddy fields. As an essential group of As, thioarsenicals have been identified as the sulfur similar to oxo-arsenicals in the organisms and these complexes where an arsinothioyl (As = S) is substituted for the arsinoyl (As = O) group (Raml et al. 2006). Herath et al. (2016) reported that this arsenic speciation occurs due to environmental (oxic/anoxic) factors, redox potential, pH in the medium and other chemical species available, microbial diversity, and oxygen concentration. Development of different thioarsenic arsenate species (—HnAs OV 4 4) 3) and arsenites (—HnAs OIII 3 n 3) have been proved experimentally (Wilkin et al. 2003; Bostick et al. 2005; Hollibaugh et al. 2005). Plants have evolved a variety of physiological and biochemical mechanisms to cope with the harmful effect that reactive oxygen species (ROS) produces. However, in various plants, the enzymatic and nonenzymatic antioxidant system converts the As(V) to As(III) followed by further reduction and methylation to a nontoxic form (Liu et al. 2015; Su et al. 2016). Uptake and accumulation of As in plants vary in the environment (Haritonidis and Malea 1999; Guo et al. 2007). Various aquatic macrophytes have traditionally been utilized in arsenic removal due to their efficiency to accumulate metals in tissues (Dhir et al. 2009; Rahman and Hasegawa 2011). Recently, several floating macrophytes have been reported as potential arsenic accumulators from contaminated sites (Xing et al. 2013; Madera-Parra et al. 2015). Similarly, different plants, particularly ferns, reported arsenic hyperaccumulators that bind arsenic in complexes usually with the phytochelatins or with the influence of rhizobacteria *Azotobactor* sp. and *Bacillus* sp. (Schmoger et al. 2000; Nearing et al. 2014; Belogova et al. 2015). Inorganic As (i-As) can be extracted from water by algae and aquatic plants, which can be utilized as a bioindicator of As in the aquatic environment for bioremediation (Hansen et al. 2006; Robinson et al. 2006).

1.8 Conclusion

Research demonstrated the current availability of sufficient analytical methods for the detection of arsenic species in the environment. An essential requirement to achieve reliable results is the quality control of the arsenic speciation process. Different mechanisms for arsenic speciation have been investigated in the past five years. Appropriate selectivity and sensitivity have been achieved with the aid of appropriate sample pretreatments in identifying and updating information on arsenic species in the environmental and biological samples. Combination of chromatography with ICP-MS/hydride generation atomic spectrometry should be more efficient for multi-As species analysis to achieve successful separation of all As species and subsequent simultaneous detection. A general finding is that there are adventitious mechanisms of arsenic uptake. For the most part, species would prefer not to absorb arsenic, but through uptake mechanisms for other nutrients and minerals such as glucose permeases and aqua glyceroporins, both As(V) and As(III) join cells. Arsenic efflux mechanisms, on the other hand, demonstrate great specificity and seem to have evolved to impart tolerance to inorganic and organic arsenic. The interactions in the rhizosphere between plant roots and microorganisms help them absorb vital mineral nutrients from the soil and prevent toxic elements from accumulating. Free-living and symbiotic nitrogen-fixing bacteria contribute a lot to meeting this demand among various rhizospheric microorganisms. The rhizospheric bacteria achieve associative nitrogen fixation in the presence of N-free or low nitrogen content and thus provide plants with essential nutrients. In addition, phosphate-solubilizing bacteria can transform the inorganic unavailable P form into plant-available soluble forms.

Acknowledgments

Authors are thankful to Manipal University, Jaipur, India, for providing facilities, supports, and encouragements.

References

Abdul, K.S.M., Jayasinghe, S.S., Chandana, E.P. et al. (2015). Arsenic and human health effects: a review. *Environ. Toxicol. Pharmacol.* 40: 828–846.

Abedin, M.J., Cotter-Howells, J., and Meharg, A.A. (2002a). Arsenic uptake and accumulation in rice (*Oryza sativa* L.) irrigated with contaminated water. *Plant Soil* 240: 311–319.

Abedin, M.J., Cresser, M.S., Meharg, A.A. et al. (2002b). Arsenic accumulation and metabolism in rice (*Oryza sativa* L.). *Environ. Sci. Technol.* 36: 962–968.

Ackerman, J.T., Eagles-Smith, C.A., Herzog, M.P. et al. (2016). Avian mercury exposure and toxicological risk across western North America: a synthesis. *Sci. Total Environ.* 568: 749–769.

Adriano, D.C. (2001). *Trace Elements in Terrestrial Environments: Biogeochemistry, Bioavailability and Risks of Metals*, 2e, 867. New York: Springer.

Adriano, D.C., Wenzel, W.W., Vangronsveld, J., and Bolan, N.S. (2004). Role of assisted natural remediation in environmental cleanup. *Geoderma* 122: 121–142.

Al-Abed, S.R., Jegadeesan, G., Purandare, J., and Allen, D. (2007). Arsenic release from iron rich mineral processing waste: influence of pH and redox potential. *Chemosphere* 66: 775–782.

Ali, H. and Khan, E. (2018). Trophic transfer, bioaccumulation, and biomagnification of non-essential hazardous heavy metals and metalloids in food chains/webs – concepts and implications for wildlife and human health. *Hum. Ecol. Risk Assess. Int. J.* 25: 1353–1376.

Anawar, H.M., Akai, J., Komaki, K. et al. (2003). Geochemical occurrence of arsenic in groundwater of Bangladesh: sources and mobilization processes. *J. Geochem. Explor.* 77: 109–131.

Anderson, L.C.D. and Bruland, K.W. (1991). Biogeochemistry of arsenic in natural waters: the importance of methylated species. *Environ. Sci. Technol.* 25: 420–427.

Asher, C.J. and Reay, P.F. (1979). Arsenic uptake by barley seedlings. *Funct. Plant Biol.* 6: 459–466.

Belogova, G.A., Sokolova, M.G., Gordeera, O.N., and Vaishlya, O.B. (2015). Speciation of arsenic and its accumulation by plants from rhizosphere soils under the influence of the *Azotobactor* and *Bacillus* bacteria. *J. Geochem. Explor.* 149: 52–58.

Benjwal, P., Kumar, M., Chamoli, P., and Kar, K.K. (2015). Enhanced photocatalytic degradation of methylene blue and adsorption of arsenic (iii) by reduced graphene oxide (rGO)–metal oxide (TiO_2/Fe_3O_4) based nanocomposites. *RSC Adv.* 5: 73249–73260.

Bentley, R. and Chasteen, T.G. (2002). Microbial methylation of metalloids: arsenic, antimony, and bismuth. *Microbiol Mol Biol Rev* 66: 250–271.

Berg, M., Tran, H.C., Nguyen, T.C. et al. (2001). Arsenic contamination of groundwater and drinking water in Vietnam, a human health threat. *Environ. Sci. Technol.* 35: 2621–2626.

Bergqvist, C., Herbert, R., Persson, I., and Greger, M. (2014). Plants influence on arsenic availability and speciation in the rhizosphere, roots and shoots of three different vegetables. *Environ. Pollut.* 184: 540–546.

Bhattacharjee, H. and Rosen, B.P. (2007). Arsenic metabolism in prokaryotic and eukaryotic microbes. In: *Molecular Microbiology of Heavy Metals*, vol. 6 (ed. I.N. DHS), 371–406. Berlin, Heidelberg: Springer.

Bienert, G.P., Schussler, M.D., and Jahn, T.P. (2008). Metalloids: essential, beneficial or toxic? Major intrinsic proteins sort it out. *Trends Biochem. Sci.* 33: 20–26.

Birnboim, H.C. and Doly, J.A. (1979). A rapid alkaline extraction procedure for screening recombinant plasmid DNA. *Nucleic Acids Res.* 7: 1513–1523.

Bissen, M. and Frimmel, F.H. (2003). Arsenic -a review. Part I: occurrence, toxicity, speciation, mobility. *Acta Hydrochim. Hydrobiol.* 31: 9–18.

Bostick, B.C., Fendorf, S., and Brown, G.E. (2005). In situ analysis of thioarsenite complexes in neutral to alkaline arsenic sulphide solutions. *Mineral. Mag.* 69: 781–795.

Braeuer, S. and Goessler, W. (2019). Arsenic species in mushrooms, with a focus on analytical methods for their determination – a critical review. *Anal. Chim. Acta.* 1073: 1–21.

Bucher, M. (2006). Functional biology of plant phosphate uptake at root and mycorrhiza interfaces. *New Phytol.* 173: 11–26.

Burger, J. and Gochfeld, M. (2005). Heavy metals in commercial fish in New Jersey. *Environ. Res.* 99: 403–412.

Cai, Q., Long, M.L., Zhu, M. et al. (2009). Food chain transfer of cadmium and lead to cattle in a lead-zinc smelter in Guizhou, China. *Environ. Pollut.* 157: 3078–3082.

Campbell, K.M. and Nordstorm, D.K. (2014). Arsenic speciation and sorption in natural environments. *Rev. Minerol. Geochem.* 79: 185–216.

Carpenter, D.O. (2006). Polychlorinated biphenyls (PCBs): routes of exposure and effects on human health. *Rev. Environ. Health* 21: 1–23.

Cascio, C., Raab, A., Jenkins, R.O. et al. (2011). The contamination in Bangladesh: effects on quality and productivity of rice impact of a rice based diet on urinary arsenic. *J. Environ. Monit.* 13: 257–265.

Chakraborti, D., Sengupta, M.K., Rahman, M.M., and Ahamed, S. (2004). Groundwater arsenic contamination and its health effects in the Ganga-Meghna-Brahmaputra plain. *J. Environ. Monit.* 6: 74N–83N.

Chakraborti, D., Das, B., Rahman, M.M. et al. (2009). Status of groundwater arsenic contamination in the state of West Bengal, India: a 20-year study report. *Mol. Nutr. Food Res.* 53: 542–551.

Chakraborty, D., Singh, S.K., Rahman, M.M. et al. (2018). Ground water arsenic contamination in the Ganga River Basin: a future health danger. *Int. J. Environ. Res. Pub. Health* 180: 2–19.

Chaudhry, T.M., Hayes, W.J., Khan, A.G., and Khoo, C.S. (1998). Phytoremediation-focusing on accumulator plants that remediate metal-contaminated soil. *Aust. J. Ecotox.* 4: 37–51.

Chen, G. and Chen, T. (2014). SPE speciation of inorganic arsenic in rice followed by hydride generation atomic fluorescence spectroscopic quantification. *Talenta* 119: 202–206.

Chen, C.-J., Chen, C.W., Wu, M.-M., and Kuo, T.-L. (1992). Cancer potential in liver, lung, bladder and kidney due to ingested inorganic arsenic in drinking water. *Br. J. Cancer* 66: 888–892.

Chen, Y., Han, Y.-H., Cao, Y. et al. (2017). Arsenic transport in rice and biological solutions to reduce arsenic risk from rice. *Front. Plant Sci.* 8: 268.

Chiavola, A., Amato, E.D., Stoller, M. et al. (2016). Application of iron based nanoparticles as adsorbents for arsenic removal from water. *Chem. Eng. Trans.* 47: 325–330.

De las Torres, A.I.G., Giráldez, I., Martínez, F. et al. (2020). Arsenic accumulation and speciation in strawberry plants exposed to inorganic arsenic enriched irrigation. *Food Chem.* 315: 126215.

Deng, H., Zhang, Z., Chang, C., and Wang, Y. (2007). Trace metal concentration in great tit (*Parus major*) and greenfinch (*Carduelis sinica*) at the western mountains of Beijing, China. *Environ. Pollut.* 148: 620–626.

Deng, Y., Li, Y., Li, X. et al. (2018). Influence of calcium and phosphate on pH dependency of arsenite and arsenate adsorption to goethite. *Chemosphere* 199: 617–624.

Dhir, B., Sharmila, P., and Saradhi, P.P. (2009). Potential of aquatic macrophytes for removing contaminants from the environment. *Crit. Rev. Environ. Sci. Technol.* 39: 754–781.

Dias, F.F., Allen, H.E., Guimarães, J.R. et al. (2009). Environmental behavior of arsenic (III) and (V) in soils. *J. Environ. Monit.* 11: 1412–1420.

Dixit, S. and Hering, J.G. (2003). Comparison of arsenic (V) and arsenic (III) sorption onto iron oxide minerals: implications for arsenic mobility. *Environ. Sci. Technol.* 37: 4182–4189.

Dos Santos, G.M., Pozebon, D., Cerveira, C., and De Moraes, D.P. (2017). Inorganic arsenic speciation in rice products using selective hydride generation and atomic absorption spectrometry (AAS). *J. Microchem.* 133: 265–271.

Fakour, H. and Lin, T.F. (2014). Experimental determination and modelling of arsenic complexation with humic and fulvic acids. *J. Hazardous Mater.* 279: 569–578.

Farid, A.T.M., Roy, K.C., Hossain, K.M., and Sen, R. (2003). A study of arsenic contaminated irrigation water and its carried over effect on vegetables. In: *Fate of Arsenic in the Environment* (ed. A.M. Feroze, A.M. Ashraf and Z. Adeel), 113–121. Dhaka: Bangladesh University of Engineering and Technology.

Fiamegkos, I., Cordeiro, F., Robouch, P. et al. (2016). Accuracy of a method based on atomic absorption spectrometry to determine inorganic arsenic in food: outcome of the collaborative trial IMEP-41. *Food Chem.* 213: 169–179.

Filali, B.K., Taoufik, J., Zeroual, Y. et al. (2000). Wastewater bacterial isolates resistant to heavy metals and antibiotics. *Curr. Microbiol.* 41: 151–156.

Fitz, W.J. and Wenzel, W.W. (2002). Arsenic transformations in the soil–rhizosphere–plant system: fundamentals and potential application to phytoremediation. *J. Biotechnol.* 99: 259–278.

Francesconi, K.A. and Kuehnelt, D. (2002). Arsenic compounds in the environment. In: *Environmental Chemistry of Arsenic* (ed. W.T. Frankenberger Jr.), 51–94. New York: Marcel Dekker.

Frankenberger, W.T. and Arshad, M. (2002). Volatilization of arsenic. *Environ. Chem. Arsenic* 363–380.

Fu, J., Wang, Q., Wang, H. et al. (2014). Monitoring of non-destructive sampling strategies to assess the exposure of avian species in Jiangsu Province, China to heavy metals. *Environ. Sci. Pollut. Res.* 21: 2898–2906.

Fulekar, M. (2010). Global status of environmental pollution and its remediation strategies. In: *Bioremediation Technology* (ed. M.H. Fulekar), 1–6. Dordrecht: Springer.

Galli, U., Schüepp, H., and Brunold, C. (1994). Heavy metal binding by mycorrhizal fungi. *Physiol. Plant.* 92: 364–368.

Garbinski, L.D., Rosen, B.P., and Chen, J. (2019). Pathways of arsenic uptake and efflux. *Environ. Int.* 126: 585–597.

Garcia-Sanchez, A., Alvarez-Ayuso, E., and Rodriguez-Martin, F. (2002). Sorption of As (V) by some oxyhydroxides and clay minerals. Application to its immobilization in two polluted mining soils. *Clay Miner.* 37: 187–194.

Giles, D.E., Mohapatra, M., Issa, T.B. et al. (2011). Iron and aluminium based adsorption strategies for removing arsenic from water. *J. Environ. Manag.* 92: 3011–3022.

Glick, B.R. (1995). The enhancement of plant growth by free-living bacteria. *Can. J. Microbiol.* 41: 109–117.

Gray, E.J. and Smith, D.L. (2005). Intracellular and extracellular PGPR: commonalities and distinctions in the plant-bacterium signaling processes. *Soil Biol. Biochem.* 37: 395–412.

Green, I.D., Diaz, A., and Tibbett, M. (2010). Factors affecting the concentration in seven spotted ladybirds (*Coccinella septempunctata* L.) of Cd and Zn transferred through the food chain. *Environ. Pollut.* 158: 135–141.

Guo, W., Zhu, Y.G., Liu, W.J. et al. (2007). Is the effect of silicon on rice uptake of arsenate (AsV) related to internal silicon concentrations, iron plaque, and phosphate nutrition? *Environ. Pollut.* 148: 251–257.

Gupta, P.K., Joshi, P., and Jahangeer (2017). In-situ observation and transport modelling of arsenic in Gangetic plain, India. *Emerg. Contam.* 3: 138–143.

Hansen, H.K., Ribeiro, A., and Mateus, E. (2006). Biosorption of arsenic (V) with *Lessonia nigrescens*. *Min. Eng.* 19: 486–490.

Haritonidis, S. and Malea, P. (1999). Bioaccumulation of metals by the green alga in *Ulva rigida* from Thermaikos Gulf Greece. *Environ. Pollut.* 104: 365–372.

Härtig, C. and Planer-Friedrich, B. (2012). Thioarsenate transformation by filamentous microbial mats thriving in an alkaline, sulfidic hot spring. *Environ. Sci. Technol.* 46: 4348–4356.

Hatje, V., Payne, T.E., Hill, D.M. et al. (2003). Kinetics of trace element uptake and release by particles in estuarine waters: effects of pH, salinity, and particle loading. *Environ. Int.* 29: 619–629.

Herath, I., Vithanage, M., Bundschuh, J. et al. (2016). Natural arsenic in global groundwaters: distribution and geochemical triggers for mobilization. *Curr. Pollut. Rep.* 2: 68–89.

Hiltner, L. (1904). Ueber neuere Erfahrungen und Probleme auf dem Gebiete der Bodenbakteriologie und unter besonderer BerUcksichtigung der Grundungung und Brache. *Arb. Deut. Landw. Gesell* 98: 59–78.

Hlavay, J. and Polyák, K. (2005). Determination of surface properties of iron hydroxide-coated alumina adsorbent prepared for removal of arsenic from drinking water. *J. Colloid Interface Sci.* 284: 71–77.

Hoang, T.H., Bang, S., Kim, K.W. et al. (2010). Arsenic in groundwater and sediment in the Mekong River delta, Vietnam. *Environ. Pollut.* 158: 2648–2658.

Hollibaugh, J.T., Carini, S., Gürleyük, H. et al. (2005). Arsenic speciation in Mono Lake, California: response to seasonal stratification and anoxia. *Geochim. Cosmochim. Acta* 69: 1925–1937.

Hu, L., Nie, Z., Wang, W. et al. (2020). Arsenic transformation behavior mediated by arsenic functional genes in landfills. *J. Hazard. Mater.* 403: 123687.

Huysmans, K. and Frankenberger, W.T. Jr. (1990). Arsenic resistant microorganisms isolated from agricultural drainage water and evaporation pond sediments. *Water Air Soil Pollut.* 53: 159–168.

International Agency for Research on Cancer (2004). Some drinking water disinfectants and contaminants, including arsenic. In: *IARC Monographs on the Evaluation of Carcinogenic Risks to Humans*, vol. 84, 269–477. Lyon, France: WHO Press.

Iriel, A., Marco-Brown, J.L., Diljkan, M. et al. (2020). Arsenic adsorption on iron-modified montmorillonite: kinetic equilibrium and surface complexes. *Environ. Eng. Sci.* 37: 22–32.

Irtelli, B. and Navari-Izzo, F. (2008). Uptake kinetics of different arsenic species by *Brassica carinata*. *Plant Soil* 303: 105–113.

Jedynak, L. and Kowalska, J. (2011). Stability of arsenic species in hydroponic media and its influence on arsenic uptake and distribution in White mustard (*Sinapis alba* L.). *Microchem. J.* 98: 163–169.

Kana, A., Koplik, R., Braeuer, S. et al. (2018). Analysis of main arsenic species in canned fish marketed in the Czech Republic and Australia. *J. Food Chem. Nanotechnol.* 4: 10–17.

Kim, M.J., Nriagu, J., and Haack, S. (2002). Arsenic species and chemistry in groundwater of southeast Michigan. *Environ. Pollut.* 120: 379–390.

Kohfahl, C., Navarro, D.S., Mendoza, J.A. et al. (2016). Algae metabolism and organic carbon in sediments determining arsenic mobilisation in ground- and surface water. A field study in Doñana National Park. *Spain. Sci. Total Environ.* 544: 874–882.

Kumar, V., Vogelsang, L., Seidel, T. et al. (2019). Interference between arsenic-induced toxicity and hypoxia. *Plant Cell Environ.* 42: 574–590.

Kumarathilaka, P., Seneweera, S., Meharg, A., and Bundschuh, J. (2018). Arsenic accumulation in rice (*Oryza sativa* L.) is influenced by environment and genetic factors. *Sci. Total Environ.* 642: 485–496.

Leyval, C., Turnau, K., and Haselwandter, K. (1997). Effect of heavy metal pollution on mycorrhizal colonization and function: physiological, ecological and applied aspects. *Mycorrhiza* 7: 139–153.

Li, M., Wang, P., Wang, J. et al. (2018). Arsenic concentrations, speciation, and localization in 141 cultivated market mushrooms: implications for arsenic exposure to humans. *Environ. Sci. Technol.* 53: 503–511.

Lin, L., Qiu, W., Wang, D. et al. (2017). Arsenic removal in aqueous solution by a novel Fe—Mn modified biochar composite: characterization and mechanism. *Ecotoxicol. Environ. Saf.* 144: 514–521.

Litter, M.I., Morgada, M.E., and Bundschuhm, J. (2010). Possible treatments for arsenic removal in Latin American waters for human consumption. *Environ. Pollut.* 158: 1105–1118.

Liu, W.J., Zhu, Y.G., Hu, Y. et al. (2006). Arsenic sequestration in iron plaque, its accumulation and speciation in mature rice plants (*Oryza sativa* L.). *Environ. Sci. Technol.* 40: 5730–5736.

Liu, X., Zhang, S., Whitworth, R.J. et al. (2015). Unbalanced activation of glutathione metabolic pathways suggests potential involvement in plant defense against the gall midge Mayetiola destructor in wheat. *Sci. Rep.* 5: 8092.

Logoteta, B., Xu, X.Y., Macnair, M.R. et al. (2009). Arsenite efflux is not enhanced in the arsenate-tolerant phenotype of *Holcus lanatus*. *New Phytol.* 183: 340–348.

Lucia, M., André, J.-M., Gontier, K. et al. (2010). Trace element concentrations (mercury, cadmium, copper, zinc, lead, aluminium, nickel, arsenic, and selenium) in some aquatic birds of the Southwest Atlantic Coast of France. *Arch. Environ. Contam. Toxicol.* 58: 844–853.

Lugtenberg, B.J., Dekkers, L., and Bloemberg, G.V. (2001). Molecular determinants of rhizosphere colonization by *Pseudomonas*. *Annu. Rev. Phytopathol.* 39: 461–490.

Ma, J., Sengupta, M.K., Yuan, D., and Dasgupta, P.K. (2014). Speciation and detection of arsenic in aqueous samples: a review of recent progress in non-atomic spectrometric methods. *Anal. Chim. Acta* 831: 1–23.

Ma, L., Yang, Z., Kong, Q., and Wang, L. (2017). Extraction and determination of arsenic species in leafy vegetables: method development and application. *Food Chem.* 217: 524–530.

Madera-Parra, C.A., Peña-Salamanca, E.J., Peña, M.R. et al. (2015). Phytoremediation of landfill leachate with *Colocasia esculenta, Gynerum sagittatum* and *Heliconia psittacorum* in constructed wetlands. *Int J Phytoremediation* 17: 16–24.

Maher, W.A., Foster, S., Krikowa, F. et al. (2013). Thio arsenic species measurements in marine organisms and geothermal waters. *Microchem. J.* 111: 82–90.

Mancinelli, R., Radicetti, E., Muleo, R. et al. (2019). Can hairy vetch cover crop affects arsenic accumulation in vegetable crops? *Agriculture* 9: 89.

Mandal, B.K. and Suzuki, K.T. (2002). Arsenic round the world: a review. *Talanta* 58: 201–235.

Mandal, B.K., Chowdhury, T.R., Samanta, G. et al. (1996). Arsenic in groundwater in seven districts of West Bengal, India – the biggest arsenic calamity in the world. *Curr. Sci.* 70: 76–86.

Marcinkowska, M. and Barałkiewicz, D. (2016). Multielemental speciation analysis by advanced hyphenated technique–HPLC/ICP-MS: a review. *Talanta* 161: 177–204.

Marin, A.R., Masscheleyn, P.H., and Patrick, W.H. (1993). Soil redox-pH stability of arsenic species and its influence on arsenic uptake by rice. *Plant Soil* 152: 245–253.

Mayorga, P., Moyano, A., Anawar, H.M., and Garcia-Sanchez, A. (2013). Uptake and accumulation of arsenic in different organs of carrot irrigated with as-rich water. *Clean – Soil, Air, Water* 41: 587–592.

Mays, D.E. and Hussam, A. (2009). Voltammetric methods for determination and speciation of inorganic arsenic in the environment – a review. *Anal. Chim. Acta* 646: 6–16.

Mazej, Z. and Germ, M. (2009). Trace element accumulation and distribution in four aquatic macrophytes. *Chemosphere* 74: 642–647.

McArthur, J.M., Banerjee, D.M., Hudson-Edwards, K.A. et al. (2004). Natural organic matter in sedimentary basins and its relation to arsenic in anoxic ground water: the example of West Bengal and its worldwide implications. *Appl. Geochem.* 19: 1255–1293.

McIntyre, D., Ranson, M.K., Aulakh, B.K., and Honda, A. (2013). Promoting universal financial protection: evidence from seven low- and middleincome countries on factors facilitating or hindering progress. *Health Res. Policy Syst.* 11: 36.

Meharg, A.A. (1994). Integrated tolerance mechanisms: constitutive and adaptive plant responses to elevated metal concentrations in the environment. *Plant Cell Environ.* 17: 989–993.

Meharg, A.A. and Hartley-Whitaker, J. (2002). Arsenic uptake and metabolism in arsenic resistant and nonresistant plant species. *New Phytol.* 154: 29–43.

Meng, X.Y., Qin, J., Wang, L.H. et al. (2011). Arsenic biotransformation and volatilization in transgenic rice. *New Phytol.* 191: 49–56.

Mishra, S., Dwivedi, S., Kumar, A. et al. (2016). Current status of ground water arsenic contamination in India and recent advancements in removal techniques from drinking water. *Int. J. Plant Environ.* 2: 1–2.

Moulin, L., Chen, W.M., Béna, G. et al. (2002). Rhizobia: the family is expanding. In: *Nitrogen Fixation. Global Perspectives* (ed. T.M. Finan, M.R. O'Brian, D.B. Layzell, et al.), 61–65. New York: CABI Publishing.

Mukhopadhyay, R., Rosen, B.P., Phung, L.T., and Silver, S. (2002). Microbial arsenic: from geocycles to genes and enzymes. *FEMS Microbiol. Rev.* 26: 311–325.

Nagajyoti, P.C., Lee, K.D., and Sreekanth, T.V.M. (2010). Heavy metals, occurrence and toxicity for plants: a review. *Environ. Chem. Lett.* 8: 199–216.

Naujokas, M.F., Anderson, B., Ahsan, H. et al. (2013). The broad scope of health effects from chronic arsenic exposure: update on a worldwide public health problem. *Environ. Health Perspect.* 121: 295–302.

Nearing, M.M., Koch, I., and Reimer, K.J. (2014). Complementary arsenic speciation methods. A review. *Spectrochim. Acta B At. Spectrosc.* 99: 150–162.

Neubauer, S.C., Toledo-Duran, G.E., Emerson, D., and Megonigal, J.P. (2007). Returning to their roots: iron-oxidizing bacteria enhance short-term plaque formation in the wetland-plant rhizosphere. *Geomicrobiol J.* 24: 65–73.

Nickson, R., McArthur, J., Burgess, W. et al. (1998). Arsenic poisoning of Bangladesh groundwater. *Nature* 395: 338.

Nikolaidis, N.P., Dobbs, G.M., Chen, J., and Lackovic, J.A. (2004). Arsenic mobility in contaminated lake sediments. *Environ. Pollut.* 129: 479–487.

Nordstrom, D.K. (2002). Worldwide occurrence of arsenic in groundwater. *Science* 296: 2143–2145.

Planer-Friedrich, B. and Wilson, N. (2012). The stability of tetrathioantimonate in the presence of oxygen, light, high temperature and arsenic. *Chem. Geol.* 322: 1–10.

Planer-Friedrich, B., Kühnlenz, T., Halder, D. et al. (2017). Thioarsenate toxicity and tolerance in the model system *Arabidopsis thaliana*. *Environ. Sci. Technol.* 51: 7187–7196.

Prathna, T.C., Sitompul, D.N., Sharma, S.K., and Kennedy, M. (2018). Synthesis, characterization and performance of iron oxide/alumina-based nanoadsorbents for simultaneous arsenic and fluoride removal. *Desalin. Water Treat.* 104: 121–134.

Quaghebeur, M., Rengel, Z., and Smirk, M. (2003). Arsenic speciation in terrestrial plant material using microwave-assisted extraction, ion chromatography and inductively coupled plasma mass spectrometry. *J. Anal. Atom. Spect.* 18: 128–134.

Raab, A., Williams, P.N., Meharg, A., and Feldmann, J. (2007). Uptake and translocation of inorganic and methylated arsenic species by plants. *Environ. Chem.* 4: 197–203.

Rahman, M.A. and Hasegawa, H. (2011). Aquatic arsenic: phytoremediation using floating macrophytes. *Chemosphere* 83: 633–646.

Rahman, M., Vahter, M., Wahed, M.A. et al. (2006). Prevalence of arsenic exposure and skin lesions. A population based survey in Matlab, Bangladesh. *J. Epidemiol. Comm. Health* 60: 242–248.

Rahman, M.A., Hasegawa, H., Ueda, K. et al. (2008). Arsenic uptake by aquatic macrophyte *Spirodela polyrhiza* L.: interactions with phosphate and iron. *J. Hazard. Mater.* 160: 356–361.

Rahman, M.S., Whalen, M., and Gagnon, G.A. (2013). Adsorption of dissolved organic matter (DOM) onto the synthetic iron pipe corrosion scales (goethite and magnetite): effect of pH. *Chem. Eng. J.* 234: 149–157.

Raml, R., Goessler, W., and Francesconi, K.A. (2006). Improved chromatographic separation of thio-arsenic compounds by reversed-phase high performance liquid chromatography-inductively coupled plasma mass spectrometry. *J. Chromatograp. A* 1128: 164–170.

Rascio, N. and Navari-Izzo, F. (2011). Heavy metal hyperaccumulating plants: how and why do they do it? And what makes them so interesting? *Plant Sci.* 180: 169–181.

Rausch, C. and Bucher, M. (2002). Molecular mechanisms of phosphate transport in plants. *Planta* 216: 23–37.

Reboredo, F., Simões, M., Jorge, C. et al. (2019). Metal content in edible crops and agricultural soils due to intensive use of fertilizers and pesticides in Terras da Costa de Caparica (Portugal). *Environ. Sci. Pollut. Res.* 26: 2512–2522.

Redman, A.D., Macalady, D.L., and Ahmann, D. (2002). Natural organic matter affects arsenic speciation and sorption onto hematite. *Environ. Sci. Technol.* 36: 2889–2896.

Robinson, B., Kim, N., Marchetti, M. et al. (2006). Arsenic hyperaccumulation by aquatic macrophytes in the Taupo Volcanic Zone, New Zealand. *Environ. Exp. Bot.* 58: 206–215.

Ronci, L., De Matthaeis, E., Chimenti, C., and Davolos, D. (2017). Arsenic-contaminated freshwater: assessing arsenate and arsenite toxicity and low-dose genotoxicity in *Gammarus elvirae* (Crustacea; Amphipoda). *Ecotoxicol.* 26: 581–588.

Rosen, B.P. and Liu, Z. (2009). Transport pathways for arsenic and selenium: a minireview. *Environ. Int.* 35: 512–515.

Sadee, B.A., Foulkes, M.E., and Hill, S.J. (2016). A study of arsenic speciation in soil, irrigation water and plant tissue: a case study of the broad bean plant, Vicia faba. *Food Chem.* 210: 362–370.

Saeki, K., Sakakibara, H., Sakai, H. et al. (2000). Arsenic accumulation in three species of sea turtles. *Biometals* 13: 241–250.

Sajnóg, A., Hanć, A., and Barałkiewicz, D. (2018). Metrological approach to quantitative analysis of clinical samples by LA-ICP-MS: a critical review of recent studies. *Talanta* 182: 92–110.

Sarkar, A. and Paul, B. (2016). The global menace of arsenic and its conventional remediation – a critical review. *Chemosphere* 158: 37–49.

Satyapal, G.K., Mishra, S.K., Srivastava, A. et al. (2018). Possible bioremediation of arsenic toxicity by isolating indigenous bacteria from the middle Gangetic plain of Bihar, India. *Biotechnol. Rep.* 17: 117–125.

Schardl, C.L., Leuchtmann, A., and Spiering, M.J. (2004). Symbioses of grasses with seedborne fungal endophytes. *Annu. Rev. Plant Biol.* 55: 315–340.

Schmoger, M.E.V., Over, M., and Grill, E. (2000). Detoxification of arsenic phytochelatins in plants. *Plant Physiol.* 122: 793–801.

Schutzendubel, A. and Polle, A. (2002). Plant responses to abiotic stresses: heavy metal-induced oxidative stress and protection by mycorrhization. *J. Exp. Bot.* 53: 1351–1365.

Shahbaz, M., Hashmi, M.Z., Malik, R.N., and Yasmin, A. (2013). Relationship between heavy metals concentrations in egret species, their environment and food chain differences from two headworks of Pakistan. *Chemosphere* 93: 274–282.

Shahid, M., Dumat, C., Khalid, S. et al. (2017). Foliar heavy metal uptake, toxicity and detoxification in plants: a comparison of foliar and root metal uptake. *J. Hazard. Mater.* 325: 36–58.

Sharma, A., Kshetrimayum, C., Sadhu, H.G., and Kumar, S. (2018). Arsenic-induced oxidative stress, cholinesterase activity in the brain of Swiss albino mice, and its amelioration by antioxidants Vitamin E and Coenzyme Q10. *Environ. Sci. Pollut. Res.* 25: 23946–23953.

Shigehiro, N., Eri, M., Kumiko, S., and Tsutomu, N. (2015). Effects of polishing cooking and storing on total arsenic and arsenic species concentrations in rice cultivated in Japan. *Food Chem.* 168: 294–301.

Singh, N.K., Raghubanshi, A.S., Upadhyay, A.K., and Rai, U.N. (2016). Arsenic and other heavy metal accumulation in plants and algae growing naturally in contaminated area of West Bengal, India. *Ecotoxicol. Environ. Saf.* 130: 224–232.

Smedley, P.L. and Kinniburgh, D.G. (2002). A review of the source, behaviour and distribution of arsenic in natural waters. *Appl. Geochem.* 17: 517–568.

Smedley, P.L., Nicolli, H.B., Macdonald, D.M.J. et al. (2002). Hydrogeochemistry of arsenic and other inorganic constituents in groundwaters from La Pampa, Argentina. *Appl. Geochem.* 17: 259–284.

Smith, A.H., Lopipero, P.A., Bates, M.N., and Steinmaus, C.M. (2002a). Arsenic epidemiology and drinking water standards. *Science* 296: 2145–2146.

Smith, E., Naidu, R., and Alston, A.M. (2002b). Chemistry of inorganic arsenic in soils: II. Effect of phosphorus, sodium, and calcium on arsenic sorption. *J. Environ. Qual.* 31: 557–563.

Srivastava, S., Srivastava, S., Bist, V. et al. (2018). *Chlorella vulgaris* and *Pseudomonas putida* interaction modulates phosphate trafficking for reduced arsenic uptake in rice (*Oryza sativa* L.). *J. Hazard. Mater.* 351: 177–187.

Su, T., Shao, Q., Wang, P., and Ma, C. (2016). Oxidative stress and its role in peroxisome homeostasis in plants. In: *Redox State as a Central Regulator of Plant-Cell Stress Responses* (ed. D.K. Gupta, J.M. Palma and F.J. Corpas), 117–136. Cham: Springer.

Takahashi, Y., Minamikawa, R., Hattori, K.H. et al. (2004). Arsenic behavior in paddy fields during the cycle of flooded and non-flooded periods. *Environ. Sci. Technol.* 38: 1038–1044.

Tang, R., Wu, G., Yue, Z. et al. (2020). Anaerobic biotransformation of roxarsone regulated by sulfate: degradation, arsenic accumulation and volatilization. *Environ. Pollut.* 267: 115602.

Terry, N. and Zayed, A.M. (1994). Selenium volatilization by plants. In: *Selenium in the Environment* (ed. W.T. Frankenberger Jr. and S. Benson), 343–369. New York: Marcel Dekker.

Tripathi, R.D., Srivastava, S., Mishra, S. et al. (2007). Arsenic hazards: strategies for tolerance and remediation by plants. *Trends Biotechnol.* 25: 158–165.

Uddin, M.J. and Jeong, Y.K. (2020). Efficiently performing periodic elements with modern adsorption technologies for arsenic removal. *Environ. Sci. Pollut. Res.* 1–25.

Ullrich-Eberius, C.I., Sanz, A., and Novacky, A.J. (1989). Evaluation of arsenate-and vanadate-associated changes of electrical membrane potential and phosphate transport in *Lemna gibba* G1. *J. Exp. Bot.* 119–128.

Ultra, V.U. Jr., Tanaka, S., Sakurai, K., and Iwasaki, K. (2007). Arbuscular mycorrhizal fungus (*Glomus aggregatum*) influences biotransformation of arsenic in the rhizosphere of sunflower (*Helianthus annuus* L.). *Soil Sci. Plant Nutr.* 53: 499–508.

Upadhyay, M.K., Shukla, A., Yadav, P., and Srivastava, S. (2019). A review of arsenic in crops, vegetables, animals and food products. *Food Chem.* 276: 608–618.

Usfar, A.A., Lebenthal, E., Atmarita, A.E., and Soekirman (2010). Obesity as a poverty-related emerging nutrition problems: the case of Indonesia. *Obesity Rev.* 11: 924–928.

Vetterlein, D., Szegedi, K., Ackermann, J. et al. (2007). Competitive mobilization of phosphate and arsenate associated with goethite by root activity. *J. Environ. Qual.* 36: 1811–1820.

Violante, A. and Pigna, M. (2002). Competitive sorption of arsenate and phosphate on different clay minerals and soils. *Soil Sci. Soc. Am. J.* 66: 1788–1796.

Vithanage, M., Herath, I., Joseph, S. et al. (2017). Interaction of arsenic with biochar in soil and water: a critical review. *Carbon* 113: 219–230.

Wang, S. and Mulligan, C.N. (2006). Effect of natural organic matter on arsenic release from soils and sediments into groundwater. *Environ. Geochem. Health* 28: 197–214.

Wang, N.X., Li, Y., Deng, X.H. et al. (2013). Toxicity and bioaccumulation kinetics of arsenate in two freshwater green algae under different phosphate regimes. *Water Res.* 47: 2497–2506.

Waxman, S. and Anderson, K.C. (2001). History of the development of arsenic derivatives in cancer therapy. *Oncologist* 6 (Suppl. 2): 3–10.

Wei, Z., Liang, K., Wu, Y. et al. (2016). The effect of pH on the adsorption of arsenic (III) and arsenic (V) at the TiO_2 anatase [1 0 1] surface. *J. Colloid Interface Sci.* 462: 252–259.

Welna, M. and Pohl, P. (2017). Potential of the hydride generation technique coupled to inductively coupled plasma optical emission spectrometry for non-chromatographic As speciation. *J. Anal. At. Spectrom.* 32: 1766–1779.

Wenzel, W.W., Brandstetter, A., Wutte, H. et al. (2002). Arsenic in field-collected soil solutions and extracts of contaminated soils and its implication to soil standards. *J. Plant Nutr. Soil Sci.* 165: 221–228.

WHO (2011). Safety evaluation of certain food contaminants in food. *WHO Food Addit. Ser.* 63: 153–316.

Wilkin, R.T., Wallschläger, D., and Ford, R.G. (2003). Speciation of arsenic in sulfidic waters. *Geochem. Trans.* 4: 1.

Williams, P.N., Price, A.H., Raab, A. et al. (2005). Variation in arsenic speciation and concentration in paddy rice related to dietary exposure. *Environ. Sci. Technol.* 39: 5531–5540.

Williams, P.N., Villada, A., Deacon, C. et al. (2007). Greatly enhanced arsenic shoot assimilation in rice leads to elevated grain levels compared to wheat and barley. *Environ. Sci. Technol.* 41: 6854–6859.

Wilson, S.C., Tighe, M., Paterson, E., and Ashley, P.M. (2014). Food crop accumulation and bioavailability assessment for antimony (Sb) compared Environ Sci Pollut Res with arsenic (As) in contaminated soils. *Environ. Sci. Pollut. Res.* 21: 11671–11681.

Wolthers, M., Charlet, L., van Der Weijden, C.H. et al. (2005). Arsenic mobility in the ambient sulfidic environment: sorption of arsenic (V) and arsenic (III) onto disordered mackinawite. *Geochim. Cosmochim. Acta* 69: 3483–3492.

World Health Organization. (2016). WHO releases country estimates on air pollution exposure and health impact. Accessed 12 February 2017.

Wu, X., Cobbina, S.J., Mao, G. et al. (2016). A review of toxicity and mechanisms of individual and mixtures of heavy metals in the environment. *Environ. Sci. Pollut. Res.* 23: 8244–8259.

Wu, X., Bowers, B., Kim, D. et al. (2019). Dissolved organic matter affects arsenic mobility and iron (III)(hydr) oxide formation: implications for managed aquifer recharge. *Environ. Sci. Technol.* 53: 14357–14367.

Xing, W., Wu, H., Hao, B. et al. (2013). Bioaccumulation of heavy metals by submerged macrophytes: looking for hyperaccumulators in eutrophic lakes. *Environ. Sci. Technol.* 47: 4695–4703.

Xu, X.Y., McGrath, S.P., and Zhao, F.J. (2007). Rapid reduction of arsenate in the medium mediated by plant roots. *New Phytol.* 176: 590–599.

Xu, X.Y., McGrath, S.P., Meharg, A.A., and Zhao, F.J. (2008). Growing rice aerobically markedly decreases arsenic accumulation. *Environ. Sci. Technol.* 42: 5574–5579.

Yamamura, S., Watanabe, M., Kanzaki, M. et al. (2008). Removal of arsenic from contaminated soils by microbial reduction of arsenate and quinone. *Environ. Sci. Technol.* 42: 6154–6159.

Yang, F., Xie, S., Liu, J. et al. (2018). Arsenic concentrations and speciation in wild birds from an abandoned realgar mine in China. *Chemosphere* 193: 777–784.

Yin, X.X., Zhang, Y.Y., Yang, J., and Zhu, Y.G. (2011a). Rapid biotransformation of arsenic by a model protozoan *Tetrahymena thermophila*. *Environ. Pollut.* 159: 837–840.

Yin, X.X., Chen, J., Qin, J. et al. (2011b). Biotransformation and volatilization of arsenic by three photosynthetic cyanobacteria. *Plant Physiol.* 156: 1631–1638.

Zavala, Y.J., Gerads, R., Gürleyük, H., and Duxbury, J.M. (2008). Arsenic in rice: II. Arsenic speciation in USA grain and implications for human health. *Environ. Sci. Technol.* 42: 3861–3866.

Zecchin, S., Colombo, M., and Cavalca, L. (2019). Exposure to different arsenic species drives the establishment of iron-and sulfur-oxidizing bacteria on rice root iron plaques. *World J. Microbiol. Biotechnol.* 35: 117.

Zhang, Y., Zhao, L., Wang, Y. et al. (2008). Enhancement of heavy metal accumulation by tissue specific co-expression of iaaM and ACC-deaminase genes in plants. *Chemosphere* 72: 564–571.

Zhang, W., Zhang, G., Liu, C. et al. (2018). Enhanced removal of arsenite and arsenate by a multifunctional Fe-Ti-Mn composite oxide: photooxidation, oxidation and adsorption. *Water Res.* 147: 264–275.

Zhang, T., Wang, W., Zhao, Y. et al. (2020). Removal of heavy metals and dyes by clay-based adsorbents: from natural clays to 1D and 2D nano-composites. *Chem. Eng. J.* 420: 127574.

Zhao, F.J., Lopez-Bellido, F.J., Gray, C.W. et al. (2007). Effects of soil compaction and irrigation on the concentrations of selenium and arsenic in wheat grains. *Sci. Total Environ.* 372: 433–439.

Zhao, F.J., Ma, J.F., Meharg, A.A., and McGrath, S.P. (2009). Arsenic uptake and metabolism in plants. *New Phytol.* 181: 777–794.

Zhao, F.J., McGrath, S.P., and Meharg, A.A. (2010). Arsenic as a food chain contaminant: mechanisms of plant uptake and metabolism and mitigation strategies. *Annu. Rev. Plant Biol.* 61: 535–559.

Zhou, J., Chen, S., Liu, J., and Frost, R.L. (2018). Adsorption kinetic and species variation of arsenic for As (V) removal by biologically mackinawite (FeS). *Chem. Eng. J.* 354: 237–244.

2

Chemistry and Occurrence of Arsenic in Water

Marta Irene Litter

IIIA (CONICET-UNSAM), Instituto de Investigación e Ingeniería Ambiental, Escuela de Hábitat y Sostenibilidad, Universidad Nacional de General San Martín, Buenos Aires Province, Argentina

2.1 Chemical Properties of Arsenic

In this chapter, a summary of the chemistry and occurrence of arsenic, mainly in waters, will be given. Due to a large number of references, only those after 2000 will be cited, although some very important papers before this date will be also referenced.

Most environmental As contamination is natural, but an important amount originates from anthropogenic activities. This issue leads millions of human beings to dangerous complications by ingesting poisoned water and meals, coming from soils contaminated with As or irrigated with water polluted with As. Arsenic pollution is nowadays a disaster, and this fact was realized during the twentieth and twenty-first centuries (da Silva et al. 2019; Hare et al. 2019). Arsenic is considered as one of the most dangerous elements for human health because of its high toxicity; this issue will be thoroughly discussed later. Consequently, it is needed to have a good knowledge and comprehension of the chemical and physical properties of the element, such as chemical forms and speciation, reaction pathways, or water solubility, mainly because removal processes can transfer an As species to another.

The metalloid As is present in approximately $5\text{-}10^{-4}\%$ in the Earth's crust (https://inchem.org/documents/jecfa/jecmono/v18je17.htm) (in the atmosphere, in the hydrosphere, and the biosphere, Basu et al. 2014; Weerasundara et al. 2021), with around $2.5\,mg\,kg^{-1}$ in the Earth's crust (Mandal and Suzuki 2002). Arsenic is the 20th, 14th, and 12th most common element in abundance in the Earth's crust, seawater, and the human body, respectively (Baig et al. 2015; Matschullat 2000; Mohan and Pittman Jr 2007; Weerasundara et al. 2021). The elemental form of As is a silver-gray, hard, and breakable crystalline solid, with the atomic number 33, the atomic weight $74922\,g\,mol^{-1}$, the specific gravity of $5.73\,g\,cm^{-3}$, a boiling point of 614 °C, and a melting point of 817 °C (Zakhar et al. 2018). The element is present in different concentrations in air, water, soil, plants, and animals; in different oxidation states (−III, 0, III, V), and inorganic and organic forms

Arsenic in Plants: Uptake, Consequences and Remediation Techniques, First Edition.
Edited by Prabhat Kumar Srivastava, Rachana Singh, Parul Parihar, and Sheo Mohan Prasad.

(Baloch et al. 2020). The organoarsenic compounds contain As bound to a carbon atom by covalent unions; the element can be either as As(III) or As(V) (Meharg and Hartley-Whitaker 2002). The organic species are present in the atmosphere, in aquatic systems, soils, sediments, and biological tissues.

It is important to cite the main arsenical species:

- Arsenate and other As(V) inorganic forms, such as arsenic acid (H_3AsO_4)
- Arsenite and other As(III) inorganic forms, such as arsenious acid ($(HO_3)As$)
- As(V) methylated forms, as dimethylarsinic acid (DMA(V), $(CH_3)_2OOH$), monomethylarsonate (MMA(V), $(CH_3)O(OH)_2$), and other As(V) organic forms
- As(III) methylated forms, such as dimethylarsenious acid (DMA(III), $(CH_3)_2AsOH$), monomethylarsonic acid (MMA(III), $CH_3AsO(OH)_2$), and other As(III) organic forms
- Trimethylarsine oxide (TMAO(V), C_3H_9AsO)
- Arsenobetaine (AB)
- Arsenocholine (AC)
- Tetramethylarsonium cation (TMA^+)
- Trimethylarsine (TMA or TMAs)
- Arsenic(0) (As(0))
- Arsine (As(-III))
- Arsenolipids
- Arsenosugars
- Other species

Relevant As species for human biomonitoring are shown in Figure 2.1 (Begerow et al. 2000).

The predominant chemical states are As(III) and As(V) and, less frequently, As(0) and As(−III). As(V) appears as arsenic acid, H_3AsO_4 ($AsO(OH)_3$), and $H_2AsO_4^-$, $HAsO_4^{2-}$, and AsO_4^{3-} as the corresponding dissociation products (pK_{a1}: 2.3; pK_{a2}: 6.8; and pK_{a3}: 11.6, respectively). It is generally retained by the soil (Alvarez Benedí et al. 2003; Calvo et al. 2003). As(III) occurs as arsenious acid, H_3AsO_3 ($As(OH)_3$), and its corresponding protolytic derivatives depending on the pH ($H_4AsO_3^+$, $H_2AsO_3^-$, AsO_3^{3-}; pK_{a1}: 9.2; pK_{a2}: 12.7) (Adriano 2001; Baig et al. 2015; Lillo 2003; Smedley and Kinniburgh 2002). Figure 2.2 shows that both acids present completely different dissociation patterns (for $100\,\mu g\,l^{-1}$ of total As). Arsenic acid is almost completely dissociated at pH > 4, whereas arsenious acid shows substantial dissociation only at pH > 8.

2.2 Worldwide Occurrence of Arsenic

At present, over 220 million people in more than 105 countries are exposed to As pollution. The most important affected countries are Argentina, Bangladesh, Cambodia, Canada, Chile, China, Germany, Hungary, India, Japan, Laos, Mexico, Nepal, Nicaragua, Pakistan, Poland, Romania, Taiwan, Thailand, United Kingdom, United States, and Vietnam, i.e. countries that present the highest As concentrations. Figure 2.3 shows the most affected regions (Shaji et al. 2021).

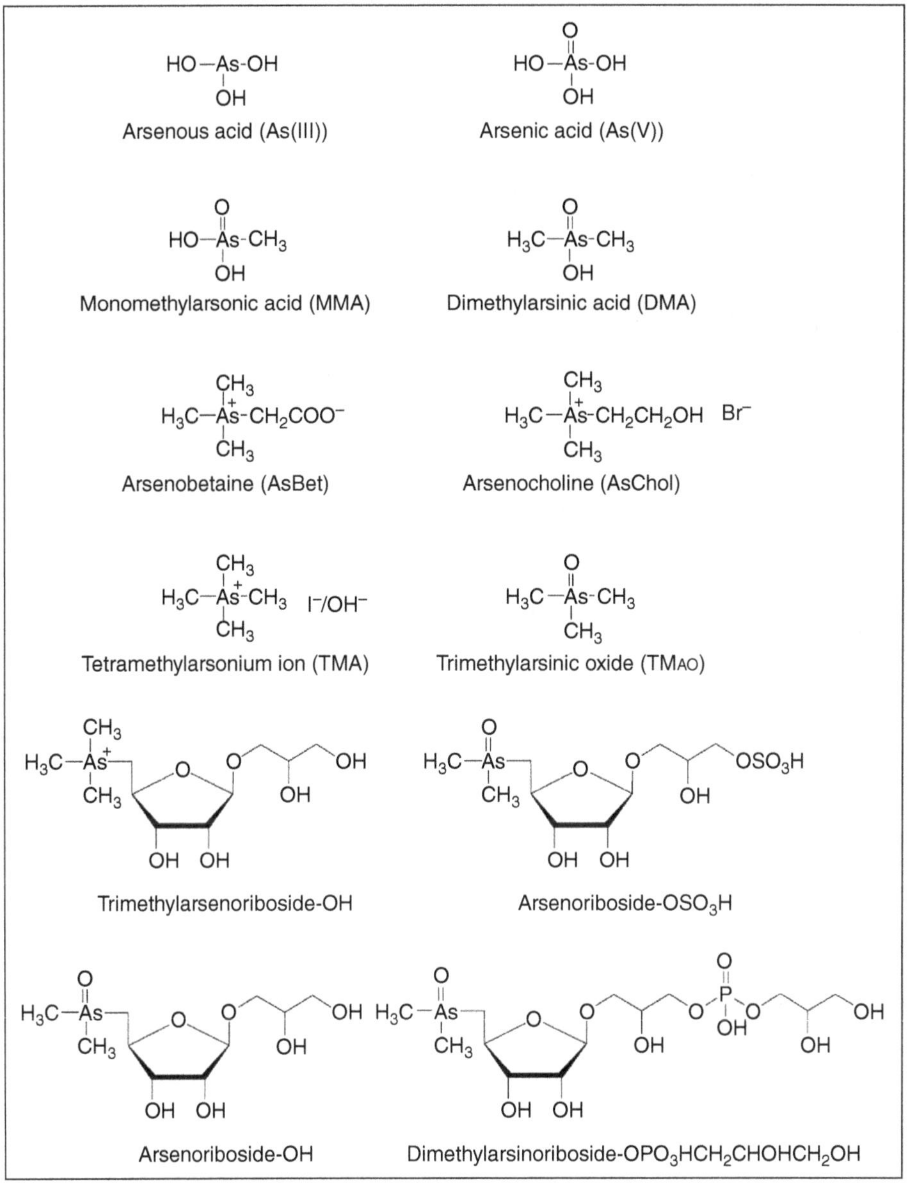

Figure 2.1 Relevant As species for human biomonitoring. *Source:* from Begerow et al. (2000) with permission.

Serious health problems due to the presence of As in groundwater affect hundreds of thousands of people, especially in Bangladesh and West Bengal, India (Huq et al. 2020; Selim Reza et al. 2010; Uppal et al. 2019; Yadav et al. 2020). Nowadays, the number of countries with large As concentrations in waters (except the geothermal ones) is constantly increasing. For this reason, public and scientific attention is put in this metalloid (Bhattacharya et al. 2009; Harvey et al. 2002; McArthur et al. 2004; Roychowdhury

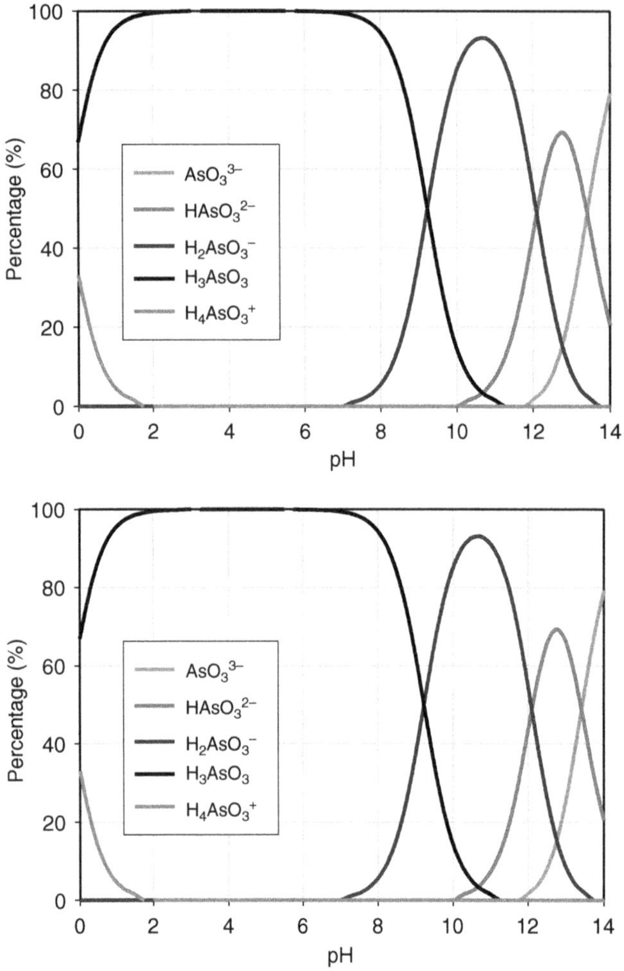

Figure 2.2 As(III) (above) and As(V) (below) speciation. Total As concentration: $100\,\mu g\,l^{-1}$ (calculated by MINEQL).

et al. 2005). As-rich aquifers were also found in China, Iran, Pakistan, and Vietnam (Dehbandi et al. 2019, Guo et al. 2008; He et al. 2020; Nickson et al. 2005; Postma et al. 2007). Since the beginning of the twentieth century, the occurrence of As in groundwater has been reported in Latin America, especially in Argentina (Bundschuh et al. 2012), and the problem is becoming more and more worrying. Recently, new zones presenting high As concentrations in groundwater have been discovered in the Chaco-Pampean plain; this is the largest area in the world containing high As concentrations in groundwater (Bundschuh et al. 2004; Smedley et al. 2000). Similarly, in other South American countries, high As levels were found, e.g. at the coastal Brazilian area as well as in marine sediments in adjacent beaches (Cagnin et al. 2017; Mirlean et al. 2014).

In the aforementioned regions, the presence of As is a consequence of the geochemical environment and particular hydrogeological conditions, with high As concentrations in

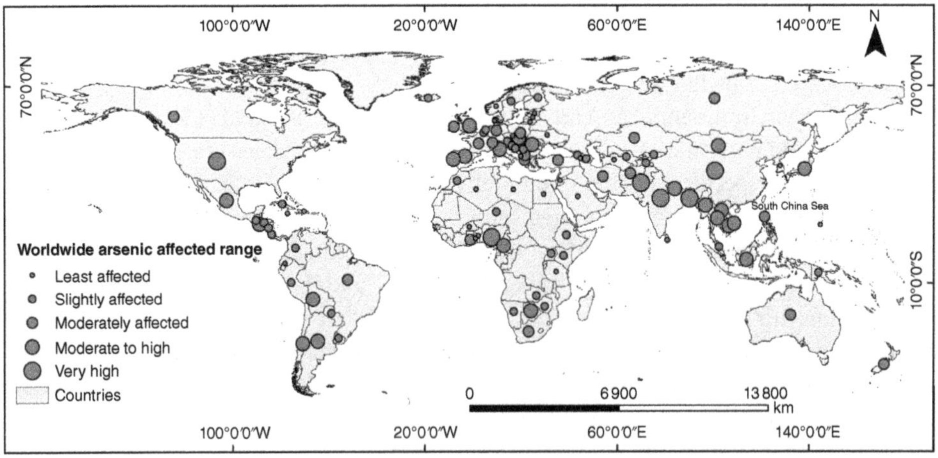

Figure 2.3 Map of worldwide affected regions. *Source:* Shaji et al. (2021), with permission.

water, in overexploited aquifers in arid or humid areas, or in free-surface and confined deep aquifers (Asere et al. 2019; Bundschuh et al. 2020; Litter et al. 2010; Luong et al. 2018; Singh et al. 2015).

Podgorski and Berg (2020) made a global map combining climate, environmental, and geological data with machine learning; in this map, it is possible to predict the location of groundwater As concentrations expectable to surpass 10 $\mu g\,l^{-1}$. The map shows many of the probably most As-contaminated places, particularly in Asia and South America (Fendorf et al. 2010), and it also points out likely high risk in other less studied regions, such as central Asia (Kazakhstan and Mongolia) and several countries close to the Sahara and in the Arctic.

2.3 Arsenic Occurrence in Natural Media

Arsenic is part of more than 200 minerals. Natural As is originated from volcanic rocks, especially their erosion products; ash, marine sedimentary rocks, hydrothermal mineral deposits, geothermal waters, and fossil fuels, including coal and oil (Smedley and Kinniburgh 2002, p. 517 and references therein), are also sources of As. The highest amounts of minerals containing As generally occur in metal ores, especially iron minerals; there, As is associated as a minor compound (e.g. pyrite [FeS_2] and sphalerite [ZnS]) or as a major compound (e.g. arsenopyrite [FeAsS], orpiment [As_2S_3], and realgar [As_4S_4]) (Baeyens et al. 2019; De Gieter and Baeyens 2005). FeAsS is the most frequent mineral, and it often contains gold; metal arsenides are also present, such as $FeAs_2$ (löllingite), NiAs (niccolite), CoAsS (brilliant cobalt), NiAsS (gersdorfite), and $CoAs_2$ (enamelite). Natural arsenates and thioarseniates are common species, and most sulfide minerals contain As (Moody 2013). It is found naturally as a cobalt mineral; it can be also usually found on the surface of rocks combined with S or metals such as Fe, Co, Mn, Sn, Ni, or Ag.

Unlike pollution due to human activities, which generally produce a localized problem, high concentrations of natural As affect large areas. The numerous cases of the presence of

As in groundwater are related to different geological environments: volcanic formations, mining districts, hydrothermal systems, and tertiary and quaternary alluvial basins (BGS and DPHE 2001; Smedley and Kinniburgh 2002).

In natural environments, As can be found in the −3, +3, and +5 oxidation states (Pirnie 2000). In natural waters, the most important oxidation states are +3 and +5, generally as dissolved species, forming oxyanions (Shakya and Ghosh 2018). In most aquifers with high As contents, not associated with geothermal processes, mineral deposits, or anthropogenic origin, the materials in solids have As contents at medium levels (0.5–1.0 $mg\,kg^{-1}$ in sandstones, 13 $mg\,kg^{-1}$ in clays, and 1–1.5 $mg\,kg^{-1}$ in carbonates). Arsenic concentration in igneous rocks is generally lower than 5 $mg\,kg^{-1}$ (Smedley and Kinniburgh 2002), but volcanic ash can release large As amounts into the water. The richest sedimentary materials in As are the carbons (maximum level of 35 000 $mg\,kg^{-1}$, Belkin et al. (2000)). Arsenic contents in sediments are similar to those in rocks, with a tendency to be higher in silt and clays.

Arsenic concentrations in water bodies are present in less than 1 $\mu g\,l^{-1}$ (in surface waters) but can attain more than 400 $\mu g\,l^{-1}$ in rivers and lakes affected by geothermal and industrial wastewater. Concentrations in aquifers range from very low values to several $mg\,l^{-1}$, naturally and anthropogenically originated.

The oxidation state of inorganic As in groundwater depends on the water chemistry; it is also affected by the exposure to air. The natural pentavalent As form is anionic and remarkably different from that of As(III), which is a neutral species: their electronic structure, dissociation properties, polarizability, sorption affinity, and toxicity are very dissimilar (Jain and Ali 2000; Shinde et al. 2013). However, As(III) and As(V) can be found together because their conversion by redox reactions is slow (Litter et al. 2010; Vatutsina et al. 2007). In freshwater environments, As content is up to an order of magnitude lower when the complexity of the trophic level increases, and there is less transfer of As from the medium to the trophic chain than in the marine environment. The arsenical profile is similar to the marine one in algae, plants, and bivalves.

Sulfide minerals are significant natural sources of As in groundwater. High As concentrations in solution can originate from arsenopyrite oxidation. The exact speciation of the soluble inorganic As, i.e. as nondissociated forms of oxyanions of the acids (H_3AsO_3 and H_3AsO_4), will depend on the prevailing Eh and pH. However, sorption of As (especially pentavalent) on ferric hydroxide produced under oxidizing conditions may limit its mobility and availability. As(III) adsorption, especially as H_3AsO_3, the predominant form under reducing conditions at $pH < 9.2$, is much less strong.

Arsenic species detected in soils are mainly inorganic and, to a lesser extent, organic compounds such as MMA, DMA, TMAO, AB, TMA^+, or AC. Methylated soil species may also result from the release of As compounds from plants or from the action of microorganisms capable of methylating As in soil.

AB and DMA(V) (see later Section 2.5) are the major compounds detected in sediments, although their concentration decreases with depth. In addition, debris found on the surface of lakes is consumed by bacteria through anaerobic respiration. Mono-methylarsonic acid (MMAA(V)) and dimethylarsinic acid (DMAA(V)), similarly to inorganic compounds, form highly insoluble complexes due to their interaction with hydrated oxides deposited on sediment clays. Their adsorption is linked to the concentration, pH, and type of clays;

however, generally, the adsorption increases with the higher iron content. The presence of organoarsenical compounds in sediments is similar to the behavior of inorganic arsenical compounds in soils. This is observed for the synthetic organoarsenicals used in agriculture, which adsorb in soils and leach in the same way as arsenates; adsorption is lower the higher the number of methyl substituents attached to the As atom.

Arsenate, the prevalent environmental species, mainly in water and soil, is present in organisms such as plants, algae, and aquatic and terrestrial animals. Arsenite (As(III)) is the reduced inorganic form and is present in a minority form in waters, soils, and living beings. It is considered the most toxic As species (Baloch et al. 2020).

2.4 Arsenic Mobilization in Natural Media

Among trace elements, As is particular because of its sensitivity to mobilize in groundwater typical pH values (6.5–8.5). The mobilization of As species is caused by the chemical properties in the sediments (pH, total iron, and free iron oxides) (Doherty et al. 2017; Rahman et al. 2009). The oxidation state and mobility of As are principally controlled by pH and redox conditions (Eh), represented in Figure 2.4 (Baloch et al. 2020; Kumar et al. 2019; Vatutsina et al. 2007).

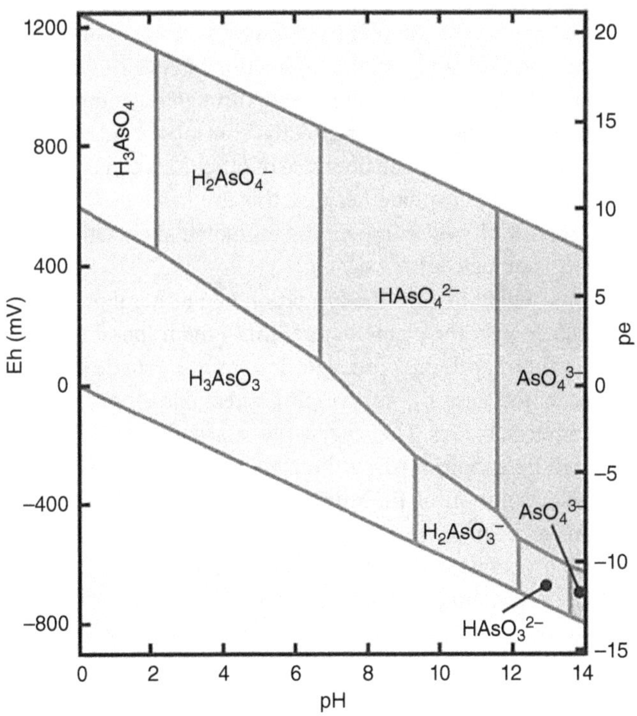

Figure 2.4 Eh-pH diagram of As species in the As–O_2–H_2O system. *Source:* modified from Smedley and Kinniburgh (2001).

The increase of As distribution in water systems can be due to the values of pH and Eh (Zakhar et al. 2018). However, arsenate highly predominates in natural environments and is stable in media-rich oxygen; in contrast, arsenite is prevalent in anaerobic conditions such as groundwater (Uppal et al. 2019).

Arsenic(III) compounds are 4 to 10 times more mobile than As(V) compounds. However, and despite this difference, and as said before, the redox kinetics is slow, and both oxidation states can be present in soils. The adsorption mechanism is specific, being the same that controls the adsorption of phosphates, and it is affected by the pH and redox conditions of the soil. At a high redox potential, As(V) predominates, and therefore, it has low mobility; at higher pH and decreased redox potential, As(III) becomes important; in this oxidation state, the element is more susceptible to leaching because of the greater mobility.

The presence of high concentrations of As in water is directly related to its release from the solid phase, mainly controlled by three factors: (i) the primary source (geosphere and atmosphere), (ii) mobilization and retention processes in the solid/liquid phase interaction, and (iii) transport in water (Smedley and Kinniburgh 2002).

The geochemical processes of water–solid phase interaction, which control the mobilization of As in water, are the sorption–desorption reactions and the precipitation–dissolution reactions of the solid phase. Both forms As(III) and As(V) can be adsorbed onto the surface of an extensive variety of oxides, especially aluminum, iron, manganese oxides, organic matter (OM), and clays. The interaction between solids and solution controls the As concentration in natural waters, a phenomenon more noticeable in the case of soils, interstitial water, and groundwater, which present high solid/solution ratio; however, this can also occur in surface water bodies where the concentration of particles (of mineral or biological origin) remains significant. In most soils and aquifers, the interaction of As with minerals probably dominates the interaction with OM, but OM can interact through reactions on the mineral surface. Oxidation–reduction reactions can indirectly control As adsorption–desorption processes by their effects on speciation. The presence of other ions can condition also As adsorption. The most important is phosphate because this ion has a geochemical behavior similar to that of As(V), and both will compete for exchange sites. Other oxyanions such as Mo, Se, and V can also compete with As(V).

Precipitation–dissolution reactions, which involve the formation of a mineral phase and its destruction, make that, theoretically, only the elements present in the mineral will participate in these processes. On the other hand, in coprecipitation, minor constituents are incorporated into the mineral, as in the case of As, which coprecipitates with pyrite. Similarly, P and As coprecipitate with iron oxides. The reverse process, codissolution, leads to As mobilization in water and will be conditioned by the stability of the mineral with which it has coprecipitated. If the dissolution of the mineral phase containing As (e.g. oxides) takes place, this mechanism can be much more effective than desorption. The most important dissolution/codissolution processes involved in As mobilization in water are the oxidation of sulfides and the reductive dissolution of oxides.

Generally, under oxidizing conditions, As(V) predominates over As(III). The species is mainly found as $H_2AsO_4^-$ at $pH < 6.9$; at higher pH values, $HAsO_4^{2-}$ predominates. In conditions of extreme acidity, the dominant species is H_3AsO_4, while in conditions of extreme basicity, the dominant species is AsO_4^{3-}. At $pH < 9.2$ and reducing conditions, the neutral species predominates (Yan et al. 2000).

Natural As is generally present as As(III) and As(V), and the speciation varies with the pH and the redox potential (Eh). Both pentavalent and trivalent As are mobile in solution, although the latter is the most mobile and toxic state. Mobilization occurs by meteorization, biological activity, and volcanic emissions, and by anthropogenic actions.

Gulens et al. (1979) conducted studies of As(III) and As(V) mobility using columns filled with sand containing Fe (0.6%) and Mn (0.01%) at different Eh and pH values, finding that:

- The mobility of As(III) in slightly acidic oxidizing conditions (pH 5.7) is 5 to 6 times faster than that of As (V);
- As(V) mobility at neutral pH values (6.9) is faster than that in the previous conditions, but always slower than that of As(III); and
- At pH 8.3, in reducing and alkaline conditions, both As(III) and As(V) move rapidly.

These phenomena would explain the high variability of the As(III)/As(V) ratio in a large number of aquifers where reducing conditions predominate.

According to published information (BGS and DPHE 2001; Lillo 2003; Smedley and Kinniburgh 2002), the main geochemical mechanisms for the release and mobilization of As into waters are as follows:

- Sulfide oxidation: an important mechanism occurring only locally and in mining areas with an abundance of sulfides; it may also involve authigenic pyrite in sedimentary succession. Sulfide oxidation takes place under oxidizing conditions. Dissolved Fe tends to precipitate as iron sulfate and iron oxides and oxyhydroxides, with As readsorption and coprecipitation. Readsorption of As means that the mechanism of arsenic mobilization by oxidation of pyrite is not considered an effective process of As mobilization.
- Dissolution of minerals: Fe and Mn oxides and oxyhydroxides are dissolved under strongly acidic conditions. Arsenic present in these materials, both adsorbed and coprecipitated, will be mobilized during dissolution. Unlike desorption, dissolution is effective in the mobilization of As embedded in the mineral structure. This explains, in part, the high As concentrations found in the acid drainage of mines or groundwater under strongly reducing conditions.
- Desorption under oxidizing conditions at high pH: under air and variable pH conditions (between neutral to acid conditions), As is strongly adsorbed onto mineral oxides in many natural environments. The adsorption isotherm (nonlinear) for As(V) indicates a relatively high amount of adsorbed As, even at low concentrations. However, under oxidizing conditions, desorption of As occurs if pH values greater than 8.5 are reached. Normally, the increase in pH results from weathering and ion exchange processes, together with an increase in salinity in arid and semiarid climates. The process of desorption under alkaline conditions is considered to be one of the most effective mechanisms in terms of As mobilization under oxidizing conditions.

The role of HCO_3^- in relation to As desorption processes is unclear.

Fulvic and humic acids at high concentrations provide dissolved organic carbon in water with high levels of As, although no direct relationship has been established between these components and desorption of As. The presence of Ca and Mg cations can facilitate the adsorption of As(V) species (negatively charged) on the solid phase. A similar effect may have Al in acidic waters and Fe^{2+} in reducing waters. Evaporative processes in arid environments allow the maintenance of high pH values as well as high concentrations of Cl, F, and U. However, this mechanism would not explain the high concentrations of As observed in reducing environments, since these usually have pH values close to neutrality.

Desorption and dissolution of As can be related to changes in reducing conditions: strongly reducing conditions capable of producing the reduction of Fe(III) and probably sulfates is another important mechanism of As mobilization in the aqueous phase. This phenomenon is mostly caused by the accumulation and burial of sediments (e.g. river basins carrying high sediment loads), where the amount of OM will determine how quickly reducing conditions will be achieved.

In the soil environment, the element is present as As(−III), As(0), As(III), or As(V), the last species in organic and inorganic forms that are subject to oxidation–reduction, precipitation–dissolution, adsorption–desorption, and biochemical and organic methylation. As already mentioned, the chemical form of As present in the medium depends on pH, redox potential, and nature and amount of sorbents and also on microbial activity (Yong and Mulligan 2004). The metallic form of As is infrequent, and As(−III) is only found in very high reducing environments. As(V) is the most common form in the most superficial area of the soil due to the aerobic conditions; in laboratory tests under aeration conditions, it has been observed to be the most stable form (Alvarez Benedí et al. 2003).

In soil and sediments, As is commonly associated with Fe, Al, or Mn oxides and oxyhydroxides, clays and sulfates, and phosphates and carbonates, which constitute an important deposit of As in soil (Foster 2003; Wang and Mulligan 2006). An important variable in the control, mobility, and As distribution is the particle size, because the smaller particles, having a larger area per unit volume, can adsorb more As. Clays and oxyhydroxides of Fe, Al, or Mn are major components of this type of particles; therefore, these compounds have a relevant role in the behavior of As in soils (Wang and Mulligan 2006). Adsorption of As defines its availability, which, in soil, is generally low (Mandal and Suzuki 2002; Huang et al. 2006). The bioavailability of As in soil is also determined by properties such as mineral composition, OM content, phosphate concentration, pH, and redox potential (Adriano, 2001; Caetano and Vale 2002; Cao and Ma 2004; Huang et al. 2006; Mandal and Suzuki 2002; Warren et al. 2003).

Arsenic levels in environmental media (air, water, soil, plants, and animals) are variable (wide range of concentrations), because of the very high mobility and capacity of transformation, being able to be adsorbed or desorbed of particles, by changing the state of oxidation by reaction with oxygen or other molecules, or by the action of microorganisms (Smedley and Kinniburgh 2002). Maity et al. (2019) indicate that, in water, 95% of As appears as inorganic species. Arsenic levels are lower in surface water (seas, rivers, and lakes) and higher in groundwater, especially in areas with deposits of volcanic rocks or As-rich minerals (Mandal and Suzuki 2002; Ng et al. 2003; Smedley and Kinniburgh 2002). Natural or anthropogenic sources of pollution increase basal As levels in soils and sediments.

The main reason for the As presence (sometimes higher than $1000\,mg\,l^{-1}$) in most groundwaters is the natural contamination by As mobilization from geogenic sources (Vatutsina et al. 2007). In groundwater, As concentration varies from less than 0.5 to $5000\,\mu g\,l^{-1}$ because concentrations of As(III) and As(V) depend on the access of As into the system, biological activity, and redox conditions. The most elevated arsenic concentrations are recorded in aquifers with oxidizing conditions and high pH, and they originate from natural geochemical processes (Smedley and Kinniburgh 2002).

In groundwaters, under oxygen, the transformation of As(V), strongly adsorbed under oxidizing conditions, to As(III), which is weakly adsorbed onto the Fe and Mn oxides and oxyhydroxides surfaces, would be one of the first reactions to occur, delivering As to the waters.

Surface waters are open systems where environmental factors vary continuously, while groundwaters are semi-closed systems, where conditions can vary discontinuously over time, allowing the solid-phase solution system to be rebalanced (Smedley and Kinniburgh 2002). In general, in surface waters, As(V) predominates over As(III), and both oxidation states can be found in groundwater. In rivers, the average concentration is generally less than $0.8\,\mu g\,l^{-1}$, although it may fluctuate depending on recharge (surface and underground), drainage of mineralized areas, climate, mining activity, and urban or industrial discharges. In lakes, As content is similar to that observed in rivers (Smedley and Kinniburgh 2002). However, in some cases, the restricted circulation of water results in intensification of evaporation processes (e.g. lakes in arid climates), or stratification of the water column, with As(V) predominance in the epilimnion and As(III) in the hypolimnion, in addition to a variation in the As concentration with depth. In seawater, the mean As concentration value is approximately $1.5\,\mu g\,l^{-1}$ (Adriano 2001). However, concentrations in estuary areas are variable, as a result of inland water supply and local variations in salinity and redox gradients, although, in the most anthropogenically unaffected cases, they remain below $4\,\mu g\,l^{-1}$.

On the other hand, plants absorb As through the roots, depending on the characteristics of the soils and plants.

In soils and groundwater, particularly in anaerobic environments, As mobility can be also associated with anthropogenic factors (Polya and Middleton 2017).

2.5 Biological Methylation of Arsenic in Organisms

The process of biological methylation is an enzymatic transfer of a methyl group from a donor to an acceptor atom within a living organism. The enzymes involved in this process are known as methyltransferases. Arsenic methylation involves the involvement of S-adenosylmethionine as a methyl group donor. Arsenic can be biotransformed in animals and humans. In mammals, inorganic As is transformed into DMA(III). MMA(III) and DMA(III) appear in the urine of humans chronically exposed to As. Only MMA(V) reductase and arsenic methyltransferase are required for biotransformation of inorganic As in humans. Figure 2.5 shows the biotransformation in mammalian systems of inorganic As(V) into As(III), followed by methylation to methyl- and dimethylarsenite, but they are easily eliminated by the organism (Vasken Aposhian et al. 2004).

Although As is metabolized by a large number of organisms, no proofs exist that this takes place in plants. In some microorganisms, biomethylation is the typical metabolic process, resulting in MMA, DMA, etc. Subsequent metabolic processes form AC, AB, and arsenosugars in several organisms and some terrestrial plants. Arsenate reduction to arsenite was absolutely demonstrated in plants, but methylation was not confirmed. Organic As

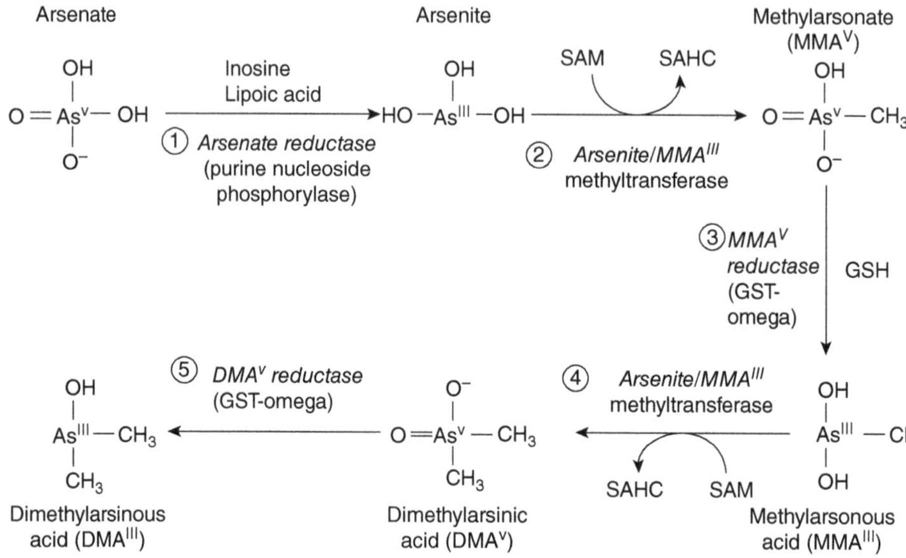

Figure 2.5 Biotransformation of As(V) in mammalians. SAM: *S*-adenosylmethionine; SAHC: *S*-adenosylhomocysteine. *Source:* From Vasken Aposhian et al. (2004), with permission.

species present in plants can be assimilated from the soil solution, and microbial activity can also cause their presence in soils. This hypothesis has been verified for MMA and DMA. Koch et al. (2000) reported that certain plant species contained MMA, DMA, TMA^+, and TMAO, but the authors did not confirm the evidence of external methylation in soils and waters. They suggest that plants could have grown in a specific microenvironment where these compounds were available (Meharg and Hartley-Whitaker 2002). These processes are shown in Figure 2.6.

In animal organisms, including humans, there is a high transformation between inorganic and organic arsenic species, regulated by environmental conditions. The different physicochemical characteristics of the arsenic forms present in the environment (As(V), As(III), MMA, DMA, TMA^+, AC, AB, arsenosugars, etc.) result in different degrees of mobility, bioavailability, and toxicity of the compounds.

The detoxification process produces harmless compounds such as AB, AC, and some arsenoribosides, and they represent the main As compounds in living organisms (Gómez Ariza et al. 2000).

Bacterial demethylation is possibly the most important route of degradation in the troposphere.

Oxidation of trimethylarsine results in partial conversion to DMA(V). In addition, hydrolysis of endocellular compounds or OM degradation can produce DMA(V) and TMAO.

In anoxic environments, rich in hydrogen sulfide and other sulfide compounds, the OM in the sediments reacts with H_2S resulting in the formation of thiols. This is of particular importance in marine environments, where the concentration of sulfates is higher than in freshwater. In this way, DMA can be converted into $[(CH_3)_2As\ (III)]_2S$, a similar reaction being proposed for MMA. $(CH_3)_2AsSR$ and $(CH_3)As(SR)_2$ formation is promoted by thiols.

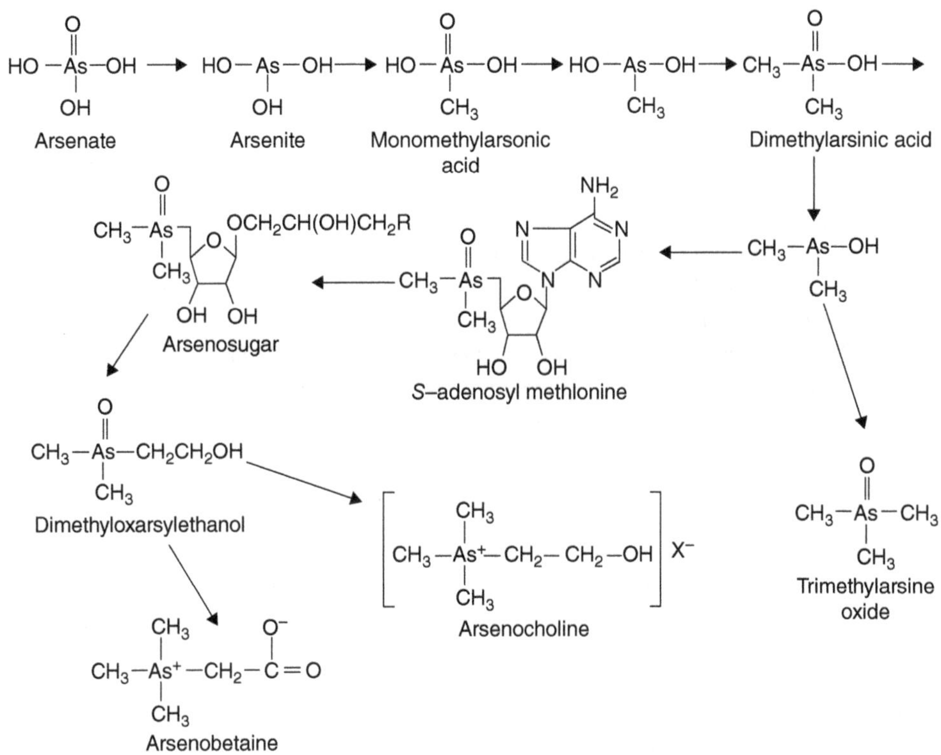

Figure 2.6 Potential pathways for reduction and methylation of As in terrestrial plants. *Source:* Meharg and Hartley-Whitaker (2002) with permission.

Organisms generally bioaccumulate arsenic as nontoxic organic compounds, suggesting that they may be the end products of metabolism as a form of detoxification of the ingested As. Organic As forms tend to appear at lower concentrations than the inorganic ones, although they may increase their proportion by methylation coming from microbial activity. Biologically mediated methylation-demethylation is important in the distribution of these species since these mechanisms are due exclusively to microorganisms present in soils, such as fungi, yeast, or bacteria (*Methanobacterium*, *Bacillus*, *Pseudomonas*, *Streptococcus*, *Staphylococcus*, *Aspergillus*, and *Penicillium*).

In plants, inorganic As and, to a lesser extent, organic compounds as MMA, DMA, TMAO, AB, TMA^+, AC, and arsenosugars can be found. The plants absorb As through the roots, in a process influenced by the characteristics of the soil (pH, OM, redox, bioavailability) and the plant itself. As(V) is picked up from the soil using the phosphorus membrane transporter, while, for As(III), a glycerol (aquaporin) transport channel is used. As(V) is reduced to As(III), which is subsequently bound to phytoquelatins. Plants are able of methylating As, although, due to the low concentrations of organic species, this is not a relevant pathway for the transformation of As species captured from the soil.

In fungi and lichens, AB and AC were detected, the first being the major species. Traces of the arsenosugar glycerol ribose were found.

In the case of terrestrial invertebrates, the ability to survive in soils with high As content depends on the metallothioneins they possess. Arsenic ingested or absorbed dermally in the trivalent state is bound to proteins rich in thiol groups and then methylated to their organic forms.

In the case of vertebrate animals, DMA was detected in blood as the major species, and in urine, DMA is the major species, followed by MMA, As(V), dimethylarsinoylethanol (DMAE), and TMA^+. Other detected compounds are DMAA, dimethylarsinothioyl acetate (DMAAS), and dimethylarsinic monosulfate (DMAS).

Organic species are found predominantly in marine organisms. In terrestrial environments, inorganic As species prevail, although small concentrations of organic species can be identified due to microbial activity (Quaghebeur and Rengel 2005).

Methylated species, which, as said, are products of degradation and/or are produced by bacterial methylation in situ, can be degraded to inorganic As in the absence of oxygen by facultative or obligate anaerobes. The highest concentrations of As methylated forms are found on the surface, in the photosynthetically active zone, caused by the greater microbial load and the presence of phytoplankton. Arsenite and arsenate inorganic compounds are converted into methylated toxic forms such as MMA(V), MMA(III), DMA(V), DMA(III), TMAO(V), and TMA^+ by biological methylation in terrestrial and marine organisms. These compounds are transformed into more complex (practically nontoxic) organic compounds, such as arsenolipids (AB, AC) and arsenosugars (Dembitsky and Levitsky 2004; Gebel 2001). MMA(V) is widely distributed as a minority species in the environment and living beings. MMA(III) is a reduced monomethylated form, a product of the metabolism of inorganic As, and is found in biological samples (urine and cells). It is currently considered the organoarsenical species with the highest degree of toxicity, and its sodium salt is used as an herbicide. DMA(V) is extensively disseminated in the environment and living beings, and in some cases, it may be the main species. It is the major metabolite of inorganic As, it is used as a herbicide and wood preservative, and it is considered teratogenic and genotoxic *in vitro* at high doses. DMA(III) is the reduced dimethylated form, a very unstable species, found in biological samples (urine and cells) as a product of the metabolism of inorganic As. It is one of the most toxic and genotoxic species. TMAO(V) is the final metabolite of inorganic As in animals such as rats and mice, although this fact has not been evidenced in humans. It is present as a minority form in some types of organisms, especially the marine ones. Its toxic effects are not yet known. AC, the trimethyl-2-hydroxyethylarsonium cation, has been found in some marine animals, mainly in white fish, shrimp, and crustaceans. AC is the precursor of the trimethyl-(carboxymethyl)arsonium zwitterion, AB, the most common arsenical species in marine organisms, of low toxicity. AB occurs principally in marine biota or marine sediments (Takeuchi et al. 2005) and surface waters (Ronkart et al. 2007). AB is the principal species in fish and mollusks, and it is present in freshwater and terrestrial organisms in varying amounts. It is almost entirely degraded to inorganic As through DMA(V). Macroscopic algae and phytoplankton, both in marine and freshwater environments, are As bioaccumulators because of the similar chemical structure of arsenate and phosphate, the essential nutrient for these microorganisms. The arsenical composition of algae is given mostly by a dimethylated arsenosugar containing the 5-deoxy-5-dimethylarsinylribofuranoside group. When the alga dies, microorganisms present in water and sediments cause the decomposition and release of the arsenosugar, most of which degrades to DMAE and then to DMA, MMA,

As(III), and As(V); the latter is the most abundant final species. In carnivorous and omnivorous zooplankton, the major species is AB; in contrast, in herbivores, the major species are arsenosugars, the As components of their basic food, algae (Ritchie et al. 2004). Aerobic microbes in marine environments quickly convert arsenosugars present in algae to DMAE, a precursor of AB. Following the death of marine organisms, AB is degraded to inorganic As via TMAO(V) or DMAA. This result is different from that of microorganisms linked to seaweed, where AB is transformed into TMAO(V) or DMA(III) but not into inorganic As. In marine animals, AB is the most abundant As compound except in bivalves, where phosphated arsenosugar is the main species; TMA^+, AC, and TMAO(V) are generally less found. In addition, many alkyl- and aryl-derivatives are commercialized for use as biocides in agriculture. After the death of marine organisms, AB decomposes by microbial action to less methylated compounds (DMA, MMA) and inorganic As. AB has low toxicity in cellular systems, and it is nontoxic *in vivo* in mammals. Therefore, moderate consumption of fishery products does not involve toxicological risk. TMA^+ is present in marine organisms (mainly bivalves and crustaceans, formed by the thermal decomposition of AB) and in terrestrial plants. It has a certain toxic character: DNA damage has been shown in human cell cultures.

TMAO is less common than MMA(V) and DMA(V), probably because of its rapid degradation into trimethylarsine or $(CH_3)_3As(V)S$; this is dependent on the relative quantities of H_2S and thiols or dithiols (cysteine, glutathione, and lipoic acid) along with numerous aerobic and anaerobic organisms.

2.6 Anthropogenic Arsenic Contamination

Although, as said, most natural sources of As are weathering, volcanic emissions, biological activity, and anthropogenic activities also greatly contribute to As pollution due to the increase of population, excessive groundwater extraction, or agricultural demand. Due to these effects, As is dissolved in the aquifers (Sarkar et al. 2012). Anthropogenic As contamination comes from industrial processes, mining (coal combustion), glass and ceramics industry (e.g. smelting of metals such as Cu), manufacturing of electronics, agriculture (e.g. pesticides, herbicides, and wood preservatives), feed additives of poultry, swine and cattle, pharmaceuticals and nutraceuticals, vitamin supplements, cosmetics, coal combustion, cigarettes, paints, dyes, and trioxide arsenals. The presence of organic As compounds in soil has been extensively studied due to the application of As compounds in agriculture and as an additive in animal food.

Among compounds used as pesticides, sodium arsenite, calcium arsenate, lead arsenate, copper acetoarsenite, and some organoarsenicals can be cited. Methylarsonic acid and DMA(V) are used as herbicides. In the farming industry, phenylarsenic compounds (arsanilic acid, i.e. aminophenylarsenic acid or aminophenylarsonic acid) are used as growth promoters in food additives for swine and poultry (Baeyens et al. 2019; Baig et al. 2015; Basu et al. 2014; Belkin et al. 2000; Goodell et al. 2007; Kumar et al. 2019; Mohan and Pittman Jr 2007; Ravenscroft et al. 2009; Smedley and Kinniburgh 2002). The chromated copper arsenate (CCA), used as a wood preservative, is prohibited nowadays in most countries (Goodell et al. 2007). Trimethyl arsine ($(CH_3)_3As$) or TMAs) was employed as a source of As in microelectronics, as a blocking component to other organoarsenical compounds, and as a ligand in coordination chemistry.

Salts of methyl- and dimethylarsinic acids were used as herbicides and pesticides. In addition, as said before, numerous synthetic As compounds are used as animal feed additives. For example, roxarsone (3-nitro-4-hydroxyphenylarsenic acid) has been used for chicken growth and control of coccidial infections. This compound is then transformed by microorganisms into 4-hydroxy-3-aminophenylarsonic acid, which contaminates soil and water. Europe banned its use in 1998.

2.7 Toxicity of Arsenic in Waters

Arsenic is a well-known poison, and it is not an essential element for the human body (Pirnie 2000). Although people are exposed to many As sources (water, food, air, and soil), the major assimilation route for toxic As is drinking water, especially where high As levels in surface and groundwaters can be found, such as Bangladesh. The relationship between clinical manifestations and the presence of As in groundwater drunk by people was first reported in 1913 in Bell Ville (Córdoba province, Argentina) (Bundschuh et al. 2012).

For total As, the major source is seafood, which fortunately has a high nontoxic AB fraction. Inorganic As(III) and As(V), MMA, and DMA are the most toxic species (Caumette et al. 2012; Santos et al. 2013). The toxic arsenic (toxic As) fraction is the sum of As(III), As(V) MMA, and DMA; this parameter is employed for risk assessment calculations for noncancer and cancer manifestations (Gao et al. 2018; Ventura-Lima et al. 2011; Xie et al. 2004).

Arsenic toxicity has been widely reported (Baig et al. 2015; Choong et al. 2007; Jain and Ali 2000; Kavcar et al. 2009). As said, inorganic As species are more toxic than most organic As compounds, As(III) being 60 times more toxic than As(V) (Jain and Ali 2000; Shinde et al. 2013; Vasken Aposhian et al. 2004).

Arsenic alters the human cells and genetic materials, causing serious diseases. The toxicity of inorganic As species takes place by two mechanisms: (i) the reaction of arsenite with protein sulfhydryl groups and (ii) the replacement of a phosphate group by arsenate. Several cancerous and noncancerous effects (either acute or nonacute) are associated with As exposure (Awual et al. 2019; Mohan and Pittman Jr 2007). Inorganic As is related to cancers as well as endocrine, immunological, and pulmonary disorders, liver diseases, cardiovascular effects, reproductive health effects, gastrointestinal and neurological disorders; genotoxicity; and dermal problems, among others (Chen et al. 2019; Liu et al. 2018; Sabir et al. 2019). Inorganic As methylation may cause cancers and dermal and gastrointestinal disorders (Alka et al. 2021; Fan et al. 2019). The International Agency for Research on Cancer has included As in group 1 (IARC 2019; Nurchi et al. 2020), i.e. the carcinogenic group. As said, As is one of the principal carcinogenic elements (Baig et al. 2015; Litter et al. 2010). For example, a $3\,mg\,l^{-1}$ concentration of As in drinking water can generate bladder and lung cancers. Furthermore, As is teratogenic, crossing the placental membrane and affecting the fetus metabolic system. Additionally, it can pass to skin, urine, fingers, hair, and nails; as a cumulative substance in hair, nails, and urine, it converts them in ideal biomarkers related to As exposure. Ischemic heart disease, blackfoot disease, cardiovascular diseases, chronic kidney disease unknown (CKDu), Bowen's disease, and Alzheimer's disease (Basu et al. 2014) are some other health problems that can be related to As consumption. Even at concentrations lower than $50\,mg\,l^{-1}$, As can play a role in birth complications, affecting neurodevelopment in early life exposure (Zheng 2020). Uterine contamination by As causes changes in pathways in gene expression associated with diabetes (Navas-Acien et al. 2019), resulting in the risk of adult diabetes. Using a Health Effects of Arsenic Longitudinal Study (HEALS) on a genome-wide DNA methylation analysis of 396 adults (Demanelis et al. 2019), it was concluded that epigenome is a general pathway of As toxicity.

As said before, biological As methylation is a metabolic pathway that takes place in most mammalians. Initially, biological methylation was assumed as a process for As detoxification because of the low toxicity of methylated As(V) species. However, it was concluded later that methylated arsenicals, especially as As(III) forms, produce carcinogenic compounds (Drobna et al. 2009), as monomethylarsonous acid (MMA(III)) as an intermediate of As metabolism. After chronic exposure to inorganic As, MMA(III) is detected in human urine, causing acute toxicity (Hughes 2002; Kumarathilaka et al. 2019).

Human As consumption is a public health problem in many geographical areas of our planet. Additionally, As is also noxious to marine animals (Alka et al. 2021; Yin et al. 2017). As stated earlier, inorganic As causes skin lesions, syndromes in vascular and nervous systems, and cancer because of its toxicity; the element is not beneficial for metabolic functions (Podgorski and Berg 2020). High As concentrations in groundwater are harmful to the users who employ this type of water as the primary source for drinking. Arsenic is detrimental also to agricultural activities by contaminating the soils. Arsenic exposure affects indirectly millions of people worldwide, particularly in partial-arid and arid lands, where groundwater is the principal source for drinking and farming (Li et al. 2020). Additionally, inorganic As significantly contaminates water involved in food, especially rice. Long exposure to water containing high inorganic As concentrations ($>100\,mg\,l^{-1}$) is associated with cancers (lung, nonmelanoma skin, and bladder; Zheng (2020)). Drinking well water with As concentrations close to $10\,mg\,l^{-1}$ poses a high risk. Particularly, in Bangladesh, approximately 20–45 million people were reported at high risk, and a death rate of 1 in every 18 adult deaths has been reported (Zheng 2020). This problem should be faced globally through multisector methodologies. As a consequence, the World Health Organization (WHO) and other health organizations proposed to lower the As guideline value in drinking water from 50 to $10\,\mu g\,l^{-1}$ (Litter et al. 2010; WHO 2011). The WHO retained the provisional value of $10\,\mu g\,l^{-1}$ for inorganic As in its 2017 edition of *Guidelines for Drinking-Water Quality*; this decision was based on the performance of the treatment and the analytical viability (World Health Organization 2017; Zheng 2020). Based on cost-benefit studies, the federal Maximum Contaminant Level (MCL) for As in the United States has been regulated also in $10\,\mu g\,l^{-1}$. However, in other countries, the allowed limit is still $50\,\mu g\,l^{-1}$. The MCL difference is worrying because new researches indicate that even $10\,\mu g\,l^{-1}$ may be not adequately safe, particularly in early and biologically vulnerable stages of life (Zheng 2020).

2.8 Conclusion

The wide distribution of As in the environment is explained by the different sources from which this element can originate and the varied inorganic and organic species in which the element can exist. Their distribution in different media is regulated by environmental factors, from which they can be mobilized and transported. This makes As enormously problematic because its mobility is relatively wide in a large range of redox conditions. It is necessary to emphasize that high concentrations of naturally occurring As in groundwater are not localized and, therefore, affect large areas of the world.

The understanding of the distribution of arsenic species in water is mandatory to understand which are the geochemical characteristics able to increase As concentration in various hydrogeological conditions. Speciation is principally controlled by the redox potential and the pH, which cause the high mobility of the element and the profound variation in its concentration in different water bodies (underground, lakes, sea, deep water, rivers, etc.).

Arsenic organic compounds, both pentavalent (majority) and trivalent, are amply distributed in waters, soils, sediments, atmosphere, and biological tissues; they are found mainly in marine organisms. As said, inorganic arsenite and arsenate are biologically converted by methylation reactions into less toxic compounds and precursors of practically nontoxic organic forms such as arsenolipids, AB, AC, and arsenosugars (see Section 2.5). The highest natural As concentrations found in groundwater result from strong interactions between water and rocks, together with favorable conditions for mobilization and accumulation. Atmospheric As, originated from volcanic emissions, volatilization of soils, wind erosion, marine aerosols, etc., deposits onto the earth and is introduced to natural waters from minerals (mainly arsenopyrite).

Arsenic tests in all domestic well water worldwide are imperative due to the known public health problem. Thus, it is needed to count on friendly, accurate, and affordable procedures for testing inorganic As in water, at least to rapidly monitor on site the element to prevent the problem in possible exposed populations. Dispersed rural or peri urban populations would be the aim of the studies on arsenic occurrence and mitigation. As described earlier, geographical sources, biological activity, and anthropogenic activities are the origins of the presence of As in the ecosystems.

References

Adriano, D. (2001). *'Arsenic' in Trace Elements in Terrestrial Environments: Biogeochemistry, Bioavailability and Risks of Metals*, 2nd.e, 220–256. New York: Springer-Verlag.

Alka, S., Shahir, S., Ibrahim, N. et al. (2021). Arsenic removal technologies and future trends: a mini review. *J. Clean. Prod.* 278: 123805.

Alvarez Benedí, J., Bolado Rodriguez, S., Cancillo Carro, I., and Calvo Revuelta, C. (2003). Adsorption – desorption dynamics of arsenic (V) in cultivated soils in Castile and Leon. In: *Studies of the Unsaturated Zone of Soil*, vol. VI (ed. J. Alvarez-Benedí and P. Marinero), 331–338. Valladolid Spain: Instituto Tecnológico Agrario de Castilla y León.

Asere, T.G., Stevens, C.V., and Du Laing, G. (2019). Use of (modified) natural adsorbents for arsenic remediation: a review. *Sci. Total Environ.* 676: 706–720.

Awual, M.R., Hasan, M.M., Asiri, A.M., and Rahman, M.M. (2019). Cleaning the arsenic(V) contaminated water for safe-guarding the public health using novel composite material. *Compos. B: Eng.* 171: 294–301.

Baeyens, W., Mirlean, N., Bundschuh, J. et al. (2019). Arsenic enrichment in sediments and beaches of Brazilian coastal waters. *Sci. Total Environ.* 681: 143–154.

Baig, S.A., Sheng, T., Hu, Y. et al. (2015). Arsenic removal from natural water using low cost granulated adsorbents: a review. *Clean-Soil, Air, Water* 43: 13–26.

Baloch, M.Y.J., Talpur, S.A., Talpur, H.A. et al. (2020). Effects of arsenic toxicity on the environment and its remediation techniques: a review. *J. Water Environ. Technol.* 18: 275–289.

Basu, A., Saha, D., Saha, R. et al. (2014). A review on sources, toxicity and remediation technologies for removing arsenic from drinking water. *Res. Chem. Intermed.* 40: 447–485.

Begerow, J., Dunemann, L., and Sur, R. (2000). Arsenic species (As(III), As(V), monomethylarsonic acid, dimethylarsinic acid), series analyses of hazardous substances in biological materials. *Biomonit. Methods* 7: 97–117.

Belkin, H., Zheng, B., Finkelman, R. (2000). Human health effects of domestic coal combustion in rural China: a causal factor for arsenic and fluoride poisoning. Second China World Conference of Geological Sciences (August 2000), Stanford University, 522–524.

BGS and DPHE (British Geological Survey/ Bangladesh Department for Public Health 407 Engineering) (2001). Arsenic contamination of groundwater in Bangladesh, Final Report. 408. In: *British Geological Survey Report WC/00/19*, vol. 2 (ed. D. Kinniburgh and P. Smedley). Keyworth: British Geological Survey.

Bhattacharya, P., Hasan, M.A., Sracek, O. et al. (2009). Groundwater chemistry and arsenic mobilization in the Holocene flood plains in south-central Bangladesh. *Environ. Geochem. Health* 31: 23–44.

Bundschuh, J., Armienta, M.A., Morales-Simfors, N. et al. (2020). Arsenic in Latin America: new findings on source, mobilization and mobility in human environments in 20 countries based on decadal research 2010–2020. *Crit. Rev. Environ. Sci. Technol.* 1–139: https://doi.org/10.1080/10643389.2020.1770527.

Bundschuh, J., Farias, B., Martin, R. et al. (2004). Groundwater arsenic in the Chaco-Pampean Plain, Argentina. *Appl. Geochem.* 19: 231–243.

Bundschuh, J., Litter, M.I., Parvez, F. et al. (2012). One century of arsenic exposure in Latin America: a review of history and occurrence from 14 countries. *Sci. Total Environ.* 429: 2–35.

Caetano, M. and Vale, C. (2002). Retention of arsenic and phosphorus in iron-rich concretions of Tagus salt marshes. *Mar. Chem.* 79: 261–271.

Cagnin, R.C., Quaresma, V.S., Chaillou, G. et al. (2017). Arsenic enrichment in sediment on the eastern continental shelf of Brazil. *Sci. Total Environ.* 607–608: 304–316.

Calvo, C., Alvarez-Benedí, J., Andrade, M. et al. (2003). Arsenic pollution in groundwater in the province of Valladolid: seasonal variations. In: *Studies of the Unsaturated Zone of Soil*, vol. VI (ed. J. Alvarez-Benedí and P. Marinero), 91–98. Valladolid Spain: Instituto Tecnológico Agrario de Castilla y León.

Cao, X. and Ma, L. (2004). Effects of compost and phosphate on the accumulation of plant arsenic in soils near pressure treated wood. *Environ. Pollut.* 132: 435–442.

Caumette, G., Koch, I., and Reimer, K.J. (2012). Arsenobetaine formation in plankton: a review of studies at the base of the aquatic food chain. *J. Environ. Monit.* 14: 2841–2853.

Chen, H., Zhao, T., Sun, D. et al. (2019). Changes of RNA N^6-methyladenosine in the hormesis effect induced by arsenite on human keratinocyte cells. *Toxicol. Vitro* 56: 84–92.

Choong, T.S.Y., Chuaha, T.G., Robiah, Y. et al. (2007). Arsenic toxicity, health hazards and removal techniques from water: an overview. *Desalination* 217: 139–166.

da Silva, E.B., Mussoline, W.A., Wilkie, A.C., and Ma, L.Q. (2019). Anaerobic digestion to reduce biomass and remove arsenic from As-hyperaccumulator *Pteris vittata*. *Environ. Pollut.* 250: 23–28.

De Gieter, M. and Baeyens, W. (2005). Arsenic in fish: implications for human health. In: *Reviews in Food and Nutrition Toxicity*, vol. 4, 57–83. Boca Raton: CRC Press.

Dehbandi, R., Abbasnejad, A., Karimi, Z. et al. (2019). Hydrogeochemical controls on arsenic mobility in an arid inland basin, Southeast of Iran: the role of alkaline conditions and salt water intrusion. *Environ. Pollut.* 249: 910–922.

Demanelis, K., Argos, M., Tong, L. et al. (2019). Association of arsenic exposure with whole blood DNA methylation: an epigenome-wide study of Bangladeshi adults. *Environ. Health Perspect.* 127: 057011.

Dembitsky, V.M. and Levitsky, D.O. (2004). Arsenolipids. *Prog. Lipid Res.* 43: 403–448.

Doherty, S.J., Tighe, M.K., and Wilson, S.C. (2017). Evaluation of amendments to reduce arsenic and antimony leaching from co-contaminated soils. *Chemosphere* 174: 208–217.

Drobna, Z., Styblo, M., and Thomas, D.J. (2009). An overview of arsenic metabolism and toxicity. *Curr. Protoc. Toxicol.* 42: https://doi.org/10.1002/0471140856.tx0431s42.

Fan, X.Y., Liu, Y.J., Cai, Y.M. et al. (2019). A mitochondria-targeted organic arsenical accelerates mitochondrial metabolic disorder and function injury. *Bioorg. Med. Chem.* 27: 760–768.

Fendorf, S., Michael, H.A., and van Geen, A. (2010). Spatial and temporal variations of groundwater arsenic in South and Southeast Asia. *Science* 328: 1123–1127.

Foster, A. (2003). Spectroscopic investigation of arsenic species in solid phases. In: *Arsenic in Groundwater: Geochemistry and Occurrence* (ed. A.H. Welch and K.G. Stollenwerk), 27–65. Kluwer Academic Publishers.

Gao, Y., Baisch, P., Mirlean, N. et al. (2018). Arsenic speciation in fish and shellfish from the North Sea (Southern Bight) and Açu Port area (Brazil) and health risks related to seafood consumption. *Chemosphere* 191: 89–96.

Gebel, T.W. (2001). Genotoxicity of arsenical compounds. *Int. J. Hyg. Environ. Health* 203: 249–262.

Gómez Ariza, J.L., Morales, E., Sánchez-Rodas, D., and Giráldez, I. (2000). Stability of chemical species in environmental matrices'. *Trends Anal.Chem.* 19 (2+3): 200–209.

Goodell, B., Merrick, P., Jellison, J., and Yuhui, Q. (2007). Marine borer degradation of treated and untreated parallel strand lumber panels. *For. **Prod. J***. 57: 72–76.

Gulens, J., Champ, D.R., and Jackson, R.E. (1979). Influence of redox environments on the mobility of arsenic in ground water. *Chem. Model. Aqueous Syst.* 81–95. doi:10.1021/bk-1979-0093.ch004.

Guo, H., Yang, S., Tang, X. et al. (2008). Groundwater geochemistry and its implications for arsenic mobilization in shallow aquifers of the Hetao Basin, Inner Mongolia. *Sci. Total Environ.* 393: 131–144.

Hare, V., Chowdhary, P., Kumar, B. et al. (2019). Arsenic toxicity and its remediation strategies for fighting the environmental threat. In: *Emerging and Eco-Friendly Approaches for Waste Management* (ed. R.N. Bharagava and P. Chowdhary), 143–170. Singapore: Springer.

Harvey, C.F., Swartz, C.H., Badruzzaman, A.B.M. et al. (2002). Arsenic mobility and groundwater extraction in Bangladesh. *Science* 298: 16602–11606.

He, X., Li, P., Ji, Y. et al. (2020). Groundwater arsenic and fluoride and associated arsenicosis and fluorosis in China: occurrence, distribution and management. *Expo. Health* https://doi.org/10.1007/s12403-020-00347-8 (accessed 3 December 2020).

Huang, R., Gao, S., Wang, W. et al. (2006). Soil arsenic availability and the transfer of soil arsenic to crops in suburban areas in Fujian Province, southeast China. *Sci. Total Environ.* 368: 531–541.

Hughes, M.F. (2002). Arsenic toxicity and potential mechanisms of action. *Toxicol. Lett.* 133: 1–16.

Huq, M.E., Fahad, S., Shao, Z. et al. (2020). Arsenic in a groundwater environment in Bangladesh: occurrence and mobilization. *J. Environ. Manage.* 262: 110318.

IARC, (2019). List of classifications, agents classified by the IARC Monographs, Volumes 1–124. IARC Monographs on the Evaluation of Risk to Humans. July 7, 2019. Retrieved July 14, 2019.

Uppal, J.S., Zheng, Q., and Le, X.C. (2019). Arsenic in drinking water – recent examples and updates from Southeast Asia. *Curr. Opin. Environ. Sci. Health* 7: 126–135.

Jain, C. and Ali, I. (2000). Arsenic: occurrence, toxicity and speciation techniques. *Water Res.* 34: 4304–4312.

Kavcar, P., Sofuoglu, A., and Sofuoglu, S.C. (2009). A health risk assessment for exposure to trace metals via drinking water ingestion pathway. *Int. J. Hyg. Environ. Health* 212: 216–227.

Koch, I., Wang, L., Ollson, C.A. et al. (2000). The predominance of inorganic arsenic species in plants from Yellowknife, Northwest Territories, Canada. *Environ. Sci. Technol.* 34: 22–26.

Kumar, R., Patel, M., Singh, P. et al. (2019). Emerging technologies for arsenic removal from drinking water in rural and peri-urban areas: methods, experience from, and options for Latin America. *Sci. Total Environ.* 694: 133427.

Kumarathilaka, P., Seneweera, S., Ok, Y.S. et al. (2019). Arsenic in cooked rice foods: assessing health risks and mitigation options. *Environ. Int.* 127: 584–591.

Li, Z., Yang, Q., Yang, Y. et al. (2020). Hydrogeochemical controls on arsenic contamination potential and health threat in an intensive agricultural area, northern China. *Environ. Pollut.* 256: 113455.

Lillo, J. (2003). Peligros geoquímicos: arsénico de origen natural en las aguas. Escuela Superior de Ciencias Experimentales y Tecnología, Universidad Rey Juan Carlos. https://www.ucm.es/data/cont/media/www/pag-15564/Peligros%20geoqu%C3%ADmicos%20del%20ars%C3%A9nico%20-%20Javier%20Lillo.pdf (accessed 3 December 2020).

Litter, M.I., Morgada, M.E., and Bundschuh, J. (2010). Possible treatments for arsenic removal in Latin American waters for human consumption. *Environ. Pollut.* 158: 1105–1118.

Liu, Q., Leslie, E.M., Moe, B. et al. (2018). Metabolism of a phenylarsenical in human hepatic cells and identification of a new arsenic metabolite. *Environ. Sci. Technol.* 52: 1386–1392.

Luong, V.T., Cañas Kurz, E.E., Hellriegel, U. et al. (2018). Iron-based subsurface arsenic removal technologies by aeration: a review of the current state and future prospects. *Water Res.* 133: 110–122.

Maity, J.P., Ho, P.-R., Huang, Y.-H. et al. (2019). The removal of arsenic from arsenic-bearing groundwater in In-situ and environment using novel natural magnetic rock material and synthesized magnetic Ex-situ material as adsorbent: a comparative assessment. *Environ. Pollut.* 253: 768–778.

Mandal, B. and Suzuki, K. (2002). Arsenic round the world: a review. *Talanta* 58: 201–235.
Matschullat, J. (2000). Arsenic in the geosphere a review. *Sci. Total Environ.* 249: 297–312.
McArthur, J.M., Banerjee, D.M., Hudson-Edwards, K.A. et al. (2004). Natural organic matter in sedimentary basins and its relation to arsenic in anoxic ground water: the example of West Bengal and its worldwide implications. *Appl. Geochem.* 19: 1255–1293.
Meharg, A. and Hartley-Whitaker, J. (2002). Arsenic uptake and metabolism in arsenic resistant and nonresistant plant species. *New Phytol.* 154: 29–43.
Mirlean, N., Baisch, P., and Diniz, D. (2014). Arsenic in groundwater of the Paraiba do Sul delta, Brazil: an atmospheric source? *Sci. Total Environ.* 482–483: 148–156.
Mohan, D. and Pittman, C.U. Jr. (2007). Arsenic removal from water/wastewater using adsorbents – a critical review. *J. Hazard. Mater.* 142: 1–53.
Moody, B. (2013). *Comparative Inorganic Chemistry*, 381. Elsevier.
Navas-Acien, A., Spratlen, M.J., Abuawad, A. et al. (2019). Early-life arsenic exposure, nutritional status, and adult diabetes risk. *Curr. Diab. Rep.* 19: 147.
Ng, J.C., Wang, J., and Shraim, A. (2003). A global health problem caused by arsenic from natural sources. *Chemosphere* 52: 1353–1359.
Nickson, R.T., McArthur, J.M., Shrestha, B. et al. (2005). Arsenic and other drinking water quality issues, Muzaffargarh District, Pakistan. *Appl. Geochem.* 20: 55–68.
Nurchi, V.M., Buha Djordjevic, A., Crisponi, G. et al. (2020). Arsenic toxicity: molecular targets and therapeutic agents. *Environ. Sci. Biomol.* 10: 235.
Pirnie, M. (2000). Technologies and costs for removal of arsenic from drinking water, EPA 815-R-00-028, 2000.
Podgorski, J. and Berg, M. (2020). Global threat of arsenic in groundwater. *Science* 368: 845–850.
Polya, D.A. and Middleton, D.R. (2017). Arsenic in drinking water: sources and human exposure. In: *Best Practice Guide on the Control of Arsenic in Drinking Water* (ed. P. Bhattacharya, D.A. Polya and D. Jovanovic), 1–23. IWA Publishing.
Postma, D., Larsen, F., Hue, N.T.M. et al. (2007). Arsenic in groundwater of the Red River floodplain, Vietnam: controlling geochemical processes and reactive transport modelling. *Geochim. Cosmochim. Acta* 71: 5054–5071.
Quaghebeur, M. and Rengel, Z. (2005). Arsenic speciation governs arsenic uptake and transport in terrestrial plants. *Microchim. Acta* 151: 141–152.
Rahman, M.M., Ng, J.C., and Naidu, R. (2009). Chronic exposure of arsenic via drinking water and its adverse health impacts on humans. *Environ. Geochem. Health* 31: 189–200.
Ravenscroft, P., Brammer, H., and Richards, K. (2009). *Arsenic Pollution, A Global Synthesis*, 1e. Chichester, West Sussex: Wiley-Blackwell.
Ritchie, A.W., Edmonds, J.S., Gössler, W., and Jenkins, R.O. (2004). An origin for arsenobetaine involving bacterial formation of an arsenic–carbon bond. *FEMS Microbiol. Lett.* 235: 95–99.
Ronkart, S.N., Laurent, V., Carbonnelle, P. et al. (2007). Speciation of five arsenic species (arsenite, arsenate, MMAAV, DMAAV and AsBet) in different kind of water by HPLC-ICP-MS. *Chemosphere* 66: 738–745.
Roychowdhury, T., Tokunaga, H., Uchino, T., and Ando, M. (2005). Effect of arsenic-contaminated irrigation water on agricultural land soil and plants in West Bengal, India. *Chemosphere* 58: 799–810.
Sabir, S., Akash, M.S.H., Fiayyaz, F. et al. (2019). Role of cadmium and arsenic as endocrine disruptors in the metabolism of carbohydrates: inserting the association into perspectives. *Biomed. Pharmacother.* 114: 108802.

Santos, C.M.M., Nunes, M.A.G., Barbosa, I.S. et al. (2013). Evaluation of microwave and ultrasound extraction procedures for arsenic speciation in bivalve mollusks by liquid chromatography inductively coupled plasma-mass spectrometry. *Spectrochim. Acta Part B.* 86: 108–114.

Sarkar, S., Greenleaf, J.E., Gupta, A. et al. (2012). Sustainable engineered processes to mitigate the global arsenic crisis in drinking water: challenges and progress. *Annu. Rev. Chem. Biomol. Eng.* 3: 497–517.

Selim Reza, A.H.M., Jean, J.-S., Yang, H.-J. et al. (2010). Occurrence of arsenic in core sediments and groundwater in the Chapai-Nawabganj District, northwestern Bangladesh. *Water Res.* 44: 2021–2037.

Shaji, E., Santosh, M., Sarath, K.V. et al. (2021). Arsenic contamination of groundwater: a global synopsis with focus on the Indian Peninsula. *Geosci. Front.* 12 (2021): 101079. https://doi.org/10.1016/j.gsf.2020.08.015 (Accessed: 13 Jan 2020).

Shakya, A.K. and Ghosh, P.K. (2018). Simultaneous removal of arsenic and nitrate in absence of iron in an attached growth bioreactor to meet drinking water standards: Importance of sulphate and empty bed contact time. *J. Clean. Prod.* 186: 304–312.

Shinde, R.N., Pandey, A., Acharya, R. et al. (2013). Chitosan-transition metal ions complexes for selective arsenic (V) preconcentration. *Water Res.* 47: 3497–3506.

Singh, R., Singh, S., Parihar, P. et al. (2015). Arsenic contamination, consequences and remediation techniques: a review. *Ecotoxicol. Environ. Saf.* 112: 247–270.

Smedley, P. and Kinninburgh, D. (2002). A review of the source, behaviour and distribution of arsenic in natural waters. *Appl. Geochem.* 17: 517–568.

Smedley, P., Macdonald, D., Nicolli, H. et al. (2000). Arsenic and other quality problems in groundwater from northern La Pampa province, Argentina. British Geological Survey Technical Report, WC/99/36. British Geological Survey, Keyworth.

Smedley, P.L. and Kinniburgh, D.G. (2001). Chapter 1. Source and behaviour of arsenic in natural waters. In: *United Nations Synthesis Report on Arsenic in Drinking-Water Water and Sanitation Protection of the Human Environment* (ed. P.L. Smedley and D.G. Kinniburgh). Wallingford, Oxon: British Geological Survey.

Takeuchi, M., Terada, A., Nanba, K. et al. (2005). Distribution and fate of biologically formed organoarsenicals in coastal marine sediment. *Appl. Organomet. Chem.* 19: 945–951.

Vasken Aposhian, H., Zakharyan, R.A., Avram, M.D. et al. (2004). A review of the enzymology of arsenic metabolism and a new potential role of hydrogen peroxide in the detoxication of the trivalent arsenic species. *Toxicol. Appl. Pharmacol.* 198: 327–335.

Vatutsina, O., Soldatov, V., Sokolova, V. et al. (2007). A new hybrid (polymer/inorganic) fibrous sorbent for arsenic removal from drinking water. *React. Funct. Polym.* 67: 184–201.

Ventura-Lima, J., Bogo, M.R., and Monserrat, J.M. (2011). Arsenic toxicity in mammals and aquatic animals: a comparative biochemical approach. *Ecotoxicol. Environ. Saf.* 74: 211–218.

Wang, S. and Mulligan, C. (2006). Occurrence of arsenic contamination in Canada: sources, behavior and distribution. *Sci. Total Environ.* 366: 701–721.

Warren, G., Alloway, B., Lepp, N. et al. (2003). Field trials to assess the uptake of arsenic by vegetables from contaminated soils and soil remediation with iron oxides. *Sci. Total Environ.* 311: 19–33.

Weerasundara, L., Ok, Y.-S., and Bundschuh, J. (2021). Selective removal of arsenic in water: a critical review. *Environ. Pollut.* 268: 115668.

WHO (2011). Arsenic in drinking-water. Background document for preparation of WHO Guidelines for drinking-water quality. Geneva, World Health Organization (WHO/SDE/WSH/03.04/75/Rev/1).

World Health Organization (2017). Guidelines for drinking-water quality: fourth edition incorporating the first addendum.

Xie, Y.X., Trouba, K.J., Liu, J. et al. (2004). Biokinetics and subchronic toxic effects of oral arsenite, arsenate, monomethylarsonic acid, and dimethylarsinic acid in v-Ha-ras transgenic (Tg.AC) mice. *Environ. Health Perspect.* 112: 1255–1263.

Yadav, S.K., Ramanathan, A.L., Kumar, M. et al. (2020). Assessment of arsenic and uranium co-occurrences in groundwater of central Gangetic Plain, Uttar Pradesh, India. *Environ. Earth Sci.* 79: 154.

Yan, X., Kerrich, R., and Hendry, M. (2000). Distribution of arsenic(III), arsenic(V) and total inorganic arsenic in pore-waters from a thick till and clay-rich aquitard sequence, Saskatchewan, Canada. *Geochim. Cosmochim. Acta* 64: 2637–2648.

Yin, H., Kong, M., Gu, X., and Chen, H. (2017). Removal of arsenic from water by porous charred granulated attapulgite-supported hydrated iron oxide in bath and column modes. *J. Clean. Prod.* 166: 88–97.

Yong, R. and Mulligan, C. (2004). *Natural Attenuation of Contaminants in Soils*. Boca Raton, FL: CRC Press.

Zakhar, R., Derco, J., and Cacho, F. (2018). An overview of main arsenic removal technologies. *Acta Chim. Slov.* 11: 107–113.

Zheng, Y. (2020). Global solutions to a silent poison. *Science* 368 (6493): 818–819.

3

Arsenic Transport and Metabolism in Plants

Gerald Zvobgo

Department of Crop Productivity and Molecular Technology, Tobacco Research Board, Harare, Zimbabwe

3.1 Introduction

Arsenic (As) is a toxic, nonessential metalloid that is ubiquitous in the environment. Absorption of As from the soil into the plant's rhizome poses significant problems to the plant and the consumer who is at the periphery of the food chain. Entry of the toxic metalloid into the environment is through leaching from geological formations. Since As is not essential for any life process and no organism known to date utilizes it for its metabolism, all uptake pathways are self-made through mimicking other essential materials such as phosphate, glucose, and glycerol (Garbinski et al. 2019). Interestingly, in a study of *Halomonadaceae*, a halophilic protobacterium, it has been argued that As can be a viable substitute for phosphorus in its biological function (Wolfe-Simon et al. 2011). However, substantial data is required to establish concrete evidence, and if this is true, it could be a groundbreaking breakthrough that will address issues of the predicted phosphorus depletion and As toxicity. Speciation studies have revealed two biologically relevant oxidation states of As: the trivalent arsenite (AsIII) and the pentavalent arsenate (AsV). Additionally, methylated species including methylarsenite (MAsIII), methylarsenate (MAsV), dimethylarsinite (DMAsIII), dimethylarsenate (DMAsV), trimethylarsine (TMAsIII), and trimethylarsine oxide (TMAs(V)O) have also been identified (Garbinski et al. 2019). The nature of the As species determines their occurrence and transportation, for instance, the pentavalent As species is more dominant in aerated conditions in contrast to the trivalent one. Generally, the trivalent aromatic arsenicals are rendered to be more toxic than the other species (Dopp et al. 2005). In food stuffs, some organic forms of As have been commonly found, which include arsenolipids, arsenochlorine, arsenobetaine, and arsenosugars.

Due to the ubiquitousness of As, excess uptake of it can present significant problems in food safety (Zhao et al. 2009). Some plants such as rice, which have an efficient As

Arsenic in Plants: Uptake, Consequences and Remediation Techniques, First Edition.
Edited by Prabhat Kumar Srivastava, Rachana Singh, Parul Parihar, and Sheo Mohan Prasad.

uptake system, present a plethora of As toxicity-related consequences, especially in half of the world's inhabitants who rely on rice as their primary source of starch. To formulate sustainable mitigation strategies for As phytotoxicity, it is imperative to understand how As is transported and metabolized in plants. Such understanding will pave the way to the formulation of countermeasures against As. In As transport, four significant elements are of crucial importance, namely, phosphorus (P), silicon (Si), sulfur (S), and iron (Fe) (Zhao et al. 2009). The transporters used by plants in accumulating P and S are shared with the inorganic As species, while the oxidized Fe can provide a covering that can help metal sequestration in plants grown under wet conditions providing a barrier and thereby changing the As uptake pathway via the roots (Azizur Rahman et al. 2011). Thus, this section will unravel the mechanisms of As transport and metabolism in plants.

3.2 Arsenite Influx and Efflux

In soils where anaerobic conditions prevail, As occurs as As(III) attributable to the vast plummeting activity of microbes (Ali et al. 2009). Unlike As(V), the pathway of As(III) uptake is less defined (Liu et al. 2002). The primary As(III) transport pathway was characterized in 1997 (Sanders et al. 1997); after that great strides have been made in identifying the proteins responsible for the uptake and sequestration of As(III). In mammals, As(III) has been shown to be transported via the glycerol channel GlpF, a member of the major intrinsic protein (MIP) or aquaporin (AQP) (Agre et al. 2002). Plants, like mammals, facilitate the uptake of As(III) through the superfamily of aquaporins. The AQPs are bidirectional transporters with small sizes, and these include the well-known water channels. On the contrary, the aquaglyceroporins are sizable to allow hefty molecules like glycerol to penetrate (Garbinski et al. 2019). The similarity of $As(OH)_3$ (form of As(III) in neutral conditions) with glycerol allows As(III) to be transported via the same channels resulting in toxicity. In plants, the AQP has been grouped into five major categories (Maurel et al. 2015).

Uptake of As(III) from the roots is necessitated by aquaporins of the NOD26-like intrinsic proteins (NIP) and plasma membrane intrinsic protein (PIP) subfamilies (Kumar et al. 2019) as evidenced in *Arabidopsis* (Xu et al. 2015) and rice (Ma et al. 2006). Functional analysis of the NP7;1 in rice demonstrated increased As tolerance in the loss of function mutant (Lindsay and Maathuis 2016). In *Arabidopsis thaliana*, several NIPs have been shown to mediate As(III) uptake into the roots (Garbinski et al. 2019), while in yeast, a strain void of As(III) transport systems exhibited the bidirectional nature of the aquaporins (Bienert et al. 2008). The NIP2;1 in rice was also shown to mediate silicon uptake (Ma et al. 2006).

It is true that As(III) transport is of significant consideration in aquatic plants like rice because of the growth of the roots in depleted oxygen conditions where As(III) prevails. Since rice plants take up Si from the soil rhizosphere in order to support its husks, the same transport pathway is utilized by As(III), hence causing toxicity (Shinde and Kumar 2021). The uptake of As(III) depends upon the varieties and the growing conditions with some varieties accumulating very high concentrations of As. The Si transporter Lsi1 located in the distal side of the plasma membrane is responsible for As(III) influx, while Lsi2 located in the neighboring sides of the identical roots is responsible for As(III) loading from the roots into the xylem (Ye et al. 2017).

3.3 Arsenate Influx and Efflux

As(V) is similar to P and is transported through the plasmalemma via the phosphate transporters (PHTs) (Wu et al. 2011). Uptake of As(V) and P is competitive via the same transport systems in As hyperaccumulators (Tu and Ma 2005), As-tolerant non-hyperaccumulators (Meharg and Macnair 1992; Bleeker et al. 2003), and nonaccumulators (Abedin et al. 2002b; Esteban et al. 2003). It is widely agreed that As toxicity is achieved by its replacement of P in chemical reactions.

3.3.1 Arsenate and Phosphate Chemistry

As(V) and P are oxyanions occurring in a similar group in the periodic table called the pnictogens. The members of this family exhibit similarity in the electron arrangement, especially in the outer shell, thereby subsequently displaying resemblances in chemical behavior (Tu and Ma 2004). Although As is highly biotoxic, P is indispensable. Apart from the four most abundant elements of life, P is one of the key life elements. Due to its major roles in the metabolism of all known lives, mainly in providing the link between the nucleotides in DNA and RNA, it is not possible for As(V) to usurp all the roles performed by P (Tawfik and Viola 2011). At the same concentration, the competition of sorption between As and P varies with the soil pH and other factors (Violante and Pigna 2002). Both As(V) and P possess the same charge on their oxygen atom and nearly identical thermochemical radii and acid dissociation constants (Kish and Viola 1999). In a study in *Pteris vittata*, PvPht1.3 had a Km value for P uptake of 22.1 μM, and the Ki of As was 14.80 μM, suggesting similar affinities for P and As (Ditusa et al. 2016).

In regard to size, the overall sizes of As(V) and P ions are similar with the thermochemical radius of P being 2.38 Å while that of As(V) being 2.48 Å (Frausto da Silva 2001). Slight variations such as bond length with oxygen occur between As(V) and P (Kish and Viola 1999). Furthermore, in regard to esterification, As esters are highly unstable in contrast to the P counterparts with the hydrolysis of As(V) esters leading to unproductive cycles (Dixon 1996).

It is also interesting to note that enzymes that utilize P have the same binding mode and kinetic parameters as As(V), which therefore inhibits metabolism when As(V) replaces P (Dixon 1996). It is worthy to note that enzymes that utilize P can imperfectly substitute As(V) as substrate, as As(V) exhibits stability problems (Figure 3.1). In addition, As(V) has been shown to be a better substrate than P in the enzyme L-aspartate-β-semialdehyde dehydrogenase, which has an identical active site for both the oxyanions (Faehnle et al. 2004).

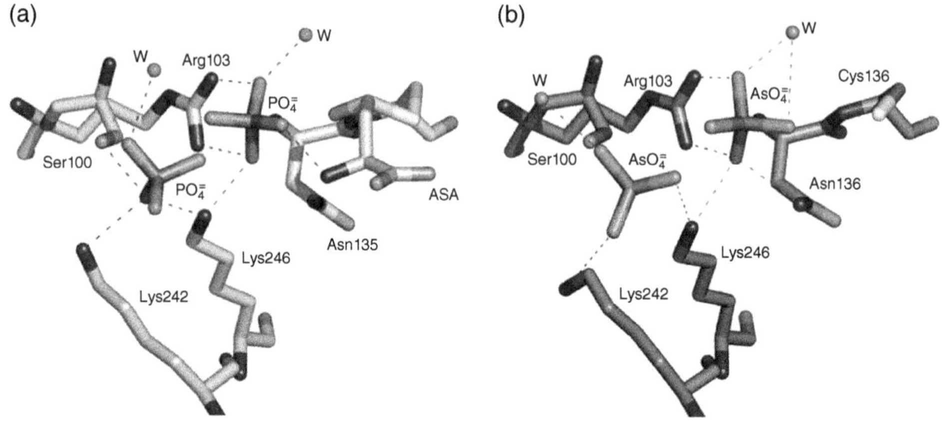

Figure 3.1 Structural similarities of As(V) and P enable them to undergo similar reactions in the presence of the enzyme L-aspartate-β-semialdehyde dehydrogenase. The structure of the enzyme-substrate complex: (a) the P-bound structure; (b) the As(V) bound structure. *Source:* Faehnle et al. (2004).

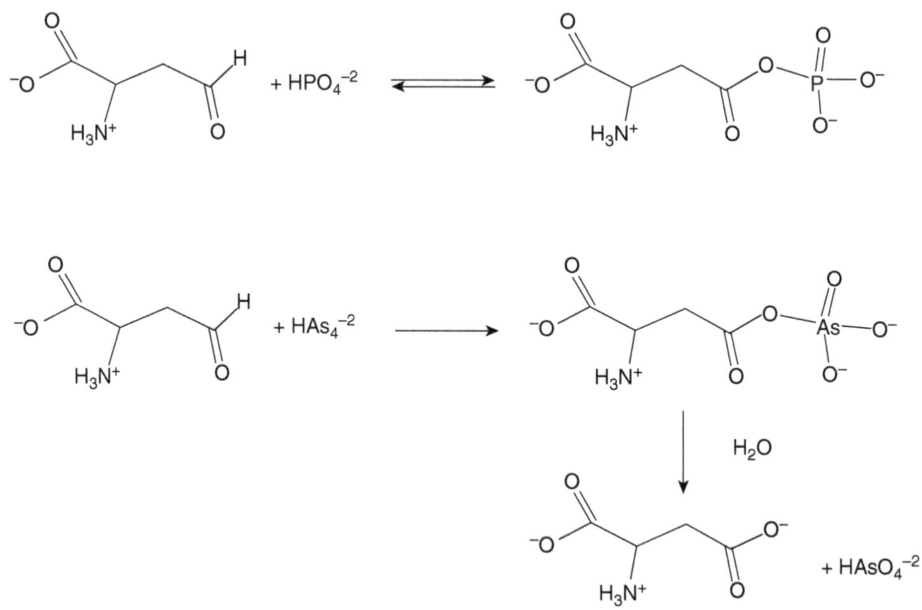

Figure 3.2 Similar positions of As(V) and P in the product formed by their reaction with L-aspartate-β-semialdehyde dehydrogenase; however, the instability of As(V) results in the production of aspartic acid and free arsenate, which is somehow different from β-aspartyl-phosphate formed by P. *Source:* Tawfik and Viola (2011)/American Chemical Society.

Although the position of As(V) and P are similar in the product, the products vary (Figure 3.2), with P yielding an unstable product that cannot easily be isolated while the product of As(V) is readily hydrolyzed to yield aspartic acid and free As(V) (Faehnle et al. 2004).

3.3.2 Effects of As and P in Plants

In limiting P environments, the uptake of As(V) is more favored than P, thus intensifying P deficiency symptoms. Contrariwise, P supplementation can harbor plants from As(V) phytotoxicity (Tu and Ma 2005). Different observations have been demonstrated in hydroponic and soil experiments, with the hydroponics experiments showing a decrease in root As(V) concentration upon P addition while the soil experiments exhibited contrary results showing an increase in root As(V) uptake upon P addition due to the competitive sorption sites in soils, thus bringing about a wider plant pool available for As(V) uptake (Fitz and Wenzel 2002). The presence of P in hydroponic-based studies has also been shown to inhibit As uptake (Huang et al. 2011). However, Purdy and Smart (2008) demonstrated contrasting results that showed a greater As uptake in the P-supplied plants as compared to the P-deprived plants, which is a phenomenon usually expected in soils rather than in hydroponic experiments. Phosphate strongly suppressed As(V) uptake in rice plants (Abedin et al. 2002a). Thus, it is clear that As and P share an antagonistic interaction in their relationship.

In As-tolerant genotypes, the interplay between As(V) uptake and P nutrition is a common mechanism responsible for their tolerance (Meharg and Hartley-Whitaker 2002). As-tolerant genotypes may exhibit lower P/As uptake capacity as compared to sensitive ones, which will result in higher As accumulation over time. However, in As-hyperaccumulators, As tolerance is not achieved by the suppression of the P/As uptake capacity since reports have shown increased As uptake by these species; thus, other mechanisms are also involved in their tolerance to As (Ditusa et al. 2016).

The means of absorption of P and As(V) by plants into the roots is active via the PHTs where it is loaded into the xylem and conveyed to the shoot. Although both analogs are cotransported into plants via the PHTs, ion movement studies have indicated a higher affinity of P in contrast to As (Ullrich-Eberius et al. 1984). Following uptake of As(V), it is converted to As(III) by arsenate reductase (Dhankher et al. 2003). The As(III) then binds to phytochelatins (PC) and is sequestrated into the vacuoles, thus initiating the process of detoxification (Schmöger et al. 2000; Figure 1.4). The enzyme-glutamylcysteine synthetase is a critical enzyme in As detoxification as it catalyzes the production of rate-limiting components during As detoxification (Dhankher et al. 2003). Thus, improved production of PCs and thiol compounds may play an important part in inducing arsenic tolerance (Liu et al. 2010). However, negligible information is available to decipher how the genes involved in As detoxification are regulated by As stress (Sung et al. 2009).

3.3.3 Nature of Phosphate Transporters in Plants

Phosphate transporters (PHTs) play a key role in the movement of As and P into the plant's protoplasm (Table 3.1). The functions and characteristics of all plant PHTs have been studied in a review by Wang et al. (2017), while their roles in *Arabidopsis* and rice have also been reviewed (Młodzińska and Zboińska 2016). The PHT genes characterized have been identified and specified based on amino acid homology with yeast (Rae et al. 2003). The high-affinity PHT in yeast, Pho84p, has been shown to cotransport P and As into the cell interior (Shen et al. 2012).

Table 3.1 The phosphate transporter families in plants.

PHT family	Function	Localization	References
PHT 1	Uptake of P from the soil and distribution within plants	Plasma membrane	Nussaume (2011)
PHT 2	H^+/Pi cotransporter in plastids of plants	Chloroplast	Versaw (2002)
PHT 3	Catalyze Pi/Pi exchange between the matrix and cytosol	Mitochondria	Stappen and Krämer (1994)
PHT 4	Pi transport in plastids and Golgi apparatus	Golgi apparatus	Guo et al. (2008)
PHO	Loading of P to the xylem	Plasma membrane	Poirier and Bucher (2002)

The differences in the localization of the PHTs suggest their varying functions in plant growth and development (Rausch and Bucher 2002; Javot et al. 2007). PHTs can generally be subcategorized into two groups, viz, the high- and the low-affinity PHTs based on the P concentrations that they operate under. Phosphate kinetic studies in suspension cell cultures have revealed that the low-affinity PHTs are expressed constitutively, while the high-affinity PHTs are induced at low P concentrations (Furihata et al. 1992). Bucher et al. (2001) suggested that the high-affinity PHT 1 family is studied more intensively in comparison to other transporters. The transporter is highly similar to the yeast PHT and possesses a conserved region GGDYPLSATlxSE in all plants (Karandashov and Bucher 2005). Various members of PHT 1 family are present in plants; for instance, rice has 13, while *A. thaliana* has 9 (Mudge et al. 2002), and *Populus* spp. has 14 (Zhang et al. 2016). Although a sum of the PHTs have been classified, not all of the PHTs have been fully functionally characterized; for example, in *Arabidopsis*, of the nine identified PHTs, only two have been functionally characterized (Misson et al. 2004), while in rice, only two PHTs also have also been functionally characterized (OsPT2 and OsPT6) (Ai et al. 2009).

Various studies have outlined greater PHT transcript abundance as a consequence of P starvation and subsequent downregulation with P resupply, which is accompanied by greater protein levels and high P transport (Ribot et al. 2008). In As tolerant genotypes, plants restrict the disastrous effects of As by restricting the expression of PHTs, which in turn enables detoxification to occur at a faster rate in contrast to influx (Puckett et al. 2012).

3.3.4 Variations in PHT upon As and P Addition

Responses of plant PHTs' genes to As(V) vary with the specific transporter, tissue type, and the concentration and extent of As(V) exposure (Muehe et al. 2014). Various studies have outlined the variation in the expression of PHTs in regard to As addition. Generally, if As is analogous to P, it is expected that the high-affinity PHT would be upregulated under As limited conditions and downregulated upon As resupply; however, this does not entirely occur across all PHTs and plant species. An addition of As under P starvation is expected to directly affect cellular metabolism due to depleted P reserves (Schachtman et al. 1998). In conditions of P stress, PHTs are upregulated following the increased capacity for As uptake. In Salix genotypes, the expressions of PHT 1;3 and PHT 1;12 were upregulated upon As exposure following As starvation (Puckett et al. 2012). Under P starvation, plants activate homeostasis responses to low P, which include upregulation of high-affinity PHTs, which consequently increase As translocation into plants, thus resulting in toxicity. Studies in plants of different mechanisms to As stress have revealed that tolerant As species restrict expression of PHT genes to a greater extent in contrast to sensitive ones, thereby limiting As(V) influx such that detoxification precedes toxicity (Meharg and Macnair 1991). The same is true for As hyperaccumulators that achieve their As tolerance by the reduction of As(V) uptake by suppressing PHTs, which allows them to accumulate high As. Due to the lower As uptake and high As accumulation, it has been suggested that the locus that controls both traits is not a simple locus but rather a complex one (Catarecha et al. 2007). However, in *A. thaliana*, the relative expression of PHT 1;1 was downregulated upon As exposure under no P, which could be due to the toxicity nature of As, thereby affecting the normal functioning of the carrier protein.

3.3.5 Gene Manipulation of PHTs and PHT Related TFs

Functional analysis of the PHTs and the related transcription factors (TFs) has been widely carried out, and it has outlined the importance of the PHTs in the cotransport of As and P. Functional analysis of PHT transporters in regulating As and P was responsible for regulating the expression of PHT 1;1 and hence alter As uptake (Gonzalez 2005). In rice OsPHF1, mutation decreased the uptake of both P and As(V) from solution, whereas overexpressed lines induced contrasting effects. It is also interesting to note that in a study on rice, PHT 1;8 overexpressing lines only increased grain P concentrations, which may indicate the involvement of OsPHT 1;8 in phloem transport of P but not As (Wu et al. 2011). Thus, it may indicate that although As and P are chemically similar, some PHTs may discriminate them (Muehe et al. 2014).

Suppression of PHTs can also confer As tolerance to plants. In *Arabidopsis*, the disruption of the root PHTs (PHO1;1 and 1;4) resulted in As tolerance (Shin et al. 2004). In rice, mutation of OsPHF1 (PHT traffic facilitator 1) resulted in As tolerance, which induced a reduction in P and As accumulation (Chen et al. 2011).

Regulation of TFs has also been shown to influence As and P uptake, which suggests that the two ions are regulated by similar TFs. WRKY45 overexpression lines increased P content and were more sensitive to As in comparison to the wild type; furthermore, RNA interference based suppression of WRKY45 lines displayed contrasting effects. The observed results were due to an epistatic regulation between WRKY45 and PHT 1;1, demonstrating that WRKY45 is involved in P starvation by the upregulation of PHT 1;1 (Wang et al. 2014). Although much of the evidence in regard to the cotransport of As(V) and P is available in plants, yeast cells lacking the high-affinity PHO84 transporter displayed higher As toxicity in comparison to the wild type, thus proving the similar mechanisms of the PHTs across plants and microbes (Shen et al. 2011). However, in barley (*Hordeum vulgare*), overexpression of the PHT 1;1 gene did not improve P uptake (Rae et al. 2003).

3.4 Transportation of Methylated As Species

Early speciation studies showed that organic As species are less taken up by plants in contrast to the inorganic species; however, recent studies have shown a higher uptake of methylated As species in contrast to the inorganic species. Source of the methylated As species MMA(V) and DMA(V) is from herbicides or pesticides; thus, their prevalence is higher in soils (Li et al. 2009). In addition, methylation by soil algae and other microbes can result in production of these methylated arsenicals. Unlike inorganic arsenicals, the transport of the methylated As forms is not well defined in plants. The uptake of MMA(V) has been postulated to follow Michaelis-Menten kinetics, while uptake of DMA(V) is unclear (Abedin et al. 2002b). The main form of methylated As is DMA(V) and accounts to almost 80% of total As (Meharg et al. 2009). In a study in rice roots, absorption rates followed the trend As(V) > MMA(V) > DMA(V), while the translocation efficiency followed the trend DMA(V) > MMA(V) > As(V) (Raab et al. 2007).

In rice, uptake of MMA(V) and DMA (V) is through the NIP and Lsi1 pathways. Mutation in the rice OsNIP2.1 resulted in the decreased uptake of MMA and DMA at 80% and 50%, respectively, indicating that the OsNIP2.1 transports As(III), DMA, MMA, and silicic acid (Li et al. 2009). Because of low pK_as of MMA and DMA, they will dissociate in pH ≤7; thus, their uptake through the aquaporins, which are solely permeable to uncharged elements, is highly dependent on the pH (Zhao et al. 2013). Thus, uptake efficiency between the organic and inorganic species varies with varying environmental pH.

3.5 Arsenic Metabolism in Plants

Because As(III) is rapidly oxidized to As(V), upon uptake of As, rapid reduction occurs (Srivastava et al. 2016). The reduction transpires via either an enzymatic process or nonenzymatically. In the nonenzymatic process, As(V) is broken down to As(III) by the oxidation of reduced glutathione (GSH) (Figure 3.3) forming oxidized GSH (Mawia et al. 2021). The oxidation is necessitated by the formation of disulfide bond resulting in a GSH dimer (GSSG), which, upon the action of GSH reductase, can be recycled to two GSH molecules (Finnegan and Chen 2012). In the enzymatic process, As(V) is converted to As(III) directly by arsenate reductase (ACR) at a much higher rate in contrast to the nonenzymatic way. Arsenate reductase was initially isolated from bacteria and yeast (Mukhopadhyay et al. 2000).

In the plant leaves, sequestration of As as free As(III) in the vacuole occurs via the ACR (Srivastava et al. 2016). In the As hypertolerant *Holcus lanatus*, *HlAsr* was shown to possess the ACR enzyme activity (Bleeker et al. 2006). In *Arabidopsis*, the high arsenic content 1 (HAC1) has been identified through Genome-Wide Association (GWA) mapping. The HAC1 catalyzes the conversion of As(V) to As(IIII) enabling the outward efflux of As(III) into the soil, thus limiting root accumulation (Chao et al. 2014). Knockout of the HAC1 induced an increase in the accumulation of As in shoot and decrease in As(III) efflux (Mawia et al. 2021). Overexpression increased As(III) outward movement into soil and decreased As uptake into the aboveground parts. Homologous protein encoding the ACR enzyme activity has been reported in rice, *P. vittata*, and *Arabidopsis* (Duan et al. 2005; Xu et al. 2017).

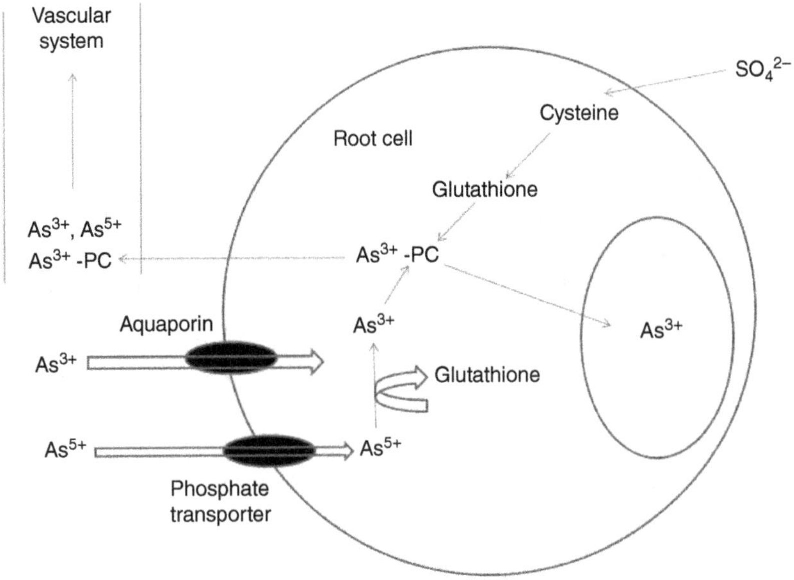

Figure 3.3 Inorganic As transport, As(III) is transported via aquaporins, while As(V) is transported via PHTs where it is converted to As(III) and is transported into the vacuole or chelated with phytochelatins (PCs) into the vacuole.

Not all of the As(III) is effluxed out of the cell (Figure 3.3); some are metabolized by forming sulfhydryl-rich components (Mawia et al. 2021). Cytosolic As increases GSH content, a predecessor of phytochelatin (PC), also a limiting component for the production of phytochelatin synthases (PCS). Upon exposure to As, sequestration of As into the vacuole is mediated by PCS and occurs via an As–PC complex. The sequestration of As into the vacuole leads to a major step in As detoxification and also reduces the translocation factor in plants. Overexpression of the PCS gene has been shown to increase As content in rice roots and shoots but a decrease in the grains (Shinde and Kumar 2021).

Besides being able to reduce inorganic As species, plants are also able to reduce the organic MMA(V) to MMA(III) taken up through the roots (Finnegan and Chen 2012). Although studies in rice have shown that MMA(V) is reduced to MMA(III) through the same principle as the challenger pathway in microbes (Li et al. 2009), the enzymes responsible for the reduction in plants are unknown.

3.6 Conclusion

It is evident that As transport, metabolism, and toxicity are major environmental problems. With research advancements, it is essential to unravel the mechanisms of reducing As uptake without affecting the uptake of other essential elements such as P and Si. In addition, the dynamics of As metabolism and detoxification still need to be unraveled so as to proffer systematic ways of controlling As toxicity. For instance, the advent of gene editing tools, such as CRISPR technology, has paved ways into smart ways of modifying plant genomes as this allows knockout and overexpression of genes of interest. Research into how gene expression of phytochelatins can be increased so as to increase As detoxification could be beneficial in phytoremediation, especially in rice-growing regions. As efflux systems can be genetically engineered so as to reduce As deposits in the above-ground parts of the plants. Although significant strides have been made in the past two decades pertaining to As uptake and metabolism, some research gaps still need to be filled, for instance, deciphering the genes responsible for As metabolism in plants. In conclusion, the studies conducted to date and the information made available can act as a benchmark toward crop improvement, thus reducing As uptake and increasing As detoxification.

References

Abedin, M.J., Cotter-Howells, J., and Meharg, A.A. (2002a). Arsenic uptake and accumulation in rice (*Oryza sativa* L.) irrigated with contaminated water. *Plant and Soil 240* (2): 311–319. https://doi.org/10.1023/A:1015792723288.

Abedin, M.J., Feldmann, J., and Meharg, A.A. (2002b). Uptake kinetics of arsenic species in rice plants. *Plant Physiology 128* (3): 1120–1128. https://doi.org/10.1104/pp.010733.

Agre, P., King, L.S., Yasui, M. et al. (2002). Aquaporin water channels – from atomic structure to clinical medicine. *The Journal of Physiology 542* (Pt 1): 3–16. https://doi.org/10.1113/jphysiol.2002.020818.

Ai, P., Sun, S., Zhao, J. et al. (2009). Two rice phosphate transporters, OsPht1;2 and OsPht1;6, have different functions and kinetic properties in uptake and translocation. *Plant Journal 57* (5): 798–809. https://doi.org/10.1111/j.1365-313X.2008.03726.x.

Ali, W., Isayenkov, S.V., Zhao, F.J., and Maathuis, F.J.M. (2009). Arsenite transport in plants. *Cellular and Molecular Life Sciences 66* (14): 2329–2339. https://doi.org/10.1007/s00018-009-0021-7.

Azizur Rahman, M., Mamunur Rahman, M., Kadohashi, K. et al. (2011). Effect of external iron and arsenic species on chelant-enhanced iron bioavailability and arsenic uptake in rice (*Oryza sativa* L.). *Chemosphere 84* (4): 439–445. https://doi.org/10.1016/j.chemosphere.2011.03.046.

Bienert, G.P., Bienert, G.P., Thorsen, M. et al. (2008). A subgroup of plant aquaporins facilitate the bi-directional diffusion of $As(OH)_3$ and $Sb(OH)_3$ across membranes. *BioMed Central Biology* 6: 26. https://doi.org/10.1186/1741-7007-6-26.

Bleeker, P.M., Hakvoort, H.W.J., Bliek, M. et al. (2006). Enhanced arsenate reduction by a CDC_{25}-like tyrosine phosphatase explains increased phytochelatin accumulation in arsenate-tolerant *Holcus lanatus*. *The Plant Journal 45* (6): 917–929. https://doi.org/10.1111/j.1365-313X.2005.02651.x.

Bleeker, P.M., Schat, H., Vooijs, R. et al. (2003). Mechanisms of arsenate tolerance in *Cytisus striatus*. *New Phytologist 157* (1): 33–38. https://doi.org/10.1046/j.1469-8137.2003.00542.x.

Bucher, M., Rausch, C., and Daram, P. (2001). Molecular and biochemical mechanisms of phosphorus uptake into plants. *Journal of Plant Nutrition and Soil Science 164* (2): 209–217. https://doi.org/10.1002/1522-2624(200104)164:2<209::AID-JPLN209>3.0.CO;2-F.

Catarecha, P., Segura, M.D., Franco-Zorrilla, J.M. et al. (2007). A mutant of the arabidopsis phosphate transporter PHT1;1 displays enhanced arsenic accumulation. *The Plant Cell Online 19* (3): 1123–1133. https://doi.org/10.1105/tpc.106.041871.

Chao, D.-Y., Chen, Y., Chen, J. et al. (2014). Genome-wide association mapping identifies a new arsenate reductase enzyme critical for limiting arsenic accumulation in plants. *PLOS Biology 12* (12): e1002009. https://doi.org/10.1371/journal.pbio.1002009.

Chen, J., Liu, Y., Ni, J. et al. (2011). OsPHF1 regulates the plasma membrane localization of low- and high-affinity inorganic phosphate transporters and determines inorganic phosphate uptake and translocation in rice. *Plant Physiology 157* (1): 269–278. https://doi.org/10.1104/pp.111.181669.

Dhankher, O.P., Shasti, N.A., Rosen, B.P. et al. (2003). Increased cadmium tolerance and accumulation by plants expressing bacterial arsenate reductase. *New Phytologist 159* (2): 431–441. https://doi.org/10.1046/j.1469-8137.2003.00827.x.

Ditusa, S.F., Fontenot, E.B., Wallace, R.W. et al. (2016). A member of the Phosphate transporter 1 (Pht1) family from the arsenic-hyperaccumulating fern Pteris vittata is a high-affinity arsenate transporter. *New Phytologist 209* (2): 762–772. https://doi.org/10.1111/nph.13472.

Dixon, H.B.F. (1996). The Biochemical action of arsonic acids especially as phosphate analogues. *Advances in Inorganic Chemistry 44* (C): 191–227. https://doi.org/10.1016/S0898-8838(08)60131-2.

Dopp, E., Hartmann, L.M., von Recklinghausen, U. et al. (2005). Forced uptake of trivalent and pentavalent methylated and inorganic arsenic and its cyto-/genotoxicity in fibroblasts and hepatoma cells. *Toxicological Sciences 87* (1): 46–56. https://doi.org/10.1093/toxsci/kfi218.

Duan, G.-L., Zhu, Y.-G., Tong, Y.-P. et al. (2005). Characterization of arsenate reductase in the extract of roots and fronds of Chinese Brake Fern, an arsenic hyperaccumulator. *Plant Physiology 138* (1): 461–469. https://doi.org/10.1104/pp.104.057422.

Esteban, E., Carpena, R.O., and Meharg, A.A. (2003). High-affinity phosphate/arsenate transport in white lupin (*Lupinus albus*) is relatively insensitive to phosphate status. *New Phytologist 158* (1): 165–173. https://doi.org/10.1046/j.1469-8137.2003.00713.x.

Faehnle, C.R., Blanco, J., and Viola, R.E. (2004). Structural basis for discrimination between oxyanion substrates or inhibitors in aspartate-β-semialdehyde dehydrogenase. *Acta Crystallographica Section D: Biological Crystallography 60* (12 II): 2320–2327. https://doi.org/10.1107/S0907444904026411.

Finnegan, P. and Chen, W. (2012). Arsenic toxicity: the effects on plant metabolism. *Frontiers in Physiology* 3: 182. https://www.frontiersin.org/article/10.3389/fphys.2012.00182.

Fitz, W.J. and Wenzel, W.W. (2002). Fundamentals and potential application to phytoremediation. *Journal of Biotechnology 99* (3): 259–278. https://doi.org/10.1016/S0168-1656(02)00218-3.

Frausto da Silva, J.J.R. and Willams, R.J.P. (2001). *The Biological Chemistry of the Elements: The Inorganic Chemistry of Life*. Oxford: Oxford University Press.

Furihata, T., Suzuki, M., and Sakurai, H. (1992). Kinetic characterization of two phosphate uptake systems with different affinities in suspension-cultured andlt;italicandgt;Catharanthus roseusandlt;/italicandgt; Protoplasts. *Plant and Cell Physiology 33* (8): 1151–1157. https://doi.org/10.1093/oxfordjournals.pcp.a078367.

Garbinski, L.D., Rosen, B.P., and Chen, J. (2019). Pathways of arsenic uptake and efflux. *Environment International 126*: 585–597. https://doi.org/10.1016/j.envint.2019.02.058.

Gonzalez, E. (2005). Phosphate transporter traffic facilitator1 is a plant-specific sec12-related protein that enables the endoplasmic reticulum exit of a high-affinity phosphate transporter in arabidopsis. *The Plant Cell Online 17* (12): 3500–3512. https://doi.org/10.1105/tpc.105.036640.

Guo, B., Jin, Y., Wussler, C. et al. (2008). Functional analysis of the *Arabidopsis* PHT4 family of intracellular phosphate transporters. *New Phytologist 177* (4): 889–898. https://doi.org/10.1111/j.1469-8137.2007.02331.x.

Huang, C.Y., Shirley, N., Genc, Y. et al. (2011). Phosphate utilization efficiency correlates with expression of low-affinity phosphate transporters and noncoding RNA, IPS1, in Barley. *Plant Physiology 156* (3): 1217–1229. https://doi.org/10.1104/pp.111.178459.

Javot, H., Penmetsa, R.V., Terzaghi, N. et al. (2007). A medicago truncatula phosphate transporter indispensable for the arbuscular mycorrhizal symbiosis. *Proceedings of the National Academy of Sciences of the United States of America 104* (5): 1720–1725. https://doi.org/10.1073/pnas.0608136104.

Karandashov, V. and Bucher, M. (2005). Symbiotic phosphate transport in arbuscular mycorrhizas. *Trends in Plant Science 10* (1): 22–29. https://doi.org/10.1016/j.tplants.2004.12.003.

Kish, M.M. and Viola, R.E. (1999). Oxyanion specificity of aspartate-β-semialdehyde dehydrogenase. *Inorganic Chemistry 38* (4): 818–820. https://doi.org/10.1021/ic981082j.

Kumar, K., Gupta, D., Mosa, K.A. et al. (2019). Arsenic transport, metabolism, and possible mitigation strategies in plants. In: *Plant-Metal Interactions* (ed. S. Srivastava, A.K. Srivastava and P. Suprasanna), 141–168. Switzerland: Springer International Publishing https://doi.org/10.1007/978-3-030-20732-8_8.

Li, R.-Y., Ago, Y., Liu, W.-J. et al. (2009). The rice aquaporin Lsi1 mediates uptake of methylated arsenic species. *Plant Physiology 150* (4): 2071–2080. https://doi.org/10.1104/pp.109.140350.

Lindsay, E.R. and Maathuis, F.J.M. (2016). Arabidopsis thaliana NIP7;1 is involved in tissue arsenic distribution and tolerance in response to arsenate. *FEBS Letters 590* (6): 779–786. https://doi.org/10.1002/1873-3468.12103.

Liu, W.J., Wood, B.A., Raab, A. et al. (2010). Complexation of arsenite with phytochelatins reduces arsenite efflux and translocation from roots to shoots in Arabidopsis. *Plant Physiology 152* (4): 2211–2221. https://doi.org/10.1104/pp.109.150862.

Liu, Z., Shen, J., Carbrey, J.M. et al. (2002). Arsenite transport by mammalian aquaglyceroporins AQP_7 and AQP_9. *Proceedings of the National Academy of Sciences of the United States of America 99* (9): 6053–6058. https://doi.org/10.1073/pnas.092131899.

Ma, J.F., Tamai, K., Yamaji, N. et al. (2006). A silicon transporter in rice. *Nature 440* (7084): 688–691. https://doi.org/10.1038/nature04590.

Maurel, C., Boursiac, Y., Luu, D.-T. et al. (2015). Aquaporins in plants. *Physiological Reviews 95* (4): 1321–1358. https://doi.org/10.1152/physrev.00008.2015.

Mawia, A.M., Hui, S., Zhou, L. et al. (2021). Inorganic arsenic toxicity and alleviation strategies in rice. *Journal of Hazardous Materials 408*: 124751. https://doi.org/10.1016/j.jhazmat.2020.124751.

Meharg, A.A. and Macnair, M.R. (1991). The mechanisms of arsenate tolerance in *Deschampsia cespitosa* (L.) Beauv. and *Agrostis capillaris* L. adaptation of the arsenate uptake system. *New Phytologist 119* (2): 291–297. https://doi.org/10.1111/j.1469-8137.1991.tb01033.x.

Meharg, A.A. and Macnair, M.R. (1992). Suppression of the high affinity phosphate uptake system: a mechanism of arsenate tolerance in *Holcus lanatus* L. *Journal of Experimental Botany 43* (4): 519–524. https://doi.org/10.1093/jxb/43.4.519.

Meharg, A.A. and Hartley-Whitaker, J. (2002). Arsenic uptake and metabolism in arsenic resistant and nonresistant plant species. *New Phytologist 154* (1): 29–43. https://doi.org/10.1046/j.1469-8137.2002.00363.x.

Meharg, A.A., Williams, P.N., Adomako, E. et al. (2009). Geographical variation in total and inorganic arsenic content of polished (white) rice. *Environmental Science and Technology 43* (5): 1612–1617. https://doi.org/10.1021/es802612a.

Misson, J., Thibaud, M.C., Bechtold, N. et al. (2004). Transcriptional regulation and functional properties of *Arabidopsis* Pht1;4, a high affinity transporter contributing greatly to phosphate uptake in phosphate deprived plants. *Plant Molecular Biology 55* (5): 727–741. https://doi.org/10.1007/s11103-004-1965-5.

Młodzińska, E. and Zboińska, M. (2016). Phosphate uptake and allocation – a closer look at *Arabidopsis thaliana* L. and *Oryza sativa* L. *Frontiers in Plant Science 7*: 1198. https://doi.org/10.3389/fpls.2016.01198.

Mudge, S.R., Rae, A.L., Diatloff, E., and Smith, F.W. (2002). Expression analysis suggests novel roles for members of the Pht1 family of phosphate transporters in *Arabidopsis*. *Plant Journal 31* (3): 341–353. https://doi.org/10.1046/j.1365-313X.2002.01356.x.

Muehe, E.M., Eisele, J.F., Daus, B. et al. (2014). Are rice (*Oryza sativa* L.) phosphate transporters regulated similarly by phosphate and arsenate? a comprehensive study. *Plant Molecular Biology 85* (3): 301–316. https://doi.org/10.1007/s11103-014-0186-9.

Mukhopadhyay, R., Shi, J., and Rosen, B.P. (2000). Purification and characterization of ACR2p, the Saccharomyces cerevisiae arsenate reductase. *Journal of Biological Chemistry 275* (28): 21149–21157. https://doi.org/10.1074/jbc.M910401199.

Nussaume, L. (2011). Phosphate import in plants: focus on the PHT1 transporters. *Frontiers in Plant Science 2*: 83. https://doi.org/10.3389/fpls.2011.00083.

Poirier, Y. and Bucher, M. (2002). Phosphate transport and homeostasis in *Arabidopsis*. *The Arabidopsis Book 1*: e0024. https://doi.org/10.1199/tab.0024.

Puckett, E.E., Serapiglia, M.J., DeLeon, A.M. et al. (2012). Differential expression of genes encoding phosphate transporters contributes to arsenic tolerance and accumulation in shrub willow (*Salix* spp.). *Environmental and Experimental Botany 75* (Supplement C): 248–257. https://doi.org/10.1016/j.envexpbot.2011.07.008.

Purdy, J.J. and Smart, L.B. (2008). Hydroponic screening of shrub willow (*Salix* spp.) for arsenic tolerance and uptake. *International Journal of Phytoremediation 10* (6): 515–528. https://doi.org/10.1080/15226510802115000.

Raab, A., Williams, P.N., Meharg, A., and Feldmann, J. (2007). Uptake and translocation of inorganic and methylated arsenic species by plants. *Environmental Chemistry 4* (3): 197–203.

Rae, A.L., Cybinski, D.H., Jarmey, J.M., and Smith, F.W. (2003). Characterization of two phosphate transporters from barley; evidence for diverse function and kinetic properties among members of the Pht1 family. *Plant Molecular Biology 53* (1–2): 27–36. https://doi.org/10.1023/B:PLAN.0000009259.75314.15.

Rausch, C. and Bucher, M. (2002). Molecular mechanisms of phosphate transport in plants. *Planta 216* (1): 23–37. https://doi.org/10.1007/s00425-002-0921-3.

Ribot, C., Wang, Y., and Poirier, Y. (2008). Expression analyses of three members of the AtPHO1 family reveal differential interactions between signaling pathways involved in phosphate deficiency and the responses to auxin, cytokinin, and abscisic acid. *Planta 227* (5): 1025–1036. https://doi.org/10.1007/s00425-007-0677-x.

Sanders, O.I., Rensing, C., Kuroda, M. et al. (1997). Antimonite is accumulated by the glycerol facilitator GlpF in *Escherichia coli*. *Journal of Bacteriology 179* (10): 3365–3367. https://doi.org/10.1128/jb.179.10.3365-3367.1997.

Schachtman, D.P., Reid, R.J., and Ayling, S.M. (1998). Phosphorus uptake by plants: from soil to cell. *Plant Physiology 116* (2): 447–453. https://doi.org/10.1104/pp.116.2.447.

Schmöger, M.E.V., Oven, M., and Grill, E. (2000). Detoxification of arsenic by phytochelatins in plants. *Plant Physiology 122* (3): 793–802. https://doi.org/10.1104/pp.122.3.793.

Shen, J., Yuan, L., Zhang, J. et al. (2011). Phosphorus dynamics: from soil to plant. *Plant Physiology 156* (3): 997–1005. https://doi.org/10.1104/pp.111.175232.

Shen, M.W.Y., Shah, D., Chen, W., and Da Silva, N. (2012). Enhanced arsenate uptake in Saccharomyces cerevisiae overexpressing the Pho84 phosphate transporter. *Biotechnology Progress 28* (3): 654–661. https://doi.org/10.1002/btpr.1531.

Shin, H., Shin, H.S., Dewbre, G.R., and Harrison, M.J. (2004). Phosphate transport in Arabidopsis: Pht1;1 and Pht1;4 play a major role in phosphate acquisition from both low- and high-phosphate environments. *Plant Journal 39* (4): 629–642. https://doi.org/10.1111/j.1365-313X.2004.02161.x.

Shinde, A. and Kumar, K. (2021). Mechanisms of arsenic transport, accumulation, and distribution in rice. In: *Arsenic Toxicity: Challenges and Solutions* (ed. N. Kumar), 279–300. Singapore: Springer https://doi.org/10.1007/978-981-33-6068-6_11.

Srivastava, S., Upadhyay, M.K., Tripathi, R.D., and Dhankher, O.P. (2016). Arsenic transport, metabolism and toxicity in plants. *International Journal of Plant and Environment 2* (1 and 2): 17–28. doi: 10.18811/ijpen.v2i1-2.6614.

Stappen, R. and Krämer, R. (1994). Kinetic mechanism of phosphate/phosphate and phosphate/OH^- antiports catalyzed by reconstituted phosphate carrier from beef heart mitochondria. *Journal of Biological Chemistry 269* (15): 11240–11246. http://europepmc.org/abstract/MED/8157653.

Sung, D.Y., Kim, T.H., Komives, E.A. et al. (2009). ARS5 is a component of the 26S proteasome complex, and negatively regulates thiol biosynthesis and arsenic tolerance in *Arabidopsis*. *Plant Journal 59* (5): 802–812. https://doi.org/10.1111/j.1365-313X.2009.03914.x.

Tawfik, D.S. and Viola, R.E. (2011). Arsenate replacing phosphate: alternative life chemistries and ion promiscuity. *Biochemistry 50* (7): 1128–1134. https://doi.org/10.1021/bi200002a.

Tu, C. and Ma, L.Q. (2005). Effects of arsenic on concentration and distribution of nutrients in the fronds of the arsenic hyperaccumulator *Pteris vittata* L. *Environmental Pollution 135* (2): 333–340. https://doi.org/10.1016/j.envpol.2004.03.026.

Tu, S. and Ma, L.Q. (2004). Comparison of arsenic and phosphate uptake and distribution in arsenic hyperaccumulating and nonhyperaccumulating fern. *Journal of Plant Nutrition 27* (7): 1227–1242. https://doi.org/10.1081/PLN-120038545.

Ullrich-Eberius, C.I., Novacky, A., and van Bel, A.J.E. (1984). Phosphate uptake in *Lemna gibba* G1: energetics and kinetics. *Planta 161* (1): 46–52. https://doi.org/10.1007/BF00951459.

Versaw, W.K. (2002). A chloroplast phosphate transporter, PHT2;1, influences allocation of phosphate within the plant and phosphate-starvation responses. *The Plant Cell Online 14* (8): 1751–1766. https://doi.org/10.1105/tpc.002220.

Violante, A. and Pigna, M. (2002). Competitive sorption of arsenate and phosphate on different clay minerals and soils. *Soil Science Society of America Journal 66* (6): 1788. https://doi.org/10.2136/sssaj2002.1788.

Wang, H., Xu, Q., Kong, Y.-H. et al. (2014). Arabidopsis WRKY45 transcription factor activates Phosphate transporter1;1 expression in response to phosphate starvation. *Plant Physiology 164* (4): 2020–2029. https://doi.org/10.1104/pp.113.235077.

Wang, D., Lv, S., Jiang, P., and Li, Y. (2017). Roles, regulation, and agricultural application of plant phosphate transporters. *Frontiers in Plant Science 8*: 817. https://doi.org/10.3389/fpls.2017.00817.

Wolfe-Simon, F., Blum, J.S., Kulp, T.R. et al. (2011). A bacterium that can grow by using arsenic instead of phosphorus. *Science 332* (6034): 1163–1166. https://doi.org/10.1126/science.1197258.

Wu, Z., Ren, H., McGrath, S.P. et al. (2011). Investigating the contribution of the phosphate transport pathway to arsenic accumulation in rice. *Plant Physiology 157* (1): 498–508. https://doi.org/10.1104/pp.111.178921.

Xu, J., Shi, S., Wang, L. et al. (2017). OsHAC4 is critical for arsenate tolerance and regulates arsenic accumulation in rice. *New Phytologist 215* (3): 1090–1101. https://doi.org/10.1111/nph.14572.

Xu, W., Dai, W., Yan, H. et al. (2015). *Arabidopsis* NIP3;1 plays an important role in arsenic uptake and root-to-shoot translocation under arsenite stress conditions. *Molecular Plant 8* (5): 722–733. https://doi.org/10.1016/j.molp.2015.01.005.

Ye, Y., Li, P., Xu, T. et al. (2017). OsPT4 contributes to arsenate uptake and transport in rice. *Frontiers in Plant Science* 8: 2197. https://www.frontiersin.org/article/10.3389/fpls.2017.02197.

Zhang, C., Meng, S., Li, M., and Zhao, Z. (2016). Genomic identification and expression analysis of the phosphate transporter gene family in poplar. *Frontiers in Plant Science 7*: 1398. https://doi.org/10.3389/fpls.2016.01398.

Zhao, F.J., Ma, J.F., Meharg, A.A., and McGrath, S.P. (2009). Arsenic uptake and metabolism in plants. *New Phytologist 181* (4): 777–794. https://doi.org/10.1111/j.1469-8137.2008.02716.x.

Zhao, F.-J., Zhu, Y.-G., and Meharg, A.A. (2013). Methylated arsenic species in rice: geographical variation, origin, and uptake mechanisms. *Environmental Science and Technology 47* (9): 3957–3966. https://doi.org/10.1021/es304295n.

4

Arsenic Induced Responses in Plants: Impacts on Different Plant Groups, from Cyanobacteria to Higher Plants

Kavita Ghosal[1], Moumita Chatterjee[2], Sharmistha Ganguly[3], Subhamita Sen Niyogi[4], and Dwaipayan Sinha[5]

[1] *Department of Botany, P.D. Women's College, Jalpaiguri, West Bengal, India*
[2] *Department of Botany, V. Sivaram Research Foundation, Bengaluru, Karnataka, India*
[3] *Department of Botany, Ranchi University, Ranchi, Jharkhand, India*
[4] *Anushka Soham Purified Water Manufacturing Co. Private Limited, Hooghly, West Bengal, India*
[5] *Department of Botany, Government General Degree College, West Bengal, India*

4.1 Introduction

Arsenic (As) is a toxic metalloid that has been considered to promote several dreadful ailments related to cardiovascular diseases, skin diseases, and cancer (Gilbert-Diamond et al. 2011; Taheri et al. 2016) and several other toxic responses (Wang et al. 2018a). Anthropogenic activities serve as the backbone and trigger water and soil contamination. Day by day the scarcity of drinking water is increasing due to As contamination (Mandal and Suzuki 2002; Sandhi et al. 2017). Arsenic exists in the biosphere in both organic oxidized and inorganic reduced forms (Sandhi et al. 2017) between which the inorganic one is very dreadful (Meharg and Hartley-Whitaker 2002; Ventura-Lima et al. 2011). Arsenate (As(V)) acts as a phosphate analogue and is absorbed by most aquatic plants, whereas arsenite (As(III)) is taken up through aquaglyceroporins and transported into the cell (Zhao et al. 2010; Chen et al. 2017; Niazi et al. 2017). There are few plants species that can alleviate As by acting as hyperaccumulators, thereby helping in phytoremediation by the process of phytofiltration (Sandhi et al. 2017). One hundred and fifty new indigenous species were identified in an As-contaminated zone in Zimbabwe (Meharg and Hartley-Whitaker 2002). Two hundred and twenty-one species belonging to five species in four different families were identified to absorb As from mercury mining site in Guizhou Province, southwestern China. The dominant species showed total As in its inorganic form, i.e. As(III) and As(V) at 87%, whereas, the organic form of As, i.e. monomethylarsonic acid (MMA) and dimethylarsinic acid (DMA) at 11.8 and 6.0%, respectively (Liang et al. 2019).

Arsenic is a hazardous metalloid element with a practically tasteless and odorless substance that is present throughout the environment (Kuivenhoven and Mason 2022). Whereas geologically occurring As cannot be eliminated, it can produce many compounds

Arsenic in Plants: Uptake, Consequences and Remediation Techniques, First Edition.
Edited by Prabhat Kumar Srivastava, Rachana Singh, Parul Parihar, and Sheo Mohan Prasad.

through reaction with several elements and organic molecules (Banerjee et al. 2019; Balali-Mood et al. 2021; Zhou et al. 2021). Arsenate and arsenite are representatives of inorganic forms, while organic molecules include methylated forms of arsenic acids and arsenobetaine (Jomova et al. 2011, Chattopadhyay et al. 2021). Semiconductors, glassware, metals, and wood preservatives all contain As (Hong et al. 2014). Arsenic has been used in medication to treat yaws, amoebic dysentery, and syphilis (Khalid et al. 2017). Plants are harmed by As, which is used as a pesticide (Bencko and Yan Li Foong 2017; Boulanger et al. 2017) and also comes from smelter waste (Zhou et al. 2021). Plant species, geographic region, soil type, and climatic circumstances all have a substantial influence on the effects, which are usually dose related. Arsenic residues and phytotoxicity are two significant dose-related effects that may be measured in plants.

Arsenic metabolism has a direct impression on its toxicity, and while the specific mode of action is unknown, there are various ideas. Organic As followed by the elemental forms are the most toxic species known (Kuivenhoven et al. 2022; Balali-Mood et al. 2021). As secondary consumers, humans consume plant products indirectly or directly as staple foods and get in contact with the chronic exposure to As without revelation of visible signs and symptoms at early stages. Inorganic forms of As are carcinogens to humans as these are the causes of malignancies of the skin (Karagas et al. 2019), lungs (Xiao et al. 2021), and bladder. There has also been evidence of a substantial association between As and liver (Hu et al. 2021), prostate, and bladder malignancies (Satyapal and Kumar 2021). A link between diabetes, neurological effects (Andersson et al. 2021), heart illness (States et al. 2009; Sobel et al. 2020), and reproductive organs (Zubair et al. 2017) has also been discovered recently, though more exploration is necessitated to corroborate these findings. Most of the research till date is based on cancer development in those who have been exposed to high doses of As. Contrarily, numerous studies have discovered that extended exposure to low amounts of As has a variety of physiological consequences (Hong et al. 2014).

Arsenate and arsenite are two inorganic As compounds that are absorbed quickly by plant roots. Arsenate is easily transformed to As(III), the more poisonous of the two forms, once within the cell. Both As(V) and As(III) impair plant metabolism although in different ways (Abbas et al. 2018). Arsenate is a phosphate analog that inhibits metabolic processes related to phosphate metabolism. Phosphate transport proteins can transfer As(V) across cellular membranes, causing phosphate resource imbalances. During phosphorylation processes, it can contest with phosphate, resulting in the creation of unstable and transient As(V) adducts. As(V)–Adenosine Di Phosphate (ADP) synthesis and rapid autohydrolysis disengage photophosphorylation and oxidative phosphorylation, resulting in a pointless cycle, thereby inhibiting the synthesis of adenosine triphosphate (ATP) (Németi et al. 2010; Finnegan and Chen 2012).

Arsenite can attach to sulfhydryl groups in a reversible manner, causing capillary dilatation and enhanced permeability (Chen et al. 2015). The discovery that As had a high affinity for vicinal dithiols led to the creation of chelating compounds, including dimercapto-propanesulfonate, dimercaptopropanol, and dimercaptosuccinic acids that bind to enzymes with narrowly located dithiol co-factors and cysteine residues and may inactivate the reactive oxygen species (ROS) generated by As as antidotes (Shen et al. 2013; Nurchi et al. 2020; Kuivenhoven and Mason 2022).

Arsenic exposure affects assimilation pathways of nitrogen, sulfur, metabolism of amino acid, carbon, and protein interactions. Arsenic exposure causes the creation of ROS, which initiates the synthesis of antioxidants. Plants with improved As tolerance have been illustrated to modify numerous metabolic pathways, such as glutathione (GSH) synthesis (de Souza Reis et al. 2020; Ahmad et al. 2021). The fact that As responsiveness varies in different species and in cultivated varieties (Mehmood et al. 2017).

4.2 Responses of Arsenic on Various Plant Groups

In the charismatic world of technology, environmental pollution is a matter of concern to many scientific researchers. In this context, "phytotoxicity" is the word that is recurring and gaining popularity due to adverse environmental pollution caused by anthropogenic and geogenic activities. Man-made pollutions are always hazardous and strikingly provoke exploitation of nature. One of such potentially toxic elements includes As contamination in the environment, which is now a global issue that is now considered as carcinogenic (Abid et al. 2016; Chandrakar et al. 2016; Mehmood et al. 2017; Niazi et al. 2017). Since As possesses combined nonmetal and metallic properties, it can take part in a plant as chemical interactions modulator (Upadhyaya et al. 2014). Herbicides, insecticides, and pesticides for cotton, rice plant, etc., which are composed of As, are extensively used nowadays, thereby causing immense contamination of domestic as well as agricultural lands (Murphy and Aucott 1998). The contamination of domestic and agricultural land causes As to enter the terrestrial food chain, which in turn dismantles the normal mechanism in both plants and animals. Being a nonessential element, it has shown phytotoxicity toward plants (Meharg and Hartley-Whitaker 2002) (Figure 4.1). Plants that are not resistant toward As have shown a considerable amount of stress on As exposure, which is exhibited by several

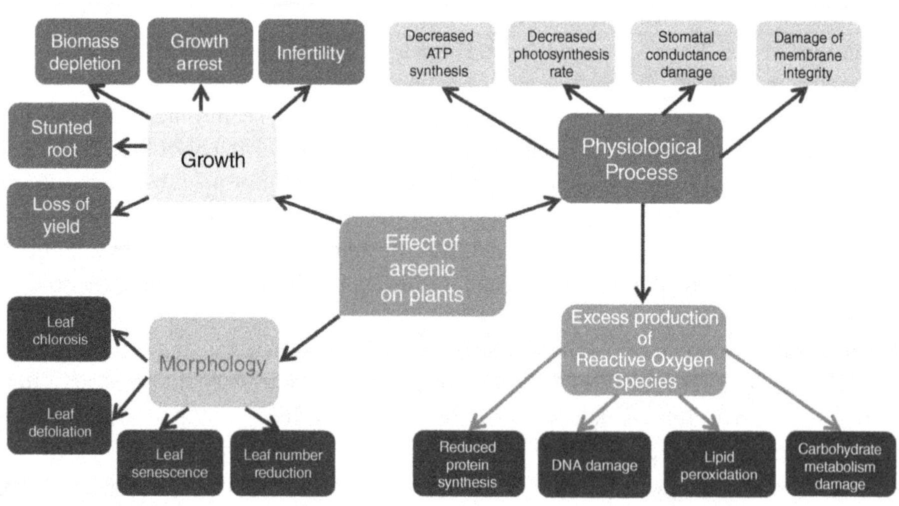

Figure 4.1 Effect on different metabolisms of plants exposed to arsenic toxicity.

abnormal symptoms such as diminished shoot growth, inhibition in growth of roots that results in the plant's death over time (Macnair and Cumbes 1987; Meharg and Macnair 1991; Paliouris and Hutchinson 1991; Carbonell-Barrachina et al. 1995).

Plants suffering from As stress have shown a considerable reduction in photosynthetic rate (Stoevaand and Bineva 2003), as seen in *Zea mays*; the functional activity of Photosystem II (PSII) is reduced in a significant amount, besides the rate of fixing CO_2 in juvenile plants is also decreased by 20% (Stoeva et al. 2005). It is assumed that stomatal and mesophyll constraints could be the probable reasons behind this reduced rate of photosynthesis. The factors that strikingly highlight the difference are chlorophyll hindrances, the light reaction, and the biochemical changes occurring during the Calvin cycle (Stoeva and Bineva 2003). Arsenic responses are also seen in seedlings of rice. In this enzyme, activity is being altered, which causes the accumulation of soluble sugars, thereby hindering carbohydrate metabolism. Under As stress, sucrose synthase plays a pragmatic role (Jha and Dubey 2004a). The oxidative stress is persuaded by As metalloid (Hartley-Whitaker et al. 2001a) and impedes the growth of root and alterations in several transporters and metabolic pathways and ultimately manifested by the development of a toxic system by hindering the electron transport system. Noxious ROS are produced in the cell that affects the cell membrane (CM) permeability, enzymatic activity, source-sink transport system, necrosis, chlorosis, and senescence and is seen in leaves due to the production of antioxidants, which convert As(V) to As(III), the dreadful one between the two (Nguyen et al. 2003). Arsenate also affects the phosphate transport system in plants as it mimics the phosphate present in ATP and forms ADP–As in the cytoplasm due to wrecking in the flow of energy inside the cell (Meharg 1994). The enzymes and proteins with –SH functional group are attacked by As(III), which retards cellular functions and in turn causes cell death (Ullrich-Eberius et al. 1989). In *Helianthus annuus*, the seed germination is dismantled and further causes plant death under As toxicity (Imran et al. 2013; Ghulam et al. 2018). *Cicer arietinum*, under As exposure, showed stunted growth of shoots and roots. Deficiency in the level of certain amino acids like lysine, methionine, proline, threonine, tryptophan, and valine was also detected (Malik et al. 2011; Ghulam et al. 2018).

4.3 Arsenic Response in Cyanophycean Algae

The As content is increasing day by day in both terrestrial and aquatic ecosystems due to anthropogenic activities such as agriculture wastes, domestic wastes, industrial wastes, and acid mine drainage. Cyanophycean algae are the potent bioaccumulator of As that have the ability to scavenge As from adjacent areas of their habitat, though it carries the risk factor of taking up As in the food chain (Yadav et al. 2021b). It was described that As level is higher in cyanophycean algae, i.e. $276.9\,\mu g\,g^{-1}$, compared to water, i.e. $244.13\,\mu g\,l^{-1}$, and soil, i.e. $19.01\,\mu g\,g^{-1}$ (Bhattacharya and Pal 2010). Arsenic toxicity can lead to cell mortality and growth inhibition in cyanophycean algae members. *Synechocystis*, *Anabaena*, *Oscillatoria*, and *Phormidium* are the dominant As-resistant genera reported from a different region of the world (Maeda and Ohki 1998; Shaheen et al. 2007; Banerjee 2008; Bhattacharya and Pal 2011; Yin et al. 2012). It is reported that As and antimony-resistant microbes have a transporter system of As(III), i.e. ars gene family (Rosen 1999). *Phormidium*, an As-resistant member, increased the growth rate against 100 mg g^{-1} of As uptake (Maeda et al. 2004). It is suggested that Ascorbate peroxidase (APX) is an essential constituent of the ascorbate–glutathione pathway, necessary for hydrogen peroxide (H_2O_2) scavenging predominantly in chloroplastid and other cell organelles to sustain cell redox level (Foyer and Noctor 2011). It is also reported that glutathione reductase is the primary factor that regulates the bacterial As(V) resistance (Mukhopadhyay et al. 2002).

The arsBHC operon includes the *ACR3* As(III) transporter gene, an *ars*H homolog that encodes Flavin Mono Nucleotide (FMN)–quinone reductase, and the *arsB* gene, and there are two more arsenate reductase genes, namely, *arsI1* and *arsI2*, that confer the As resistance property in *Synechocystis* sp. strain PCC 6803 (Lopez-Maury et al. 2003; Hervas et al. 2012). The As(V) and As(III) induce related genomic responses in *Synechocystis* sp. PCC 6803, which influences strong stress responses like upregulation of chaperones and redox scavenging properties and inhibits growth and photosynthesis. If *arsB* gene is absent in *Synechocystis* sp. PCC 6803 strain, then As(III) detoxification is not possible (Sánchez-Riego et al. 2014). As As(V) is a phosphate analogue, it enters inside the cell via phosphate transporters; thus, its toxicity replaces the phosphate in many biochemical pathways like glycolysis and oxidative phosphorylation (Fekry et al. 2011; Tawfik and Viola 2011; Kamerlin et al. 2013). Arsenite enters the cell via aquaglyceroporins, and as a result, toxicity increases than its forms of arsenothiols after binding the cellular dithiols. Arsenothiols block the protein function as a ROS (Bernstam and Nriagu 2000; Wysocki et al. 2001; Liu et al. 2002; Meng et al. 2004). Cyanophycean members adopted a mechanism to combat ROS load by modulating several antioxidant enzymes. The oxidative outburst, disruption of metal homeostasis, and reduction in GSH pool are associated with As content (Pandey et al. 2012; Sánchez-Riego et al. 2014). Phosphate content hampers the As(V) transport, and as a result, it regulates the As resistance potential in Cyanophyceae. If the phosphate content declines, then As(V) transport system is upregulated, and as a result, transport of As(V) is increased (Thiel 1988; Takahashi et al. 1990; Takahashi et al. 2001; Markley 2004; Pitt et al. 2010; Yin et al. 2012; Zhang et al. 2014). *arsB* gene expression is upregulated in *Crocosphaera watsonii* (Dyhrman and Haley 2011) due to low levels of phosphate, whereas phosphate transport gene expression is inhibited by As in *Synechocystis* (Sánchez-Riego et al. 2014). According to Elias et al. (2012), the selectivity and specificity of the phosphate transport pathway are regulated by As(V) transport. Reduction of active arsenate level in cytoplasm effectively decreases As(III) formation rate from As(V). Arsenate is not able to take entry inside the cell once it is formed by the oxidation of As(III); if the phosphate level is declined, then only As(V) produced outside of the cell enters inside the cytoplasm (Zhang et al. 2014). The polymers of GSH formed phytochelatins, which have a vital role in As(III) resistance (Wysocki et al. 2003). Homologous gene of phytochelatin synthase occurs in a few cyanophycean members (Harada et al. 2004), and also As content induced elevation of phytochelatin (Pandey et al. 2012). In *Microcystis aeruginosa*, phosphorus plays an important role in As efflux for upregulating bio-oxidation of As(III). Not only methylation and reduction but also efflux of As is elevated by low phosphorus content in algal cells (Yan et al. 2014). *Aphanothece halophytica*, a halotolerant cyanophycean member, tolerates 40 mg l^{-1} of As concentration (Laloknam et al. 2009).

Shilpa and Roychoudhury (2017) suggested that *Nostoc*, *Leptolyngbya*, and *Anabaena* are potent As bioaccumulators; these three species have a vital role in plant growth in As-rich habitat. Sharma and Rathore (2018) reported elevation of As concentrations that reduced frequency of heterocyst and growth rate in three cyanophycean members, namely, *Nostoc muscorum*, *Anabaena doliolum*, and *Aulosira fertilissima*. Uncoupled phosphorylation is associated with As(V), and as a result, ADP for coupled phosphorylation is eliminated from the cell that leads to ATP deficiency (Singh et al. 2016). Adenosine triphosphate deficiency is a major cause of the declining growth, and if it continues, the organism will die. *Microcystis novacekii* has a high As(III) scavenging property (21.2%) from the culture medium containing 14.7 $mg\,l^{-1}$ concentration of As(III) (Ferreira et al. 2018). According to Dutta and Bhadury (2020), in *Nostoc ellipsosporum*, growth rate decreases whereas exopolysaccharide, i.e. EPS, production increases with increasing As content.

4.4 Responses on Other Groups of Algae (Chlorophyceae, Phaeophyceae, Rhodophyceae, Diatoms, Xanthophyceae, Charophyceae, etc.)

4.4.1 Chlorophyceae

Chlorophyceae are eukaryotic freshwater green algae that are important both ecologically and scientifically and have the potential to take up toxic As from the environment (Senthilkumar et al. 2020). Arsenate competitively slows down phosphate uptake by green algae *Euglena gracilis* (Miot et al. 2009). Chlorophyceae absorb As by passive surface adsorption followed by electrostatic attraction, coordination, ion exchange, chelation, complexation, and finally microprecipitation (Suresh Kumar et al. 2015). The surface characteristics of the cell wall (CW) play a role in As absorption in *Chlorella vulgaris*. Many algae species have the potentiality to reduce As(V). *Chlamydomonas reinhardtii* reduces 18–24% of the total As (Yin et al. 2011), while *Chlorella salinacan* reduces up to 32% of the As intracellularly when exposed to As(V) for 72 hours (Karajova et al. 2008). At high As concentrations, the development of *Chlorella vulgaris* isolated from an As-polluted environment was unaffected (Maeda et al. 1985). According to Hofslagare et al. (1994), As(III) is less toxic than As(V) in *Scenedesmus obliquus*; the carbon uptake potential is reduced by 29 and 41% due to the presence of As(III) and As(V), respectively. Arsenate and As(III) alter the extracellular polymeric substances, i.e. EPS production potential of *Chlorella pyrenoidosa* (Zhanga et al. 2020).

Arsenite oxidation by the biological method has been characterized in *Alcaligenes faecalis*. MMA, DMA, and various trimethyl forms of As are the most prevalent stable organic As forms found inside freshwater algae (Murray et al. 2003). The most frequent methylation As forms are less toxic than As(V) and As(III) (Hirano et al. 2004). As a result, methylation is thought to be a component of the detoxification process. In a series of reduction/oxidation reactions, inorganic As is converted to organic methylated As. The addition of methyl groups generates MMA, DMA, and, to a lesser extent, trimethyl arsenic and lipoic acid, and in the reduction stage, GSH has been discovered to behave as a reducing agent (Sun et al. 2016b). S-adenosylmethionine (SAM) serves as a methyl donor in oxidative methylation and is transformed to *S*-adenosylhomocysteine (Bentley and Chasteen 2002).

4.4.2 Phaeophyceae

Phaeophyceae or brown algae are characterized by the presence of fucoxanthin (Sheath and Wehr 2003). Methylated As compounds are considered as toxic arsenolipids and arsenosugars, and formation of these toxic substances is regulated by phosphate availability (Hughes et al. 2011). Deficiency of phosphate in the habitat of *Ectocarpus* species upregulates the phospholipids biosynthesis (Pétursdóttir et al. 2015). *Fucus serratus* sequestrated arsenate in a very high rate and produced toxic arsenosugars (Geiszinger et al. 2001). *Hijikia fusiforme* contains huge amount of As, which is about 50% of their biomass content (Ichikawa et al. 2010), whereas *Laminaria digitata* possesses As of about 2.8 to 20 $mg\,kg^{-1}$ of dry weight (Taylor and Jackson 2016).

4.4.3 Rhodophyceae

Rhodophyceae are marine multicellular algae in which red Phycoerythrin and blue Phycocyanin mask the chlorophyll (Cotas et al. 2020). Arsenic buildup and transformation mechanisms in macroalgae are poorly understood. The phosphate transport system actively absorbs As(V) in marine algae. As(V) is reduced to As(III) in the cells, which is then methylated to produce variants of arsenic acids. Microalgae take up more As(V) when phosphorus is scarce and reduce it to As(III), and afterward, methylation occurs for the production of organo-arsenic acid and ultimately is excreted out of the cell (Levy et al. 2005). Under phosphorus sufficient condition, As(V) is converted to As(III). Brown algae generally accumulate more sulfonate as well as sulfate arsenoribosides. Red algae usually contain glycerol and phosphate. Most seaweeds convert As to arsenosugars (Taylor et al. 2017). Gametophytes of the giant kelp *Macrocystis pyrifera* when exposed to a higher concentration of As affect nuclear migration (Gayle et al. 1994).

4.4.4 Diatoms

Diatoms belonging to Bacillariophyceae are microscopic photosynthetic algae mainly found in aquatic habitats. Heavy metals such as As are degraded, speciated, and detoxified in large part by diatoms (Kiran Marella et al. 2020). The diatom community showed a strong reduction in cell bio volume. Diatoms sequestrate heavy metals like As by passive adsorption and active assimilation from the aqueous environment. Frustules prevent heavy metal toxicity (Kiran Marella et al. 2020). For detoxification process, a study by Hasegawa et al. (2001) had shown that microalgae *Closterium acicolare* convert As(V) mainly about 80% into pentavalent methylated intermediate, which is less harmful (Murray et al. 2003).

Phytochelatins are GSH derivatives that are good chelators for heavy metals like As and lower their toxicity. Phytochelatin complexes with As are restricted into vacuoles. Papry et al. (2019) reported that with salinity and temperature, marine diatom species, namely, *Skeletonema* sp., *Chaetoceros didymus*, *Asteroplanus karianus*, *Nitzschia longissima*, *Ditylum brightwellii*, and *Thalassionema nitzschioides* alter their growth, pattern of speciation, and also As biotransformation.

4.5 Responses on Moss

Mosses have always served as bioindicators because they are devoid of roots, protective epidermis, and vascular system; thereby, their whole thalli are directly exposed to the contaminated environment, so their cells can uptake heavy metals quite easily (Donovan et al. 2016). The first research work done on mosses used as bioindicators was done in Sweden (Ruhling and Tyler 1968). Aquatic plants are more efficient than terrestrial plants in absorbing As (Bergqvist and Greger 2012). Submerged bryophytes are the ones that not only resist As but also accumulate it (Díaz et al. 2012). *Drepanocladus* sp. are aquatic moss that accumulated As from mining sites (Koch et al. 2000). In Skellefteå field area, the aquatic moss *Warnstorfia fluitans* was found to accumulate As from As-contaminated water (Sandhi et al. 2017). Phytofiltration is a method where pollutants are removed from the environment using plants. It is a cost-effective and environmentally friendly procedure that aids in phytoremediation (Islam et al. 2013). Phytofiltration has the potential to help in the stabilization, extraction, degradation, and volatilization of pollutants. The hyperaccumulators are the plants that not only tolerate extreme As stress but also accumulate it by absorption (Claire-Lise and Nathalie 2012). The heavy metals accumulation by bryophytes is done by several mechanisms, mainly by irreversible absorption of the metals on the cellular surface (Gonzalez and Pokrovsky 2014). The metals that are absorbed are imprisoned as particulate matters that are either dissolved in liquids or deposited intercellularly on the surrounding cells, attached to the CW's chelating sites and CM extracellularly or transported from cell to cell intracellularly (Vazquez et al. 1999, Salemaa et al. 2004; Gonzalez and Pokrovsky 2014). The ion exchange system has a crucial role in the extracellular accumulation of these heavy metals, resulting in complex formation between the functional groups of the bryophyte's CW and the metals (Shakya et al. 2008). The CW contains the functional groups of polygalacturonic acid, which contribute toward the binding capacities of the mosses (Tipping et al. 2008). The mosses' acid–base characteristics can be another reason for the binding that involves several functional groups like polyphenols, carboxyl, amine phosphodiester, etc. Out of which the phosphoryl and carboxyl groups are considered to be dominant while binding with the metal because of the organic composition of the CW (Stanković et al. 2018). Under extreme pH conditions, other groups like amines and sulfhydryl groups can be present in very low amounts (Gonzalez and Pokrovsky 2014). The presence of an abundant amount of uronic acid rather than cellulose and hemicellulose can also be another reason behind the binding capabilities of mosses with heavy metals (Wells and Brown 1987). The higher amount of carboxyl and phosphoryl moieties in the bryophyte CW can be another explanation for heavy metal uptake (Vazquez et al. 1999; Tremper et al. 2004; Gonzalez and Pokrovsky 2014). Different mosses have different uptaking capacities toward the same metal (Shakya et al. 2008). Therefore, As-stressed mosses can potentially be utilized as an indicator of oxidative stress even when the morphological characters of the plant remain the same (Stanković et al. 2018).

4.6 Arsenic Response on Pteridophyte

Plants' adopted strategies to overcome As stress by accumulating this potentiality are known as phytoremediation. Phytoremediation process helps in removal of As from the soil as well as water; there some fern species have hyperaccumulation property, namely, *Pteris vittata, Pteris quadriavrita, Didymochlaena truncatula, Pityrogramma calomelanos, Dryopteris atrata, Pteris multifida, Dryopteris erythrosora, Pteris ensiformis, Cyrtomium falcatum, Arachnoides simplicor, Pteris cretica* var. *nervosa, Adiantum hispidulum, Pteris dentata*, and *Pteris oshimensis*. The absorption process of pollutants from water, known as rhizofiltration, is used widely for cleaning up As-rich water, and also these plants are used for removal of As from soil (Wongkongkatep et al. 2003; Luongo and Ma 2005; Wang et al. 2006; Danh et al. 2014). Wongkongkatep et al. (2003) reported that the hyperaccumulation property of *P. calomelanos*, also known as silver back fern, elevated in the presence of a chelating compound, namely, ethylenediaminetetraacetic acid (EDTA), whereas the presence of dimercaptosuccinic acid (DMSA), a chelating compound, decreases the As accumulation property of the fern in their aerial portion. Arsenite and As(V), both the As forms, are absorbed by *P. calomelanos*. *P. vittata* is a potent As bioaccumulator as it shows storage of 503 and 1748 $mg\,kg^{-1}$ As in their roots and shoots, respectively, if the fern grows at 10 $mg\,l^{-1}$ of As concentration level (Luongo and Ma 2005). According to Wang et al. (2002), *P. vittata* absorbed As(V) through the phosphate transport system, and as a result, As(III) accumulated in the foliar region, and As(V) influx decreased if the phosphate concentration is more. Arsenate absorbed by the root of *P. vittata* translocates to not only the cell vacuoles of both lower and upper epidermis but also the trichromes of the foliage at a very higher rate (Su et al. 2008). Arsenic is able to inhibit cellular metabolism and produce oxidative stress responses and change in the DNA (Ushijima and Okochi-Takada 2005; Bona et al. 2010; Campos et al. 2016; He et al. 2017; Praveen et al. 2018). Zemanová et al. (2020) reported that *P. cretica* var. *Albo-lineata*, an As hyperaccumulator pteridophyte, sequestrated As more in older frond compared to younger frond. The accumulation of As induced methylation of DNA and reduction of chlorophylls; root morphology also altered as CW lignified and elevated the nutrients such as Zn, Cu, S, Mg, Mn, etc. *Pteris umbrosa*, a native of Australia, show As sequestration potential reduced in aged fern individuals; adventitious roots accumulated less As compared to rhizome, whereas senescing fronds accumulate less As compared to mature fronds, and increased As sequestration rate leads to the phytotoxicity, and as a result, biomass is reduced (Koller et al. 2007). Campos et al. (2016) reported that *P. calomelanos* can increase the rate of amino acids as well as antioxidant synthesis with disturbing major carbon metabolism, which altogether gives the As tolerance potentiality. ROS production is elevated at high As concentration which leads to the necrosis of the frond. Indole-3-acetic acid (IAA) application increased Catalase (CAT) and Superoxide dismutase (SOD) activities in *P. cretica* var. *nervosa* and *P. ensiformis*. Peroxidase (POD) activity and As sequestration rate increased in *P. cretica* var. *nervosa* under the stress of As (He et al. 2017). Arsenic is mobilized in the rhizome as well as the root of *P. vittata* and is translocated via xylem tissue to the leaf in the form of As(III) at a very high rate (Lombi et al. 2002; Pickering et al. 2006; Su et al. 2008). Indriolo et al. (2010) reported that for the vacuolar As(III) accumulation, PvACR3 protein is responsible. PvACR3 in *P. vittata* acts as As(III) efflux, though it does not transport As(III) from the cell to outside as PvACR3 is present in the membrane of vacuole, so it translocates As(III) toward the vacuole. Lycophytes, mosses, ferns, and gymnosperms possess a single copy of *ACR3* genes responsible for As tolerance; on the other hand, angiosperms lack *ACR3* gene, which makes them As susceptible. Plants have two types of metal uptake systems: (i) the highly inducible high-affinity system operated at low content, for example, uptake system of phosphate operated at low concentrations of phosphate; and (ii) the constitutive low-affinity system operated at high content, in which transpiration and root pressure regulate the As transport system (Bondada and Ma 2003). According to Marin et al. (2002), As moves from the root to the shoot by the formation of dimethyl arsenic acid, accumulated inside the tissue as arsenite-tris-thiolate complex (Hartley-Whitaker et al. 2001). Fronds of *P. vittata* have higher As accumulator property; fronds are able to store 93%, whereas rhizome shows only 8% of the concentrated As (Ma et al. 2001). *P. vittata* by the phytoextraction process eliminates As from habitat and accumulates in their fronds as well as roots (Mirza et al. 2014), and then the accumulated As is harvested from fronds and disposed of as toxic waste (Xie et al. 2009). According to Anjum et al. (2015), phytoremediation by *P. vittata* is an environmentally friendly and simple process, and also *P. vittata* is not an invasive plant, so it can be widely used to remove As from polluted areas. *P. vittata* in association with Arbuscular mycorrhizal fungi (AMF) enhances the As sequestration rate up to 20 cm soil depth (Cantamessa et al. 2020).

4.7 Responses in Angiosperms

Arsenic toxicity, as a result of irrigation with As-contaminated groundwater that is posing threats to human health, is the most common hazard for rice plants, particularly in Bangladesh and West Bengal (India) (Majumder et al. 2019). Arsenic has been detected in rice grains from North America, Asia, and Europe, posing a future hazard to food surety. Because of its abundance, phytotoxicity and mobility As^{3+} is the most detrimental species for rice crops (Kalita et al. 2018).

A recent report (Samal et al. 2021) revealed that high yield rice varieties (HYV) are more likely to acquire As in comparison to local varieties (LVs), thus affecting all the age ranges crossing estimated daily intakes. The calculated Cumulative Hazard Index and Individual Excess Lifetime Cancer Risk values revealed that As co-exposures via drinking water and rice provide extremely serious carcinogenic and noncancerous dangers to the study population.

Besides, according to new research (Farhat et al. 2021), increased warmth can raise As concentrations in rice tissue. Temperature and rice grain As (As(III)) had a positive linear relationship. The gene encoding the OsABCC1 transporter, which is thought to be critical for As sequestration in vacuoles and hence inhibiting As transfer to grains, showed no increase. The finding suggests that temperature affects As sorption to root plaque where higher growth temperatures may raise the risk of dietary As exposure in rice systems previously thought to be low risk.

The reactions of oxidative stress indices, metabolites, antioxidant enzymes, and ions were likewise tissue and time dependent, with a distinct pattern in the two kinds of rice varieties: tolerant Pooja and vulnerable CO-50 found in a study (Yadav et al. 2021). Salicylic acid and abscisic acid, among others, displayed varied responses in the two varieties. The investigation disclosed that resistance comes up due to early detection of As stress, optimal synchronization of hormones, biochemical responses, and ionic and metabolic profiles. The iron sulfate amendment also resulted in a considerable reduction in grain As contents in both rice genotypes of Basmati-385 and BR-1, where Bas-385 was naturally more productive and tolerant than the rest and was also supported by regression analysis (Irem et al. 2019).

Rice plant absorption at the root surface is influenced by factors such as radial oxygen loss (ROL) and iron plaque. Several transporters, like aquaglyceroporins and phosphate transporters, are involved in the uptake and transport of various As species, as well as their translocation to subcellular compartments. Sulfur availability, and hence the level of thiol (-SH)-containing As binding peptides such as phytochelatins and GSH, affects these activities. It's also hypothesized that phloem has a function in As transport to the grains (Awasthi et al. 2017, Kumarathilaka et al. 2018).

Affecting phosphorylation metabolism and interfering with the flow electron leakage that occurs during As(V) to As(III) reduction cause ROS to develop, and the redox-led methylation that follows adds to the ROS production. The As-generated ROS interactions with macromolecules, such as protein and nucleic acid damage and lipid peroxidation, are the signatures of As stress. The activation of MAPK signaling cascades by AS-induced ROS leads to the targeting of transcription factors and gene expression. The enzymatic antioxidant system is influenced by redox imbalances, which activate the cell to generate low-molecular-weight antioxidants, which are critical in the protection of ROS-induced damage (Kostecka-Gugała and Latowski 2018).

In contrast to the As-sensitive genotype *Nicotiana sylvestris*, total photosynthesis proportions and saccharide quantities were not affected by As exposure in tolerant genotype *Nicotiana tabacum, cv. "Wisconsin"* (Kofroňová et al. 2020; Naeem et al. 2020). Tolerant type holds higher total antioxidant capacity, as evidenced by higher antioxidant levels (e.g. ascorbate, phenolics, GSH) and activities and/or proper localizations of antioxidant enzymes where a lipid peroxidation marker, malondialdehyde levels, improved solely in sensitive tobacco, demonstrating that As-tolerant species have effective membrane protection. The researchers believe that strong metabolism of carbohydrate based on uninterrupted photosynthesis is critical not only for supplying carbon and energy for defense but also for directly quenching (ROS) (Irato and Santovito 2021).

When compared to sensitive cultivars, As-tolerant cultivars are more efficient in antioxidant metabolism (Kalita et al. 2018; Das et al. 2021). The oxidative stress generated by As has been discovered to be alleviated by iron and selenium. Rice roots absorbed more As(III) and As(V) after being exposed to selenite. Selenite altered the localization and speciation of As in the shoot, lowering its content. Selenite inhibited As translocation more effectively than Selenite nanoparticles (SeNPs) (Camara et al. 2019; Wang et al. 2021).

Microbes in the plant rhizosphere may provide As-stress resistance in plants cultivated in As-contaminated environments. The sensitivity of rice cultivars to As was assessed using oxidative stress indicators, antioxidant enzyme activities, and genotoxicity levels. Under

As(V)-provoked conditions, the levels of oxidative stress markers such as malondialdehyde (MDA), H_2O_2, and proline, as well as the actions of antioxidant enzymes such as CAT and SOD, as well as the level of genotoxicity as measured by Random Amplified Polymorphic DNA (RAPD) profiling, were all altered to varying degrees in all rice cultivars tested and exhibited a considerable change in banding patterns in cultivated varieties seedlings treated with As(V) (Majumder et al. 2019; Majumder et al. 2020a; Chattopadhyay et al. 2021).

The effect of maize seedlings to exposure of inorganic under hydroponic conditions displayed differential expression in their sensitivity to the metalloid (Requejo and Tena 2014). The findings revealed that maize is relatively resistant to As toxicity like soybean plants where researchers focused on the root structure, plant development, photosynthetic pigment substance, and antioxidant system, which were all assessed after As(V) and As(III) treatments (Armendariz et al. 2016). In both plants under As(III) therapy, the negative effect was more pronounced than under As(V) therapy in all of the investigated parameters. After two and five days of treatment, antioxidant enzyme activity (SOD and total peroxidases) was increased in roots and shoots taken from soybean seedlings treated with As. The maize seedling disclosed that the metalloid is less toxic to roots than it is to shoots, thereby producing a large accumulation of nonprotein thiols (probably phytochelatins) in roots without a significant effect on GSH content. Maize can store a lot of As in its roots and only transfers a small amount to the shoots. As a result of its low proportion of acropetal passage of As from roots to shoots, maize appears to be a risk-free crop in terms of metalloid entry into the food chain, and as a major bioenergy harvest competent of presenting extreme degrees of As curb and accretion in roots, maize may demonstrate a fascinating opportunity for cultivation. In soybeans, there was a decrease in chlorophyll concentration and an upsurge in peroxidation of membrane phospholipids. It's possible that As-induced root structural changes, such as a specific array of dark acquisitions in the vascular system, are linked to an adaptation or detoxifying mechanism that prevents As from reaching aboveground tissues like maize. When plant–water relations and heavy metals' effect concerned scientists, another study reported that root absorption rate was significantly reduced by As(V) and As(III), perhaps reducing metalloid uptake in soybean plants, but restored water absorption after few days. Due to the accumulation of proline and soluble carbohydrates, this was linked to reduction in stomatal conductance as well as a decrease in leaf water and osmotic potential. As stress modulates lipid signaling causing stomatal closure of soybean plants (Armendariz et al. 2016; Abbas et al. 2018; Zhao et al. 2018). In addition, plants exposed to the therapy had reduced leaf xylem vessels and aberrant stomata. These systems improved the plant's ability to retain water and avoid dehydration as a result.

In a recent case study, the impact of 24-Epibrassinolide (EBL) on the antioxidant complex in *Arabidopsis thaliana* under As(V) stress was investigated (Surgun-Acara and Zemheri-Navruz 2019). The degree of expression of CAT (CAT3, CAT2,CAT1), SOD (Cu-ZnSODs, FeSODs, and MnSOD) isoforms, proline biosynthesis (*P5CS2* and *P5CS1*), and heat shock protein (*Hsp90–1* and *Hsp70–4* genes), as well as concentration of MDA, content of proline and total antioxidant status, was verified in rosette leaves of 2-month-old plants exposed to As(V) and/or EBL for 24 hours. According to the findings, As(V) exposure stimulates antioxidative resistance in diverse cellular parts, and combined EBL and stress treatment causes considerable alterations in levels of transcript of multiple SOD and CAT isoforms. The overexpression of several *HSPs* and *P5CS* genes in leaves subjected to

As(V) demonstrated that Brassinosteroids (BRs) play a role in stress tolerance induction. Studies on the molecular underpinnings of BRs' actions will aid in the development of improved ways to protect plants from metal toxicity by approving a better vision of this phytohormone's anti-stress activities.

4.8 Perception of Arsenic Stress by Plants and Triggering of Signaling Cascades

Understanding the movement and detoxification of hazardous components in plants necessitates a study of ionomic variation. Phosphorous (Pi) and Potassium (K^+) are two of the most critical nutrients that plants require, and researchers are interested in learning more about how they are obtained. Most of the research focused on determining the mechanisms of Pi and K^+ absorption and circulation inside plants, in addition to the regulatory routes involved. Proof is accumulating that suggests connections in the feeding of different nutrients, as well as cross talk in the signaling pathways that control their acquisition.

In *Arabidopsis* plants, elevated concentrations of K^+ in the external medium slow down uptake of Pi and hindered nutrition, leading to the development of the phosphate starvation response (PSR) and activation of genes programming root phosphate uptake techniques. The PHL1 and PHR1 transcription factors, which are vital components of Pi signaling in *Arabidopsis*, are required for the high K^+-induced PSR (Bustos et al. 2010; Sun et al. 2016a; Wang et al. 2018c). Importantly, high K^+ levels in plants decrease the As buildup and its harmful consequences. The findings could aid in the development of solutions to address both Pi deficit and As accumulation in crops (Rodenas et al. 2019; Liang et al. 2020). *Vallisneria natans, Potamageton crispus, Myriophyllum spicatum, Hydrilla verticillate*, and *Ceratophyllum demersum* (Abu Bakar et al. 2013) are the examples of common submerged plants that show significant potential of As phytofiltration for bioremediation as they accumulate As in the shoot and immobilize As below ground in roots due to As acquisition more in a basipetal manner than acropetal fashion (Xue and Yan 2011). The ability of submerged plants to concentrate As is influenced by As concentration in the water, as well as coexistence of ions like PO_4^{3-} and Zn^{2+} (Ran and Chen 2020). In rice, nodulin 26-like intrinsic channels and phosphate transporters transfer As(V) and As(III) to the root. The entry of MMA, DMA into the root may be aided by the silicic acid transporter. DMA(V) is one of the most transportable As species in plants, quickly transferring from root to shoot. The gene *OsPTR7* plays a role in DMA transport between the vascular bundle (Abedi and Mojiri 2020). Rice roots treated with any of the As species showed positive correlations between P and Fe, showing that P and Fe share biological activities. The principal component analysis (PCA) results in ionomic distinction between shoots and roots agreed with the heatmap from the hierarchical cluster analysis (HCA). The ionomic profiles in rice plants are sufficient to detect reactions to environmental disturbances (Du et al. 2019, 2020).

In rice roots, As(V) is able to intracellular ROS manufacture and calcium oscillation. Abiotic stress, secondary metabolic processes, and detoxification pathways were all activated by As(V). Cell cycle and oligopeptide transport and Secondary CW biogenesis genes were all downregulated. The expression of genes encoding signaling components such as receptor-like cytoplasmic kinases, protein kinase, heat shock factor, myeloblastosis transcription factor (MYB TF), APETALA2/ethylene response factor, and zinc-finger protein expressed in inflorescence meristem transcription factors was upregulated (Chi et al. 2013). As(V) stress selectively influenced C3H and GARP-G2-like transcription expression factors (Huang et al. 2012). The telurite-resistance/dicarboxylate and ATP-binding cassette superfamily transporters were the most common As(V)-regulated transporter families. Ca^{2+} signaling may be involved in the stress-induced generation of jasmonates (JAs). JA transporters in Subclass 4 of ABCGs may also play a role in heavy metal detoxification using the ATP-Binding Cassette G (ABCG) transporter subfamilies. They together limit the buildup of hazardous substances and improve tolerance to them by harmonizing the ion transport system, antioxidant enzyme activity, and chelating capability in plants (Chen et al. 2021). Under As stress, several signaling proteins also increase, including MAPK kinase, calcium-dependent protein kinase (CDPK), and MAPK. Furthermore, As(V) significantly boosts the activity of CDPK-like kinases and MAPKs, and As-induced MAPK activation implicates CDPK and Nicotinamide Adenine Dinucleotide Phosphate (NADPH) oxidases (Huang et al. 2012).

Checking of As-induced stress upon two popular rice varieties, namely, Tarom hashemi and Fajr, in Iran displayed that As lowered fluorescence, chlorophyll content, and biomass production, while methyl jasmonate (MJ) increased photosynthetic pigments and plant development according to the findings. In both rice types, As caused oxidative stress (H_2O_2 and MDA); however, MJ decreased As-induced oxidative stress by moderating antioxidant enzyme activity and the ASA–GSH cycle (Jung et al. 2019). By lowering the manifestation of *Lsi1*, *Lsi2*, and *Lsi6* genes induced by As treatment, MJ lowered As accumulation. MJ boosted Fe accumulation in the leaves through escalating the expression of YSL2 and FRDL1 transporters even under As toxicity. Administration of MJ amplified the vigor and yield of both rice varieties under As toxicity by reducing As accumulation by regulating As transporters and alleviating oxidative stress by boosting the activity of antioxidant enzymes and the ASA–GSH cycle (Mousavi et al. 2020). Manifestation of corresponding thiol-metabolic genes (*OsPCS1* and *OsMT2*) downregulated with respect to As toxicity, and the role of melatonin seems to be more important in Khitish, an As-sensitive rice variety, as evidenced by better usage of thiol components like GSH, cysteine nonprotein thiols with higher GSH/GSSG ratio, and also phytochelatins, following exogenous Melatonin application in rice (Samanta et al. 2021).

In *Arabidopsis*, the discovery of the three hypersensitive arsenate (aos) mutants (ptlpd1–1, ptlpd2, and ptlpd1–4) marked their identical phenotype by the interference of plastidial LIPOAMIDE DEHYDROGENASE1 (*ptLPD1*), a gene encoding one of the two E3 isoforms located in the plastidial pyruvate dehydrogenase complex (PDC) that equally contributes in catabolic and anabolic metabolism. Among them, ptlpd1–1 plants displayed comparatively diminished root and shoot growth and increased anthocyanin buildup in the presence of arsenate but accumulating the same quantity of As as wild type demonstrating the innate sensitivity to the poison rather than its escalation. These data imply that the ptLPD isoforms play a role in As(III)-mediated As sensitivity in *Arabidopsis in vivo* and could be used to increase As tolerance (Chen et al. 2010). Following As treatment, transcriptional alterations have been found in two *Arabidopsis* accessions, Col-0, the tolerant, and Ws-2, the sensitive type exhibiting a faster root elongation rate with lower As accumulation than

Ws-2. Intensive studies revealed their shared genes related to heat sensitivity and oxidative stress, indicating possessing common but conserved As stress-related responses. HSPs and factors, ubiquitin, and aquaporin transporters were mostly coded by the responsive genes. Genes coding for ethylene-signaling components were elevated in Col-0 after exposure to As. Col-0-specific, tolerance-associated, linked with innate immunity responsible *LRR-RLK VIII* gene's genetic loss-of-function research indicated decreased metal buildup sensitivity toward As in roots. Thus, ethylene-related pathways, protein structure preservation, and *LRR-RLK VIII*-mediated signaling may be essential way outs for the species' toxicity and tolerance (Fu et al. 2014). Accessions were divided into two groups based on root length reduction: tolerant and sensitive to As (V). Col-0 (with 10% reduction) and Slavi-1 (with >60% reduction) were found to be the most tolerant and sensitive accessions, respectively, based on root length analysis. Again 168 and 548 genes were found to have differential expressions in a comparative global expression study, with 120 common differentially expressed genes. Differential expression in diverse accessions was linked to defense and stress responses, transport, regulatory systems, and metabolic processes (Shukla et al. 2018). Similar reports on the activation of stress-induced genes were also unearthed in As-stressed mustard plants, while GSH metabolism, carbon metabolism, and plant hormone signal transduction are the top pathways with the most KEGG orthology hits (Thakur et al. 2019).

Most genes concerned in developmental processes, growth of cells, and photosynthesis were found to be downregulated in tissues treated with As, while 10870 considerably differentially expressed genes mostly related to metabolic processes, stress, signal transduction, transcription regulator activity, and transporter activity were found to be highly expressed. To understand the As stress management beyond phytochelatins, As(V)-induced chlorosis1 (aic1), an As(V)-sensitive *Arabidopsis* mutant, in the backdrop of the phytochelatin's synthase-defective mutant cadmium-sensitive1–3 (cad1–3), has been introduced and checked. Both mutants displayed lower chlorophyll content, fewer and bigger chloroplasts, and a substantially deformed chloroplast structure than cad1–3. TRANSLOCON PRESENT AT THE OUTER MEMBRANE OF CHLOROPLAST132 (TOC132) was recognized as the altered gene using genome complementation and resequencing in both the mutant types. Under arsenate stress, the double mutant aic1 cad1-3 had fewer proteins associated to chloroplast ribosomes, photosynthesis, compound synthesis, and thioredoxin systems than *cad1–3*. These findings suggest that chloroplasts are a particularly vulnerable target for As harmfulness and that AIC1/TOC132 plays a vital function in preventing As toxicity in chloroplasts (Wang et al. 2018b).

Recently, 841 shared genes were found to respond to both As^{3+} and Cd^{+2} stimuli in 30-day rice seedlings, grown in hydroponic cultures for up to one week for transcriptome analysis. Redox control, transcriptional regulation, stress response, signal transduction, transmembrane transport, and biosynthesis and metabolism of macromolecules and sulfur compounds were among the genes shared by Cd and As stress (Li et al. 2019). The GSH metabolism-linked gene *Os09g0367700* was considerably upregulated in plants exposed to both methods compared to unexposed controls and exhibited a positive synergistic effect under combined Cd and As stress, according to qRT-PCR validation (Wang et al. 2018c). Furthermore, after seven days of combined exposure to Cd and As, the redox control genes *Os07g0638300, Os06g0216000*, and *Os01g0294500*, as well as the CW biogenesis genes *Os05g0247800, Os11g0592000*, and *Os03g0416200*, the GSH metabolism gene, the transmembrane transport associated genes Os04g0524500 differed considerably from an unexposed control and showed synergistic effects (Huang et al. 2019).

When two varieties of *Populus deltoides* were compared to "I-45/51," "zhonglin 2025," showed less morphological and structural harm and reduced ROS and MDA buildup (Kofroňová et al. 2020) and greater photosynthesis and ROS scavenging abilities under As(V) stress due to the presence of defense and stress-related proteins mostly. PdDRT102 was discovered to be the only one of these proteins that were highly elevated in "zhonglin 2025" in As(V) stress. In *Arabidopsis*, heterologous overexpression of PdDRT102 resulted in increased As(V) and salt chloride tolerance. In *Arabidopsis*, fluorescent-tagged PdDRT102 protein signals in the nucleus and CM were observed to be significantly elevated under As stress, demonstrating that the DRT102 protein may play a role in shielding poplar from this stress (Liu et al. 2017).

4.9 Mechanistic Aspects of Responses Related to Arsenic (Effect on ATP Synthesis, Photosynthesis, DNA, Protein, Cell Membrane, Carbohydrate, and Lipid Metabolism)

4.9.1 Effect of Arsenic on ATP Synthesis

Arsenic is accumulated by plants through phosphate transporters (Awasthi et al. 2017). This is because As and Pi are analogues to each other and hence can use the same transporter to enter the plant body (Abbas et al. 2018). It is for the same reason that As is able to disrupt phosphate-dependent metabolic pathways related to ATP synthesis (Finnegan and Chen 2012). During phosphorylation processes, As can compete with phosphate resulting in the formation of As adducts. The production and rapid autohydrolysis of As(V)–ADP initiate a futile cycle in which photophosphorylation and oxidative phosphorylation are decoupled, lowering the ability of cells to produce ATP and carry out regular metabolism. Arsenate toxicity in plants is reported to be caused by the replacement of Pi in essential metabolic pathways, thereby disrupting key biochemical processes (Finnegan and Chen 2012). Purine nucleoside phosphorylase converts inosine to hypoxanthine and ribose-1-arsenate by cleaving it with As(V). Thiols rapidly convert the resultant As(V) ester to As(III) in a nonenzymatic manner (Breydo 2013). Arsenic can affect a wide range of metabolic pathways related to DNA (Srivastava and Sharma 2014; Zemanová et al. 2020), RNA (Pandey et al. 2020), oxidative phosphorylation (Hughes et al. 2011), and induction of signaling pathways related to phosphorylation (Huang et al. 2012). In plants, ATP is generated from ADP and inorganic phosphate in the thylakoid membrane (Yin et al. 2010). Pentavalent As decouples the oxidative phosphorylation pathway, in which a phosphate molecule is replaced from ATP resulting in formation of ADP–As(V) complex. This perturbs the energy flow in cellular level and suppresses overall ATP synthesis (Chandrakar et al. 2016). In addition to it, As is recognized by mitochondrial enzymes, which results in disturbance of ATP synthesis (Abbas et al. 2018).

4.9.2 Arsenic's Effect on Photosynthesis

Arsenic creates toxicity to the plants and inhibits the photosynthetic process (Gusman et al. 2013; Singh et al. 2019). It has also been linked to a reduction in photosynthetic pigment (Srivastava and Sharma 2014). Arsenic damages the integrity of chlorophylls and carotenoids of photosynthetic system and reduction in activities of antenna complex and electron transport chain. As a result, ATP and NADPH synthesis are inhibited, and energy is dissipated as fluorescence or heat (Kofroňová et al. 2020). Arsenic also negatively affects photosystem II as a result of a decrease in values of Relationship between photochemical efficiency (JIP-parameters), namely, ϕPO, FV/F0, ϕE0,Ψ0, and PIABS (Patel et al. 2018). The decrease in pigments upon As state is due to lack of adaptive adjustments in relation to synthesis of pigments (Srivastava and Sharma, 2013). Investigations on As toxicity on *C. demersum* revealed that inhibition of tetrapyrrole biosynthetic pathway is responsible for reduction of chlorophyll content. It was also observed that As inhibits chlorophyll biosynthetic pathway upstream of coproporphyrinogen III (Mishra et al. 2016).

4.9.3 Effect of Arsenic on Cell Membrane

Arsenic poisoning in plants results in the production of oxidative stress (Sharma 2012; Piacentini et al. 2020). As a result, there is an overabundance of reactive oxygens and consequent oxidation of lipid layers of CM (Álvarez-Robles et al. 2020). This is evident by increase in MDA content, which is a by-product of lipid peroxidation of CM (Farooq et al. 2016). In rice seedling, As induces damage to the CM due to lipid peroxidation as evident from increase in MDA concentration (Choudhury et al. 2010). Oxidation of lipid results in electrolyte leakage by the cells (Talukdar 2013). Arsenic toxicity also results in disruption of nutrient uptake and nutrient balance (Li et al. 2019). Arsenic toxicity also results in decrease in stomatal conductance (Maghsoudi et al. 2020). All these also cumulate to membrane damage (Abbas et al. 2018). Arsenic results in alteration of lipid signaling and induces closure of stomata possibly to reduce water loss through transpiration (Armendariz et al. 2016). It was further observed that As toxicity results in decrease in membrane stability index in *P. vittata* and *P. ensiformis* (Singh et al. 2006).

4.9.4 Arsenic Induced Oxidative Stress

Several studies have demonstrated that As toxicity results in enhanced ROS generation in plants (Hossain et al. 2018; Majumder et al. 2020b). Electron leakage during reduction of pentavalent As to trivalent As causes the formation of ROS. Reactive oxygen species is further contributed by redox-driven methylation reactions (Kostecka-Gugała and Latowski 2018). In plants, As is also available in methylated forms such as MMA, and in most of the cases, these methylated forms are of microbial origin and are absorbed by the plants (Lomax et al. 2012; Di et al. 2019). These methylated forms readily react with molecular oxygen to give rise to ROS within the cells. Additionally, iron-dependent oxidative stress is caused by DMA, which is related to iron released from ferritin, whereas dimethyl arsenate-induced ROS may directly cause oxidation of DNA (Sharma 2012). Cytochrome oxidase also reduces pentavalent form of As to trivalent form using oxygen as terminal electron acceptor and ultimately results in formation of ROS (Abbas et al. 2018).

4.9.5 Effect of Arsenic on Carbohydrate Metabolism

Arsenic has a negative impact on metabolism of plant. The metabolism of carbohydrates is affected by As stress (Majumder et al. 2020). Arsenic stress results in increase of soluble sugar in plants (Vezza et al. 2018). Arsenic stress also increases the concentration of

Table 4.1 Arsenic on different plant enzymes.

S.No.	Enzyme	Effect	Plants affected	Reference
1.	α-Amylase	Inhibition	Rice	Jha and Dubey (2004)
2.	α-Amylase	Inhibition	Wheat	Liu et al. (2005)
3.	Starch phosphorylase	Increased activity	Rice	Choudhury et al. (2010)
4.	Sucrose synthase	Increased activity	Wheat	Sil et al. (2019)
5.	Starch phosphorylase	Increased activity	Wheat	Sil et al. (2019)
6.	Acid invertase	Increased activity	Rice	Jha and Dubey (2005)

reducing sugars in plants (Choudhury et al. 2011). In addition to it, there have been variations of activity of numbers of carbohydrate metabolizing enzymes during As stress. Table 4.1 illustrates the As influence on a variety of plant enzymes involved in metabolism.

4.9.6 Effect of Arsenic on Lipid Metabolism

One of the most deleterious effects on plants during As stress is oxidation of lipids. The oxidation of lipids of cellular membrane results in leakage of the cells (Talukdar 2013). Three steps are involved in the lipid peroxidation reaction, namely, initiation, progression, and termination. During the initiation step, the allylic hydrogen is abstracted by prooxidants such as the hydroxyl radical, resulting in the carbon-centered lipid radical ($L^{\cdot}$). The lipid radical reacts swiftly with oxygen to form a lipid peroxyl radical at the time of propagation ($LOO^{\cdot}$), which then removes a hydrogen atom from another lipid molecule, forming a new lipid radical ($L^{\bullet}$) and lipid hydroperoxide (LOOH). In termination reaction, an antioxidant donates a hydrogen atom to the $LOO^{\cdot}$, generating a radical of antioxidant that interacts with an additional $LOO^{\cdot}$ to produce nonradical compounds (Ayala et al. 2014). During oxidative stress, increased oxidation of PUFA occurs, which ultimately results in increase of membrane permeability (Sharma et al. 2012). It is observed that lipid hydroperoxides and various products of peroxidation of lipids, such as malondialdehyde, form adducts with biomolecules of the cell, thereby altering their function (Blair 2008).

4.9.7 Effect of Arsenic on Protein

Arsenic is highly toxic to plants, and one of the major reasons of this toxicity is its high affinity with sulfhydryl group of protein (Shen et al. 2013). This results in deterioration of membrane and ultimately death of cells (Abbas et al. 2018). In addition, As plays a role in the inhibition of enzymes related to protein metabolism. A study reports that As inhibits nitrate reductase, nitrite reductase, and glutamine synthetase in seeds of rice (Jha and Dubey 2004). It was also documented in another investigation. That in rice seedlings, As exposure inhibits RNA and protein hydrolysis, which is due to suppression of activities of RNAase and protease (Mishra and Dubey 2006). Arsenic reacts with the thiol and sulfhydryl groups of enzymes involved in synthesis of genetic material of plant (Agarwal and Khan 2020).

4.9.8 Effect of Arsenic on DNA

An oxidative burst is one of the first signs of As toxicity. The generated ROS causes a number of changes in the nucleic acid structure. Arsenic-mediated oxidative stress results in the formation of DNA-protein adducts (Chandrakar et al. 2016) and oxidizes the base pairs and damages sugar residues, which leads to fragmentation and strand breakage (Roldán-Arjona and Ariza 2009). It also initiates single and double stranded breakage of DNA, chromosomal aberration in plants (Dutta et al. 2018). Arsenic-induced ROS also modifies bases and generates 8-hydroxydeoxyguanasine and 8-oxoguanine (Tripathi et al. 2020). The DNA lesion 8-oxoguanine is highly mutagenic and causes mutation of various degrees (Ohno et al. 2014).

4.10 Future Prospects and Conclusion

Volcanic eruption and man-made attributions for As pollution present themselves the talk of decades for perturbation of homeostasis. Soil and water are getting affected equally due to the escalation of As salts uncompromisingly. Access to safe drinking water is laden with political, socioeconomic, and cultural disparities. Research data have shown that interdisciplinary perspective of water purification is the main issue to consider, and long-running improvements involve larger viewpoints than strict scientific approaches. A statistical survey, conducted by Food and Agriculture Organization (FAO) in 2014, exhibits that over 56% of the population on the earth is active in agricultural practice. As a result, As pollution of water and soil poses a serious threat to everyone's survival. Direct and indirect consumption of As through flora and fauna of biome causes eventual biomagnification in physiology. Plants have acquired a variety of escapism routes and heavy metal tolerance in the course of evolution.

Plants may withstand As by a variety of processes, including phytochelatins, vacuole sequestration, and antioxidant defense system activation. To cope with As poisoning, plants have evolved several signaling pathways that utilize proteins, ROS, hormones, nitric oxide, and calcium ions as signaling molecules. Plants can endure under metal stress, thanks to these mechanisms, which activate their defensive systems. The routes through which these stress signals are interpreted and responded to are a largely unexplored area of research with many unanswered questions. Greater knowledge of these signaling routes can aid in the development of plants that can thrive in As-adulterated soil and water. The routes through which these stress signals are interpreted and responded are a largely unexplored area of research with many unanswered questions. There is a considerable urge to distinguish relevant gene targets that may be adjusted as per needs of accomplices utilizing diverse biotechnological approaches to boost the survival of plants in contaminated environments. Arsenic-induced toxicity in macroalgae is poorly understood. Detailed study about the relationship of the macroalgae with the absorption and metabolism of As is necessary for the future. More studies on the mechanism of As incorporation, biotransformation, and emission at the biochemical and molecular levels are important.

A transcriptome study of As accumulation and detoxification will not only aid in improving its phytoremediation effectiveness but also help with management measures needed to prevent As bio-deposit in the food chain. Bio-phytoremediation is another approach appreciated and practiced by cultivars across the world to boost up crop production against As stress, which is being assisted by plant growth-promoting bacteria (PGPB) or bacteria consortium (BC). Under PGPB ambiance, plants not only accumulate biomass but also show minimal visible signs of As stress.

Responses to the stress induction by inorganic and organic As salts in land and aquatic plants exhibit their tolerance and vulnerability toward As accumulation. Stunted growth and anomalies in structural protein–lipid composition directly affect photosynthetic production, whereas there are some aquatic and semiaquatic plants securely storing As salts without letting any alteration in the metabolic network. Understanding and discoveries of switching on the stress-induced genes provide a necessary shield against the impact of As. Conjoint physiological and molecular studies are the pillars to take a stand against the stress influence.

Industrial waste, unrestricted mining, and inadequate knowledge about the future consequences are intensifying more severity to the already-prevailed issues. Therefore, the governments' intervention in constraining utilization of land and water is of utmost importance to resist such practice legally. In many ways, plants pave the way for stress management by their broad-spectrum modes in cleansing As load of the spheres. Therefore, studying plants in As management and diverse applications according to the need and specificities may open up the doors of possibilities for future implications on a large scale.

References

Abbas, G., Murtaza, B., Bibi, I. et al. (2018). Arsenic uptake, toxicity, detoxification, and speciation in plants: physiological, biochemical, and molecular aspects. *Int. J. Environ. Res. Public Health* 15 (1): 59. https://doi.org/10.3390/ijerph15010059.

Abedi, T. and Mojiri, A. (2020). Arsenic uptake and accumulation mechanisms in rice species. *Plants (Basel)* 9 (2): 129. https://doi.org/10.3390/plants9020129.

Abid, M., Niazi, N.K., Bibi, I. et al. (2016). Arsenic (V) biosorption by charred orange peel in aqueous environments. *Int. J. Phytoremed.* 18: 442–449.

Abu Bakar, A.F., Yusoff, I., Fatt, N.T. et al. (2013). Arsenic, zinc, and aluminium removal from gold mine wastewater effluents and accumulation by submerged aquatic plants (*Cabomba piauhyensis*, *Egeria densa*, and *Hydrilla verticillata*). *Biomed. Res. Int.* 2013: 890803. https://doi.org/10.1155/2013/890803.

Agarwal, S. and Khan, S. (2020). Heavy metal phytotoxicity: DNA damage. In: *Cellular and Molecular Phytotoxicity of Heavy Metals. Nanotechnology in the Life Sciences* (ed. M. Faisal, Q. Saquib, A.A. Alatar and A.A. Al- Khedhairy), 157–177. Cham: Springer https://doi.org/10.1007/978-3-030-45975-8_10.

Ahmad, A., Khan, W.U., Ali Shah, A. et al. (2021). Synergistic effects of nitric oxide and silicon on promoting plant growth, oxidative stress tolerance and reduction of arsenic uptake in *Brassica Juncea*. *Chemosphere* 262: 128384. https://doi.org/10.1016/j.chemosphere.2020.128384.

Álvarez-Robles, M.J., Bernal, M.P., Sánchez-Guerrero, A. et al. (2020). Major as species, lipid peroxidation and protein carbonylation in rice plants exposed to increasing as (V) concentrations. *Heliyon* 6 (8): e04703. https://doi.org/10.1016/j.heliyon.2020.e04703.

Andersson, S., Romero, A., Rodrigues, J.I. et al. (2021). Genome-wide imaging screen uncovers molecular determinants of arsenite-induced protein aggregation and toxicity. *J. Cell Sci.* May 18:jcs.258338. https://doi.org/10.1242/jcs.258338.

Anjum, N.A., Hasanuzzaman, M., and Hossain, M.A. (2015). Jacks of metal/metalloid chelation trade in plants-an overview. *Front. Plant Sci.* 6: 192.

Armendariz, A.L., Talano, M.A., Villasuso, A.L. et al. (2016). Arsenic stress induces changes in lipid signalling and evokes the stomata closure in soybean. *Plant Physiol. Biochem.* 103: 45–52. https://doi.org/10.1016/j.plaphy.2016.02.041.

Awasthi, S., Chauhan, R., Srivastava, S., and Tripathi, R.D. (2017). The journey of arsenic from soil to grain in rice. *Front. Plant Sci.* 8: 1007. https://doi.org/10.3389/fpls.2017.01007.

Ayala, A., Muñoz, M.F., and Argüelles, S. (2014). Lipid peroxidation: production, metabolism, and signaling mechanisms of malondialdehyde and 4-hydroxy-2-nonenal. *Oxidative Med. Cell. Longev.* 2014: 360438. https://doi.org/10.1155/2014/360438.

Balali-Mood, M., Naseri, K., Tahergorabi, Z. et al. (2021). Toxic mechanisms of five heavy metals: mercury, lead, chromium, cadmium, and arsenic. *Front. Pharmacol.* 12: 643972. https://doi.org/10.3389/fphar.2021.643972.

Banerjee, A., Samanta, S., Singh, A., and Roychoudhury, A. (2019). Deciphering the molecular mechanism behind stimulated co-uptake of arsenic and fluoride from soil, associated toxicity, defence and glyoxalase machineries in arsenic-tolerant rice. *J. Hazard. Mater.* 390: 121978. https://doi.org/10.1016/j.jhazmat.2019.121978.

Banerjee, M. (2008). Arsenic a threatening environmental issue and role of cyanobacteria in toxicity mitigation. *J. Ecophysiol. Occup. Health* 8: 153–159.

Bencko, V. and Yan Li Foong, F. (2017). The history of arsenical pesticides and health risks related to the use of agent blue. *Ann. Agric. Environ. Med.* 24 (2): 312–316. https://doi.org/10.26444/aaem/74715.

Bentley, R. and Chasteen, T.G. (2002). Microbial methylation of metalloids arsenic, antimony, and bismuth. *Microbiol. Mol. Biol. Rev.* 66: 250–271.

Bergqvist, C. and Greger, M. (2012). Arsenic accumulation and speciation in plants from different habitats. *Appl. Geochem.* 27: 615–622.

Bernstam, L. and Nriagu, J. (2000). Molecular aspects of arsenic stress. *J. Toxicol. Environ. Health. B Crit. Rev.* 3 (4): 293–322.

Bhattacharya, P. and Pal, R. (2010). Response of cyanobacteria to arsenic toxicity. *J. Appl. Phycol.* 23: 293–299.

Bhattacharya, P. and Pal, R. (2011). Response of cyanobacteria to arsenic toxicity. *J. Appl. Phycol.* 23: 293–299.

Blair, I.A. (2008). DNA adducts with lipid peroxidation products. *J. Biol. Chem.* 283 (23): 15545–15549. https://doi.org/10.1074/jbc.R700051200.

Bona, E., Cattaneo, C., Cesaro, P. et al. (2010). Proteomic analysis of *Pteris vittata* fronds: two arbuscular mycorrhizal fungi differentially modulate protein expression under arsenic contamination. *Proteomics* 10 (21): 3811–3834.

Bondada, B.R. and Ma, L.Q. (2003). Tolerance of heavy metals in vascular plants: arsenic hyperaccumulation by chinese, brake fern (*Pterzs vzttata* L.). In: *Pteridology in the New Millennium* (ed. S. Chandra and M. Srivastava), 397–420. Dordrecht: Springer https://doi.org/10.1007/978-94-017-2811-9_28.

Boulanger, M., Tual, S., Lemarchand, C. et al. (2017). Agricultural exposure and risk of bladder cancer in the AGRIculture and CANcer cohort. *Int. Arch. Occup. Environ. Health* 90 (2): 169–178. https://doi.org/10.1007/s00420-016-1182-y.

Breydo, L. (2013). Arsenic, biologically active compounds. In: *Encyclopedia of Metalloproteins* (ed. R.H. Kretsinger, V.N. Uversky and E.A. Permyakov), 143–149. New York: Springer https://doi.org/10.1007/978-1-4614-1533-6_484.

Bustos, R., Castrillo, G., Linhares, F. et al. (2010). A central regulatory system largely controls transcriptional activation and repression responses to phosphate starvation in *Arabidopsis*. *PLoS Genet.* 6 (9): e1001102. https://doi.org/10.1371/journal.pgen.1001102.

Camara, A.Y., Wan, Y., Yu, Y. et al. (2019). Effect of endogenous selenium on arsenic uptake and antioxidative enzymes in as-exposed rice seedlings. *Int. J. Environ. Res. Public Health* 16 (18): 3350. https://doi.org/10.3390/ijerph16183350.

Campos, N.V., Araújo, T.O., Arcanjo-Silva, S. et al. (2016). Arsenic hyperaccumulation induces metabolic reprogramming in *Pityrogramma calomelanos* to reduce oxidative stress. *Physiol. Plant* 157 (2): 135–146.

Cantamessa, S., Massa, N., Gamalero, E., and Berta, G. (2020). Phytoremediation of a highly arsenic polluted site, using *Pteris vittata* L. and arbuscular mycorrhizal fungi. *Plan. Theory* 9: 1211.

Carbonell-Barrachina, A.A., Burlo, F., and Mataix, J. (1995). Arsenic uptake, distribution and accumulation in tomato plants: effect of arsenite on plant growth and yield. *J. Plant Nutr.* 18: 1237–1250. https://doi.org/10.1080/01904169509364975.

Chandrakar, V., Naithani, S.C., and Keshavkant, S. (2016). Arsenic-induced metabolic disturbances and their mitigation mechanisms in crop plants: a review. *Biologia* 71 (1): 367–377. https://doi.org/10.1515/biolog-2016-0052.

Chattopadhyay, A., Singh, A.P., Kasote, D. et al. (2021). Effect of phosphorus application on arsenic species accumulation and co-deposition of polyphenols in rice grain: phyto and food safety evaluation. *Plants (Basel)* 10 (2): 281. https://doi.org/10.3390/plants10020281.

Chen, B., Liu, Q., Popowich, A. et al. (2015). Therapeutic and analytical applications of arsenic binding to proteins. *Metallomics* 7 (1): 39–55. https://doi.org/10.1039/c4mt00222a.

Chen, W., Chi, Y., Taylor, N.L. et al. (2010). Disruption of ptLPD1 or ptLPD2, genes that encode isoforms of the plastidial lipoamide dehydrogenase, confers arsenate hypersensitivity in *Arabidopsis*. *Plant Physiol.* 153 (3): 1385–1397. https://doi.org/10.1104/pp.110.153452.

Chen, X., Jiang, W., Tong, T. et al. (2021). Molecular interaction and evolution of jasmonate signaling with transport and detoxification of heavy metals and metalloids in plants. *Front. Plant Sci.* 12: 665842. https://doi.org/10.3389/fpls.2021.665842.

Chen, Y., Han, Y.H., Cao, Y. et al. (2017). Arsenic transport in rice and biological solutions to reduce arsenic risk from rice. *Front. Plant Sci.* 8: 268.

Chi, W.C., Chen, Y.A., Hsiung, Y.C. et al. (2013). Autotoxicity mechanism of *Oryza sativa:* transcriptome response in rice roots exposed to ferulic acid. *BMC Genomics* 14: 351. https://doi.org/10.1186/1471-2164-14-351.

Choudhury, B., Chowdhury, S., and Biswas, A.K. (2011). Regulation of growth and metabolism in rice (*Oryza sativa* L.) by arsenic and its possible reversal by phosphate. *J. Plant Interact.* 6 (1): 15–24.

Choudhury, B., Mitra, S., and Biswas, A.K. (2010). Regulation of sugar metabolism in rice (*Oryza sativa* L.) seedlings under arsenate toxicity and its improvement by phosphate. *Physiol. Mol. Biol. Plants* 16 (1): 59–68.

Claire-Lise, M. and Nathalie, V. (2012). The use of the model species *Arabidopsis halleri* towards phytoextraction of cadmium polluted soils. *New Biotechnol.* 30 (1): 9–14. https://doi.org/10.1016/j.nbt.2012.07.009.

Cotas, J., Leandro, A., Pacheco, D. et al. (2020). A comprehensive review of the nutraceutical and therapeutic applications of red seaweeds (Rhodophyta). *Life* 10 (3): 19. https://doi.org/10.3390/life10030019.

Danh, L.T., Truong, P., Mammucari, R., and Foster, N. (2014). A critical review of the arsenic uptake mechanisms and phytoremediation potential of *Pteris vittata*. *Int. J. Phytoremediation* 16 (5): 429–453.

Das, P., Khare, P., Singh, R.P. et al. (2021). Arsenic-induced differential expression of oxidative stress and secondary metabolite content in two genotypes of *Andrographis paniculata*. *J. Hazard. Mater.* 406: 124302. https://doi.org/10.1016/j.jhazmat.2020.124302.

Di, X., Beesley, L., Zhang, Z. et al. (2019). Microbial arsenic methylation in soil and uptake and metabolism of methylated arsenic in plants: a review. *Int. J. Environ. Res. Public Health* 16 (24): 5012. https://doi.org/10.3390/ijerph16245012.

Díaz, S., Villares, R., and Carballeira, A. (2012). Uptake kinetics of As, Hg, Sb, and Se in the aquatic moss *Fontinalis antipyretica* Hedw. *Water Air Soil Pollut.* 223: 3409–3423. https://doi.org/10.3390/ijerph15010059.

Donovan, G.H., Jovan, S.E., Gatziolis, D. et al. (2016). Using an epiphytic moss to identify previously unknown sources of atmospheric cadmium pollution. *Sci. Total Environ.* 559: 84–93.

Du, F., Liu, P., Wang, K. et al. (2019). Influence of different arsenic species on uptake, speciation and efflux of arsenic in hydroponic rice plants. *Ecotoxicol. Environ. Saf.* 186: 109791. https://doi.org/10.1016/j.ecoenv.2019.109791.

Du, F., Liu, P., Wang, K. et al. (2020). Ionomic responses of rice plants to the stresses of different arsenic species in hydroponics. *Chemosphere* 243: 125398. https://doi.org/10.1016/j.chemosphere.2019.125398.

Dutta, S. and Bhadury, P. (2020). Effect of arsenic on exopolysaccharide production in a diazotrophic cyanobacterium. *J. Appl. Phycol.* 32: 2915–2926.

Dutta, S., Mitra, M., Agarwal, P. et al. (2018). Oxidative and genotoxic damages in plants in response to heavy metal stress and maintenance of genome stability. *Plant Signal. Behav.* 13 (8): e1460048. https://doi.org/10.1080/15592324.2018.1460048.

Dyhrman, S.T. and Haley, S.T. (2011). Arsenate resistance in the unicellular marine diazotroph *Crocosphaera watsonii*. *Front. Microbiol.* 2: 214.

Elias, M., Wellner, A., Goldin-Azulay, K. et al. (2012). The molecular basis of phosphate discrimination in arsenate-rich environments. *Nature* 491 (7422): 134–137.

Farhat, Y.A., Kim, S.H., Seyfferth, A.L. et al. (2021). Altered arsenic availability, uptake, and allocation in rice under elevated temperature. *Sci. Total Environ.* 763: 143049. https://doi.org/10.1016/j.scitotenv.2020.143049.

Farooq, M.A., Gill, R.A., Islam, F. et al. (2016). Methyl jasmonate regulates antioxidant defense and suppresses arsenic uptake in *Brassica napus* L. *Front. Plant Sci.* 7: 468. https://doi.org/10.3389/fpls.2016.00468.

Fekry, M.I., Tipton, P.A., and Gates, K.S. (2011). Kinetic consequences of replacing the internucleotide phosphorus atoms in DNA with arsenic. *ACS Chem. Biol.* 18 6 (2): 127–130.

Ferreira, F.A.G., Franco, M.W., Oliveira, D.M.D. et al. (2018). Arsenic tolerance of *Microcystis novacekii* (KomárekCompère, 1974) and its arsenic decontamination potential. *Braz. Arch. Biol. Technol.* 61: e18160644.

Finnegan, P.M. and Chen, W. (2012). Arsenic toxicity: the effects on plant metabolism. *Front. Physiol.* 3: 182. https://doi.org/10.3389/fphys.2012.00182.

Foyer, C.H. and Noctor, G. (2011). Ascorbate and glutathione: the heart of the redox hub. *Plant Physiol.* 155 (1): 2–18. https://doi.org/10.1104/pp.110.167569.

Fu, S.F., Chen, P.Y., Nguyen, Q.T. et al. (2014). Transcriptome profiling of genes and pathways associated with arsenic toxicity and tolerance in *Arabidopsis*. *BMC Plant Biol.* 14: 94. https://doi.org/10.1186/1471-2229-14-94.

Gayle, D.G., Murali, C.P., and Gary, N.C. (1994). Inhibition of cellular events during early algal gametophyte development: effects of select metals and an aqueous petroleum waste. *Aquat. Toxicol.* 28 (1–2): 127–144. https://doi.org/10.1016/0166-445X(94)90025-6.

Geiszinger, A., Goessler, W., Pedersen, S.N., and Francesconi, K.A. (2001). Arsenic biotransformation by the brown macroalga *Fucus serratus*. *Environ. Toxicol.* 20 (10): 2255–2262.

Ghulam, A., Behzad, M., Irshad, B. et al. (2018). Arsenic uptake, toxicity, detoxification, and speciation in plants: physiological, biochemical, and molecular aspects. *Int. J. Environ. Res. Public Health* 15 (59): 1–45.

Gilbert-Diamond, D., Cottingham, K.L., Gruber, J.F. et al. (2011). Rice consumption contributes to arsenic exposure in US women. *Proc. Natl. Acad. Sci. U.S.A.* 108: 20656–20660.

Gonzalez, A.G. and Pokrovsky, O.S. (2014). Metal adsorption on mosses: toward a universal adsorption model. *J. Colloid Interface Sci.* 415: 169–178.

Gusman, G.S., Oliveira, J.A., and Farnese, F.S. (2013). Arsenate and arsenite: the toxic effects on photosynthesis and growth of lettuce plants. *Acta Physiol. Plant.* 35: 1201–1209. https://doi.org/10.1007/s11738-012-1159-8.

Harada, E., von Roepenack-Lahaye, E., and Clemens, S. (2004). A cyanobacterial protein with similarity to phytochelatin synthases catalyzes the conversion of glutathione to gamma-glutamylcysteine and lacks phytochelatin synthase activity. *Phytochemistry* 65 (24): 3179–3185.

Hartley-Whitaker, J., Ainsworth, G., and Meharg, A. (2001a). Copper- and arsenic-induced oxidative stress in *Holcus lanatus* L. Clones with differential sensitivity. *Plant Cell Environ.* 24: 713–722. https://doi.org/10.1046/j.0016-8025.2001.00721.x.

Hasegawa, H., Sohrin, Y., Seki, K. et al. (2001). Biosynthesis and release of methylarsenic compounds during the growth of fresh water algae. *Chemosphere.* 43 (3): 265–272.

Hartley-Whitaker, J., Ainsworth, G., Vooijs, R. et al. (2001b). Phytochelatins are involved in differential arsenate tolerance in *Holcus lanatus*. *Plant Physiol.* 126 (1): 299–306.

He, S., Hu, Y., Wang, H. et al. (2017). Effects of indole-3-acetic acid on arsenic uptake ad antioxidative enzymes in *Pteris cretica* var. *nervosa* and *Pteris ensiformis*. *Int. J. Phytoremediation* 19 (3): 231–238.

Hervas, M., Lopez-Maury, L., Leon, P. et al. (2012). ArsH from the cyanobacterium *Synechocystis* sp. PCC 6803 is an efficient NADPH-dependent quinone reductase. *Biochemistry* 51: 1178–1187.

Hirano, S., Kobayashi, Y., Cui, X. et al. (2004). The accumulation and toxicity of methylated arsenicals in endothelial cells: important roles of thiol compounds. *Toxicol. Appl. Pharmacol.* 198 (3): 458–467. https://doi.org/10.1016/j.taap.2003.10.023.

Hofslagare, O., Sjöberg, S., and Samuelsson, G. (1994). The effect of arsenate and arsenite on photosynthesis in *Scenedesmus obliquus*. A potentiometric study in a closed CO_2-system. *Chem. Speciat. Bioavailab.* 6 (4): 95–102.

Hong, Y.S., Song, K.H., and Chung, J.Y. (2014). Health effects of chronic arsenic exposure. *J. Prev. Med. Public Health* 47 (5): 245–252. https://doi.org/10.3961/jpmph.14.035.

Hossain, M.M., Khatun, M.A., Haque, M.N. et al. (2018). Silicon alleviates arsenic-induced toxicity in wheat through vacuolar sequestration and ROS scavenging. *Int. J. Phytoremediation* 20 (8): 796–804. https://doi.org/10.1080/15226514.2018.1425669.

Hu, Y., Xiao, T., and Zhang, A. (2021). Associations between and risks of trace elements related to skin and liver damage induced by arsenic from coal burning. *Ecotoxicol. Environ. Saf.* 208: 111719. https://doi.org/10.1016/j.ecoenv.2020.111719.

Huang, T.L., Nguyen, Q.T., Fu, S.F. et al. (2012). Transcriptomic changes and signalling pathways induced by arsenic stress in rice roots. *Plant Mol. Biol.* 80 (6): 587–608. https://doi.org/10.1007/s11103-012-9969-z.

Huang, Y., Chen, H., Reinfelder, J.R. et al. (2019). A transcriptomic (RNA-seq) analysis of genes responsive to both cadmium and arsenic stress in rice root. *Sci Total Environ.* 666: 445–460. https://doi.org/10.1016/j.scitotenv.2019.02.281.

Hughes, M.F., Beck, B.D., Chen, Y. et al. (2011). Arsenic exposure and toxicology: a historical perspective. *Toxicol. Sci.* 123 (2): 305–332.

Ichikawa, S., Nozawa, S., Hanaoka, K., and Kaise, T. (2010). Ingestion and excretion of arsenic compounds present in edible brown algae, Hijikia fusiforme, by mice. *Food Chem. Toxicol.* 48 (2): 465–469.

Imran, M.A., Khan, R.M., Ali, Z., and Mahmood, T. (2013). Toxicity of arsenic (As) on seed germination of sunflower (*Helianthus annuus* L.). *Int. J. Phys. Sci.* 8: 840–847.

Indriolo, E., Na, G., Ellis, D. et al. (2010). A vacuolar arsenite transporter necessary for arsenic tolerance in the arsenic hyperaccumulating fern *Pteris vittata* is missing in flowering plants. *Plant Cell* 22 (6): 2045–2057.

Irato, P. and Santovito, G. (2021). Enzymatic and non-enzymatic molecules with antioxidant function. *Antioxidants* 10 (4): 579. https://doi.org/10.3390/antiox10040579.

Irem, S., Islam, E., Maathuis, F.J.M. et al. (2019). Assessment of potential dietary toxicity and arsenic accumulation in two contrasting rice genotypes: effect of soil amendments. *Chemosphere* 225: 104–114. https://doi.org/10.1016/j.chemosphere.2019.02.202.

Islam, M.S., Ueno, Y., Sikder, M.T., and Kurasaki, M. (2013). Phytofiltration of arsenic and cadmium from the water environment using micranthemum umbrosum (J.F. Gmel) S. F Blake as a hyperaccumulator. *Int. J. Phytoremediation* 15: 1010–1021.

Jha, A.B. and Dubey, R.S. (2004a). Carbohydrate metabolism in growing rice seedlings under arsenic toxicity. *J. Plant Physiol.* 161 (7): 867–872. https://doi.org/10.1016/j.jplph.2004.01.004.

Jha, A.B. and Dubey, R.S. (2004b). Effect of arsenic on nitrogen assimilatory enzymes in germinating rice seeds. *Indian J. Plant Physiol.* 9: 438–441.

Jha, A.B. and Dubey, R.S. (2005). Effect of arsenic on behaviour of enzymes of sugar metabolism in germinating rice seeds. *Acta Physiol. Plant.* 27: 341–347. https://doi.org/10.1007/s11738-005-0010-x.

Jomova, K., Jenisova, Z., Feszterova, M. et al. (2011). Arsenic: toxicity, oxidative stress and human disease. *J. Appl. Toxicol.* 31 (2): 95–107. https://doi.org/10.1002/jat.1649.

Jung, H.I., Kong, M.S., Lee, B.R. et al. (2019). Exogenous glutathione increases arsenic translocation into shoots and alleviates arsenic-induced oxidative stress by sustaining ascorbate-glutathione homeostasis in rice seedlings. *Front. Plant Sci.* 10: 1089. https://doi.org/10.3389/fpls.2019.01089.

Kalita, J., Pradhan, A.K., Shandilya, Z.M., and Tanti, B. (2018). Arsenic stress responses and tolerance in rice: physiological, cellular and molecular approaches. *Rice Sci.* 25 (5): 235–249. https://doi.org/10.1016/j.rsci.2018.06.007.

Kamerlin, S.C., Sharma, P.K., Prasad, R.B., and Warshel, A. (2013). Why nature really chose phosphate. *Q. Rev. Biophys.* 46 (1): 1–132.

Karagas, M.R., Punshon, T., Davis, M. et al. (2019). Rice intake and emerging concerns on arsenic in rice: a review of the human evidence and methodologic challenges. *Curr. Environ. Health Rep.* 6 (4): 361–372. https://doi.org/10.1007/s40572-019-00249-1.

Karajova, I.B., Slaveykova, V.I., and Tsalev, D.L. (2008). The bio uptake and toxicity of arsenic species on the green microalga *Chlorella salina* in seawater. *Aquat. Toxicol.* 87: 264–271.

Khalid Wang, S., Shahid, M., Niazi, N.K. et al. (2017). Arsenic behavior in soil-plant system: biogeochemical reactions and chemical speciation influences. In: *Enhancing Cleanup of Environmental Pollutants*, vol. 2 (ed. Anjum et al.), 97–140. Berlin, Germany: Springer.

Kiran Marella, T., Saxena, A., and Tiwari, A. (2020). Diatom mediated heavy metal remediation: a review. *Bioresour. Technol.* 305: 123068. https://doi.org/10.1016/j.biortech.2020.123068.

Koch, I., Wang, L., Ollson, C.A. et al. (2000). The predominance of inorganic arsenic species in plants from yellowknife, Northwest territories, Canada. *Environ. Sci. Technol.* 34: 22–26.

Kofroňová, M., Hrdinová, A., Mašková, P. et al. (2020). Multi-component antioxidative system and robust carbohydrate status, the essence of plant arsenic tolerance. *Antioxidants* 9 (4): 283. https://doi.org/10.3390/antiox9040283.

Koller, C.E., Patrick, J.W., Rose, R.J. et al. (2007). *Pteris umbrosa* R. Br. As an arsenic hyperaccumulator: accumulation, partitioning and comparison with the established as hyperaccumulator *Pteris vittata*. *Chemosphere* 66 (7): 1256–1263.

Kostecka-Gugała, A. and Latowski, D. (2018). Arsenic-induced oxidative stress in plants. In: *Mechanisms of Arsenic Toxicity and Tolerance in Plants* (ed. M. Hasanuzzaman, K. Nahar and M. Fujita). Singapore: Springer https://doi.org/10.1007/978-981-13-1292-2_4.

Kuivenhoven, M. and Mason, K. (2022). Arsenic toxicity. [Updated 2020 Jun 26]. In: *StatPearls [Internet]*. Treasure Island (FL): StatPearls Publishing Available from: https://www.ncbi.nlm.nih.gov/books/NBK541125/.

Kumarathilaka, P., Seneweera, S., Meharg, A., and Bundschuh, J. (2018). Arsenic accumulation in rice (*Oryza sativa* L.) is influenced by environment and genetic factors. *Sci. Total Environ.* 642: 485–496. https://doi.org/10.1016/j.scitotenv.2018.06.030.

Laloknam, S., Sirisopana, S., Phornphisutthimas, S. et al. (2009). Removal of mercury, arsenic, and cadmium in synthetic wastewater by cyanobacterium *Aphanothece halophytica*. *NU Sci. J.* 6 (S1): 96–104.

Levy, J.L., Stauber, J.L., Adams, M.S. et al. (2005). Toxicity, biotransformation, and mode of action of arsenic in two freshwater microalgae (*Chlorella* sp. and *Monoraphidium arcuatum*). *Environ. Toxicol. Chem.* 24 (10): 2630–2639. https://doi.org/10.1897/04-580r.1.

Li, J., Zhao, Q., Xue, B. et al. (2019). Arsenic and nutrient absorption characteristics and antioxidant response in different leaves of two ryegrass (*Lolium perenne*) species under arsenic stress. *PLoS One* 14 (11): e0225373. https://doi.org/10.1371/journal.pone.0225373.

Liang, G., Song, H., Xiao, Y., and Zhang, Z. (2020). Ammonium accumulation caused by reduced tonoplast V-ATPase activity in *Arabidopsis thaliana*. *Int. J. Mol. Sci.* 22 (1): 2. https://doi.org/10.3390/ijms22010002.

Liang, L., Xu, Z., Qiu, G. et al. (2019). Arsenic accumulation and speciation in epilithic moss collected from an abandoned mercury mining area, south-western China. *Atmos. Poll. Res.* 10 (2): 455–461. https://doi.org/10.1016/j.apr.2018.09.005.

Liu, X., Zhang, S., Shan, X., and Zhu, Y.G. (2005). Toxicity of arsenate and arsenite on germination, seedling growth and amylolytic activity of wheat. *Chemosphere* 61 (2): 293–301.

Liu, Y., Damaris, R.N., and Yang, P. (2017). Proteomics analysis identified a DRT protein involved in arsenic resistance in *Populus*. *Plant Cell Rep.* 36 (12): 1855–1869. https://doi.org/10.1007/s00299-017-2199-8.

Liu, Z., Shen, J., Carbrey, J.M. et al. (2002). Arsenite transport by mammalian aquaglyceroporins AQP7 and AQP9. *Proc. Natl. Acad. Sci. U.S.A.* 99 (9): 6053–6058.

Lomax, C., Liu, W.J., Wu, L. et al. (2012). Methylated arsenic species in plants originate from soil microorganisms. *New Phytol.* 193 (3): 665–672. https://doi.org/10.1111/j.1469-8137.2011.03956.x.

Lombi, E., Zhao, F.J., Fuhrmann, M. et al. (2002). Arsenic distribution and speciation in the fronds of the hyperaccumulator *Pteris vittata*. *New Phytol.* 156: 195–203.

Lopez-Maury, L., Florencio, F.J., and Reyes, J.C. (2003). Arsenic sensing and resistance system in the cyanobacterium *Synechocystis* sp. strain PCC 6803. *J. Bacteriol.* 185: 5363–5371.

Luongo, T. and Ma, L. (2005). Characteristics of arsenic accumulation by *Pteris* and non-*Pteris* ferns. *Plant Soil* 277: 117.

Ma, L., Komar, K., and Tu, C. (2001). A fern that hyperaccumulates arsenic. *Nature* 409: 579.

Macnair, M.R. and Cumbes, Q. (1987). Evidence that arsenic tolerance in *Holcus lanatus* L. Is caused by an altered phosphate uptake system. *New Phytol.* 107: 387–394. https://doi.org/10.1111/j.1469-8137.1987.tb00190.x.

Maeda, S., Nakashima, S., Takeshita, T., and Higashi, S. (1985). Bioaccumulation of arsenic by freshwater algae and the application to the removal of inorganic arsenic from an aqueous phase. Part II. By *Chlorella vulgaris* isolated from arsenic-polluted environment. *Sep. Sci. Technol.* 20 (2–3): 153–161.

Maeda, S. and Ohki, A. (1998). Bioaccumulation and biotransformation of arsenic, antimony and bismuth compounds by freshwater algae. In: *Wastewater Treatment with Algae*, 2nde (ed. Y.S. Wong and N.F.Y. Tam), 73–91. Berlin Heidelberg: Springer-Verlag.

Maeda, S., Wada, H., Kumeda, K. et al. (2004). Methylation of inorganic arsenic by arsenic-tolerantfreshwater algae. *Appl. Organomet. Chem.* 1: 465–472.

Maghsoudi, K., Arvin, M.J., and Ashraf, M. (2020). Mitigation of arsenic toxicity in wheat by the exogenously applied salicylic acid, 24-epi-brassinolide and silicon. *J. Soil Sci. Plant Nutr.* 20: 577–588. https://doi.org/10.1007/s42729-019-00147-3.

Majumder, B., Das, S., Biswas, S. et al. (2020a). Differential responses of photosynthetic parameters and its influence on carbohydrate metabolism in some contrasting rice (*Oryza sativa* L.) genotypes under arsenate stress. *Ecotoxicology* 29 (7): 912–931. https://doi.org/10.1007/s10646-020-02241-0.

Majumder, B., Das, S., Mukhopadhyay, S., and Biswas, A.K. (2019). Identification of arsenic-tolerant and arsenic-sensitive rice (*Oryza sativa* L.) cultivars on the basis of arsenic accumulation assisted stress perception, morpho-biochemical responses, and alteration in genomic template stability. *Protoplasma* 256 (1): 193–211. https://doi.org/10.1007/s00709-018-1290-5.

Majumder, B., Das, S., Pal, B., and Biswas, A.K. (2020b). Evaluation of arsenic induced toxicity based on arsenic accumulation, translocation and its implications on physio-chemical changes and genomic instability in indica rice (*Oryza sativa* L.) cultivars. *Ecotoxicology* 29 (1): 13–34. https://doi.org/10.1007/s10646-019-02135-w.

Malik, J.A., Goel, S., Sandhir, R., and Nayyar, H. (2011). Uptake and distribution of arsenic in chickpea: effects on seed yield and seed composition. *Commun. Soil Sci. Plant Anal.* 42: 1728–1738.

Mandal, B. and Suzuki, K.T. (2002). Arsenic round the world: a review. *Talanta* 58: 201–235.

Marin, A.R., Pezashkip, S.R., Masschelen, H., and Choi, H.S. (2002). Effect of dimethylarsenic acid on growth, tissue arsenic, and photosynthesis of rice plants. *J. Plant Nutr.* 16: 865–880.

Markley, C.T. (2004). Arsenate uptake, sequestration and reduction by a freshwater cyanobacterium: a potential biologic control of arsenic in South Texas. Master's thesis. Texas A&M University, College Station.

Meharg, A. (1994). Integrated tolerance mechanisms – constitutive and adaptive plant responses to elevated metal concentrations in the environment. *Plant Cell Environ.* 17: 989–993. https://doi.org/10.1111/j.1365-3040.1994.tb02032.x.

Meharg, A.A. and Hartley-Whitaker, J. (2002). Arsenic uptake and metabolism in arsenic resistant and nonresistant plant species. *New Phytol.* 154: 29–43. https://doi.org/10.1046/j.1469-8137.2002.00363.x.

Meharg, A.A. and Macnair, M.R. (1991). The mechanisms of arsenate tolerance in *Deschampsia cespitosa* (L.) beauvand *Agrostis capillaris* L. *New Phytol.* 119: 291–297. https://doi.org/10.1111/j.1469-8137.1991.tb01033.x.

Mehmood, T., Bibi, I., Shahid, M. et al. (2017). Effect of compost addition on arsenic uptake, morphological and physiological attributes of maize plants grown in contrasting soils. *J. Geochem. Explor.* 178: 83–91. https://doi.org/10.1016/j.gexplo.2017.03.018.

Meng, Y.L., Liu, Z., and Rosen, B.P. (2004). As(III) and Sb(III) uptake by GlpF and efflux by ArsB in *Escherichia coli*. *J. Biol. Chem.* 279 (18): 18334–18341.

Miot, J., Morin, G., Skouri-Panet, F. et al. (2009). Speciation of arsenic in *Euglena gracilis* cells exposed to As(V). *Environ. Sci. Technol.* 43 (9): 3315–3321. https://doi.org/10.1021/es802833s.

Mirza, N., Mahmood, Q., Maroof Shah, M. et al. (2014). Plants as useful vectors to reduce environmental toxic arsenic content. *Sci. World J.* 2014: 1–11.

Mishra, S., Alfeld, M., Sobotka, R. et al. (2016). Analysis of sublethal arsenic toxicity to *Ceratophyllum demersum*: subcellular distribution of arsenic and inhibition of chlorophyll biosynthesis. *J. Exp. Bot.* 67 (15): 4639–4646.

Mishra, S. and Dubey, R.S. (2006). Inhibition of ribonuclease and protease activities in arsenic exposed rice seedlings: role of proline as enzyme protectant. *J. Plant Physiol.* 163 (9): 927–936. https://doi.org/10.1016/j.jplph.2005.08.003.

Mousavi, S.R., Niknejad, Y., Fallah, H., and Tari, D.B. (2020). Methyl jasmonate alleviates arsenic toxicity in rice. *Plant Cell Rep.* 39 (8): 1041–1060. https://doi.org/10.1007/s00299-020-02547-7.

Mukhopadhyay, R., Rosen, B.P., Phung, L.T., and Silver, S. (2002). Microbialarsenic: from geocycles to genes and enzymes. *FEMS Microbiol. Rev.* 26: 311–325.

Murphy, E.A. and Aucott, M. (1998). An assessment of the amounts of arsenical pesticides used historically in a geographic area. *Sci. Total Environ.* 218: 89–101. https://doi.org/10.1016/S0048-9697(98)00180-6.

Murray, L.A., Raab, A., Maar, L.L., and Feldmann, J. (2003). Biotransformation of arsenate to arsenosugars by *Chlorella vulgaris*. *Appl. Organomet. Chem.* 17: 669–674.

Naeem, M., Nabi, A., Aftab, T., and Khan, M.M.A. (2020). Oligomers of carrageenan regulate functional activities and artemisinin production in *Artemisia annua* L. exposed to arsenic stress. *Protoplasma* 257 (3): 871–887. https://doi.org/10.1007/s00709-019-01475-y.

Németi, B., Regonesi, M.E., Tortora, P., and Gregus, Z. (2010). Polynucleotide phosphorylase and mitochondrial ATP synthase mediate reduction of arsenate to the more toxic arsenite by forming arsenylated analogues of ADP and ATP. *Toxicol. Sci.* 117 (2): 270–281. https://doi.org/10.1093/toxsci/kfq141.

Nguyen, T., Sherratt, P.J., Huang, H.C. et al. (2003). Increased protein stability as a mechanismthat enhances Nrf2-mediated transcriptional activation of the antioxidant response element. degradation of Nrf2 by the 26 S proteasome. *J. Biol. Chem.* 278: 4536–4541. https://doi.org/10.1074/jbc.M207293200.

Niazi, N.K., Bibi, I., Fatimah, A. et al. (2017). Phosphate-assisted phytoremediation of arsenic by *Brassica napus* and *Brassica juncea*: morphological and physiological response. *Int. J. Phytoremed.* 19: 670–678. https://doi.org/10.1080/15226514.2016.1278427.

Nurchi, V.M., Djordjevic, A.B., Crisponi, G. et al. (2020). Arsenic toxicity: molecular targets and therapeutic agents. *Biomolecules* 10 (2): 235. https://doi.org/10.3390/biom10020235.

Ohno, M., Sakumi, K., Fukumura, R. et al. (2014). 8-oxoguanine causes spontaneous de novo germline mutations in mice. *Sci. Rep.* 4 (1): 1–9.

Paliouris, G. and Hutchinson, T.C. (1991). Arsenic, cobalt and nickel tolerances in two populations of *Silene vulgaris* (Moench) Garcke from Ontario, Canada. *New Phytol.* 117: 449–459. https://doi.org/10.1111/j.1469-8137.1991.tb00009.x.

Pandey, A.K., Gedda, M.R., and Verma, A.K. (2020). Effect of arsenic stress on expression pattern of a rice specific *miR156j* at various developmental stages and their allied co-expression target networks. *Front. Plant Sci.* 11: 752. https://doi.org/10.3389/fpls.2020.00752.

Pandey, S., Rai, R., and Rai, L.C. (2012). Proteomics combines morphological, physiological and biochemical attributes to unravel the survival strategy of *Anabaena* sp. PCC7120 under arsenic stress. *J. Proteome* 75 (3): 921–937.

Papry, R.I., Ishii, K., Mamun, M.A.A. et al. (2019). Arsenic biotransformation potential of six marine diatom species: effect of temperature and salinity. *Sci. Rep.* 9: 10226.

Patel, A., Tiwari, S., and Prasad, S.M. (2018). Toxicity assessment of arsenate and arsenite on growth, chlorophyll a fluorescence and antioxidant machinery in *Nostoc muscorum*. *Ecotoxicol. Environ. Saf.* 157: 369–379. https://doi.org/10.1016/j.ecoenv.2018.03.056.

Pétursdóttir, Á.H., Fletcher, K., Gunnlaugsdóttir, H. et al. (2015). Environmental effects on arsenosugars and arsenolipids in *Ectocarpus* (Phaeophyta). *Environ. Chem.* 13 (1): 21–33.

Piacentini, D., Corpas, F.J., D'Angeli, S. et al. (2020). Cadmium and arsenic-induced-stress differentially modulates *Arabidopsis* root architecture, peroxisome distribution, enzymatic activities and their nitric oxide content. *Plant Physiol. Biochem.* 148: 312–323. https://doi.org/10.1016/j.plaphy.2020.01.026.

Pickering, I.J., Gumaelius, L., Harris, H.H. et al. (2006). Localizing the biochemical transformations of arsenate in a hyperaccumulating fern. *Environ. Sci. Technol.* 40 (16): 5010–5014.

Pitt, F.D., Mazard, S., Humphreys, L., and Scanlan, D.J. (2010). Functional characterization of *Synechocystis* sp. strain PCC 6803 pst1 and pst2 gene clusters reveals a novel strategy for phosphate uptake in a freshwater cyanobacterium. *J. Bacteriol.* 192 (13): 3512–3523.

Praveen, A., Mehrotra, S., and Singh, N. (2018). Nutrient constraints in arsenic phytoremediation. *Russ. J. Plant Physiol.* 65: 15–22.

Ran, Y. and Chen, G. (2020). Arsenic accumulation by submerged plants: a review. *Sheng Wu Gong Cheng Xue Bao* 36 (3): 407–415. https://doi.org/10.13345/j.cjb.190237.

Requejo, R. and Tena, M. (2014). Intra-specific variability in the response of maize to arsenic exposure. *Environ. Sci. Pollut. Res. Int.* 21 (18): 10574–10582. https://doi.org/10.1007/s11356-014-3097-z.

Rodenas, R., Martínez, V., Nieves-Cordones, M., and Rubio, F. (2019). High external K^+ concentrations impair Pi nutrition, induce the phosphate starvation response, and reduce arsenic toxicity in arabidopsis plants. *Int. J. Mol. Sci.* 20 (9): 2237. https://doi.org/10.3390/ijms20092237.

Roldán-Arjona, T. and Ariza, R.R. (2009). Repair and tolerance of oxidative DNA damage in plants. *Mutat. Res.* 681 (2–3): 169–179. https://doi.org/10.1016/j.mrrev.2008.07.003.

Rosen, B.P. (1999). Families of arsenic transporters. *Trends Microbiol.* 7: 207–212.

Ruhling, A. and Tyler, G. (1968). An ecological approach to lead problem. *Bot. Notiser* 121: 21.

Salemaa, M., Derome, J., Helmisaari, H.S. et al. (2004). Element accumulation in boreal bryophytes, lichens and vascular plants exposed to heavy metal and sulphur deposition in Finland. *Sci. Total Environ.* 324: 141–160.

Samal, A.C., Bhattacharya, P., Biswas, P. et al. (2021). Variety-specific arsenic accumulation in 44 different rice cultivars (*O. sativa* L.) and human health risks due to co-exposure of arsenic-contaminated rice and drinking water. *J. Hazard. Mater.* 407 (124804): https://doi.org/10.1016/j.jhazmat.2020.124804.

Samanta, S., Banerjee, A., and Roychoudhury, A. (2021). Exogenous melatonin regulates endogenous phytohormone homeostasis and thiol-mediated detoxification in two indica rice cultivars under arsenic stress. *Plant Cell Rep.* https://doi.org/10.1007/s00299-021-02711-7.

Sa'nchez-Riego, A.M., Lo'pez-Maury, L., and Florencio, F.J. (2014). Genomic responses to arsenic in the cyanobacterium *Synechocystis* sp. PCC 6803. *PLoS One* 9 (5): e96826.

Sandhi, A., Landberg, T., and Greger, M. (2017). Phytofiltration of arsenic by aquatic moss (*Warnstorfia fluitans*). *Environ. Pollut.* 37: 1098–1105. https://doi.org/10.1016/j.envpol.2017.11.038.

Satyapal, G.K. and Kumar, N. (2021). Arsenic: source, distribution, toxicity and bioremediation. In: *Arsenic Toxicity: Challenges and Solutions* (ed. N. Kumar). Singapore: Springer https://doi.org/10.1007/978-981-33-6068-6_6.

Senthilkumar, R., Reddy, P.D.M., Govindarajan, L. et al. (2020). Synthesis of green marine algal-based biochar for remediation of arsenic(V) from contaminated waters in batch and column mode of operation. *Int. J. Phytoremediation* 22 (3): 279–286. https://doi.org/10.1080/15226514.2019.1658710.

Shaheen, R., Mahmud, R., and Sen, J. (2007). A study on arsenic decontaminating cyanobacteria of an arsenic affected soil. *J. Soil Nature* 1: 23–29.

Shakya, K., Chettri, M.K., and Sawidis, T. (2008). Impact of heavy metals (copper, zinc, and lead) on the chlorophyll content of some mosses. *Arch. Environ. Contam. Toxicol.* 54: 412–421.

Sharma, I. (2012). Arsenic induced oxidative stress in plants. *Biologia* 67: 447–453. https://doi.org/10.2478/s11756-012-0024-y.

Sharma, P., Jha, A.B., Dubey, R.S., and Pessarakli, M. (2012). Reactive oxygen species, oxidative damage, and antioxidative defense mechanism in plants under stressful conditions. *J. Bot.* 2012.

Sharma, P. and Rathore, D.K. (2018). Toxicity of arsenic on nitrogen-fixing cyanobacteria. *J. Algal Biomass. Utln.* 9 (1): 68–71.

Sheath, R.G. and Wehr, J.D. (2003). Introduction to freshwater algae. In: *Aquatic Ecology, Freshwater Algae of North America* (ed. J.D. Wehr and R.G. Sheath), 1–9. Academic Press.

Shen, S., Li, X.F., Cullen, W.R. et al. (2013). Arsenic binding to proteins. *Chem. Rev.* 113 (10): 7769–7792. https://doi.org/10.1021/cr300015c.

Shilpa, P. and Roychoudhury, K. (2017). Effect of arsenic on plant growth and remediation of arsenic by Cyanobacterial species Mankar. *ESSENCE Int. J. Env.* VIII (1): 36–45.

Shukla, T., Khare, R., Kumar, S. et al. (2018). Differential transcriptome modulation leads to variation in arsenic stress response in *Arabidopsis thaliana* accessions. *J. Hazard. Mater.* 351: 1–10. https://doi.org/10.1016/j.jhazmat.2018.02.031.

Sil, P., Das, P., and Biswas, S. (2019). Modulation of photosynthetic parameters, sugar metabolism, polyamine and ion contents by silicon amendments in wheat (*Triticum aestivum* L.) seedlings exposed to arsenic. *Environ. Sci. Pollut. Res.* 26: 13630–13648. https://doi.org/10.1007/s11356-019-04896-7.

Singh, N., Ma, L.Q., Srivastava, M., and Rathinasabapathi, B. (2006). Metabolic adaptations to arsenic-induced oxidative stress in *Pteris vittata* L and *Pteris ensiformis* L. *Plant Sci.* 170 (2): 274–282.

Singh, R., Jha, A.B., Misra, A.N., and Sharma, P. (2019). Differential responses of growth, photosynthesis, oxidative stress, metals accumulation and NRAMP genes in contrasting Ricinus communis genotypes under arsenic stress. *Environ. Sci. Pollut. Res.* 26 (30): 31166–31177. https://doi.org/10.1007/s11356-019-06243-2.

Singh, S., Parihar, P., Singh, R. et al. (2016). Heavy metal tolerance in plants: role of transcriptomics, proteomics, metabolomics, and ionomics. *Front. Plant Sci.* 6: 1143. https://doi.org/10.3389/fpls.2015.01143.

Sobel, M.H., Sanchez, T.R., Jones, M.R. et al. (2020). Rice intake, arsenic exposure, and subclinical cardiovascular disease among US adults in MESA. *J. Am. Heart Assoc.* 9 (4): e015658. https://doi.org/10.1161/JAHA.119.015658.

de Souza Reis, I.N.R., de Alves, Oliveira, J., Ventrella, M.C. et al. (2020). Involvement of glutathione metabolism in *Eichhornia crassipes* tolerance to arsenic. *Plant Biol.* 22 (2): 346–350. https://doi.org/10.1111/plb.12988.

Srivastava, S. and Sharma, Y.K. (2013). Impact of arsenic toxicity on black gram and its amelioration using phosphate. *Int. Sch. Res. Notices* 2013. https://doi.org/10.1155/2013/340925.

Srivastava, S. and Sharma, Y.K. (2014). Arsenic induced changes in growth and metabolism of black gram seedlings (*Vigna Mungo* L.) and the role of phosphate as an ameliorating agent. *Environ. Process.* 1: 431–445. https://doi.org/10.1007/s40710-014-0035-5.

Stanković, J.D., Sabovljević, A.D., and Sabovljević, M.S. (2018). Bryophytes and heavy metals: a review. *Acta Bot. Croat.* 77 (2): 109–118.

States, J.C., Srivastava, S., Chen, Y., and Barchowsky, A. (2009). Arsenic and cardiovascular disease. *Toxicol. Sci.* 107 (2): 312–323. https://doi.org/10.1093/toxsci/kfn236.

Stoeva, N., Berova, M., and Zlatev, Z. (2005). Effect of arsenic on some physiological parameters in bean plants. *Plant Biol.* 49: 293–296. https://doi.org/10.1007/s10535-005-3296-z.

Stoeva, N. and Bineva, T. (2003). Oxidative changes and photosynthesis in oat plants grown in as-contaminated soil. *Bulg. J. Plant Physiol.* 29: 87–95.

Su, Y.H., McGrath, S.P., Zhu, Y.G., and Zhao, F.J. (2008). Highly efficient xylem transport of arsenite in the arsenic hyperaccumulator *Pteris vittata*. *New Phytol.* 180 (2): 434–441.

Sun, L., Song, L., Zhang, Y. et al. (2016a). *Arabidopsis* PHL2 and PHR1 act redundantly as the key components of the central regulatory system controlling transcriptional responses to phosphate starvation. *Plant Physiol.* 170 (1): 499–514. https://doi.org/10.1104/pp.15.01336.

Sun, Y., Liu, G., and Cai, Y. (2016b). Thiolated arsenicals in arsenic metabolism: occurrence, formation, and biological implications. *J. Environ. Sci.* 49: 59–73. https://doi.org/10.1016/j.jes.2016.08.016.

Suresh Kumar, K., Dahms, H.U., Won, E.J. et al. (2015). Microalgae – a promising tool for heavy metal remediation. *Ecotoxicol. Environ. Saf.* 113: 329–352. https://doi.org/10.1016/j.ecoenv.2014.12.019.

Surgun-Acar, Y. and Zemheri-Navruz, F. (2019). 24-epibrassinolide promotes arsenic tolerance in *Arabidopsis thaliana* L. by altering stress responses at biochemical and molecular level. *J. Plant Physiol.* 238: 12–19. https://doi.org/10.1016/j.jplph.2019.05.002.

Taheri, M., Mehrzad, J., MahmudyGharaie, M.H. et al. (2016). High soil and groundwater arsenic levels induce high body arsenic loads, health risk and potential anemia for inhabitants of northeastern Iran. *Environ. Geochem. Health* 38: 469–482.

Takahashi, A., Kawakami, H., Bada, A. et al. (1990). Effects of phosphate on arsenate inhibition in a marine cyanobacterium, *Phormidium* sp. *Appl. Organomet. Chem.* 4: 269–279.

Takahashi, A., Kawakami, H., Iwakiri, K., and Matsuto, S. (2001). Some characteristics of arsenate transport in a marine cyanobacterium, *Synechococcus* sp. *Appl. Organomet. Chem.* 15: 291–298.

Talukdar, D. (2013). Arsenic-induced oxidative stress in the common bean legume, *Phaseolus vulgaris* L. seedlings and its amelioration by exogenous nitric oxide. *Physiol. Mol. Biol. Plants* 19 (1): 69–79. https://doi.org/10.1007/s12298-012-0140-8.

Tawfik, D.S. and Viola, R.E. (2011). Arsenate replacing phosphate: alternative life chemistries and ion promiscuity. *Biochemistry* 50 (7): 1128–1134.

Taylor, V.F. and Jackson, B.P. (2016). Concentrations and speciation of arsenic in New England seaweed species harvested for food and agriculture. *Chemosphere* 163: 6–13.

Taylor, V.F., Li, Z., Sayarath, V. et al. (2017). Distinct arsenic metabolites following seaweed consumption in humans. *Sci. Rep.* 7 (1): 3920. https://doi.org/10.1038/s41598-017-03883-7.

Thakur, S., Choudhary, S., Dubey, P., and Bhardwaj, P. (2019). Comparative transcriptome profiling reveals the reprogramming of gene networks under arsenic stress in Indian mustard. *Genome* 62 (12): 833–847. https://doi.org/10.1139/gen-2018-0152.

Thiel, T. (1988). Phosphate transport and arsenate resistance in the cyanobacterium *Anabaena variabilis*. *J. Bacteriol.* 170 (3): 1143–1147.

Tipping, E., Vincent, C.D., Lawlor, A.J., and Lofts, S. (2008). Metal accumulation by stream bryophytes, related to chemical speciation. *Environ. Pollut.* 156: 936–943.

Tremper, A.H., Agneta, M., Burton, S., and Higgs, D.E. (2004). Field and laboratory exposures of two moss species to low level metal pollution. *J. Atmos. Chem.* 49: 111–120.

Tripathi, D., Nam, A., Oldenburg, D.J., and Bendich, A.J. (2020). Reactive oxygen species, antioxidant agents, and DNA damage in developing maize mitochondria and plastids. *Front. Plant Sci.* 11: 596.

Ullrich-Eberius, C.I., Sanz, A., and Novacky, A.J. (1989). Evaluation of arsenate- and vanadate-associated changes of electrical membrane potential and phosphate transport in *Lemna Gibba* G1. *J. Exp. Bot.* 40: 119–128. https://doi.org/10.1093/jxb/40.1.119.

Upadhyaya, H., Shome, S., Roy, D., and Bhattacharya, M.K. (2014). *Am. J. Plant Sci.* 5: 3609–3618. https://doi.org/10.4236/ajps.2014.524377.

Ushijima, T. and Okochi-Takada, E. (2005). Aberrant methylations in cancer cells: where do they come from? *Cancer Sci.* 96 (4): 206–211.

Vazquez, M.D., Lopez, J., and Carballeira, A. (1999). Uptake of heavy metals to the extracellular and intracellular compartments in three species of aquatic bryophyte. *Ecotoxicol. Environ. Saf.* 44: 12–24.

Ventura-Lima, J., Bogo, M.R., and Monserrat, J.M. (2011). Arsenic toxicity in mammals and aquatic animals: a comparative biochemical approach. *Ecotox. Environ. Safe.* 74: 211–218.

Vezza, M.E., Llanes, A., Travaglia, C. et al. (2018). Arsenic stress effects on root water absorption in soybean plants: physiological and morphological aspects. *Plant Physiol. Biochem.* 123: 8–17. https://doi.org/10.1016/j.plaphy.2017.11.020.

Wang, H.B., Ye, Z.H., Shu, W.S. et al. (2006). Arsenic uptake and accumulation in fern species growing at arsenic-contaminated sites of southern China: field surveys. *Erratum in: Int. J. Phytoremediation* 8 (2): 185.

Wang, J., Zhao, F.J., Meharg, A.A. et al. (2002). Mechanisms of arsenic hyperaccumulation in *Pteris vittata*. Uptake kinetics, interactions with phosphate, and arsenic speciation. *Plant Physiol.* 130 (3): 1552–1561.

Wang, K., Wang, Y., Wan, Y. et al. (2021). The fate of arsenic in rice plants (*Oryza sativa* L.): influence of different forms of selenium. *Chemosphere* 264 (1): 128417. https://doi.org/10.1016/j.chemosphere.2020.128417.

Wang, P., Chen, X., Xu, X. et al. (2018a). Arsenate induced chlorosis 1/translocon at the outer envolope membrane of chloroplasts 132 protects chloroplasts from arsenic toxicity. *Plant Physiol.* 178 (4): 1568–1583. https://doi.org/10.1104/pp.18.01042.

Wang, Q.Y., Sun, J.Y., Xu, X.J., and Yu, H.W. (2018b). Integration of chemical and toxicological tools to assess the bioavailability of copper derived from different copper-based fungicides in soil. *Ecotoxicol. Environ. Saf.* 161 (1): 662–668.

Wang, Z., Zheng, Z., Song, L., and Liu, D. (2018c). Functional characterization of *Arabidopsis* PHL4 in plant response to phosphate starvation. *Front. Plant Sci.* 9: 1432. https://doi.org/10.3389/fpls.2018.01432.

Wells, J.M. and Brown, D.H. (1987). Factors affecting the kinetics of intra-and extracellular cadmium uptake by the moss *Rhytidiadelphus squarrosus*. *New Phytol.* 105: 123–137.

Wongkongkatep, J., Fukushi, K., Parkpian, P. et al. (2003). Arsenic uptake by native fern species in Thailand: effect of chelating agents on hyperaccumulation of arsenic by *Pityrogramma calomelanos*. *J. Environ. Sci. Health A Tox. Hazard. Subst. Environ. Eng.* 38 (12): 2773–2784.

Wysocki, R., Chéry, C.C., Wawrzycka, D. et al. (2001). The glycerol channel Fps1p mediates the uptake of arsenite and antimonite in *Saccharomyces cerevisiae*. *Mol. Microbiol.* 40 (6): 1391–1401.

Wysocki, R., Clemens, S., Augustyniak, D. et al. (2003). Metalloid tolerance based on phytochelatins is not functionally equivalent to the arsenite transporter Acr3p. *Biochem. Biophys. Res. Commun.* 304 (2): 293–300.

Xiao, T., Zou, Z., Xue, J. et al. (2021). LncRNA H19-mediated M2 polarization of macrophages promotes myofibroblast differentiation in pulmonary fibrosis induced by arsenic exposure. *Environ. Pollut.* 268 (115810): https://doi.org/10.1016/j.envpol.2020.115810.

Xie, Q.E., Yan, X.L., Liao, X.Y., and Li, X. (2009). The arsenic hyperaccumulator fern *Pteris vittata* L. *Environ. Sci. Technol.* 43: 8488–8495.

Xue, P.Y. and Yan, C.Z. (2011). Arsenic accumulation and translocation in the submerged macrophyte *Hydrilla verticillata* (L.f.). *Chemosphere* 85 (7): 1176–1181. https://doi.org/10.1016/j.chemosphere.2011.09.051.

Yadav, A.P.S., Dwivedi, V., Kumar, S. et al. (2021a). Cyanobacterial extracellular polymeric substances for heavy metal removal: a mini review. *J. Compos. Sci.* 5 (1): 1. https://doi.org/10.3390/jcs5010001.

Yadav, P., Srivastava, S., Patil, T. et al. (2021b). Tracking the time-dependent and tissue-specific processes of arsenic accumulation and stress responses in rice (*Oryza sativa L.*). *J. Hazard. Mater.* 406 (124307): https://doi.org/10.1016/j.jhazmat.2020.124307.

Yan, C., Wang, Z., and Luo, Z. (2014). Arsenic efflux from *Microcystis aeruginosa* under different phosphate regimes. *PLoS One* 9 (12): e116099.

Yin, L., Lundin, B., Bertrand, M. et al. (2010). Role of thylakoid ATP/ADP carrier in photoinhibition and photoprotection of photosystem II in *Arabidopsis*. *Plant Physiol.* 153 (2): 666–677.

Yin, X., Wang, L., Duan, G., and Sun, G. (2011). Characterization of arsenate transformation andidentification of arsenate reductase in a green alga *Chlamydomonas reinhardtii*'. *J. Environ. Sci.* 23: 1186–1193.

Yin, X., Wang, L.H., Bai, R. et al. (2012). Accumulation and transformation of arsenic in the blue-green alga *Synechocysis* sp. PCC6803. *Water Air Soil Pollut.* 223: 1183–1190.

Zemanová, V., Popov, M., Pavlíková, D. et al. (2020). Effect of arsenic stress on 5-methylcytosine, photosynthetic parameters and nutrient content in arsenic hyperaccumulator *Pteris cretica* (L.) var. Albo-lineata. *BMC Plant Biol.* 20 (1): 1–10.

Zhang, S., Rensing, C., and Zhu, Y.G. (2014). Cyanobacteria-mediated arsenic redox dynamics is regulated by phosphate in aquatic environments. *Environ. Sci. Technol.* 48 (2): 994–1000.

Zhanga, J., Zhou, F., Liu, Y. et al. (2020). Effect of extracellular polymeric substances on arsenic accumulation in *Chlorella pyrenoidosa*. *Sci. Total Environ.* 704: 135–368.

Zhao, F., McGrath, S.P., and Meharg, A.A. (2010). Arsenic as a food chain contaminant: mechanisms of plant uptake and metabolism and mitigation strategies. *Annu. Rev. Plant Biol.* 61: 535–559.

Zhao, Z., Zhang, H., Fu, Z. et al. (2018). Genetic-based dissection of arsenic accumulation in maize using a genome-wide association analysis method. *Plant Biotechnol. J.* 16 (5): 1085–1093. https://doi.org/10.1111/pbi.12853.

Zhou, H., Liu, G., Zhang, L. et al. (2021). Strategies for arsenic pollution control from copper pyrometallurgy based on the study of arsenic sources, emission pathways and speciation characterization in copper flash smelting systems. *Environ. Pollut.* 270: 116203. https://doi.org/10.1016/j.envpol.2020.116203.

Zubair, M., Ahmad, M., and Qureshi, Z.I. (2017). Review on arsenic-induced toxicity in male reproductive system and its amelioration. *Andrologia* 49 (9): https://doi.org/10.1111/and.12791.

5

Arsenic-Induced Responses in Plants

Impacts on Morphological, Anatomical, and Other Quantitative and Qualitative Characters

Sumaya Farooq[1], *Simranjeet Singh*[2], *Vijay Kumar*[3], *Daljeet Singh Dhanjal*[1], *Praveen C. Ramamurthy*[2], *and Joginder Singh*[1]

[1] *Department of Biotechnology, School of Bioengineering and Biosciences, Lovely Professional University, Phagwara, Punjab, India*
[2] *Interdisciplinary Centre for Water Research (ICWaR), Indian Institute of Science, Bengaluru, Karnataka, India*
[3] *Department of Chemistry, Central Ayurveda Research Institute, Jhansi, Uttar Pradesh, India*

5.1 Introduction

Contamination of heavy metals in the soil has become the major environmental concern, which has adversely affected plants' normal cellular functioning (Kumar et al. 2015a, 2015b; Singh et al. 2019a). Among heavy metals (HM), arsenic (As) is stated to be a carcinogenic and nonessential metalloid that is held responsible for food safety issues (Zhao et al. 2010). Metals possess potential health effects on animals and humans, and environmental As contamination has become a matter of public concern (Kumar et al. 2015a; Singh et al. 2019a, 2020b, 2020c). Soil to plant transmission is considered the main pathways for animal and human exposure (Türkdoğan et al. 2003). Mineralization and weathering of the earth's surface, herbicides, mining operations, insecticides, utilization of As-based preservatives for wood, and irrigation of As-loaded groundwater crops are the main sources of As; (Mirza et al. 2014; Chandrakar et al. 2016). The widespread application of arsenic-containing pesticides has led to the contamination of agricultural and domestic land (Kumar et al. 2014; Singh et al. 2016, 2017a, 2020a; Sidhu et al. 2019). Moreover, the inorganic form of As is phytotoxic to plants and a variety of animals. Depending on the soil's redox state, As is mainly found in soil as inorganic arsenate (As(V)) or arsenite (As(III)). Humans get directly exposed to As via the consumption of water contaminated with As and indirectly to As via the consumption of crops grown in As-contaminated field (Rosas-Castor et al. 2014).

Arsenate is absorbed by root phosphate transporters and hampersthe phosphate metabolism during ATP(Adenosine triphosphate) synthesis and phosphorylation, leading to inhibited growth and oxidative stress (Ronzan et al. 2018). These challenges and issues associated with As have received increased attention, especially in nations like India, Bangladesh, and China, due to high prevalence, chronic toxicity, and growing

Arsenic in Plants: Uptake, Consequences and Remediation Techniques, First Edition.
Edited by Prabhat Kumar Srivastava, Rachana Singh, Parul Parihar, and Sheo Mohan Prasad.

environmental feedback worldwide (Meharg 2004). Anthropogenic practices involving the use of As-containing herbicides, mining operations, insecticides, and even irrigation of crops with polluted groundwater are linked to As contamination in the soil (Pandey et al. 2016; Mishra et al. 2017; Singh et al. 2017b). In the terrestrial world, the most common and toxic types include oxidation states of As(III) and As(V) (Zhao et al. 2009). As has been reported, a number of symptoms of toxicity in plants, ranging from the root and shoot biomass reduction, are strongly phytotoxic and exhibit necrosis, wilting of leaf blades, low fruit and grain yield, and lessening in the photosynthesis process and chlorophyll content. Additionally, the effect of irrigation with highly contaminated water on crops and soil has attracted considerable interest because of their transition in food chain through the groundwater-plant–soil system (Mukherjee and Bhattacharya 2001; Tripathi et al. 2007).

In many metabolic processes, plant cells allow arsenic entry through essential element channels and result in adverse effects like reduced germination, restricted growth, and low yield of certain crops. Soybeans (*Glycine max* L.) are highly valuable crops worldwide because of their high nutritional value and application in biofuel production. It is often grown in stressful environments and also in soil contaminated with a high concentration of As (Miransari 2016). The presence of As can lead to potentially dangerous human and animal ingestion, as in edible parts of plants. Under such circumstances, the detection of As in plants is highly important to minimize animal and human exposure chances.

5.2 Impact of Arsenic on the Morphological Characters of Plants

The morphological study conducted on *Brassica juncea* var. *varuna* roots showed a reduction in the length of roots of treated plants and reduced the number of lateral roots (Pandey et al. 2016). The growth of *Arabidopsis* pathogenesis-related genes was affected differently by arsenic exposure, exhibiting similar effects on lateral root density as well as lateral root length (LRL) (Piacentini et al. 2020). The permissible limit of arsenic ranges, i.e. $20\,mg\,kg^{-1}$ in soil and $10\,\mu g\,L^{-1}$ in groundwater, results in obstruction in metabolic processes in a plant cell, resulting in wilting and curling of the plant cells (Figure 5.1), leaf blade necrosis, reduction in the leaf area and a number of leaves, loss of mineral content, proliferation without elongation, reduction in photosynthesis rate, reduced rate of elongation, and biomass accumulation (Talukdar 2013a; Chandrakar et al. 2017a). Arsenite binding to both enzymes and proteins in sulfhydryl groups contributes to distortion in cell membranes, resulting in the loss of turgor and rigidity, causing wilting in plant sections. Additionally, the absorption of essential minerals and water via root cells, As in the soil, limits the cell's dehydration condition, which eventually results in curling of leaves. A study reported that exposure of $50\,\mu M$ As on mung bean seedlings caused a substantial reduction in both shoot and root length. The inhibitory effect on root length was more noticeable in contrast to shoot length. In answer to $50\,\mu M$ As, the root length declined by over 49.3%. The reduction in shoot length, on the other hand, was relatively smaller (15.13%), which was also apparent from a decrease in the root–shoot ratio for As treatment (Upadhyaya et al. 2014). Fenugreek plants subjected to treatments showed substantial differences ($P < 0.05$) in their growth and seed yield components between the treatments (Talukdar 2013a). The mean

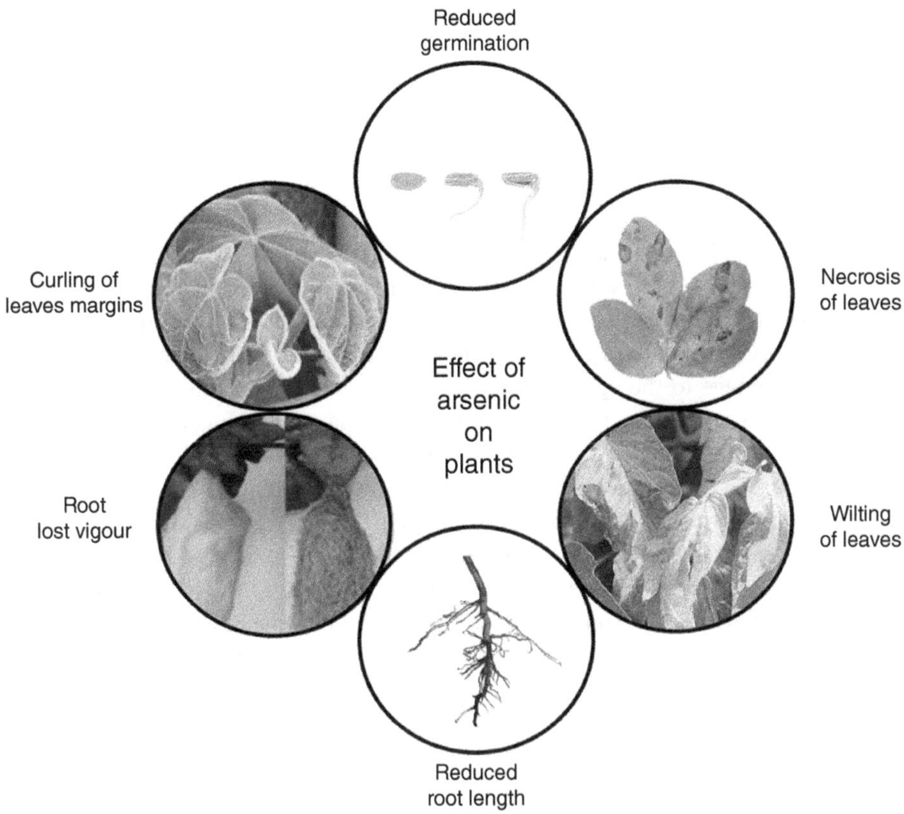

Figure 5.1 The potent effect of Arsenic on plant growth and development.

value of the physical parameters such as dry and fresh weight and length of root shoot and fresh root decreases with an increased arsenic concentration in *Trigonella foenum-graecum* L. and *Lathyrus sativis* L. crops (Talukdar 2013b).

Essential nutrients deficiency also leads to the development of toxicity symptoms like necrosis, which causes the browning or blackening of the affected parts of the plant, especially the green leaves. As a result, the formation of circular necrotic areas causes the death of tissues in a specific part, which subsequently progresses and leads to necrosis. As shown by ultrastructural analysis, during necrosis cytoplasmic leakage, cell disorganization and membrane vesiculation occur (Farnese et al. 2017). It was recorded that As has negative consequences on cell elongation as a reduction in root length was recorded after performing As-stressed bioassay on *Brassica napus* L. seedling roots (Farooq et al. 2018).

5.3 Impact of Arsenic on the Anatomical Characters of Plants

Increasing evidence is elucidating that heavy metals exposure causes structural changes that can affect water relationships with potential functional consequences. Arsenic effect on soybean plants has been observed in radical structure (Armendariz et al. 2016). Even reduction in stomatal index of *Beta vulgaris* and treatment of *Sorghum bicolor* Cu and Cd have been recorded (Kasim 2006; Sagardoy et al. 2010). The increase in the number of stomata was recorded in *Helianthus annuus* inoculated with metals like Cu, Pb, Zn, or Cd and *Vignaradiata* exposed to As (Kastori et al. 1992; Gupta and Bhatnagar 2015).

Inoculation of As(V) in soybean plants increases the frequency of anomalous stomata by 15%, although no significant variations in the number of stomata were observed. Other abnormalities like rounded stomata were observed. Also, stomata developed from guard cells (undifferentiated) and smaller ostioles because there was no distinct separation between the two guard cells. Moreover, the formation of nonfunctional and anomalous stomata has been reported in plants exposed to various types of heavy metals (Gupta and Bhatnagar 2015). Certain studies have reported that As interferes with microtubules assembly and cause damage to the cytoskeleton (Dho et al. 2010). In plants treated with As, microtubule failure occurs due to nonfunctional and anomalous stomata. The thickening in the guard cells of plants with the treated periclinal wall is reported to be responsible for keeping ostiole partially closed, which defines the cause responsible for reduced water loss in As(V)-treated soybean plants (Gupta and Bhatnagar 2015).

Additionally, As(III)- and As(V)-treated plants have shown a reduction in xylem diameter in leaves' vascular bundles. Similar variations are observed in the roots of As-treated soybean plants (Armendariz et al. 2016). Changes in both tissue and stomatal structure have been reported to affect the water relationships by affecting transpiration process as it alters the water flow from root to shoot. In another study, As treatment showed no changes in the shape and size of the spinach leaf's mesophyll cells, although the substantial change was observed in intercellular spaces and leaf thickness (Bandaru et al. 2010).

Arsenic is well remembered for undermining the plasma membrane structure. Any disruption of the outer membrane of the plant cells results in disproportionate absorption and transportation of ions, nutrients, and water that eventually reduce conductance of stomatal structure (Päivöke and Simola 2001; Wani et al. 2017). Arsenic-induced chloroplastic degradation and internal membranes modification have a sturdy effect on photosynthesis, associated with decreased carbon assimilation rate and photosynthetic pigment concentration (Farnese et al. 2017).

5.4 Effect of As on stem Anatomy of Plants

The epidermis, cortex, trichomes, xylem, and phloem cells among the various stem tissues showed maximum alterations after control care. As-treated plants display epidermal cells that were enlarged than the control cells. Trichome apical cells (the site of biosynthesis of artemisinin) were turgid with possible accumulation of artemisinin. Although the phloem cells and xylem cells were loaded densely with stained arsenic deposits, no deposits were observed in the control (Farnese et al. 2014).

Compared with regulation, As exposure causes crystals' development in a high number in the cortex, epidermal layer, stem pith, and vascular bundles. They also found crystal deposits in the petiole. The SEM (Scanning emission microscopy) analysis revealed the presence of surface layer crystals and the formation of druses, crystal and prismatic sands in petioles and stems. In SEM leaves, both druse- as well as needle-like were observed.

Vascular bundle disintegration in different parts of the plant and cell loosening in the absence of root hairs are common abnormalities observed in As exposed parts of roots (Talukdar 2013b).

5.4.1 Effect of Arsenic on Anatomy of Plants Roots

Tissues were also nearly normal in the roots; however, thickening in cell walls was observed in As + SNP exposed plant. As-induced anomalies have increased with the inclusion of 2-phenyl-4,4,5-tetramethylimidazole-1-oxyl 3-oxide (PTIO) in the media, having a more severe and noticeable effect on the roots in contrast to shoots. Alteration in roots has also shown the influence on permeability and selectivity of the cell membrane, which has resulted in lower uptake of nutrients (Gusman et al. 2013).

5.5 Impacts of Arsenic on Quantitative Characters of Plants

Arsenic is a nonessential and poisonous metalloid that has a serious influence on productivity and plant growth. The multiple transporters contained in the root cells take As from the substrate (Chandrakar et al. 2016).

5.5.1 Root Plasmolysis

Roots are generally the first part of the plant that get exposed to As, where cell decoloration and plasmolysis are caused by this metalloid (Shaibur and Kawai 2011). Most of the hyperosmotic solutions containing sorbitol or mannitol or sucrose cause the extruding of vacuole from the tonoplast resulting in an imbalance of turgor pressure. If this ailment continues for a longer time, it causes protoplast retraction, which aids in the separation of plasma membrane from the solid cell wall. An active process, plasmolysis is well known to function in viable cells; it is frequently used under different environmental stresses to test cell viability (Lang et al. 2014). Plant root exposed to As causes a sharp increase in the width of the cell, root lignification, and microtubule disintegration, eventually causing an increase in diameter of the root (Lou et al. 2015).

5.5.2 Cell Division

Development reduction/inhibition is associated with the declined mitotic activity, due to As exposure on the meristematic zone of roots, which reduced the rate of cell differentiation in the apical meristem and minimized the elongation and expansion of freshly formed cells (Farooq et al. 2015). With therapy, cellular turgor pressure decreases, preventing elongation of the root cells. Arsenic is also recognized to inhibit the amino acid synthesis required in the cell cycle (Reichard and Puga 2010). Arsenic seriously affected the cell division machinery of plant systems (Sharma 2012). Arsenic binds to thiol groups and contributes to the dysfunction of enzyme activities involved in the DNA repair mechanism (Patra et al. 2004). Genomic studies of two genes, *Os04g46650* and *Os01g14660*, showed that their expression was downregulated in the presence of As in rice plants (Sharma 2012). It also binds to the tubulin protein thiol group, contributing to the disruption of cell division in the spindles. The transcript abundance in *Oryza sativa* L. under As stress was strongly repressed by two each tubulin genes (*Os03g45920* as well as *Os03g56810*) and microtubule genes (*Os03g13460* as well as *Os09g27700*) (Norton et al. 2008). In reaction to As, chromosome aberrations lower the mitotic index in various plants (Patra et al. 2004). Another detrimental effect of As is chromosome stickiness, owing to the formation of the bridge. The chromosomal anomalies caused by As are categorized into two categories, i.e. clastogenic effects like micronuclei, ring chromosome bridges, fragments, and the precocious chromosome motion anaphase because of chromosome stickiness and premature terminalization (Mumthas et al. 2010).

5.5.3 Biomass

The fact that As causes a major decrease in biomass production has been well established in plants, owing to the fact that the point of contact for As is the roots (Chandrakar et al. 2018). Arsenic has a high affinity in the radicular membranes for protein and enzyme sulfhydryl groups, which causes obstruction in root functioning and death of cells (Shaibur and Kawai 2011). The negative effect of As on cell activity of the plant is that most of the available energy is used for the synthesis of stress-related essentials like phytochelatins, antioxidants, etc. may be due to reduced germination and growth responses. Another reason for decreases in biomass accumulation is reduced water absorption via root cells (Gomes et al. 2013). Furthermore, As exposure leads to protein degradation because of carbohydrate deficiency, cell adaptation leading to accretion of biomass, and reduction of protein content (Agnihotri and Seth 2016). Accumulation of As has been reported to affect plant growth severely (Chandrakar et al. 2016). It has been found that these enzymes are used by plants for reducing the As concentration in tissues of the plant (Meadows 2014). Further, genetic engineering approaches might have the application in manipulating the pattern of protein expression so that plants sustain and grow in As-contaminated regions (Farooq et al. 2016).

5.5.4 Energy Flow

As a phosphate analogue, As(V) uncouples the oxidative phosphorylation mechanism by dislocating one molecule of phosphate from ATP, creating an unstable, volatile complex, i.e. adenosine diphosphate–As(V) (Geng et al. 2006). In the ATP synthesis rate, this leads to suppression and disruption in cellular processes' energy flow (Meharg and Hartley-Whitaker 2002). The aqueous medium hydrolyzes this ADP–As(V) complex and reuses ADP as well as As, which again react at different coupling sites.

5.5.5 Photosynthetic Pigments

The exposure of As on curcumin has been reported to alter the photosynthetic pigments (Chl a/b), carotenoid and total chlorophyll content in comparison to controls, and reduced intake potential of developing mung seedlings. However, in the mung bean shoot, curcumin induced an increase of photosynthesis pigments 25, 50, and 100 μM and its association with As treatment in contrast to As-treated plants, respectively. On analysis, 50 μM As and 25 μM curcumin-treated mung seedlings showed the highest decrease in chlorophyll ratio. Results show thousandfold increases in total carotenoid content compared to only As-treated plants, respectively (Upadhyaya et al. 2014). This effect of curcumin may be due to its ability, as mentioned elsewhere, to act as an antioxidant in animal cells (Banerjee et al. 2010).

5.6 Impact of Arsenic on the Qualitative Characters of Plants

5.6.1 Cellular Membrane Damage

Cell membranes in plants are very susceptible to damage caused by an attack (Ismail 2012). Increased accumulation and consequent loss of membrane integrity were caused by exposure of plants or their parts to As (Kaur et al. 2012). Two mechanisms are reported to the sensitivity of cellular membranes to toxicity: As for the thiol groups of membrane protein groups, specifically binding, and As-induced reactive oxygen species (ROS) production disintegrates the membrane's stability (Karam et al. 2016). The binding properties change the proper functioning of proteins as far as membrane proteins are concerned, thereby breaking the membrane. Furthermore, ROS-regulated lipid peroxidation of the membrane is apprehended for loss of integrity of the membrane (Talukdar 2014). Furthermore, free radicals are responsible for the concussion of plasma and vacuolar membrane by peroxidation of polyunsaturated fatty acids from cell membranes. Moreover, ROS attacks disrupt the membrane structure and elevate membrane permeability (Kapoor et al. 2019; Singh et al. 2019b). Arsenic-induced ROS are found to be involved in the cell signaling cascade of vacuolar membranes', which leads to the inactivation of vacuolar enzymes causing structural changes in the vacuolar membrane of cell (Farnese et al. 2017). Cytosol denaturation increased permeability, and electrolyte and ion loss are the main outcomes of As-induced stress injuries. The depth of injury can be directly linked to the measurement of the membrane stability index (Pathare et al. 2016). Another study reported that presence of As in the cells alters the interaction of lipid-protein in the cellular membranes of plant and amends the normal functioning of constituents of metabolic processes, involving the key H + ATPase enzymes, which increases the permeability of membrane as well as the loss of electrolytes (Chandrakar et al. 2017b).

5.6.2 Leaf Reflectance

With increased As-treated plant reflectance, As-treated plant leaf reflectance decreases at infrared wavelengths (above 750 nm) and increases at visible wavelengths (400–700 nm). As low light absorption increases reflectance in visible spectra with decreased leaf chlorophyll content, these findings are consistent with plant health (Horler et al. 1983). The leaf structure gets primarily influenced by LR in the Near-infrared area (Slaton et al. 2001). Due to changes in refraction indices of leaf components, multiple reflections near mesophyll cells and interference in the air are held responsible for reflectance in this region (Gates et al. 1965). Higher SM/PM ratio (ratio of palisade mesophyll to spongy mesophyll thickness) is accorded in As-treated plants due to a decrease in SM cells. Moreover, reduced SM thickness is claimed to be responsible for less scattering, which results in decreased NIR reflectance. Furthermore, the decrease in Aspaces/S supports the evidence that reduced intercellular air spaces via As minimize manifold SM dispersions and increase absorption (Bandaru et al. 2010).

5.6.3 Water Loss

The As presence in plant cells alters the expression of different transporter genes of the outer membrane, disrupting soil water absorption and the flow of water and nutrients (Finnegan and Chen 2012). During abiotic stresses, reduced water intake and water loss are associated with reduced plant biomass (Agnihotri and Seth 2016). Because of turbulences in nutrient and water absorption and passage in the cell, stomatal conductance and transpiration rates also get disrupted during As exposure (Stoeva et al. 2003a). The relative water content of As-treated plants has decreased marginally in relation to their respective controls (Stoeva et al. 2003b). It also affects the functioning of aquaporins that are involved in soil water absorption, in the PM intrinsic protein class (Siddiqui et al. 2015). Under stress, intrinsic PM protein gene expression gets downregulated and causes a reduction in total water content, which eventually disturbs water balance and obstructs the growth of the seedling (Srivastava et al. 2013a). Amendment in endogenous factors like anatomy and root morphology is claimed to indirectly affect water absorption by plant root cells (Rucińska-Sobkowiak 2016). For a fact, water balance inside the cells is highly important for plants' growth, and any disturbance in water homeostasis often impacts the different characteristics of plant physiology (Srivastava et al. 2013b).

5.7 Conclusion

Owing to its deleterious effects on animals and plants, As has attracted scientists' significant attention. The presence of As has a harmful effect on plant morphology, anatomy, and physiology. Awareness of physiological and morphological processes of plant responses exposed to As toxicity will allow us to remediate via new deliberate and controlled methods. Toxicity symptoms of As involve growth attribute inhibition, photosynthetic apparatus distortion, biomass accumulation inhibition, etc. While significant attempts have been made in different phases of plant growth to unveil the mechanisms of As-induced injury symptoms, our understanding of the topic is still far from comprehensive. Via enhanced ROS production, As imposes oxidative stress and ultimately oxidizes cellular macromolecules like lipids, carbohydrates, amino acids, and nucleic acids. The indirect mechanism relies on reducing decreasing the phytotoxicity of As in the soil and exposes plants to less harmful compounds of As. A direct mechanism focused on stimulating the growth and activity of indigenous soil microbiota, which can have a positive impact on the general fitness of plants and increase the efficacy of the absorption of As. This knowledge of various morphological, physiological, and biochemical parameters may pave the way for the metalloid stress enhancement of fenugreek genotypes.

References

Agnihotri, A. and Seth, C.S. (2016). Exogenously applied nitrate improves the photosynthetic performance and nitrogen metabolism in tomato (*Solanum lycopersicum* L. cv Pusa Rohini) under arsenic (V) toxicity. *Physiol. Mol. Biol. Plants* 22: 341–349. https://doi.org/10.1007/s12298-016-0370-2.

Armendariz, A.L., Talano, M.A., Travaglia, C. et al. (2016). Arsenic toxicity in soybean seedlings and their attenuation mechanisms. *Plant Physiol. Biochem.* 98: 119–127. https://doi.org/10.1016/j.plaphy.2015.11.021.

Bandaru, V., Hansen, D.J., Codling, E.E. et al. (2010). Quantifying arsenic-induced morphological changes in spinach leaves: implications for remote sensing. *Int. J. Remote Sens.* 31: 4163–4177. https://doi.org/10.1080/01431161.2010.498453.

Banerjee, S., Bandyopadhyaya, G., Chattopadhyay, K., and Chattopadhyay, B.D. (2010). Amelioration of nicotine-induced damage of blood cells in protein malnourished female rats by curcumin. *Int. J. Pharmacol.* 6: 409–420. https://doi.org/10.3923/ijp.2010.444.455.

Chandrakar, V., Dubey, A., and Keshavkant, S. (2018). Modulation of arsenic-induced oxidative stress and protein metabolism by diphenyleneiodonium, 24-epibrassinolide and proline in *Glycine max* L. *Acta Bot. Croat.* 77: 51–61. https://doi.org/10.2478/botcro-2018-0004.

Chandrakar, V., Naithani, S.C., and Keshavkant, S. (2016). Arsenic-induced metabolic disturbances and their mitigation mechanisms in crop plants: a review. *Biol.* https://doi.org/10.1515/biolog-2016-0052.

Chandrakar, V., Parkhey, S., Dubey, A., and Keshavkant, S. (2017a). Modulation in arsenic-induced lipid catabolism in *Glycine max* using proline, 24-epibrassinolide and diphenylene iodonium. *Biol.* 72: 292–299. https://doi.org/10.1515/biolog-2017-0033.

Chandrakar, V., Yadu, B., Meena, R.K. et al. (2017b). Arsenic-induced genotoxic responses and their amelioration by diphenylene iodonium, 24-epibrassinolide and proline in *Glycine max* L. *Plant Physiol. Biochem.* 112: 74–86. https://doi.org/10.1016/j.plaphy.2016.12.023.

Dho, S., Camusso, W., Mucciarelli, M., and Fusconi, A. (2010). Arsenate toxicity on the apices of *Pisum sativum* L. seedling roots: effects on mitotic activity, chromatin integrity and microtubules. *Environ. Exp. Bot.* 69: 17–23. https://doi.org/10.1016/j.envexpbot.2010.02.010.

Farnese, F.S., Oliveira, J.A., Gusman, G.S. et al. (2014). Effects of adding nitroprusside on arsenic stressed response of *Pistia stratiotes* L. under hydroponic conditions. *Int. J. Phytoremediation* 16: 123–137. https://doi.org/10.1080/15226514.2012.759532.

Farnese, F.S., Oliveira, J.A., Paiva, E.A.S. et al. (2017). The involvement of nitric oxide in integration of plant physiological and ultrastructural adjustments in response to arsenic. *Front. Plant Sci.* 8: https://doi.org/10.3389/fpls.2017.00516.

Farooq, M.A., Islam, F., Ali, B. et al. (2016). Arsenic toxicity in plants: cellular and molecular mechanisms of its transport and metabolism. *Environ. Exp. Bot.* https://doi.org/10.1016/j.envexpbot.2016.08.004.

Farooq, M.A., Islam, F., Yang, C. et al. (2018). Methyl jasmonate alleviates arsenic-induced oxidative damage and modulates the ascorbate–glutathione cycle in oilseed rape roots. *Plant Growth Regul.* 84: 135–148. https://doi.org/10.1007/s10725-017-0327-7.

Farooq, M.A., Li, L., Ali, B. et al. (2015). Oxidative injury and antioxidant enzymes regulation in arsenic-exposed seedlings of four *Brassica napus* L. cultivars. *Environ. Sci. Pollut. Res.* 22: 10699–10712. https://doi.org/10.1007/s11356-015-4269-1.

Finnegan, P.M. and Chen, W. (2012). Arsenic toxicity: the effects on plant metabolism. *Front. Physiol.* 3: 182. https://doi.org/10.3389/fphys.2012.00182.

Gates, D.M., Keegan, H.J., Schleter, J.C., and Weidner, V.R. (1965). Spectral properties of plants. *Appl. Opt.* 4: 11–20.

Geng, C.N., Zhu, Y.G., Tong, Y.P. et al. (2006). Arsenate (As) uptake by and distribution in two cultivars of winter wheat (*Triticum aestivum* L.). *Chemosphere* 62: 608–615. https://doi.org/10.1016/j.chemosphere.2005.05.045.

Gomes, M.P., Carneiro, M.M.L.C., Nogueira, C.O.G. et al. (2013). The system modulating ROS content in germinating seeds of two Brazilian savanna tree species exposed to As and Zn. *Acta Physiol. Plant.* 35: 1011–1022. https://doi.org/10.1007/s11738-012-1140-6.

Gupta, P. and Bhatnagar, A.K. (2015). Spatial distribution of arsenic in different leaf tissues and its effect on structure and development of stomata and trichomes in mung bean, *Vigna radiata* (L.) Wilczek. *Environ. Exp. Bot.* 109: 12–22. https://doi.org/10.1016/j.envexpbot.2014.08.001.

Gusman, G.S., Oliveira, J.A., Farnese, F.S., and Cambraia, J. (2013). Mineral nutrition and enzymatic adaptation induced by arsenate and arsenite exposure in lettuce plants. *Plant Physiol. Biochem.* 71: 307–314. https://doi.org/10.1016/j.plaphy.2013.08.006.

Horler, D.N.H., Dockray, M., and Barber, J. (1983). The red edge of plant leaf reflectance. *Int. J. Remote Sens.* 4: 273–288. https://doi.org/10.1080/01431168308948546.

Ismail, G.S. (2012). Protective role of nitric oxide against arsenic-induced damages in germinating mung bean seeds. *Acta Physiol. Plant.* 34: 1303–1311. https://doi.org/10.1007/s11738-012-0927-9.

Kapoor, D., Singh, S., Kumar, V. et al. (2019). Antioxidant enzymes regulation in plants in reference to reactive oxygen species (ROS) and reactive nitrogen species (RNS). *Plant Gene* 19: 100182. https://doi.org/10.1016/j.plgene.2019.100182.

Karam, E.A., Keramat, B., Asrar, Z., and Mozafari, H. (2016). Triacontanol-induced changes in growth, oxidative defense system in Coriander (*Coriandrum sativum*) under arsenic toxicity. *Indian J. Plant Physiol.* 21: 137–142. https://doi.org/10.1007/s40502-016-0213-8.

Kasim, W.A. (2006). Changes induced by copper and cadmium stress in the anatomy and grain yield of *Sorghum bicolor* (L.) moench. *Int. J. Agric. Biol.* 8: 123–128.

Kastori, R., Petrović, M., and Petrović, N. (1992). Effect of excess lead, cadmium, copper, and zinc on water relations in sunflower. *J. Plant Nutr.* 15: 2427–2439. https://doi.org/10.1080/01904169209364485.

Kaur, S., Singh, H.P., Batish, D.R. et al. (2012). Arsenic (As) inhibits radicle emergence and elongation in phaseolus aureus by altering starch-metabolizing enzymes vis-à-vis disruption of oxidative metabolism. *Biol. Trace Elem. Res.* 146: 360–368. https://doi.org/10.1007/s12011-011-9258-8.

Kumar, V., Singh, S., Bhadrecha, P. et al. (2015a). Bioremediation of heavy metals by employing resistant microbial isolates from agricultural soil irrigated with industrial waste water. *Orient. J. Chem.* 31: 357–361. https://doi.org/10.13005/ojc/310142.

Kumar, V., Singh, S., Singh, J., and Upadhyay, N. (2015b). Potential of plant growth promoting traits by bacteria isolated from heavy metal contaminated soils. *Bull. Environ. Contam. Toxicol.* 94: 807–814. https://doi.org/10.1007/s00128-015-1523-7.

Kumar, V., Upadhyay, N., Kumar, V. et al. (2014). Environmental exposure and health risks of the insecticide monocrotophos-a review. *J. Biodivers. Environ. Sci.* 2014: 111–120.

Lang, I., Sassmann, S., Schmidt, B., and Komis, G. (2014). Plasmolysis: loss of turgor and beyond. *Plants* 3: 583–593. https://doi.org/10.3390/plants3040583.

Lou, L.Q., Shi, G.L., Wu, J.H. et al. (2015). The Influence of phosphorus on arsenic uptake/efflux and as toxicity to wheat roots in comparison with sulfur and silicon. *J. Plant Growth Regul.* 34: 242–250. https://doi.org/10.1007/s00344-014-9460-y.

Meadows, R. (2014). How plants control arsenic accumulation. *PLoS Biol.* 12: e1002008. https://doi.org/10.1371/journal.pbio.1002008.

Meharg, A.A. (2004). Arsenic in rice – understanding a new disaster for South-East Asia. *Trends Plant Sci.* 9: 415–417. https://doi.org/10.1016/j.tplants.2004.07.002.

Meharg, A.A. and Hartley-Whitaker, J. (2002). Arsenic uptake and metabolism in arsenic resistant and nonresistant plant species. *New Phytol.* 154: 29–43. https://doi.org/10.1046/j.1469-8137.2002.00363.x.

Miransari, M. (2016). *Abiotic and Biotic Stresses in Soybean Production*. Amsterdam: Elsevier Inc. https://doi.org/10.1016/c2014-0-00087-1.

Mirza, N., Mahmood, Q., Maroof Shah, M. et al. (2014). Plants as useful vectors to reduce environmental toxic arsenic content. *Sci. World J.* 2014: 921581. https://doi.org/10.1155/2014/921581.

Mishra, A., Kumar, J., and Melo, J.S. (2017). An optical microplate biosensor for the detection of methyl parathion pesticide using a biohybrid of *Sphingomonas* sp. cells-silica nanoparticles. *Biosens. Bioelectron.* 87: 332–338. https://doi.org/10.1016/j.bios.2016.08.048.

Mukherjee, A.B. and Bhattacharya, P. (2001). Arsenic in groundwater in the Bengal Delta Plain: slow poisoning in Bangladesh. *Environ. Rev.* 9: 189–220. https://doi.org/10.1139/a01-007.

Mumthas, S., Chidambaram, A.A., Sundaramoorthy, P., and Ganesh, S.K. (2010). Effect of arsenic and manganese on root growth and cell division in root tip cells of green gram (*Vigna radiata* L.). *Emir. J. Food Agric.* 22: 285–297. https://doi.org/10.9755/ejfa.v22i4.4876.

Norton, G.J., Lou-Hing, D.E., Meharg, A.A., and Price, A.H. (2008). Rice-arsenate interactions in hydroponics: whole genome transcriptional analysis. *J. Exp. Bot.* 59: 2267–2276. https://doi.org/10.1093/jxb/ern097.

Päivöke, A.E.A. and Simola, L.K. (2001). Arsenate toxicity to pisum sativum: mineral nutrients, chlorophyll content, and phytase activity. *Ecotoxicol. Environ. Saf.* 49: 111–121. https://doi.org/10.1006/eesa.2001.2044.

Pandey, C., Khan, E., Panthri, M. et al. (2016). Impact of silicon on Indian mustard (*Brassica juncea* L.) root traits by regulating growth parameters, cellular antioxidants and stress modulators under arsenic stress. *Plant Physiol. Biochem.* 104: 216–225. https://doi.org/10.1016/j.plaphy.2016.03.032.

Pathare, V., Srivastava, S., Sonawane, B.V., and Suprasanna, P. (2016). Arsenic stress affects the expression profile of genes of 14-3-3 proteins in the shoot of mycorrhiza colonized rice. *Physiol. Mol. Biol. Plants* 22: 515–522. https://doi.org/10.1007/s12298-016-0382-y.

Patra, M., Bhowmik, N., Bandopadhyay, B., and Sharma, A. (2004). Comparison of mercury, lead and arsenic with respect to genotoxic effects on plant systems and the development of genetic tolerance. *Environ. Exp. Bot.* 52: 199–223. https://doi.org/10.1016/j.envexpbot.2004.02.009.

Piacentini, D., Corpas, F.J., D'Angeli, S. et al. (2020). Cadmium and arsenic-induced-stress differentially modulates *Arabidopsis* root architecture, peroxisome distribution, enzymatic activities and their nitric oxide content. *Plant Physiol. Biochem.* 148: 312–323. https://doi.org/10.1016/j.plaphy.2020.01.026.

Reichard, J.F. and Puga, A. (2010). Effects of arsenic exposure on DNA methylation and epigenetic gene regulation. *Epigenomics.* 2: 87–104. https://doi.org/10.2217/epi.09.45.

Ronzan, M., Piacentini, D., Fattorini, L. et al. (2018). Cadmium and arsenic affect root development in *Oryza sativa* L. negatively interacting with auxin. *Environ. Exp. Bot.* 151: 64–75. https://doi.org/10.1016/j.envexpbot.2018.04.008.

Rosas-Castor, J.M., Guzmán-Mar, J.L., Hernández-Ramírez, A. et al. (2014). Arsenic accumulation in maize crop (*Zea mays*): a review. *Sci. Total Environ.* https://doi.org/10.1016/j.scitotenv.2014.04.075.

Rucińska-Sobkowiak, R. (2016). Water relations in plants subjected to heavy metal stresses. *Acta Physiol. Plant.* 38: 1–13. https://doi.org/10.1007/s11738-016-2277-5.

Sagardoy, R., Vázquez, S., Florez-Sarasa, I.D. et al. (2010). Stomatal and mesophyll conductances to CO_2 are the main limitations to photosynthesis in sugar beet (*Beta vulgaris*) plants grown with excess zinc. *New Phytol.* 187: 145–158. https://doi.org/10.1111/j.1469-8137.2010.03241.x.

Shaibur, M.R. and Kawai, S. (2011). Arsenic toxicity in Akitakomachi rice in presence of Fe^{3+}–citrate. *Adv. Environ. Biol.* 5: 1411–1422. https://doi.org/10.3329/bjar.v36i4.11742.

Sharma, I. (2012). Arsenic induced oxidative stress in plants. *Biologia* 67: 447–453. https://doi.org/10.2478/s11756-012-0024-y.

Siddiqui, F., Tandon, P.K., and Srivastava, S. (2015). Analysis of arsenic induced physiological and biochemical responses in a medicinal plant, Withania somnifera. *Physiol. Mol. Biol. Plants* 21: 61–69. https://doi.org/10.1007/s12298-014-0278-7.

Sidhu, G.K., Singh, S., Kumar, V. et al. (2019). Toxicity, monitoring and biodegradation of organophosphate pesticides: a review. *Crit. Rev. Environ. Sci. Technol.* 49: 1135–1187. https://doi.org/10.1080/10643389.2019.1565554.

Singh, S., Kumar, V., Chauhan, A. et al. (2017a). Toxicity, degradation and analysis of the herbicide atrazine. *Environ. Chem. Lett.* 161 (16): 211–237. doi: 10.1007/S10311-017-0665-8.

Singh, S., Kumar, V., Datta, S. et al. (2020a). Glyphosate uptake, translocation, resistance emergence in crops, analytical monitoring, toxicity and degradation: a review. *Environ. Chem. Lett.* 18: 663–702. https://doi.org/10.1007/s10311-020-00969-z.

Singh, S., Kumar, V., Sidhu, G.K. et al. (2019a). Plant growth promoting rhizobacteria from heavy metal contaminated soil promote growth attributes of *Pisum sativum* L. *Biocatal. Agric. Biotechnol.* 17: 665–671. https://doi.org/10.1016/j.bcab.2019.01.035.

Singh, S., Kumar, V., Singh, D. et al. (2020b). A sustainable paradigm of sewage sludge biochar: valorization, opportunities, challenges and future prospects. *J. Clean. Prod.* 269: 122259. https://doi.org/10.1016/j.jclepro.2020.122259.

Singh, S., Kumar, V., Singla, S. et al. (2020c). Kinetic study of the biodegradation of acephate by indigenous soil bacterial isolates in the presence of humic acid and metal ions. *Biomol. Ther.* 10: 433. https://doi.org/10.3390/biom10030433.

Singh, S., Kumar, V., Upadhyay, N. et al. (2017b). Efficient biodegradation of acephate by *Pseudomonas pseudoalcaligenes* PS-5 in the presence and absence of heavy metal ions [Cu(II) and Fe(III)], and humic acid. *3 Biotech.* 7: 1–10. https://doi.org/10.1007/s13205-017-0900-9.

Singh, S., Singh, N., Kumar, V. et al. (2016). Toxicity, monitoring and biodegradation of the fungicide carbendazim. *Environ. Chem. Lett.* 14: 317–329. https://doi.org/10.1007/s10311-016-0566-2.

Singh, S., Kumar, V., Kapoor, D. et al. (2019b). Revealing on hydrogen sulfide and nitric oxide signals co-ordination for plant growth under stress conditions. *Physiol. Plant.* 168: 301–317. https://doi.org/10.1111/ppl.13002.

Slaton, M.R., Raymond Hunt, E., and Smith, W.K. (2001). Estimating near-infrared leaf reflectance from leaf structural characteristics. *Am. J. Bot.* 88: 278–284. https://doi.org/10.2307/2657019.

Srivastava, S., Srivastava, A.K., Singh, B. et al. (2013a). The effect of arsenic on pigment composition and photosynthesis in *Hydrilla verticillata*. *Biol. Plant.* 57: 385–389. https://doi.org/10.1007/s10535-012-0288-7.

Srivastava, S., Srivastava, A.K., Suprasanna, P., and D'Souza, S.F. (2013b). Quantitative real-time expression profiling of aquaporins-isoforms and growth response of *Brassica juncea* under arsenite stress. *Mol. Biol. Rep.* 40: 2879–2886. https://doi.org/10.1007/s11033-012-2303-7.

Stoeva, N., Berova, M., and Zlatev, Z. (2003a). Physiological response of maize to arsenic contamination. *Biol. Plant.* 47: 449–452. https://doi.org/10.1023/B:BIOP.0000023893.12939.48.

Stoeva, N., Bineva, T., and Pushkarov, N. (2003b). Oxidative changes and photosynthesis in oat plants grown in As-contaminated soil. *Bulg. J. Plant Physiol.* 29: 87–95.

Talukdar, D. (2013a). Arsenic-induced changes in growth and antioxidant metabolism of fenugreek. *Russ. J. Plant Physiol.* 60: 652–660. https://doi.org/10.1134/S1021443713050130.

Talukdar, D. (2013b). Arsenic-induced oxidative stress in the common bean legume, *Phaseolus vulgaris* L. seedlings and its amelioration by exogenous nitric oxide. *Physiol. Mol. Biol. Plants* 19: 69–79. https://doi.org/10.1007/s12298-012-0140-8.

Talukdar, D. (2014). Arsenic-induced oxidative stress and its reversal by thiourea in mung bean (*Vigna radiata* (L.) Wilczek.) genotype. *Eur. J. Exp. Biol.* 3: 13–18.

Tripathi, R.D., Srivastava, S., Mishra, S. et al. (2007). Arsenic hazards: strategies for tolerance and remediation by plants. *Trends Biotechnol.* 25: 158–165. https://doi.org/10.1016/j.tibtech.2007.02.003.

Türkdoğan, M.K., Kilicel, F., Kara, K. et al. (2003). Heavy metals in soil, vegetables and fruits in the endemic upper gastrointestinal cancer region of Turkey. *Environ. Toxicol. Pharmacol.* 13: 175–179. https://doi.org/10.1016/S1382-6689(02)00156-4.

Upadhyaya, H., Shome, S., Roy, D., and Bhattacharya, M.K. (2014). Arsenic induced changes in growth and physiological responses in *Vigna radiata* seedling: effect of curcumin interaction. *Am. J. Plant Sci.* 05: 3609–3618. https://doi.org/10.4236/ajps.2014.524377.

Wani, A.B., Chadar, H., Wani, A.H. et al. (2017). Salicylic acid to decrease plant stress. *Environ. Chem. Lett.* 15: 101–123. https://doi.org/10.1007/s10311-016-0584-0.

Zhao, F.J., Ago, Y., Mitani, N. et al. (2010). The role of the rice aquaporin Lsi1 in arsenite efflux from roots. *New Phytol.* 186: 392–399. https://doi.org/10.1111/j.1469-8137.2010.03192.x.

Zhao, F.J., Ma, J.F., Meharg, A.A., and McGrath, S.P. (2009). Arsenic uptake and metabolism in plants. *New Phytol.* 181: 777–794. https://doi.org/10.1111/j.1469-8137.2008.02716.x.

6

Arsenic-Induced Responses in Plants

Impacts on Biochemical Processes

Sanjay Kumar[1], *Varsha Rani*[2], *Simranjeet Singh*[3], *Dhriti Kapoor*[4], *Daljeet Singh Dhanjal*[5], *Ankita Thakur*[4], *Mamta Pujari*[4], *Praveen C. Ramamurthy*[3], *and Joginder Singh*[5]

[1] *Institute of Multidisciplinary Research for Advanced Materials (IMRAM), Tohoku University, Sendai, Japan*
[2] *Department of Pharmaceutical Engineering and Technology, Indian Institute of Technology, Banaras Hindu University, Varanasi, Uttar Pradesh, India*
[3] *Interdisciplinary Centre for Water Research (ICWaR), Indian Institute of Science, Bengaluru, Karnataka, India*
[4] *Department of Botany, School of Bioengineering and Biosciences, Lovely Professional University, Phagwara, Punjab, India*
[5] *Department of Biotechnology, School of Bioengineering and Biosciences, Lovely Professional University, Phagwara, Punjab, India*

6.1 Introduction

Arsenic (As) is an extremely hazardous toxic heavy metal for both animals and plants, which is ubiquitously distributed. It is accumulated in groundwater and a number of edible as well as medicinal plants. The crust of the earth is the primary natural reserve of arsenic. It is one of the abundant elements found in the crust of the earth. The average availability of arsenic ranges from 1 mg to 100 g kg^{-1} (Mirza et al. 2014). The overaccumulation of arsenic into plants is because of the mineralization, mining activities, irrigation with arsenic-contaminated groundwater, utilization of arsenic-containing wood preservatives, herbicides, insecticides, and fertilization with solid municipal wastes (Mirza et al. 2014; Kumar et al. 2015b; Singh et al. 2016, 2017a; Wani et al. 2017; Sidhu et al. 2019). There are four oxidative states of arsenic, such as As III, As V, As-III, and As 0. Out of these four states, arsenate and arsenite are highly toxic compared to organic ones due to their chemical analogy with essential ions such as borate and phosphate (Zhao et al. 2009). The state of both arsenic (III) and (V) triggers an overproduction of reactive oxygen species (ROS) like hydroxyl radical ($OH^{\cdot}$), oxide radical ($O_2^{\cdot}-$), and hydrogen peroxide (H_2O_2) (Nath et al. 2014; Kapoor et al. 2019). Excessive production of ROS causes oxidative damage of various biochemicals like nucleic acids, proteins, and lipids (Parkhey et al. 2012, 2014). They also stimulate various enzymes like CAT, ascorbate peroxidase (APX), glutathione reductase (GR), superoxide dismutase (SOD), guaiacol peroxidase (POX), and nonenzymatic components like glutathione (GSH), ascorbate, carotenoid, and α-tocopherol inside the plants that

Arsenic in Plants: Uptake, Consequences and Remediation Techniques, First Edition.
Edited by Prabhat Kumar Srivastava, Rachana Singh, Parul Parihar, and Sheo Mohan Prasad.

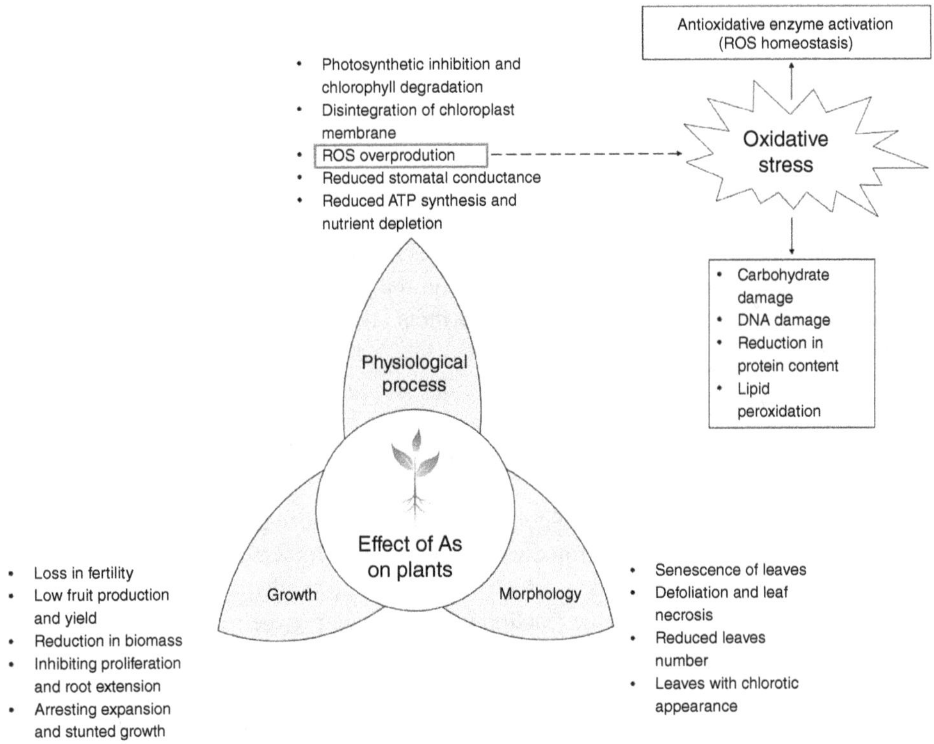

Figure 6.1 Effect of arsenic (As) on plants.

act as antioxidants (Singh et al. (2007) Mishra et al. 2016; Singh et al. 2017b; Kaur et al. 2018; Singh et al. 2020b, 2020c, 2020d, 2020e, 2020f). observed arsenic toxicity induces by SOD, guaiacol POX, and GR, in mung bean (*Phaseolus aureus* Roxb). They observed 50.0 μM arsenic elevates the response of SOD as well as GR by 60 and 90%, respectively. They also observed enhanced APX activity with 83% at 10.0 μM arsenic concentration, while at 50.0 μM, a significant decline is observed in mung bean (Singh et al. 2006). Stoeva et al. observed the increase in peroxidase activity as well as lipid peroxidation after exposure of 5 mg dm^{-3} arsenic in *P. vulgaris* L (Stoeva et al. 2005). The effect of arsenic on the plant is illustrated in Figure 6.1.

As toxicity affects the biochemical process of the plant by affecting several pathways inside the plant cell. The present paper combined the main pathways (either enzymatic or nonenzymatic) affected by arsenic toxicity.

6.2 Arsenic Effect on Biochemical Process in Plants

Accumulation of arsenic beyond the limit in soil or water interferes with plant metabolism, which causes curling, wilting, and necrosis of leaf blades, and reduction of leaves number and leaf area, which eventually reduces photosynthesis activity and biomass accumulation, reduces mineral content level, curbs elongation of the root, nodulation, and proliferation, and results in poor yield (Kumar et al. 2014b; Kumar et al. 2015a, 2015b; Mishra et al. 2016; De Mandal et al. 2021; Singh et al. 2020a, 2021). Singh et al. reported arsenic toxicity causes a reduction in root elongation in mung bean by inducing oxidative stress, which significantly elevates the electrolyte leakage, lipid peroxidation, and oxidizability (Singh et al. 2007). Stoeva et al. observed a reduction in water potential, protein content, biomass accumulation, and leaf gas exchange in *Phaseolus vulgaris* L. after a higher dose of arsenic ($5\,mg\,dm^{-3}$) (Stoeva et al. 2005). Gusman et al. reported a decrease in the photosynthesis rate in arsenic-induced plants because of variations in chlorophyll a (fluorescence parameters) and gas exchange and the content of the chloroplastic pigment. They also observed a decrease in the relative growth rate (RGR) of roots and leaves in arsenic-induced plants (Gusman et al. 2013).

6.3 Oxidative Stress on the Arsenic-Induced Plant

Accumulation of arsenic leads to ROS generation in plants due to inhibition of key enzymes as well as the release of a free electron during conversion of As V to As III (Figure 6.2). This reduction processes a redox reaction, which also occurs, i.e. methylation of inorganic arsenic, followed by methylated arsenic reaction with molecular oxygen, and evoking the ROS generation. Biomethylation of arsenic gives dimethylarsinic acid (DMA), trimethylarsonium oxide (TMAO), monomethylarsonic acid (MMA), and tetramethylarsonium ion (TETRA). Furthermore, arseno-sugars, arsenocholine, and arsenobetaine are limited forms of metabolized arsenic that have been reported in different plant species (Koch et al. 2000).

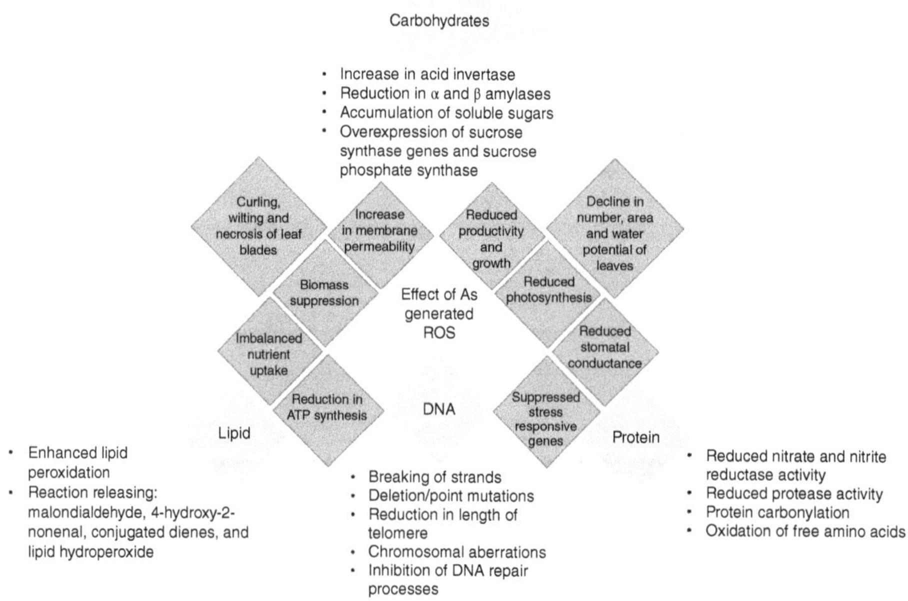

Figure 6.2 Effect of arsenic-generated reactive oxygen species.

DMA has been reported to cause oxidative stress (iron-dependent), which occurs due to the release of iron from ferritin, whereas ROS directly generated by DMA^{3+} causes DNA damage (Shi et al. 2004).

Most of the ROS are free radicals such as $O_2^{\cdot}-$, H_2O_2, $O\dot{H}$, NO, dimethyl arsenic radical $[(CH_3)_2A\dot{s}]$, and dimethylarsinic peroxyl radical $[(CH_3)_2AsO\dot{O}]$. ROS are mainly generated in the chloroplast on the site of the electron transport chain. ROS are also generated in the cell wall, plasma membranes, peroxisomes, and endoplasmic reticulum. These free radicals are highly unstable and highly reactive. Therefore, it reacts with non-radical molecules and generates new free radicals, resulting in high ROS production within plant cells, which eventually causes oxidative stress.

As-induced oxidative stress-inducing glycolate oxidase, lowering the concentration of antioxidant compounds (ascorbate, carotenoids, and GSH), chlorophyll, total protein content, and carotenoids, decreases the activity of SOD, CAT, and APX, (Gupta et al. 2013; Singh et al. 2006). Whereas it increases the levels of total sugar, Pro, and oxidative stress markers like electrolyte leakage, protein-lipid-oxidized products, carbonyl hydroperoxides, ROS, proteasome, and protease. This stress results into release of accruals of malondialdehyde (MDA) and 4-hydroxy-2-nonenal (HNE) along with certain lipid peroxidation biomarkers (Chandrakar et al. 2018). Talukdar et al. (Talukdar 2013) observed a decrease in the dry weights and lengths of roots and shoots, photosynthetic traits as well as a decrease in the grain yield in arsenic-induced fenugreek (*Trigonella foenumgraecum* L. cv. Azad) plants due to a decrease in the response of antioxidant enzymes like APX, SOD, and CAT.

As-induced oxidative stress can be modulated by the application of other heavy metals, biomolecules, and antioxidant agents in plants. Kumar et al. (2016) observed a decrease in the accumulation of arsenic in shoot and root by twofold and threefold, respectively, in the presence of Se in rice (*Oryza sativa* L.). They observed increased activities of CAT, glutathione peroxidase (GPx), and APX and induced levels of GSH, phytochelatins (PCs), and nonprotein thiols (NPTs) after exposure of Se in arsenic-treated plants. Selenium supplementation also modulated the thiol metabolism enzymes, namely glutathione-S-transferase (GST), phytochelatin synthase (PCS), and γ-glutamylcysteine synthetase (γ-ECS,) in arsenic-treated plants (Kumar et al. 2014a). NO donor and sodium nitroprusside (SNP) can reduce the arsenic-induced oxidative effect in plants. Hasanuzzaman et al. observed that SNP supplementation increases the relative water content (RWC), Pro and chlorophyll contents, ascorbate (AsA) and reduced GSH contents, and GSH/glutathione disulfide (GSSG) ratio in arsenic-treated wheat (Hasanuzzaman and Fujita 2013). Jin et al. also observed the improvement in MDA content and arsenic-induced electrolyte leakage after application of 100 μM SNP in arsenic-induced tall fescue leaves. Reduction in H_2O_2 and $O_2^{\cdot-}$ and enhancement in activity of antioxidant enzymes like CAT, APX, and SOD was also observed. Proline (Pro), diphenyleneiodonium (DPI), and 24-epibrassinolide (EBL) play a protective role against arsenic-induced oxidative stress in the plant (Figure 6.3) (Jin et al. 2010). Chandrakar et al. observed EBL, DPI, and Pro enhance accruals of total protein, viability, endogenous Pro, total sugar, and antioxidant activity in arsenic-induced *Glycine max* L (Chandrakar et al. 2018). Salicylic acid (SA) also plays a protective role against arsenic-induced oxidative stress in the plant. Saidi et al. observed SA improves GPx, APX, and CAT activity, alongside decreased POX and SOD activities in arsenic-induced *Helianthus annuus* (Saidi et al. 2017).

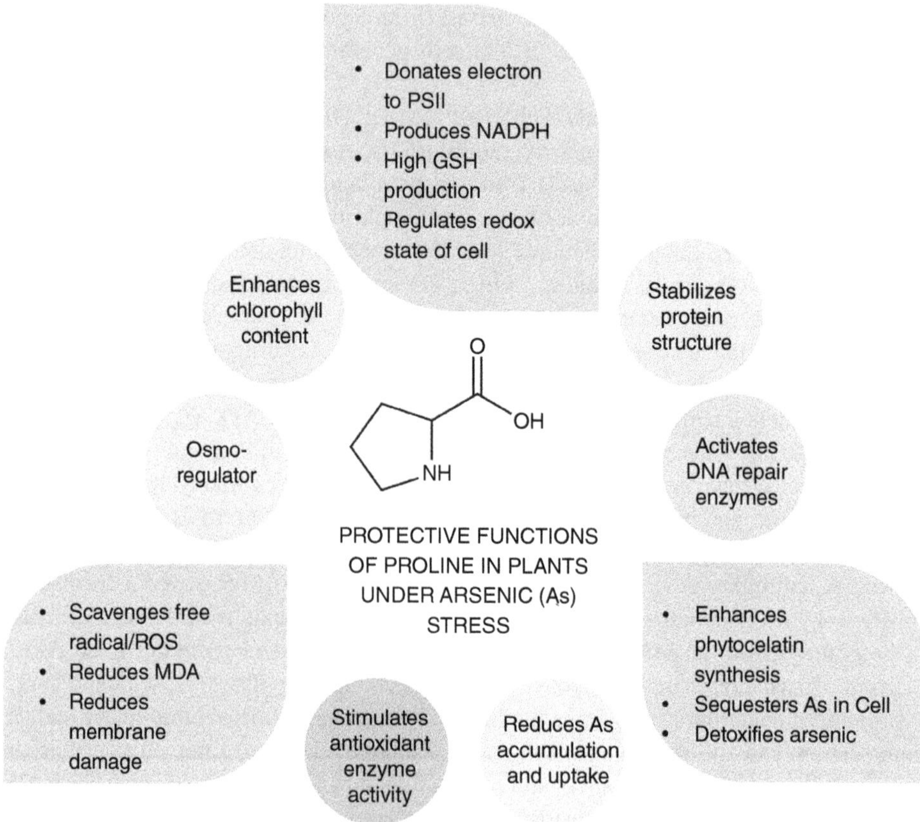

Figure 6.3 Protective functioning of Proline (Pro) in plants under arsenic stress.

6.4 Carbohydrate Metabolism in the Arsenic-Induced Plant

As is one of the most poisonous and lethal metalloids found on the earth's surface. This environmental toxin is harmful to all living organisms, including plants, animals, and humans too. Arsenic can cause various morphological, biochemical, and physiological alterations in plants, even at minute concentrations. The lethal effects of arsenic play a major role in plant metabolism. arsenic affecting the food chain is a serious matter of global concern. Apart from water contamination, rice is identified as the most affected crop by the arsenic toxicity. Arsenic affects the root part of the plant initially and gradually interferes with the metabolic process of the plant, which can lead to death (Garg and Singla 2011). During the metabolism of inorganic arsenic, arsenic (V) is reduced to arsenic (III) by losing two electrons and then transforming to pentavalent organic arsenic by oxidative methylation (Hughes 2002).

Various plant physiological activities are impacted by arsenic toxicity. Plant cellular membrane is exposed to arsenic, which ultimately causes electrolyte leakage (Singh et al. 2006). Arsenic stress is also known to adversely affect primary carbohydrate metabolisms such as sugars and starch levels. A significant reduction in the proportion of reducing to nonreducing sugar was observed within the stems of *O. sativa* under arsenic stress (Jha and Dubey 2004), indicating that arsenic toxicity affects carbohydrate metabolism in rice seedlings that started with the aggregation of solvent sugars. Sucrose synthase is recognized to perform as a beneficial moiety in the organization of the sucrose under arsenic toxicity. The change of nonreducing sugar, chiefly sucrose, to reducing sugar is a distinguished feature of arsenic stress. It may further destruct the sucrose assembly by affecting the activity of hexose monophosphate. Further decline in the proportion of sucrose synthesis was also noticed, which is applicable for the convenient hexose monophosphate units. The activity of sucrose phosphate synthase, alpha, and beta-amylase was diminished, while enzymes such as starch phosphorylase, sucrose synthase, and acid invertase were improved. Acid invertase is the chief determinant, which is required for the production of fructose and glucose from sucrose. With this regard, an immediate association that links its activity and hexose level was measured (Kaur et al. 2012). Hexoses comprised of sucrose synthase and acid invertase that are linked to the glycolytic pathway later on. However, this metabolic activity was impaired under the elevated levels of arsenic.

Kofroňová et al. (2020) recently observed the carbohydrate metabolism and antioxidative defense of plants under arsenic stress. They found that application of arsenic in the tolerant genotype of *Nicotiana tabacum* cv Wisconsin showed the same amount of carbohydrates in soluble as well as form of starch. At the same time, there was a decline in sugar content and no change in the starch level in the case of the sensitive genotype of *Nicotiana sylvestris* (Kofroňová et al. 2020). Carbohydrates play a significant role in protecting plants from stress conditions by supplying carbon and energy source and also help in declining the level of ROS. Sucrose, along with glucose recognized to control gene expression to provide stress resistance. Gupta and Seth (2019) obtained similar results in relative arsenic-sensitive tomatoes while the level of carbohydrate was not affected in the tolerant genotype (Gupta and Seth 2019). On the other hand, the tolerant tobacco was observed to grow well even under the arsenic-treatment, which was proposed to be a result of effective and complex simulation of a broad range of defense mechanisms. In plant species like citruses or rice, a link between carbohydrate accumulation and heavy metal resistance was reported (Jiang et al. 2015).

As-induced changes in growth and metabolism in *Vigna mungo* seedlings to different levels of arsenic (V) were investigated (Srivastava and Sharma 2014). Parameters studied for the carbohydrate metabolism in seedlings were total sugars (reducing and nonreducing), phenol, and starch content, along with amylase enzyme activity. Accumulation of starch and sugar was seen in plant tissue due to arsenic exposure. A significant increase in total sugar level was seen. Yet there was more increase in the level of nonreducing sugar than the reducing sugar. With an increase in the dose of arsenic, starch content also increased along with sugar. A decline in the activity of the amylase enzyme (responsible for starch hydrolysis) was seen on arsenic application. Arsenic exposure resulted in the decrease of amylase activity in rice seedlings also (Choudhury et al. 2010). Arsenic application increased the enzymes associated with carbohydrate metabolism, for instance, sucrose synthase, neutral invertase, sucrose phosphate synthase, fructokinase, hexokinase, phosphofructokinase, and pyruvate kinase except for acid invertase activity in *Solanum tuberosum* and decreased the amounts of sugars and starch (Shahid et al. 2019).

6.5 Lipid Metabolism in the Arsenic-Induced Plant

Lipid oxidation is a well-known impact in arsenic-stressed plants (Bhowmik et al. 2015). The cell membrane is degraded by arsenic stress, which leads to an increase in the electrolyte leakage and oxidative stress markers (Gill and Tuteja 2010). In addition to this, lipid peroxidation occurs in both organelle membranes and the cellular membrane due to the overaccumulation of ROS because of which the formation of lipid-derived cytotoxic radicals changes the regular activity of the cell (Sytar et al. 2013). Significantly, three essential phases, initiation, progression, and termination, are included in the cascade of lipid peroxidation reactions (Parkhey et al. 2012) (Table 6.1).

Lipid peroxidation is caused by the excessive generation of $OH^{\cdot}$ radicals in cellular membranes. During arsenic poisoning, this process can produce ROS, reacting with proteins, lipids, nucleic acids, and pigments. An increase in the level of lipid peroxidation is cytotoxic and can influence healthy cellular functioning (Li et al. 2007). $OH^{\cdot}$ is mainly responsible for the initiation of lipid peroxidation, which transfers one H-atom from polyunsaturated fatty acid (PUFA) moieties of the membrane (Gill and Tuteja 2010). The carbon-centered lipid radical attaches O_2 to the fatty acid under aerobic conditions and structures $ROO^{\cdot}$. One H-atom from the adjoining PUFA is preoccupied with the same $ROO^{\cdot}$ radical; thus, there is an improvement in the peroxidation reaction. Linoleic acid and linolenic acid are highly inclined to $^1O_2$ and OH intrusion as a by-product, and deliver a mixture of lipid hydroperoxides (Das and Roychoudhury, 2014). There is a decrease in membrane fluidity due to an increase in the peroxidation of PUFA, which causes significant damage to protein synthesis. MDA, 4-hydroxy-2-nominal, keto fatty acids, and hydroxyl are major by-products of lipid peroxidation. MDA is the most notable by-product of lipid peroxidation that is produced after ROS exposure. A conjugate can be formed by these products with both protein and DNA. In the mitochondria of *Zea mays*, aldehydes are produced and are displayed to inflict cytoplasmic male sterility. Lately, researchers have revealed that arsenic-mediated lipid peroxidation reactions were due to the notable escalation in MDA level (Gusman et al. 2013). MDA-level production is strongly associated with membrane fluidity due to the peroxidation of PUFAs, which ultimately elevates the level of ROS to cause oxidative stress.

Lipid catabolism was induced with arsenic in *G. max*, which was mitigated by the use of diphenylene iodonium, 24-EBL, and Pro. The outcomes revealed that arsenic caused a decline in the plant morphological characteristics like dry mass, radical length, germination percentage, etc. (Chandrakar et al. 2017). Arsenic significantly elevated the level of oxidative stress markers and sharply lessened the total lipid content. Interestingly, as protective agents, 24-EBL, diphenylene iodonium, and Pro performed their roles and reduced arsenic accumulation and enhanced plant growth. The level of oxidative stress markers was limited by the application of these moieties and reduced the functioning of lipoxygenase and lipase enzymes to boost the total lipid content.

Effect of arsenic on the growth and metabolism in *O. sativa* seedlings was observed (Choudhury et al. 2010). A gradual increase in MDA was seen in rice seedlings exposed to arsenic, which indicated an enhanced rate of lipid peroxidation. Similar kinds of results were seen in H_2O_2 content on exposure to arsenic treatment. Generally, at a low concentration of arsenic, H_2O_2 helps in providing stress tolerance, but at high concentration, it results in cellular damage, which ultimately causes cell death. Arsenic treatment leads to membrane

Table 6.1 Effect of arsenic on lipid metabolism in plants.

S. No.	Plant species	Plant part	As concentration	Results	Reference
1.	*Phaseolus aureus*	Roots	10 and 50 μM	Increased electrolyte and leakage lipid peroxidation was observed due to elevation in the MDA under arsenic stress.	Singh et al. (2007)
2.	*Oryza sativa*	Seedlings	50 μM	Arsenic induced the overaccumulation of MDA.	Smith et al. (2010)
3.	*Brassica juncea*	Roots and leaves	0.0, 0.1, 0.2, and 0.3 mM	Arsenic treatment adversely affected the antioxidative enzymes, lipid peroxidation, and protein content.	Kanwar et al. (2015)
4.	*Pisum sativum*	Leaves and pods	50 μM	Arsenic-induced, increasing levels of ROS caused damage to the membranes, lipids, and proteins.	Singh et al. (2015)
5.	*Lettuce sativa*	Roots and shoots	50 μM	Arsenic treatment increased the oxidative damage in *Lactuca sativa* by elevating the concentrations of H_2O_2 and MDA.	Silveira et al. (2015)
6.	*Glycine max*	Seedlings	25 μM	Arsenic treatment induced the phosphatidic acid (PA) signaling by phospholipase D (PLD) enzyme and also by inducing the activity of diacylglycerol kinase (DGK)/ phospholipase C (PLC).	Armendariz et al. (2016)
7.	*Vigna mungo*	Leaves	100 and 200 μM	Arsenic application adversely affected the plant growth by elevating the level of lipid peroxidation.	Srivastava et al. (2017)
8.	*Vigna unguiculata*	Seedlings	25, 50, 75, and 100 μM	MDA amount was observed to be reduced in seedlings of cowpea with increasing the concentration of arsenic.	Khan (2020)
9.	*Oryza sativa*	Seedlings	50 μM	Arsenic application altered the membrane permeability in rice seedlings by increased production of ROS and elevated level of lipid peroxidation via overaccumulation in thiobarbituric acid-reactive substances.	Gautam et al. (2020)
10.	*Gycine max*	Leaves	As(III)	Level of H_2O_2, MDA, and electrolyte leakage was observed to be elevated in arsenic-treated plants.	Ahmad et al. (2020)

damage due to lipid peroxidation by changing plant antioxidant machinery, resulting in stunted growth in plants. Due to oxidative stress, cell damage in plants is examined as the most antagonistic effect of arsenic or any other heavy metal.

6.6 Protein Metabolism in the Arsenic-Induced Plant

As is a naturally occurring toxic metalloid that adversely impacts plants by inhibiting plant growth, metabolic alterations, oxidative stress, and, ultimately, can result in cell death. Arsenic stress induces the expression of several stress-associated proteins such as ubiquitin, heat shock proteins (HSPs), and metallothioneins (MTs) by activating a network of complex enzymes. Arsenic (III) and arsenic (V) are two inorganic forms of arsenic, which are abundantly present in the environment. Both arsenic (V) and arsenic (III) can disturb plant metabolism through distinct mechanisms. At the same time, arsenic (III) is a hundred times more toxic than arsenic (V) in plants (Mishra and Dubey 2006). The reason for the high noxiousness of arsenic (III) is its association in proteins with sulfhydryl groups, which results in cell death because of membrane deterioration (Meharg and Hartley-Whitaker 2002). In experiments performed on different plant species, including slender brake fern and ladder brake fern (Singh et al. 2006), red clover (Mascher et al. 2002), rice (Singh et al. 2009), maize, and mung bean (Stoeva et al. 2005), there was a decrease in the quantity of total plant proteins due to exposure of arsenic to these plants.

The arsenic-induced decline in the functioning of both nitrite and nitrate reductase enzymes has resulted in lessened protein content in rice (*O. sativa*) plants. Proteins are hydrolyzed into short peptides and free amino acids by the activity of peptidases and proteases. A diminished level of proteases in arsenic-exposed mung bean cotyledons and red clover seedlings was defended (Ismail 2012; Mascher et al. 2002). Similarly, in some arsenic-stressed plants, like wheat, ladder brake fern, slender brake fern, rice, and mung bean seeds, deterioration in proteases was documented. The proteins are quite susceptive to ROS-induced oxidative stress, which, ultimately, causes enhancement in their hydrolysis, like lipid moieties (Parkhey et al. 2014). The modifications induced in proteins because of ROS could also be inaugurated through the auto-oxidation of lipids and carbohydrates and the free flow of electrons throughout metal ion-dependent reactions. Amino acids like Pro, Arg, His, Lys, and Trp formed proteins that become oxidized by forming a free carboxyl group, which dramatically can hinder or change the activities of protein (Hawkins and Davies 2019). The protein consecutively becomes much sensitive to proteolytic damage (Møller et al. 2007). The ratio of protein carbonylation may become apparent because of the immediate loss of electrons from the side chains of the amino acid (Dar et al. 2017). Proteins with thiol groups and sulfur-containing amino acids are comparatively much sensitive to the ROS attack (Oracz et al. 2007). To this date, no inquiries or research has proved whether arsenic-induced ROS can lead to the initiation of oxidation and carbonylation in a cell's protein, especially in plants. Many researchers have concentrated on metals like zinc, cobalt, and copper (Karuppanapandian et al. 2011; Mustafa and Komatsu 2016; Ramakrishna and Seeta Ram Rao 2013) and under salinity stress in plants (Naliwajski and Skłodowska 2014). Loss of amino acids is consequently caused by the interruption in nutrient metabolism caused by arsenic, as unveiled by amino acid profiling (Kumar et al. 2014a).

In black gram seedlings, arsenic-induced changes in growth and metabolism were observed (Srivastava and Sharma 2014), which showed that on the arsenic application, there was a decrease in protein and nucleic acid content. In a plant cell, arsenic (V) is effortlessly converted to arsenic (III). This results in alteration of protein function by binding with a sulfhydryl group. ROS generated by arsenic has the capability to damage both proteins and nucleic acids, causing various lesions in DNA (Panda et al. 2010).

By delivering reactive carbonyl species, proteins are also altered or destroyed covalently by arsenic-induced ROS (Parkhey et al. 2014). The amino acids, specifically Lys, Arg, Pro, Trp, His, and Thr, get oxidized to aminoadipic semialdehyde, γ-glutamyl semialdehyde, and aminoketobutyrate by ROS, which inhibits the function of proteins (Gill and Tuteja 2010). Both Met and Cys are quite sensitive to $^1O_2$ and $OH^{\cdot}$ radicals (Parkhey et al. 2014). Although, in animals, arsenic and carbonyl formation was reviewed, and a close connection was seen (Flora 2011). Consequently, the monitoring of protein carbonyl was justified to use as an arsenic-induced oxidative stress marker.

During the initial degrees of seedling development, arsenic was seen to reduce and/or inhibit the root/shoot growth, seed germination rate, and several other developmental processes (Li et al. 2007). Various mechanisms have been developed in plants over a certain period of time to manage the toxic consequences of arsenic. Such mechanism includes synthesis of arsenic-binding proteins, compartmentation of metal and accumulation of osmoprotectants like glycinebetaine, Pro, and mannitol. It is obvious that arsenic exposure significantly influences various plant processes at a molecular and biochemical level, including a reduction in photosynthetic efficiency, hindrance of overall plant growth, biomass reduction, imbalance of mineral nutrients, etc. As a consequence, arsenic-induced oxidative stress damages biomolecules like lipids, proteins, carbohydrates, etc., and alters their metabolic pathways, either indirectly or directly by altering the functioning of certain key enzymes (Abbas et al. 2018).

14-3-3 protein group is considered a beneficial moiety for regulating plant metabolism and enhancing stress tolerance in plants. Arsenic treatment caused a significant alteration in 14-3-3 protein genes functioning in both non-colonized and mycorrhiza-colonized rice genotypes (Pathare et al. 2016). Arsenic treatment in *G. max* decreased the total protein content and elevated the level of protein carbonyls, hydroperoxides, protease, and proteasome enzyme activities (Chandrakar et al. 2018). Similar results were reported by Mallick et al. (2013) in *Z. mays* where reduced protein amount with an enhanced level of protein carbonyls was found (Mallick et al. 2013). A decrease of 44, 42, 40, and 38% in the level of essential amino acids, i.e. Phe, Leu, Val, and Lys, respectively, was observed in arsenic III-treated rice plants (Kumar et al. 2016). This arsenic-induced decrease in essential amino acids was due to the influence on amino acid formation, nitrogen metabolism, and degradation routes (Finnegan and Chen 2012).

6.7 Conclusion

Even though substantial steps have been taken to unveil the mechanism of arsenic-induced injury at different plant development stages, continuous efforts are still being made to unravel it. It cannot be neglected that exposure to arsenic adversely affects the plant by inhibiting its growth responses and processes, biomass accumulation, and photosynthetic efficacy. It has been comprehended that exposure to arsenic induces oxidative stress due to ineffective elimination and overproduction of ROS, which subsequently damages nucleic acids, proteins, and lipids. Additionally, the oxidized by-product formed interferes with a different metabolic process by either directly acting as a competitive phosphate inhibitor or indirectly obstructing the normal functioning of main enzymes. Therefore, the toxicity of arsenic is stated to negatively affect the productivity as well as the quality of plants and impact the well-being and health of animals and humans consuming it. With passing time, plants have started evolving and developing strategies to curb the toxic effect induced by arsenic via mechanisms like accumulation of compatible solutes (mannitol, glycine betaine, sugar, and Pro), biosynthesis of polyphenols, compartmentalization, and metal-binding proteins. Moreover, exogenous application of NO, phosphate, Pro, and potassium has been reported to reduce arsenic toxicity substantially. Besides the availability of extensive literature on the transportation and uptake of arsenic in plants, the defined mechanism involved in amending the different biochemical reactions within the antioxidant system of the plant against arsenic toxicity is still under investigation. Considering this, it has become essential to understand how the minute concentration of arsenic elicits different chemical reactions in a plant and demands instant attention. Still, mechanisms or a mixture of mechanisms, such as DNA oxidation, lipid oxidation, protein oxidation, and ROS generation, which are operative in arsenic-induced toxicity, are not well-comprehended. Thus, it has become important to understand the extent of arsenic-induced injury and mitigation strategies. The existing (NO, phosphate, Pro, polyphenol, and potassium) and new compounds that are involved in mitigation during arsenic-induced injury need to be explored and extreme changes occurring at biochemical, molecular, and physiological levels need to be recognized. Other than this, the involvement of important stress markers (protein carbonyl, 8-OHdG, and 4-hydroxy-2-nonenal) that are involved in arsenic toxicity in different crop plants need to be assessed. Thus, continuous effort and research are required in this direction to uncover the precise molecular mechanism involved in ameliorating arsenic-induced toxicity responses in different plants.

References

Abbas, G., Murtaza, B., Bibi, I. et al. (2018). Arsenic uptake, toxicity, detoxification, and speciation in plants: physiological, biochemical, and molecular aspects. *Int. J. Environ. Res. Public Health* 15: 59. https://doi.org/10.3390/ijerph15010059.

Ahmad, P., Alyemeni, M.N., Al-Huqail, A.A. et al. (2020). Zinc oxide nanoparticles application alleviates arsenic (As) toxicity in soybean plants by restricting the uptake of as and modulating key biochemical attributes, antioxidant enzymes, ascorbate-glutathione cycle and glyoxalase system. *Plants* 9: 1–18. https://doi.org/10.3390/plants9070825.

Armendariz, A.L., Talano, M.A., Villasuso, A.L. et al. (2016). Arsenic stress induces changes in lipid signalling and evokes the stomata closure in soybean. *Plant Physiol. Biochem.* 103: 45–52. https://doi.org/10.1016/j.plaphy.2016.02.041.

Bhowmik, A.K., Alamdar, A., Katsoyiannis, I. et al. (2015). Mapping human health risks from exposure to trace metal contamination of drinking water sources in Pakistan. *Sci. Total Environ.* 538: 306–316. https://doi.org/10.1016/j.scitotenv.2015.08.069.

Chandrakar, V., Dubey, A., and Keshavkant, S. (2018). Modulation of arsenic-induced oxidative stress and protein metabolism by diphenyleneiodonium, 24-epibrassinolide and proline in *Glycine max* L. *Acta Bot. Croat.* 77: 51–61. https://doi.org/10.2478/botcro-2018-0004.

Chandrakar, V., Parkhey, S., Dubey, A., and Keshavkant, S. (2017). Modulation in arsenic-induced lipid catabolism in *Glycine max* using proline, 24-epibrassinolide and diphenylene iodonium. *Biologia* 72: 292–299. https://doi.org/10.1515/biolog-2017-0033.

Choudhury, B., Mitra, S., and Biswas, A.K. (2010). Regulation of sugar metabolism in rice (*Oryza sativa* L.) seedlings under arsenate toxicity and its improvement by phosphate. *Physiol. Mol. Biol. Plants* 16: 59–68. https://doi.org/10.1007/s12298-010-0008-8.

Dar, M.I., Naikoo, M.I., Khan, F.A. et al. (2017). An introduction to reactive oxygen species metabolism under changing climate in plants. In: *Reactive Oxygen Species and Antioxidant Systems in Plants: Role and Regulation under Abiotic Stress* (ed. M.I.R. Khan and N.A. Khan), 25–52. Singapore: Springer https://doi.org/10.1007/978-981-10-5254-5_2.

Das, K. and Roychoudhury, A. (2014). Reactive oxygen species (ROS) and response of antioxidants as ROS-scavengers during environmental stress in plants. *Front. Environ. Sci.* 2: 53. https://doi.org/10.3389/fenvs.2014.00053.

De Mandal, S., Sonali, Singh, S. et al. (2021). Plant–microbe association for mutual benefits for plant growth and soil health. In: *Current Trends in Microbial Biotechnology for Sustainable Agriculture* (ed. A.N. Yadav, J. Singh, C. Singh and N. Yadav), 95–121. Singapore: Springer https://doi.org/10.1007/978-981-15-6949-4_5.

Finnegan, P.M. and Chen, W. (2012). Arsenic toxicity: the effects on plant metabolism. *Front. Physiol.* 3: 182. https://doi.org/10.3389/fphys.2012.00182.

Flora, S.J.S. (2011). Arsenic-induced oxidative stress and its reversibility. *Free Radic. Biol. Med.* 51: 257–281. https://doi.org/10.1016/j.freeradbiomed.2011.04.008.

Garg, N. and Singla, P. (2011). Arsenic toxicity in crop plants: physiological effects and tolerance mechanisms. *Environ. Chem. Lett.* 9: 303–321. https://doi.org/10.1007/s10311-011-0313-7.

Gautam, A., Pandey, A.K., and Dubey, R.S. (2020). *Azadirachta indica* and *Ocimum sanctum* leaf extracts alleviate arsenic toxicity by reducing arsenic uptake and improving antioxidant system in rice seedlings. *Physiol. Mol. Biol. Plants* 26: 63–81. https://doi.org/10.1007/s12298-019-00730-z.

Gill, S.S. and Tuteja, N. (2010). Reactive oxygen species and antioxidant machinery in abiotic stress tolerance in crop plants. *Plant Physiol. Biochem.* 48: 909–930. https://doi.org/10.1016/j.plaphy.2010.08.016.

Gupta, D.K., Inouhe, M., Rodríguez-Serrano, M. et al. (2013). Oxidative stress and arsenic toxicity: role of NADPH oxidases. *Chemosphere* 90: 1987–1996. https://doi.org/10.1016/j.chemosphere.2012.10.066.

Gupta, P. and Seth, C.S. (2019). Nitrate supplementation attenuates As(V) toxicity in *Solanum lycopersicum* L. cv Pusa Rohini: insights into As(V) sub-cellular distribution, photosynthesis, nitrogen assimilation, and DNA damage. *Plant Physiol. Biochem.* 139: 44–55. https://doi.org/10.1016/j.plaphy.2019.03.007.

Gusman, G.S., Oliveira, J.A., Farnese, F.S., and Cambraia, J. (2013). Arsenate and arsenite: the toxic effects on photosynthesis and growth of lettuce plants. *Acta Physiol. Plant.* 35: 1201–1209. https://doi.org/10.1007/s11738-012-1159-8.

Hasanuzzaman, M. and Fujita, M. (2013). Exogenous sodium nitroprusside alleviates arsenic-induced oxidative stress in wheat (*Triticum aestivum* L.) seedlings by enhancing antioxidant defense and glyoxalase system. *Ecotoxicology* 22: 584–596. https://doi.org/10.1007/s10646-013-1050-4.

Hawkins, C.L. and Davies, M.J. (2019). Detection, identification, and quantification of oxidative protein modifications. *J. Biol. Chem.* 294: 19683–19708. https://doi.org/10.1074/jbc.REV119.006217.

Hughes, M.F. (2002). Arsenic toxicity and potential mechanisms of action. *Toxicol. Lett.* 133: 1–16. https://doi.org/10.1016/S0378-4274(02)00084-X.

Ismail, G.S.M. (2012). Protective role of nitric oxide against arsenic-induced damages in germinating mung bean seeds. *Acta Physiol. Plant.* 34: 1303–1311. https://doi.org/10.1007/s11738-012-0927-9.

Jha, A.B. and Dubey, R.S. (2004). Carbohydrate metabolism in growing rice seedlings under arsenic toxicity. *J. Plant Physiol.* 161: 867–872. https://doi.org/10.1016/j.jplph.2004.01.004.

Jiang, H.X., Yang, L.T., Qi, Y.P. et al. (2015). Root iTRAQ protein profile analysis of two *Citrus* species differing in aluminum-tolerance in response to long-term aluminum-toxicity. *BMC Genomics* 16: https://doi.org/10.1186/s12864-015-2133-9.

Jin, J.W., Xu, Y.F., and Huang, Y.F. (2010). Protective effect of nitric oxide against arsenic-induced oxidative damage in tall fescue leaves. *Afr. J. Biotechnol.* 9: 1619–1627. https://doi.org/10.5897/ajb10.1442.

Kanwar, M.K., Poonam, and Bhardwaj, R. (2015). Arsenic induced modulation of antioxidative defense system and brassinosteroids in *Brassica juncea* L. *Ecotoxicol. Environ. Saf.* 115: 119–125. https://doi.org/10.1016/j.ecoenv.2015.02.016.

Kapoor, D., Singh, S., Kumar, V. et al. (2019). Antioxidant enzymes regulation in plants in reference to reactive oxygen species (ROS) and reactive nitrogen species (RNS). *Plant Gene* 19: 100182. https://doi.org/10.1016/j.plgene.2019.100182.

Karuppanapandian, T., Moon, J.C., Kim, C. et al. (2011). Reactive oxygen species in plants: their generation, signal transduction, and scavenging mechanisms. *Aust. J. Crop. Sci.* 5: 709–725.

Kaur, P., Singh, S., Kumar, V. et al. (2018). Effect of rhizobacteria on arsenic uptake by macrophyte *Eichhornia crassipes* (Mart.) Solms. *Int. J. Phytoremediation* 20: 114–120. https://doi.org/10.1080/15226514.2017.1337071.

Kaur, S., Singh, H.P., Batish, D.R. et al. (2012). Arsenic (As) inhibits radicle emergence and elongation in phaseolus aureus by altering starch-metabolizing enzymes vis-à-vis disruption of oxidative metabolism. *Biol. Trace Elem. Res.* 146: 360–368. https://doi.org/10.1007/s12011-011-9258-8.

Khan, F. (2020). Effect of arsenic on growth, lipid peroxidation and antioxidant defence system in cowpea. *Legum. Res.* 43: 200–205. https://doi.org/10.18805/LR-457.

Koch, I., Wang, L., Ollson, C.A. et al. (2000). The predominance of inorganic arsenic species in plants from Yellowknife, Northwest Territories, Canada. *Environ. Sci. Technol.* 34: 22–26. https://doi.org/10.1021/es9906756.

Kofroňová, M., Hrdinová, A., Mašková, P. et al. (2020). Multi-component antioxidative system and robust carbohydrate status, the essence of plant arsenic tolerance. *Antioxidants* 9: 283. https://doi.org/10.3390/antiox9040283.

Kumar, A., Dixit, G., Singh, A.P. et al. (2016). Selenate mitigates arsenite toxicity in rice (*Oryza sativa* L.) by reducing arsenic uptake and ameliorates amino acid content and thiol metabolism. *Ecotoxicol. Environ. Saf.* 133: 350–359. https://doi.org/10.1016/j.ecoenv.2016.06.037.

Kumar, A., Singh, R.P., Singh, P.K. et al. (2014a). Selenium ameliorates arsenic induced oxidative stress through modulation of antioxidant enzymes and thiols in rice (*Oryza sativa* L.). *Ecotoxicology* 23: 1153–1163. https://doi.org/10.1007/s10646-014-1257-z.

Kumar, V., Singh, S., Bhadrecha, P. et al. (2015a). Bioremediation of heavy metals by employing resistant microbial isolates from agricultural soil irrigated with industrial waste water. *Orient. J. Chem.* 31: 357–361. https://doi.org/10.13005/ojc/310142.

Kumar, V., Singh, S., Singh, J., and Upadhyay, N. (2015b). Potential of plant growth promoting traits by bacteria isolated from heavy metal contaminated soils. *Bull. Environ. Contam. Toxicol.* 94: 807–814. https://doi.org/10.1007/s00128-015-1523-7.

Kumar, V., Upadhyay, N., Kumar, V. et al. (2014b). Environmental exposure and health risks of the insecticide monocrotophos-a review. *J. Biodivers. Environ. Sci.* 2014: 111–120.

Li, C.X., Feng, S.L., Shao, Y. et al. (2007). Effects of arsenic on seed germination and physiological activities of wheat seedlings. *J. Environ. Sci.* 19: 725–732. https://doi.org/10.1016/S1001-0742(07)60121-1.

Mallick, S., Kumar, N., Singh, A.P. et al. (2013). Role of sulfate in detoxification of arsenate-induced toxicity in *Zea mays* L.(SRHM 445): nutrient status and antioxidants. *J. Plant Interact.* 8: 140–154.

Mascher, R., Lippmann, B., Holzinger, S., and Bergmann, H. (2002). Arsenate toxicity: effects on oxidative stress response molecules and enzymes in red clover plants. *Plant Sci.* 163: 961–969. https://doi.org/10.1016/S0168-9452(02)00245-5.

Meharg, A.A. and Hartley-Whitaker, J. (2002). Arsenic uptake and metabolism in arsenic resistant and nonresistant plant species. *New Phytol.* 154: 29–43. https://doi.org/10.1046/j.1469-8137.2002.00363.x.

Mirza, N., Mahmood, Q., Maroof Shah, M. et al. (2014). Plants as useful vectors to reduce environmental toxic arsenic content. *Sci. World J.* 2014: 921581. https://doi.org/10.1155/2014/921581.

Mishra, S. and Dubey, R.S. (2006). Inhibition of ribonuclease and protease activities in arsenic exposed rice seedlings: role of proline as enzyme protectant. *J. Plant Physiol.* 163: 927–936. https://doi.org/10.1016/j.jplph.2005.08.003.

Mishra, V., Gupta, A., Kaur, P. et al. (2016). Synergistic effects of *Arbuscular mycorrhizal* fungi and plant growth promoting rhizobacteria in bioremediation of iron contaminated soils. *Int. J. Phytoremediation* 18: 697–703. https://doi.org/10.1080/15226514.2015.1131231.

Møller, I.M., Jensen, P.E., and Hansson, A. (2007). Oxidative modifications to cellular components in plants. *Annu. Rev. Plant Biol.* 58: 459–481. https://doi.org/10.1146/annurev.arplant.58.032806.103946.

Mustafa, G. and Komatsu, S. (2016). Toxicity of heavy metals and metal-containing nanoparticles on plants. *Biochim. Biophys. Acta – Proteins Proteomics* 1864: 932–944. https://doi.org/10.1016/j.bbapap.2016.02.020.

Naliwajski, M.R. and Skłodowska, M. (2014). The oxidative stress and antioxidant systems in cucumber cells during acclimation to salinity. *Biol. Plant.* 58: 47–54. https://doi.org/10.1007/s10535-013-0378-1.

Nath, S., Panda, P., Mishra, S. et al. (2014). Arsenic stress in rice: redox consequences and regulation by iron. *Plant Physiol. Biochem.* 80: 203–210. https://doi.org/10.1016/j.plaphy.2014.04.013.

Oracz, K., Bailly, C., Gniazdowska, A. et al. (2007). Induction of oxidative stress by sunflower phytotoxins in germinating mustard seeds. *J. Chem. Ecol.* 33: 251–264. https://doi.org/10.1007/s10886-006-9222-9.

Panda, S.K., Upadhyay, R.K., and Nath, S. (2010). Arsenic stress in plants. *J. Agron. Crop Sci.* 196: 161–174.

Parkhey, S., Naithani, S.C., and Keshavkant, S. (2012). ROS production and lipid catabolism in desiccating *Shorea robusta* seeds during aging. *Plant Physiol. Biochem.* 57: 261–267. https://doi.org/10.1016/j.plaphy.2012.06.008.

Parkhey, S., Naithani, S.C., and Keshavkant, S. (2014). Protein metabolism during natural ageing in desiccating recalcitrant seeds of *Shorea robusta*. *Acta Physiol. Plant.* 36: 1649–1659. https://doi.org/10.1007/s11738-014-1540-x.

Pathare, V., Srivastava, S., Sonawane, B.V., and Suprasanna, P. (2016). Arsenic stress affects the expression profile of genes of 14-3-3 proteins in the shoot of mycorrhiza colonized rice. *Physiol. Mol. Biol. Plants* 22: 515–522. https://doi.org/10.1007/s12298-016-0382-y.

Ramakrishna, B. and Seeta Ram Rao, S. (2013). 24-Epibrassinolide maintains elevated redox state of AsA and GSH in radish (*Raphanus sativus* L.) seedlings under zinc stress. *Acta Physiol. Plant.* 35: 1291–1302. https://doi.org/10.1007/s11738-012-1168-7.

Saidi, I., Yousfi, N., and Borgi, M.A. (2017). Salicylic acid improves the antioxidant ability against arsenicinduced oxidative stress in sunflower (*Helianthus annuus*) seedling. *J. Plant Nutr.* 40: 2326–2335. https://doi.org/10.1080/01904167.2017.1310888.

Shahid, M.A., Balal, R.M., Khan, N. et al. (2019). Selenium impedes cadmium and arsenic toxicity in potato by modulating carbohydrate and nitrogen metabolism. *Ecotoxicol. Environ. Saf.* 180: 588–599. https://doi.org/10.1016/j.ecoenv.2019.05.037.

Shi, H., Shi, X., Liu, K.J., and Liu, K.J. (2004). Oxidative mechanism of arsenic toxicity and carcinogenesis. *Mol. Cell. Biochem.* 255: 67–78.

Sidhu, G.K., Singh, S., Kumar, V. et al. (2019). Toxicity, monitoring and biodegradation of organophosphate pesticides: a review. *Crit. Rev. Environ. Sci. Technol.* 49: 1135–1187. https://doi.org/10.1080/10643389.2019.1565554.

Silveira, N.M., De Oliveira, J.A., Ribeiro, C. et al. (2015). Nitric oxide attenuates oxidative stress induced by arsenic in Lettuce (*Lactuca sativa*) leaves. *Water Air Soil Pollut.* 226: 226–379. https://doi.org/10.1007/s11270-015-2630-0.

Singh, H.P., Batish, D.R., Kohli, R.K., and Arora, K. (2007). Arsenic-induced root growth inhibition in mung bean (*Phaseolus aureus* Roxb.) is due to oxidative stress resulting from enhanced lipid peroxidation. *Plant Growth Regul.* 53: 65–73. https://doi.org/10.1007/s10725-007-9205-z.

Singh, H.P., Kaur, S., Batish, D.R. et al. (2009). Nitric oxide alleviates arsenic toxicity by reducing oxidative damage in the roots of *Oryza sativa* (rice). *Nitric Oxide – Biol. Chem.* 20: 289–297. https://doi.org/10.1016/j.niox.2009.02.004.

Singh, N., Ma, L.Q., Srivastava, M., and Rathinasabapathi, B. (2006). Metabolic adaptations to arsenic-induced oxidative stress in *Pteris vittata* L. and *Pteris ensiformis* L. *Plant Sci.* 170: 274–282. https://doi.org/10.1016/j.plantsci.2005.08.013.

Singh, S., Kumar, V., Chauhan, A. et al. (2017a). Toxicity, degradation and analysis of the herbicide atrazine. *Environ. Chem. Lett.* 161 (16): 211–237. https://doi.org/10.1007/S10311-017-0665-8.

Singh, S., Kumar, V., Datta, S. et al. (2020a). Plant disease management by bioactive natural products. In: *Natural Bioactive Products in Sustainable Agriculture*, 15–29. Singapore: Springer https://doi.org/10.1007/978-981-15-3024-1_2.

Singh, S., Kumar, V., Datta, S. et al. (2020b). Glyphosate uptake, translocation, resistance emergence in crops, analytical monitoring, toxicity and degradation: a review. *Environ. Chem. Lett.* https://doi.org/10.1007/s10311-020-00969-z.

Singh, S., Kumar, V., Dhanjal, D.S., and Singh, J. (2020c). Biological control agents: diversity, ecological significances, and biotechnological applications. In: *Natural Bioactive Products in Sustainable Agriculture* (ed. J. Singh and A.N. Yadav), 31–44. Switzerland: Springer https://doi.org/10.1007/978-981-15-3024-1_3.

Singh, S., Kumar, V., Dhanjal, D.S., and Singh, J. (2020d). Herbicides and plant growth regulators: current developments and future challenges. In: *Natural Bioactive Products in Sustainable Agriculture*, 67–81. Switzerland: Springer https://doi.org/10.1007/978-981-15-3024-1_5.

Singh, S., Kumar, V., Dhanjal, D.S. et al. (2021). Rhizosphere biology: a key to agricultural sustainability. In: *Current Trends in Microbial Biotechnology for Sustainable Agriculture* (ed. A.N. Yadav, J. Singh, C. Singh and N. Yadav), 161–182. Singapore: Springer https://doi.org/10.1007/978-981-15-6949-4_7.

Singh, S., Kumar, V., Singla, S. et al. (2020e). Kinetic study of the biodegradation of acephate by indigenous soil bacterial isolates in the presence of humic acid and metal ions. *Biomolecules* 10: 433. https://doi.org/10.3390/biom10030433.

Singh, S., Kumar, V., Upadhyay, N. et al. (2017b). Efficient biodegradation of acephate by *Pseudomonas pseudoalcaligenes* PS-5 in the presence and absence of heavy metal ions [Cu(II) and Fe(III)], and humic acid. *3 Biotech* 7: 262. https://doi.org/10.1007/s13205-017-0900-9.

Singh, S., Singh, N., Kumar, V. et al. (2016). Toxicity, monitoring and biodegradation of the fungicide carbendazim. *Environ. Chem. Lett.* 14: 317–329. https://doi.org/10.1007/s10311-016-0566-2.

Singh, S., Kumar, V., Singh, S. et al. (2020f). Global scenario of plant–microbiome for sustainable agriculture: current advancements and future challenges. In: *Plant Microbiomes for Sustainable Agriculture* (ed. A.N. Yadav, J. Singh, A.A. Rastegari and N. Yadav), 425–443. Springer https://doi.org/10.1007/978-3-030-38453-1_14.

Singh, V.P., Singh, S., Kumar, J., and Prasad, S.M. (2015). Hydrogen sulfide alleviates toxic effects of arsenate in pea seedlings through up-regulation of the ascorbate-glutathione cycle: possible involvement of nitric oxide. *J. Plant Physiol.* 181: 20–29. https://doi.org/10.1016/j.jplph.2015.03.015.

Smith, S.E., Christophersen, H.M., Pope, S., and Smith, F.A. (2010). Arsenic uptake and toxicity in plants: integrating mycorrhizal influences. *Plant Soil* 327: 1–21. https://doi.org/10.1007/s11104-009-0089-8.

Srivastava, S. and Sharma, Y.K. (2014). Arsenic induced changes in growth and metabolism of black gram seedlings (*Vigna mungo* L.) and the role of phosphate as an ameliorating agent. *Environ. Process.* 1: 431–445. https://doi.org/10.1007/s40710-014-0035-5.

Srivastava, S., Sinha, P., and Sharma, Y.K. (2017). Status of photosynthetic pigments, lipid peroxidation and anti-oxidative enzymes in *Vigna mungo* in presence of arsenic. *J. Plant Nutr.* 40: 298–306. https://doi.org/10.1080/01904167.2016.1240189.

Stoeva, N., Berova, M., and Zlatev, Z. (2005). Effect of arsenic on some physiological parameters in bean plants. *Biol. Plant.* 49: 293–296. https://doi.org/10.1007/s10535-005-3296-z.

Sytar, O., Kumar, A., Latowski, D. et al. (2013). Heavy metal-induced oxidative damage, defense reactions, and detoxification mechanisms in plants. *Acta Physiol. Plant.* 35: 985–999. https://doi.org/10.1007/s11738-012-1169-6.

Talukdar, D. (2013). Arsenic-induced changes in growth and antioxidant metabolism of fenugreek. *Russ. J. Plant Physiol.* 60: 652–660. https://doi.org/10.1134/S1021443713050130.

Wani, A.B., Chadar, H., Wani, A.H. et al. (2017). Salicylic acid to decrease plant stress. *Environ. Chem. Lett.* 15: 101–123. https://doi.org/10.1007/s10311-016-0584-0.

Zhao, F.J., Ma, J.F., Meharg, A.A., and McGrath, S.P. (2009). Arsenic uptake and metabolism in plants. *New Phytol.* 181: 777–794. https://doi.org/10.1111/j.1469-8137.2008.02716.x.

7

Photosynthetic Responses of Two Salt-Tolerant Plants, *Tamarix gallica* and *Arthrocnemum indicum* Against Arsenic Stress

A Case Study

Dhouha Belhaj Sghaier[1], *Sílvia Pedro*[2], *Bernardo Duarte*[2], *Isabel Caçador*[2,3], *and Noomene Sleimi*[1]

[1] *UR MaNE, Faculté des Sciences de Bizerte, Université de Carthage, Tunis, Tunisia*
[2] *MARE – Marine and Environmental Sciences Centre, Faculty of Sciences, University of Lisbon, Lisbon, Portugal*
[3] *Department of Plant Biology, Faculty of Sciences, University of Lisbon, Lisbon, Portugal*

7.1 Introduction

Over the last decades, we have noticed a drastic rise in environmental contamination. It is noteworthy that any pollutant, heavy metals, in particular, have the ability to provoke harmful effects on the plants (González et al. 2014). Additionally, according to their oxidation degree, heavy metals (HMs) could be awfully reactive, causing plant cell toxicity in several ways. Heavy metal ions may induce deteriorations at the cellular and molecular stages via substituting functional components of biomolecules, stopping essential groups and enzymes' active regions, hampering membranes, and altering plants' transcription models (Guo et al. 2016, 2017). The generation of ROS by heavy metals not only leads to oxidative impairment to different biomolecules but also overlaps with the electron movement system that occurs in the leaves (Guo et al. 2016).

HM toxicity leads to alterations in various plant physiological processes, which involve inactivated and denatured enzymes, proteins, blocking functional groups of metabolically important molecules (Khan et al. 2018). HM toxicity also leads to displace/substitute metal ions essential from biomolecules and functional cell units, conformational variation, and disruption of membrane integrity (Khan et al. 2019).

It is well documented that the main entryway for As (V) in plants is realized across the Pi transport proteins. As (V) and Pi are competing for the equivalent access scheme (Finnegan and Chen 2012). When the arsenate As (V) enters the roots, a rise in its concentration results. A fraction of As (V) is transported to the outer environment (Zhao et al. 2009), whereas, the other arsenate fraction is quickly transformed to As (III) in the belowground parts. Among the effects of arsenic toxicity, there is a regression in the chlorophyll tenor (Chl) and/or the disturbance of the photosystem II activity (Duman et al. 2010).

Arsenic in Plants: Uptake, Consequences and Remediation Techniques, First Edition.
Edited by Prabhat Kumar Srivastava, Rachana Singh, Parul Parihar, and Sheo Mohan Prasad.

The incidence of As toxicity on plants is well established since As (V) could replace the inorganic phosphate in biochemical reactions, overgenerate reactive oxygen species (ROS), while As (III) can fix thiol-reactive composite (Flora 2011). As easily gets accumulated in the plants and can be removed to other parts of plants which are usually used as the source of food, and easily enters the food chain (Mecwan Neha et al. 2018).

As subjection to plants has as consequence development suppression, physiological waste and therefore cell mortality (Stoeva et al. 2003). As gives rise to the ROS, which impair the membranes and many functional groups in the cell (Meharg and Hartley-Whitaker 2002). To know the overall aftereffect of As, numerous physiological and biochemical indicators in various plants species have been figured out (Dave et al. 2013; Gangwar et al. 2014; Rai et al. 2016). It has been recognized that elevated concentration As causes many chemical and physiological dysfunctions including water and nutrients uptake, unsettle enzyme and chloroplast building, minimizing photosynthesis competence, necrosis, and eventually death (Finnegan and Chen 2012; Khalid et al. 2016; Mecwan Neha et al. 2018). The acuteness of photosynthetic machinery destruction is mainly linked to the As concentrations and the As supplied form (Kofroňová et al. 2020).

Some plants are capable to sustain elevated concentrations of metalloids without demonstrating lethal signs (Van Der Ent et al. 2013). Other specialized species of plants metal tolerant are named hyperaccumulators. Those groups of plants are able to accumulate internally elevated concentrations of the metal, without altering the growth. While some halophytes are capable to withstand metal/metalloid toxicity to some level, a high endogenous level of metal in plants may induce a harsh breakdown of basic processes related to metabolism (Patra et al. 2004; Liu et al. 2013).

The physiological adjustment of halophytes to support salt plays a crucial role in surviving in this habitat and in overcoming adverse conditions (Hasanuzzaman et al. 2014). In some countries, salt marshes or Sebkha like Tunisia, this habitat is used as a sink for industrial waste. As a result, prior attention is paid to the exploitation of salt marshes and to the halophytes' resistance to soil containments. The rehabilitation of degraded lands and removal of pollutants are other emerging areas of application of these halophytes (Duarte and Cacador 2021). Rising agro-management systems and halophytes revegetation are required for landscaping and phytoremediation in the last years (Duarte and Cacador 2021).

The chapter is addressed to the salt marsh plants *Tamarix gallica* and *Arthrocnemum indicum* and sought to ensure a vital view on (i) processes of absorption, displacement, (ii) chlorophyll degradation, (iii) PS II electron movement, and (iv) electron transport chain (ETC) performance in leaves. These salt marsh plants were chosen because there were abundant halophyte species that occupy salt marshes, increase size well from small pieces, and have two different ways to tolerate saline soils and metal contamination. Based on these facts, plants' development in synthetic soils polluted with arsenic (As) were used to reach their phytoremediation efficiency. The chapter aims to verify and compare the As exposure of *T. gallica* and *A. indicum* and its impact on photosynthetic pigments and apparatus.

7.2 Metal Uptake

In *T. gallica,* results showed that the accumulation of As was influenced by the metal quantity in the irrigation medium (Figure 7.1a). The As content was around 92 and 94 $\mu g\,g^{-1}$ DW at 500 and 800 µM, respectively in leaves and roots. The append of NaCl in the solution affected the concentration of As in the leaves, whereas, in the roots, the addition of salt did

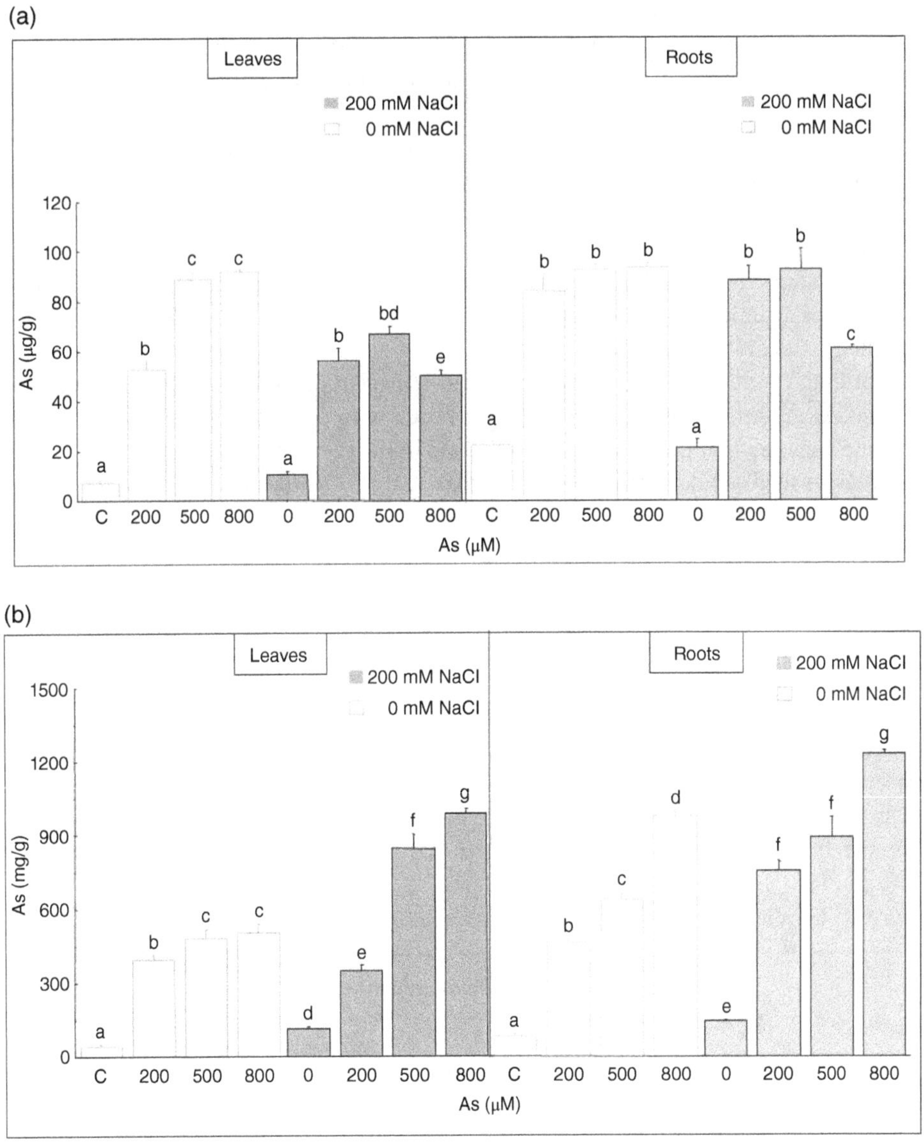

Figure 7.1 As uptake in (a) *T. gallica* (the concept of this figure is based on Sghaier et al. (2015)) and (b) *A. indicum*. Statistical significance at $p < 0.05$ between control and treated plants. Bars marked with the same letter are not significantly different at $p = 0.05$.

not reduce its content except for the 800 μM As dose (54 $\mu g g^{-1}$ DW), where there was an estimated decrease in accumulation of 35%. The same effect was observed in the leaves, at a concentration of 500 μM (Figure 7.1a).

In *A. indicum*, the accumulation of As in the leaves increased with a rising concentration in the medium. This increase was amplified by salt supplementation to the medium. The foliar As content was estimated at 450 $\mu g g^{-1}$ DW in plants subjected to 200 μM; then at doses of 500 and 800 μM, it reached 550 $\mu g g^{-1}$ DW (Figure 7.1b).

The same behavior is observed at the root level which accumulates large amounts of As up to 900 and 1200 $\mu g g^{-1}$ DW in plants treated with NaH_2AsO_4 alone or combined with a salt (Figure 7.1b). The As accumulation by *A. indicum* was higher than *T. gallica* (Figure 7.1).

The results presented in Table 7.1 show that the roots of both species constitute the major site of accumulation of As. The removal of this ion from the below to the above part is evidenced through the factor of translocation (FT).

Indeed, the FTs recorded in *T. gallica* fluctuated according to the dose of As. The application of salt stress dramatically increased the translocation factor. Thus, at a concentration of 800 μM of arsenic, the *Tamarix* translocation factor was only 0.11, whereas it reached 0.19 after the addition of salt.

Regarding *A. indicum*, we noted values that vary from 0.45 ± 0.04 to 0.26 ± 0.01. Compared to *T. gallica, A. indicum* exhibited FTs which raised with increasing concentration of As. It should be noted that the FT values determined in the two species are less than 1.

As previously discussed in Sghaier et al. (2015, 2019), *T. gallica* could not be considered as an accumulator plant with a small amount of metal uptake but an As tolerant plant (see detailed results in Sghaier et al. (2015)), whereas *A. indicum* could be defined as an accumulator plant especially in a salty medium as shown in Figure 7.1a.

In literature, hyperaccumulator plants are defined as plants that accumulate at least 1000 $\mu g g^{-1}$ of Co, Cu, Cr, Pb, and As in the dry weight (Reeves and Baker 2000; Van der Ent et al. 2013).

The low As translocation factor demonstrated that these halophytes behaved as an As-tolerant plant. Li et al. (2016) reported that the accumulator plants were storing high concentrations of toxic metals in the shoots.

The arsenic accumulation in the belowground part is a dominant way in tolerant plants with the transport of small amounts to the aboveground parts (Caçador et al. 2009).

Table 7.1 Translocation factor in two halophyte species *A. indicum* and *T. gallica*.

	Translocation factor					
	200 μM As	**500 μM As**	**800 μM As**	**200 μM As + NaCl**	**500 μM As + NaCl**	**800 μM As + NaCl**
T. gallica	0.104 ± 0.001	0.07 ± 0.01	0.11 ± 0.04	0.09 ± 0.04	0.13 ± 0.01	0.19 ± 0.03
A. indicum	0.43 ± 0.03	0.37 ± 0.03	0.45 ± 0.04	0.26 ± 0.01	0.36 ± 0.07	0.34 ± 0.02

Mean values of six repetitions ± SE.

Thus, the limitation of translocation could be defined by As blocking transport through immobilizing them in the compartment by complexing with organic acids as malate, citrate, and oxalate or complexation through phytochelatins or metallothioneins followed by sequestration in the vacuolar system as discussed in our previous studies in Sghaier et al. (2016, 2020).

7.3 Impact of Arsenic on Photosynthetic Pigments

After exposing the plants to different arsenic concentrations (200, 500, and 800 μM), *T. gallica* showed the presence of high levels of chlorophylls a and b, reaching $100\,\text{mg}\,\text{g}^{-1}$ fresh weight (FW) for chlorophyll a and $70\,\text{mg}\,\text{g}^{-1}$ FW for chlorophyll b. On the other hand, depending on the increasing concentrations of As, there was no effect on the chlorophyll content or on its degradation, which was observed except for a nonsignificant fluctuation (Figure 7.2a).

In *A. indicum*, the results obtained revealed a decline in the level of chlorophylls a and b at high concentrations of As (800 μM). The levels recorded are in order of $220\,\text{mg}\,\text{g}^{-1}$ FW in untreated plants and in plants treated with 200 and 500 μM, while at 800 μM, the chlorophyll content reached $120\,\text{mg}\,\text{g}^{-1}$ FW (Figure 7.2b).

The salt supplementation to the solution did not alter the contents of chlorophylls a and b. Constant values were recorded at different concentrations of element trace metals (ETM). Regardless of the metallic stress alone or combined with NaCl, *T. gallica* did not show significant variations in the content of chlorophylls a and b (Figure 7.2a). However, *A. indicum* showed after the salt supplementation a marked rise in the chlorophyll a level, particularly at 800 μM As which reached $280\,\text{mg}\,\text{g}^{-1}$ FW (Figure 7.2b).

In *T. gallica*, metal stress did not alter the levels of carotenoids and total chlorophylls (Figure 7.3a). At high concentrations of As (800 μM), the content of carotenoids was above $150\,\text{mg}\,\text{g}^{-1}$ FW. This content is recorded in untreated plants. These contents did not vary after the application of the salt stress (Figure 7.3a). Contrarily, in *A. indicum*, we observed a reduction in the level of carotenoids and total chlorophylls at high concentration of As used alone. However, a linear increase in the contents of total chlorophylls and carotenoids was obtained following metal stress combined with salt stress. The values recorded at 200 and 800 μM were around 50 and $65\,\text{mg}\,\text{g}^{-1}$ FW, respectively (Figure 7.3b).

The effect of the stress induced by As on xanthophyll cycle and their degradation products in *T. gallica* is presented in Table 7.2. The content of violaxanthin in the absence of As was registered as $12\,\text{mg}\,\text{g}^{-1}$ FW. This content remained constant regardless of the amount of metal ion used.

Likewise, the plants subjected to the salt had values close to that recorded in the untreated plants. The same level was noticed in *T. gallica* plants subjected to combined stress of As and NaCl. With the exception of 200 μM, there was a decrease in the violaxanthin content (Table 7.2). Violaxanthin degradation products such as antheraxanthin and zeaxanthin showed similar values in the presence of arsenic alone or combined with salt. The same results were also revealed in plants subjected to salt in comparison to the untreated plants (Table 7.2).

(a)

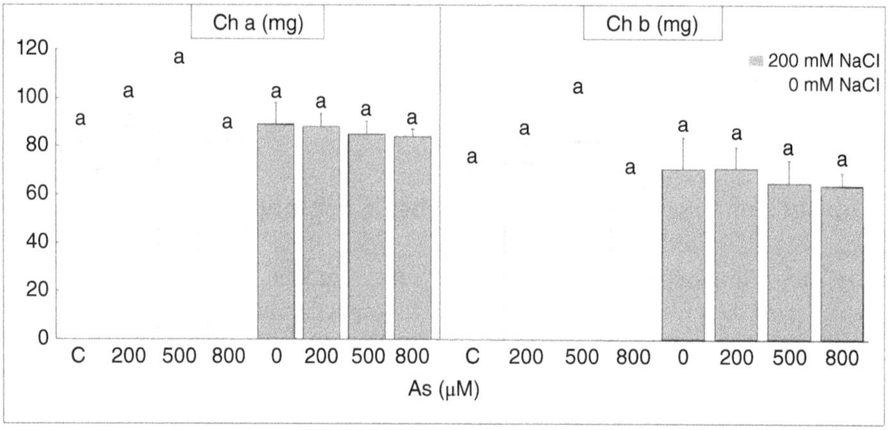

(b)

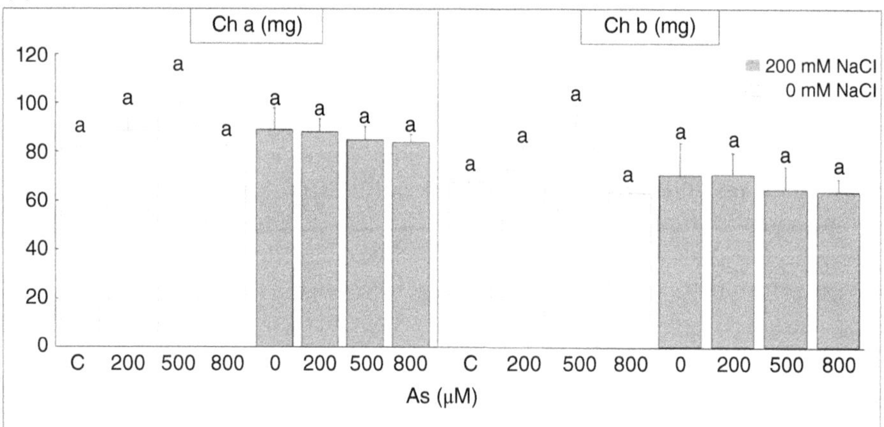

Figure 7.2 Chlorophylls a and b and total carotenoid content of (a) *T. gallica* (the concept of this figure is based on Sghaier et al. (2015)) and (b) *A. indicum* under combined stresses with As and NaCl. Bars marked with the same letter are not significantly different at $p = 0.05$.

On the other hand, in *A. indicum*, an augmentation in the content of violaxanthin and antheraxanthin was noted after intoxication with As (Table 7.3). By increasing the concentration of arsenic in the medium, the violaxanthin and antheraxanthin content increased for plants subjected to 500 and 800 μM As in comparison to untreated plants. Salt supplementation substantially amplified the violaxanthin and antheraxanthin content as evidenced by the plants grown in the presence of 200 mM NaCl compared to the control plants (4.5/6.1 and 1.4/3.1 $mg\,g^{-1}$ FW, respectively) (Table 7.3). The same result was observed in plants subjected to combined stress. We illustrated a rise in the content of violaxanthin and antheraxanthin especially at 500 and 800 μM of NaH_2AsO_4 and NaCl, respectively, compared to those treated with arsenic. In contrast, there was no change in the zeaxanthin content in the existence of As, salt, or As added with salt.

The quantification of chlorophylls makes it possible to know the physiological state of a plant. Chlorophyll was part of the reaction center (RC) which allows the

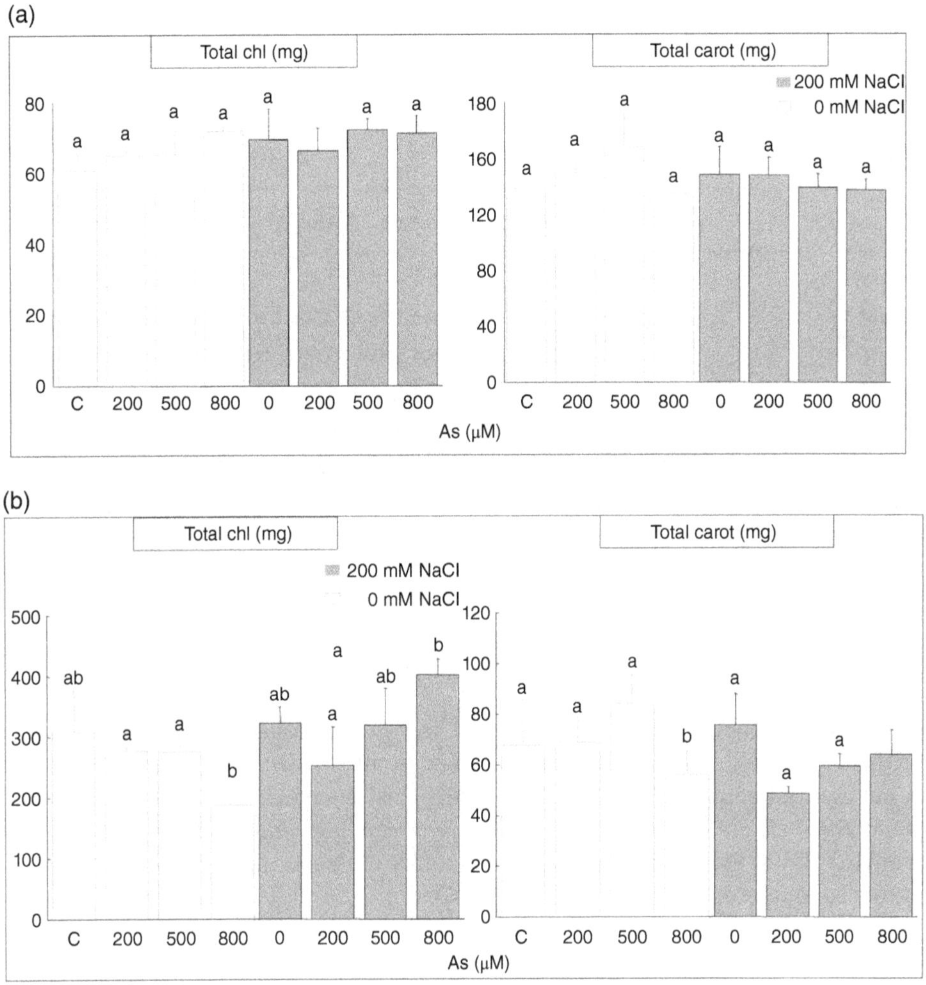

Figure 7.3 Total chlorophyll and total carotenoid content of (a) *T. gallica* (the concept of this figure is based on Sghaier et al. (2015)) and (b) *A. indicum* under combined stresses with As and NaCl. Bars marked with the same letter are not significantly different at $p = 0.05$.

Table 7.2 Chlorophyll degradation products, carotenoids, and xanthophyll cycle contents ($\mu g\, g^{-1}$ FW) in *T. gallica* exposed to increasing As levels.

	Pheo[a]	β-Carotène	Lutein	Viola	Anthera	Zea
Control	41.2 ± 8.9	5.4 ± 1.3	7.3 ± 1.8	12.1 ± 3.5	7.8 ± 4.2	9.7 ± 3.1
200 μM As	22.50 ± 4.2	6.7 ± 1.4	7.4 ± 1.3	10.1 ± 1.8	8.3 ± 0.8	10.7 ± 4.7
500 μM As	15.8 ± 4.4	10.0 ± 1.2[a]	9.2 ± 1.2[a]	12.1 ± 3.1	9.1 ± 1.4	10.9 ± 1.6
800 μM As	10.7 ± 1.7	7.8 ± 1.2	8.6 ± 2.2	6.8 ± 2.5	5.3 ± 0.1	22.6 ± 4.1
200 mM NaCl	19.80 ± 5.8	6.5 ± 0.7	11.4 ± 2.4	11.1 ± 2.5	8.2 ± 1.4	14.0 ± 4.7
200 mM NaCl/200 μM As	19.6 ± 5.3	5.0 ± 0.4[a]	4.7 ± 0.4[a]	6.3 ± 0.9	6.2 ± 0.5	5.9 ± 0.7
200 mM NaCl/500 μM As	39.9 ± 17.4	5.1 ± 1.3	6.3 ± 1.3	9.6 ± 1.1	7.1 ± 1.7	6.8 ± 1.3
200 mM NaCl/800 μM As	51.44 ± 13.3	5.5 ± 1.7	8.8 ± 3.2	13,3 ± 3.7	8.6 ± 1,5	10.4 ± 1.4

Anthera, antheraxanthin; βCar, β-carotene; Lut, lutein; Pheo, pheophytin; Viola, violaxanthin; Zea, zeaxanthin. Mean values of six repetitions ± SE.

[a] represent statistically differences ($p < 0.05$).

Table 7.3 Chlorophyll degradation products, carotenoids, and xanthophyll cycle contents ($\mu g\, g^{-1}$ FW) in *A. indicum* exposed to increasing As levels.

	Pheo[a]	β-Carotène	Lutein	Viola	Anthera	Zea
Control	51.01 ± 3.7	12.02 ± 0.3	6.7 ± 1.4	1.40 ± 0.5	3.10±0.7	28.5 ± 3.1
200 μM As	54.90 ± 4.2	12.20 ± 1.3	8.1 ± 0.5	2.3 ± 1.8	4.30±0.6	29.80 ± 0.6
500 μM As	58.03 ± 2.9	11.50 ± 0.8	9.1 ± 0.4	5.80 ± 0.4	5.0 4±1.0	27.30 ± 3.1
800 μM As	52.35 ± 1.3	11.66 ± 0.3	7.7 ± 1.1	3.1 ± 0.9	4.8±0.5	31.50 ± 2.2
200 mM NaCl	55.90 ± 6.9	13.6 ± 3.3	8.1 ± 0.9	4.50 ± 0.9	6.10±1.3	30.9 ± 3.2
200 mM NaCl/200 μM As	54.97 ± 1.7	12.40 ± 0.9	8.5 ± 0.2	4.60 ± 0.6	5.50±0.6	30.4 ± 1.6
200 mM NaCl/500 μM As	54.04 ± 4.7	12.00 ± 1.4	9.0 ± 0.5	4.40 ± 0.8	4.80±0.7	31.60 ± 0.4
200 mM NaCl/800 μM As	53.44 ± 1.1	12.08 ± 0.6	8.1 ± 0.7	4.9 ± 0.3	4.60±0 8	31.70 ± 0.5

Anthera, antheraxanthin; βCar, β-carotene; Lut, lutein; Pheo, pheophytin; Viola, violaxanthin; Zea, zeaxanthin. Mean values of six repetitions ± different.
[a] Pheo, pheophytin.

modification of photon energy into chemical energy. The collecting antenna, grouping together chlorophylls a and b, absorbs photons (Vincent and Zivy 2007).

Chlorosis and stunted plant growth are usually evidenced as a consequence of metal toxicity that indicates a perturbation of photosynthetic pigment formation path, thereby modifying plastid processing and photosynthetic forcefulness. Heavy metals further mitigate the buildup of chlorophylls (Rai et al. 2016).

Thus, chlorophyll had a main contribution to the photosynthesis pathway through its light absorption capability. ALA transformation to porphobilinogen is realized via ALA dehydratase (ALAD, porphobilinogen synthase) (Rai et al. 2016). ALAD is a regulatory enzyme in chlorophyll biosynthesis and is a metal-sensitive enzyme which requires Mg^{2+} or Zn^{2+} for its activity. It has been suggested that ALAD inhibition could be induced by the interconnection realized by heavy metals and the SH enzyme associations (Rai et al. 2016).

Furthermore, it has been revealed that cadmium (Cd) may induce inhibition of protochlorophyllide reductase activity of the enzyme key of the protochlorophyllide conversion to chlorophyllide, thus by blocking the pathway (Sharma et al. 2020). Endowed with an inhibitory activity similar to Cd, Cr is able to alter the enzymatic activity of the enzymes responsible for the biosynthesis path. Cr could induce the reduction of magnesium (Mg) and nitrogen (N) uptake, leading to a chlorophyll content decline (Singh et al. 2013). On the other hand, Pb leads to a stimulation in the chlorophyllase activity, inducing chlorophyll degradation (Sharma and Dubey 2005).

Heavy metals produce ultrastructural modifications in plant leaves cultivated in a heavy metal-polluting zone, plenty of studies have been revealed that metal-provoke inhibition of chlorophyll biosynthesis and regression of Chl a and Chl b amount, and its product degradation (Ali et al. 2014; Mei et al. 2015; Khan et al. 2016). It has been reported that these metals can lead to the inhibition of the activity of certain enzymes interconnected to biosynthesis, or by modifying certain necessary groups like the Mg in chlorophyll and, consequently, altering the biosynthesis of chlorophylls. In addition, the pigment–protein union could be altered by these metals.

Additionally, Zhang et al. (2018) reported that other heavy metals like Pb and Mn induced inhibition in chlorophyll biosynthesis in *Suaeda glauca* and *Arabidopsis thaliana*. Similar to our results obtained in *T. gallica*, the total chlorophyll content in the rice leaves presented no variations in the presence of arsenite and arsenate concentrations compared to the control. Elevated chlorophyll content was obtained in untreated plants which demonstrated a decline with rising arsenic concentration. Deeper, the top reduction in total chlorophyll content was noted at the treatment with $15\,\text{mg}\,\text{l}^{-1}$ As (Gaikwad et al. 2020).

Underneath heavy metal stress, overabundance of ROS is formed owing to excessive reduction of the electron movement sequence. ROS is destructive to chlorophylls and metabolic reactions. The uncontrolled H_2O_2 produces oxidation of Calvin cycle enzymes and leads to a drop in photosynthesis rate (Hajiboland 2014).

During various environmental stresses, such as the presence of ETM and extreme salinity, the transport of electrons via photosystem II (PSII) and photosystem I (PSI) and the processes of CO_2 fixation in sugars are reduced. Under these conditions, to protect the photosynthetic apparatus, an expansion in energy waste as heat form by the activation of the xanthophyll cycle has been observed (Hendrickson et al. 2003; Heber et al. 2006). *The non-photochemical waste of energy* is regulated by the xanthophyll cycle, which allows the synthesis of zeaxanthin when plants are subjected to huge light power and thus protects against the formation of ROS (Anjum et al. 2016).

Violaxanthin is reversibly converted to zeaxanthin; thereby the transitional pigment antheraxanthin (Demmig-Adams and Adams 1996). This regulatory role of the xanthophyll cycle is conditional on the transthylakoidal proton slope, which is itself closely related to the redox state of electron transport of PSII-PSI. Zeaxanthin can also be found in free form in thylakoid membranes, where it acts as an antioxidant, protecting the membranes against lipid peroxidation induced by ROS (Moronsito et al. 2003). For example, when *Jungermannia exsertifolia* sub sp. Cordifolia was exposed to Cd, they noted that the ratio of zeaxanthin to violaxanthin and antheraxanthin increased (Otero et al. 2006).

7.4 Effect of Arsenic on Photosynthetic Apparatus

The consequences of As on photosynthetic efficiency were determined via various physiological factors. The putative As effects on photosynthetic apparatus was represented in Figures 7.4–7.7.

Figures showed the parameters of the fluorescence of chlorophyll a (Fv/Fm, F0, Fm, F′0, F′m, F′v, and Fv) of both halophytes *T. gallica* and *A. indicum* allowing the evaluation of the effects generated by As on the capacity and efficiency of photosystems to convert light energy into chemical energy. The evolution of the Fv/Fm ratio in the presence of inorganic As demonstrated stability of the photon energy absorbed in *T. gallica*. It is around 0.71 even when the arsenic content in the nutrient solution is $800\,\mu\text{M}$ (Figure 7.4a).

In addition, the presence of arsenic at high concentration did not affect the photochemical performance of PSIIs, described by the variables F′v/F′m and ϕPSII (Figure 7.4b). These results implied that arsenic did not alter the photochemical efficacy of PSIIs and their

Tamarix gallica

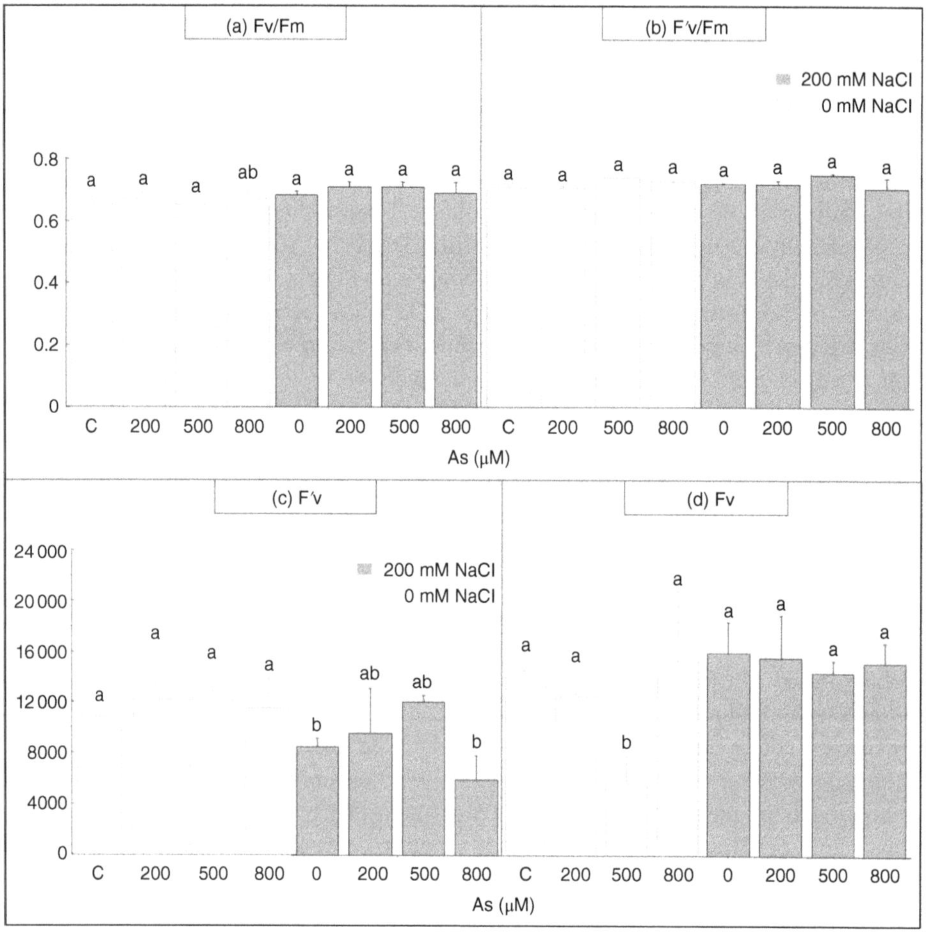

Figure 7.4 PSII quantum yield (a, b), variable fluorescence (c, d) in light and dark-adapted leaves, and non-photochemical quenching in *T. gallica*. Bars marked with the same letter are not significantly different at $p = 0.05$. The concept of this figure is based on Sghaier et al. (2015).

ability to reduce the primary acceptor QA (determined by the Fv/Fm ratio). The measurements taken demonstrated a low impact of As on the functioning of PSII. In accordance with the small incidence on photosynthetic efficiency, leaf content persisted unchanged all over the investigation. In addition, we have observed that the addition of NaCl had a weak effect on ϕPo which varies between 0.7 and 0.72 (Figure 7.4a).

For fluorescence analysis, variable fluorescence values at light-adapted leaf levels (F′v) and dark-adapted ones (Fv) showing that there were no effects of stress observed on performance of operational or quantum PSII maximal regardless of the rise and fall of the lower rate F′v (Figure 7.4c, d).

In this study, the effect of the stress applied on *A. indicum* showed a small difference in the maximum and operational quantum yield of the primary photochemistry φPo and

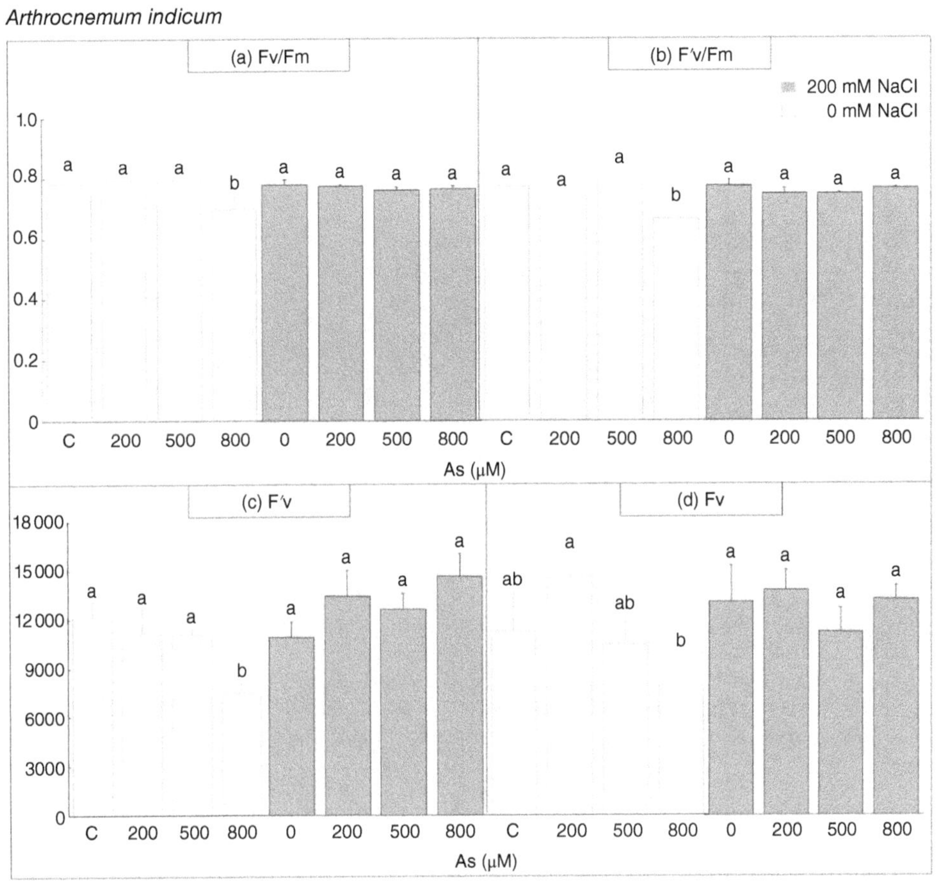

Figure 7.5 PSII quantum yield (a, b), variable fluorescence (c, d) in light and dark-adapted leaves, and non-photochemical quenching in *A. indicum*. Bars marked with the same letter are not significantly different at $p = 0.05$.

φP′o in comparison to the untreated plants and this at a high concentration of metal applied (800 μM) (Figure 7.5a, b). A value around 0.8 was recorded in untreated plants. The same value was noted in plants treated with arsenic at increasing concentrations up to 500 μM. While at 800 μM, there was a decline, although the value recorded, which is equal to 0.74, indicating a good functioning of the photosystems. This drop was as consequent of a reduction in the fluorescence variable at leaves adapted to light and/or darkness (F′v and Fv) (Figure 7.5c, d). However, only a decrease was observed in plants treated with As at very high concentrations (800 μM).

Contrarily, the addition of the salt to the medium improved the efficiency of the operational or maximum quantum PSII. Constant values similar to that registered in control plants and which were around 0.8 were noticed even at high concentrations (800 μM). Same effect of salt was observed on the fluorescence variable, the addition of NaCl in the medium attenuated the effect of the toxicity of As and no resulting significant variation was detected in the treated and untreated plants (Figure 7.5a, b). Therefore, there was no

Tamarix gallica

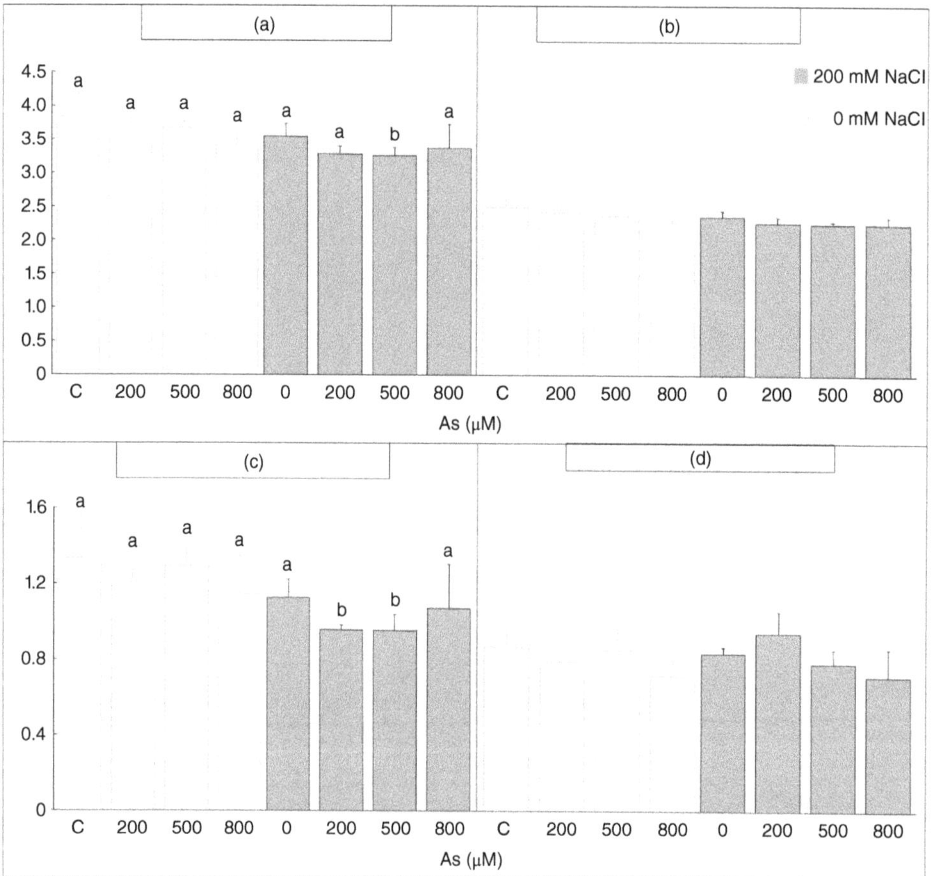

Figure 7.6 Energy fluxes in *T. gallica* under combined stresses with As and NaCl: (a) absorbed energy flux, (b) trapped energy flux, (c) dissipated energy flux, (d) transported energy flux. Statistical significance at $p < 0.05$ between control and treated plants. Bars marked with the same letter are not significantly different at $p = 0.05$. The concept of this figure is based on Sghaier et al. (2015).

loss in the efficiency of the fundamental photochemistry of PSII, thus confirming the high stability of the photochemical efficiency of PSII to metallic stress and combined stress (ETM, NaCl) (Figure 7.5).

A representation of the specific activity of a photosynthetic unit across four streams for each sample is given in Figures 7.6 and 7.7. The accurate fluxes at the stage of the RCs, namely ABS/RC absorption, TRo/RC trapping, dissipate energie flux (DIo)/RC dissipation and electron transport energy flux (ETo)/RC electron transport by the active RC (the centers which reduce QA to QA−).

RC absorption (ABS/RC) and TRo/RC trapping were not altered by either As or salt. Likewise, the electron transport (ETo/RC) for treated *T. gallica* demonstrated a nonsignificant fluctuation and presented values close to the control (Figure 7.6d). However, energy

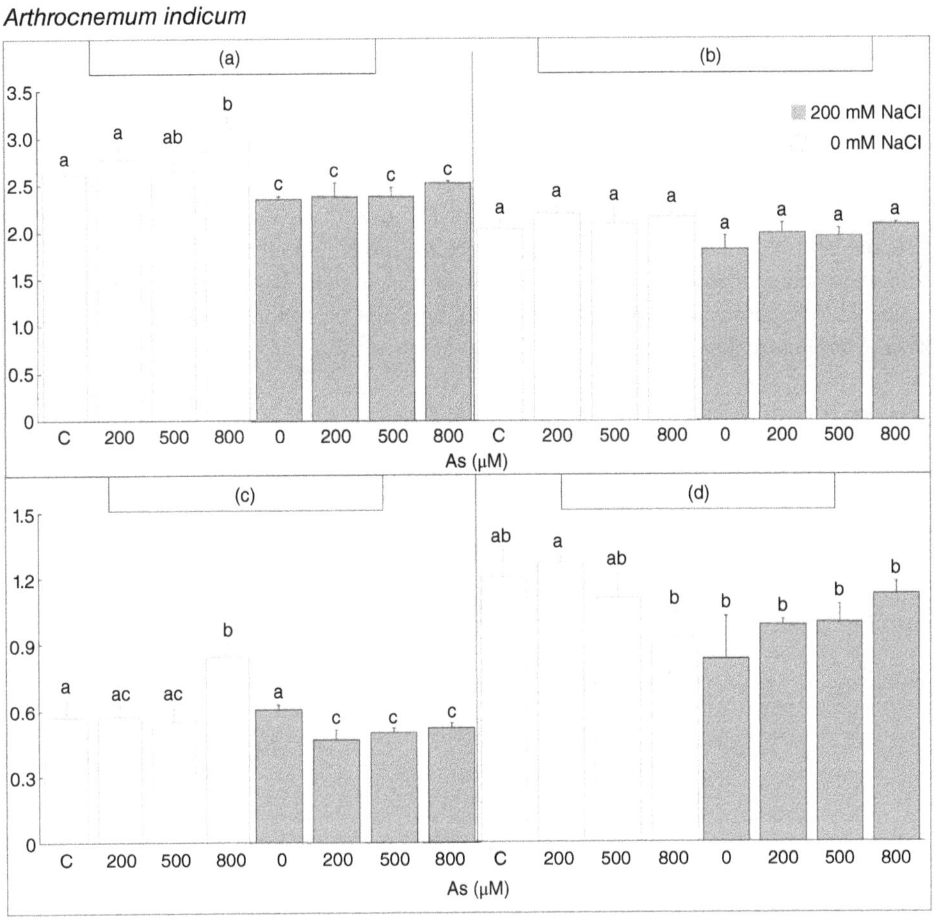

Figure 7.7 Energy fluxes in *A. indicum* under combined stresses with As and NaCl: (a) absorbed energy flux, (b) trapped energy flux, (c) dissipated energy flux, (d) transported energy flux. Bars marked with the same letter are not significantly different at $p = 0.05$.

dissipation (DIo) from the RC appeared to be impaired during combined stress (As, NaCl). A drop was observed at concentrations of 200 and 500 μM As (Figure 7.6c).

In *A. indicum*, a decline was observed in the energy fluxes transported at 800 μM As concentration (Figure 7.7d). Indeed, these observations were in accordance with our previous results and which allowed us to understand the decrease observed in plants treated at very high concentrations of As. However, at this same concentration, there was an increase in flux at the level of energy absorbed and dissipated (Figure 7.7a, c). So, for the other control and treated plants, the same rate of dissipation and energy uptake was registered (Figure 7.7c). For the fourth "trapped energy" signal transduction stream, no difference was illustrated (Figure 7.7b). Plants trap light in the same way whatever concentration of ETM was applied and this explained the small difference in operational or quantum performance. However, the addition of 200 mM NaCl improved the ability of plants to signal transduction. No difference was found for the four energy streams and this clarified the stability of fluorescence and yield recorded (Figure 7.7).

The performance index (PI) is a variable employed to differentiate the response of plants to stress. The PI mirrored the situation of PSI and II and afford further specific points about plant efficiency, notably underneath stress. Therefore, the stress impact on these components will be visualized in the PI. In the current research, *T. gallica* exhibited few PI changing, although they were not significant (Figure 7.8a). Finally, the application of de-epoxidation State (DES), an additional decisive index related to the efficiency of light utilization and energy waste, indicates that there was no light disturbance. *T. gallica* leaves subjected to As alone, or combined with NaCl, with the exception of samples treated with 500 μM As (Figure 7.8a).

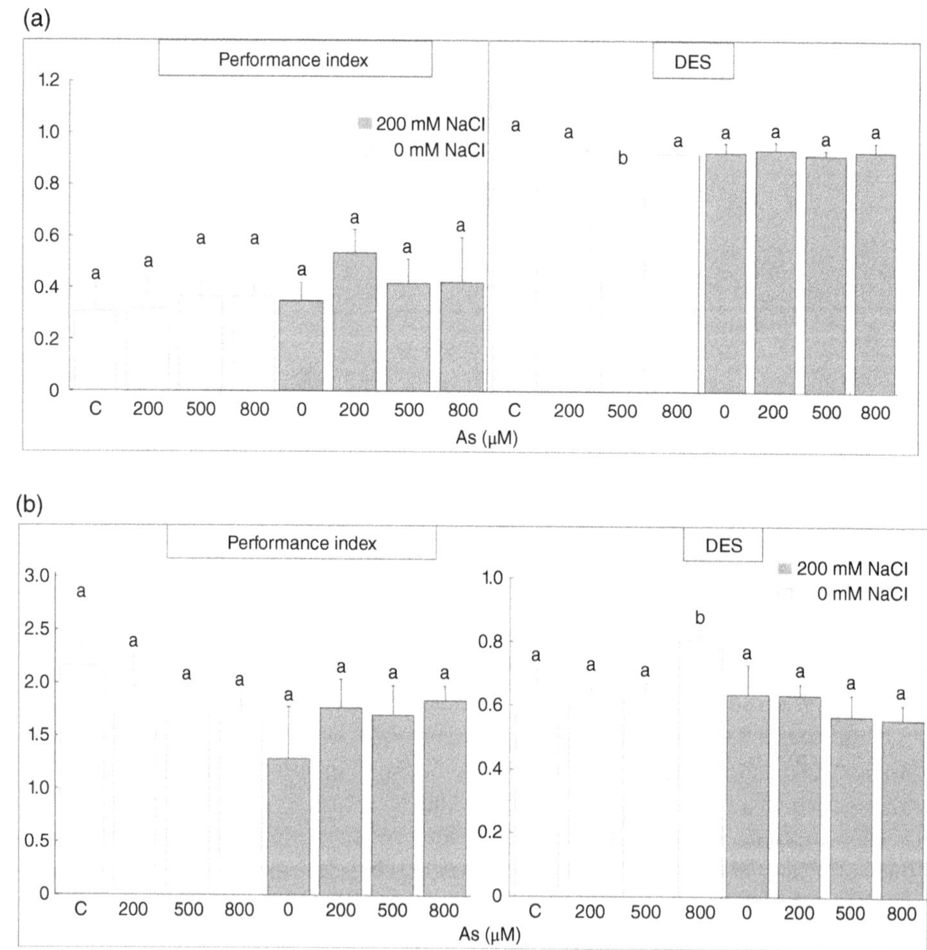

Figure 7.8 The performance index (PI) and DES values in (a) *T. gallica* (the concept of this figure is based on Sghaier et al. (2015)) and (b) *A. indicum*. Bars marked with the same letter are not significantly different at $p = 0.05$.

The overall energetic flux efficiency in *A. indicum* was determined by the PI showing a nonsignificant decrease as the concentration of As increased in the medium. This drop was not significant. However, the opposite effect was determined by the salt supplementation. However, this increase was not significant. The PI gave us all the information concerning the path of energy transduction. The latter allowed us to conclude the photochemical efficiency of PSII and, at the same time, the role played by salt in reducing the toxic effect of ETM (Figure 7.8b). The last parameter was DES. The latter represented the dissipation of energy by the conversion and degradation of the photosynthetic pigments previously analyzed. The graph showed the same result obtained downstream, a significant elevation at a very high concentration of As. Even observation could be reported that the addition of salt in the medium had a depressive effect on the toxicity of ETM. No significant difference was detected between the treatments (Figure 7.8b).

The OJIP polyphasic curve of fluorescence was measured and plotted on a logarithmic time scale (Figures 7.9 and 7.10). The rapid increase in the yield of chlorophyll fluorescence between Fo and FM during the first second of intense illumination was used to analyze electron transport in the PSII. The 2 ms stage J gave information on simple turnover of the events of the primary reactions of photochemistry, principally the reduction of QA. The time interval of 2 to ~200 ms (the J–P step) reflects multiple separation charges where the redox composant of the ETC gets reduced. The various phases of the fluorescence curve are marked in alphabetical order.

Taking into account the thermal phase, in comparison to the donor (J–I) and the acceptor (O–J), there were a few changes in the fluorescence rate. This may be as a result of the inactivation of the oxygen complex (OEC). In addition, *T. gallica*, in the high concentration of As (800 μM), had the ability to dissipate excess energy to overcome the buildup and reduction of ROS, thus avoiding photo-inhibition of photosynthesis. However, no notable changes were reported in plants subjected to combined stress (Figure 7.9).

The representation of the different oxidation–reduction states of the electron transporters associated with the PSII makes it possible to define the Kautsky curve. These transitions show that at the level of the intervals (OJ) and (JI), *A. indicum* exhibited the same fluorescence values and a similar capacity for reducing the primary electron acceptors QA QB in the presence of high concentrations of As (Figure 7.10). Only at 200 μM of As, values higher than those observed in the plants treated at the different intervals of the curve are recorded. However, concerning the segment (I–P), the variation was observed especially at 800 μM As where a decrease in reduction at the PQH_2 pool was noted (Figure 7.10). All of these behavioral differences allowed us to understand the differences recorded for the energy flows previously discussed.

The addition of salt in the irrigation solution improved the levels of reduction of QA QB until the reduction of the entire PQH_2 pool and which is represented by the intervals (OJ), (JI), and (IP). Some of the concentrations of ETM are applied. The salt effect was clearly seen by comparing two groups of plants, one for control and the other treated with 200 mM NaCl. We noticed a decline in the level of reduction of QA- QB- to QA- QB2- but, finally, the same pool PQH_2 was reached.

Heavy metals can alter the performance of PS II through the plastoquinone pool or by feedback control through alteration of enzymes (Rai et al. 2016). These metals may condition the role of the photosynthetic system by interacting with the photosynthetic system at

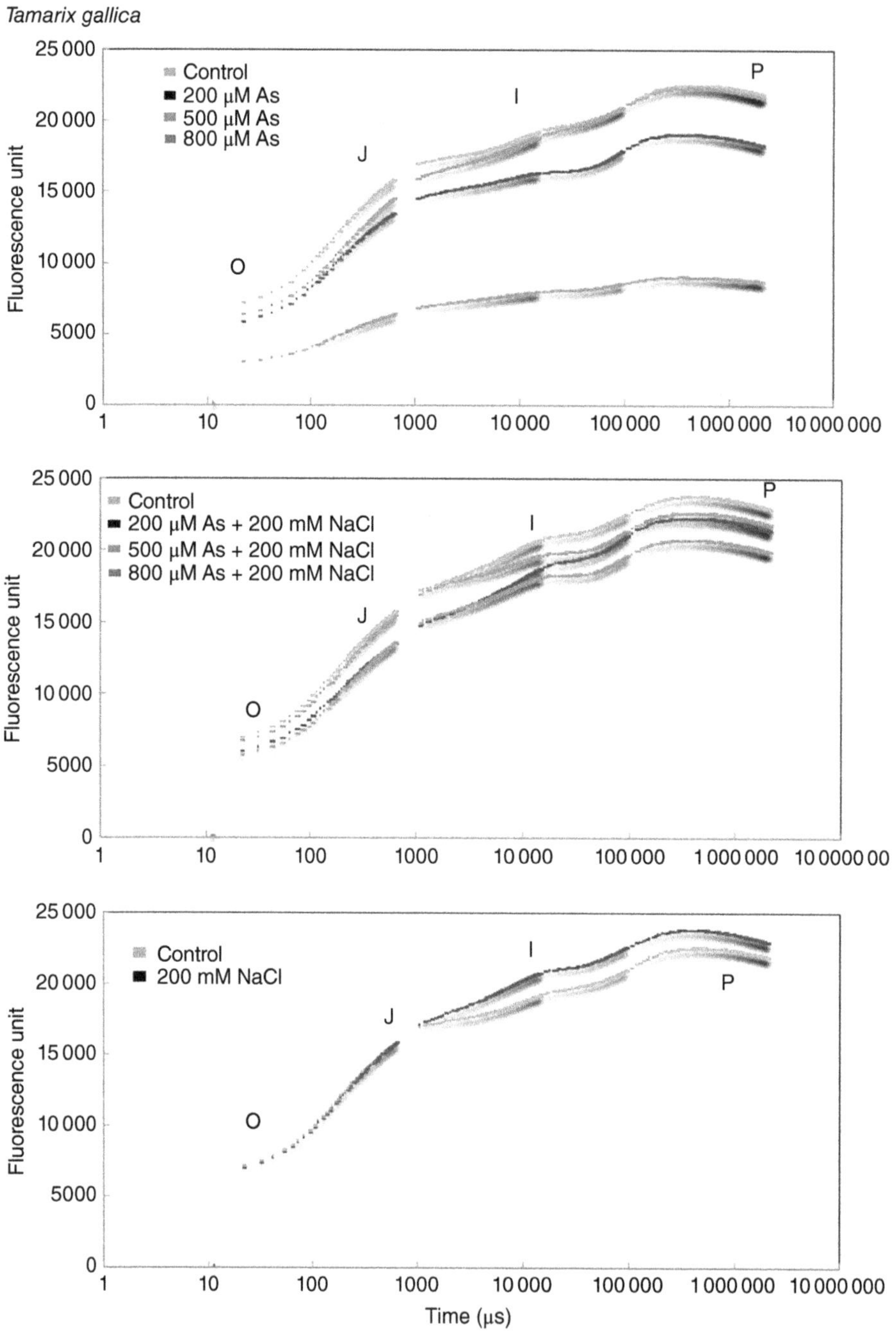

Figure 7.9 Average values of the Kautsky curves in dark-adapted leaves of *T. gallica* under combined stresses with As and NaCl. Bars marked with the same letter are not significantly different at $p = 0.05$. The concept of this figure is based on Sghaier et al. (2015).

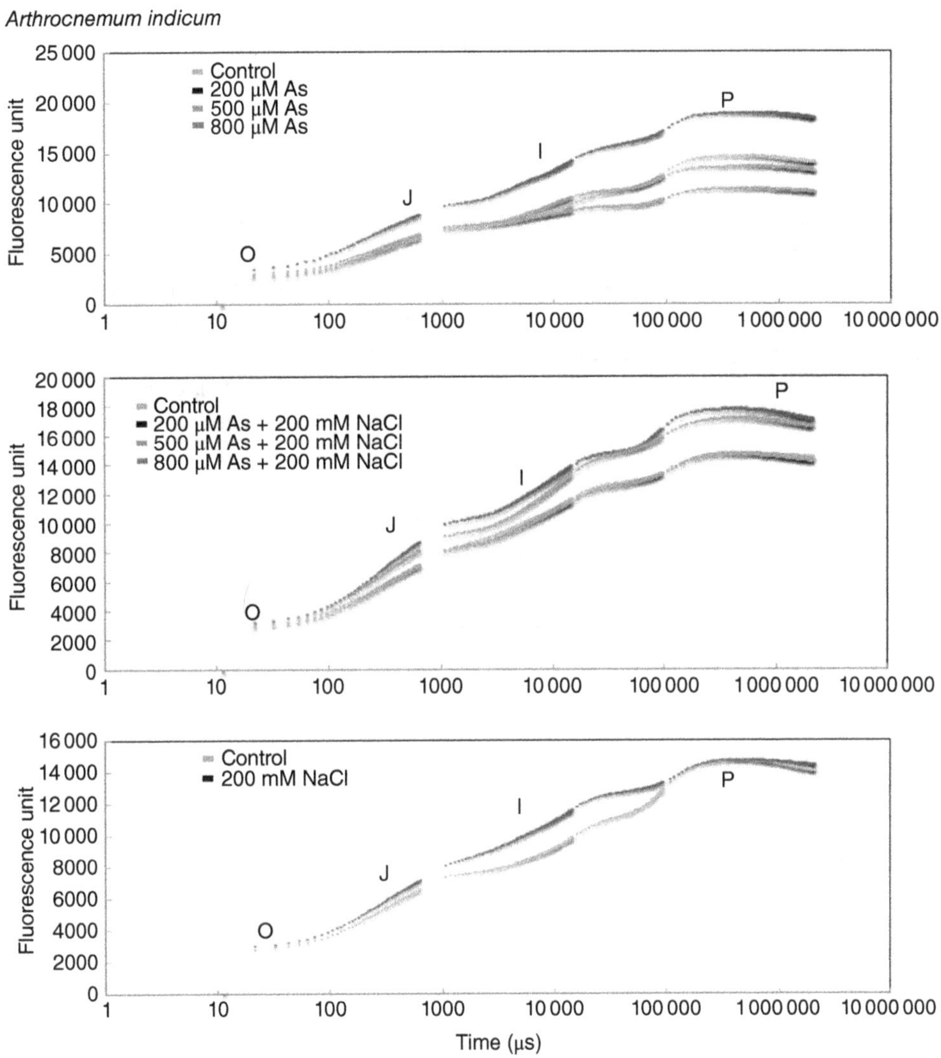

Figure 7.10 Average values of the Kautsky curves in dark-adapted leaves of *A. indicum* under combined stresses with As and NaCl. Bars marked with the same letter are not significantly different at $p = 0.05$.

different stages of building design and structure altering the composition and/or structure chlorophyll, thereby interfering with the organization of pigment–protein groups key for the perfect role of PS II (Rai et al. 2016). Therefore, extensive research had reported that the heavy metals incidence on the light period of photosynthesis is focused on the responses interconnected to PSII. In several species, metal stress has usually been associated with decreased fluorescence (Zou et al. 2009). Indeed, photosynthesis could be disrupted via a diminishment in plastid pigment content as suggested by Mishra et al. (2017) in rice, Malik et al. (2011) in chickpeas, and Naeem et al. (2019) in *Artemisia annua*.

The decreasing proportion of RCs active under heavy metal treatment as in the case of As can reduce the transport of photochemical electrons, leading to diminishing of the peak fluorescence capacity of chlorophyll a (Reboredo et al. 2021). In fact, the application of different doses of Hg in *Halimione portulacoides* did not induce symptoms of toxicity over the early hours of subjection, as evidenced through a powerful ET/CS energy flow and a poor energy waste. Nevertheless, 120 hours later to the subjection, few signals of stress were observed. The form of the oxidized quinone pool showed an expansion, revealing a few rates of electron transport from PS II to PS I, moreover noticeable through a small ET/CS (Cabrita et al. 2019), showing a characteristic stress mark under subjection to metals/metalloids as a result of the redox intrusion with cations inner to the chloroplasts and the oxidative explosion provoked through the ROS and, therefore, the peroxidation of membrane lipids, where quinones and all of ETC are holed (Anjum et al. 2016), conducting to an expansion in waste energy, making the chloroplast minor powerful and then conducting a fewer quantity of energy to the entire plant (Santos et al. 2015; Anjum et al. 2016).

The unvaried parameters F0 and Fv/Fm reveal that the light-collecting complexes II (LHCII) of this plant were not damaged by TME (Gusman et al. 2013), indicating the conservation of the total efficacy of PSII underneath exposure to arsenic alone or in combination with salt. This can be figured as the main scheme for *T. gallica* and *A. indicum* rising in polluted soil. In this study, we demonstrate that in these two species, to avoid metallic stress, plants used energy dissipation to overcome toxicity. The increase in non-photochemical quenching associated with the reduction in photochemical efficiency can be elucidated as a protected measure to prevent the phenomena of irreparable impairment to photosystems once photochemical reactions and/or carbon uptake are damaged (Müller et al. 2001; Lambrev et al. 2012; Arena et al. 2017). Yet, it is worth noting that in *A. indicum*, the presence of As generates a major rise in non photochemical quenching (NPQ) in comparison to the unpolluted area, manifesting non-photochemical waste procedure as a major means to counter the hazard of damage to photosystems. These results deal with prior research in which it was shown that although Pb did not alter ΦPSII, there was an increase in NPQ in parallel (Figlioli et al. 2019).

Likewise, our research is also coherent with the finding of Reboredo et al. (2021) who describes that in populations of hyperaccumulators of *Isatis cappadocica*, harvested in mine cuttings polluted by As, unchanged values of Fv/Fm and transfer rates of electrons once subjected to As in the 5–800 μM scale were presented, demonstrating usual photosynthetic powerful and plant resistance with the existence of a huge amount of As (Karimi et al. 2013). Furthermore, it was noted that a decrease in F0 was observed (and not in Fv/Fm) without any major incidence perceived on photosynthetic factors interconnected to photochemical performance PSII (Fv′/Fm′) and the consumption of photochemical energy (Y (II), qL). Moreover, once the single incidence on Y (II) was noted which was offset through an expansion in Y (NPQ), revealing the setting off of the photoprotective waste process. This purpose is related to the non-photochemical extinction imputed to photo-inactivation and the waste of unregulated energy (heat and fluorescence) in PSII (Krause and Jahns 2007; Busch et al. 2009) and its fewer changes in dealing plant with PSII. As indicated by the keeping of PSII function, and also by the no existence of major changes in factors revealing the PSII photoinhibition situation (Roboredo et al. 2021).

In addition, in the *Nicotiana tabacum* tolerant group, the net estimate of photosynthesis did not change mainly among the controls and As compared to the susceptible *Nicotiana sylvestris* group, where it significantly decreased under As (Kofroňová et al. 2020). In this research, the optimal fluorescence power of the supporting group was not altered under As, unlike the non-supporting group, where a major decline was presented (Kofroňová et al. 2020).

Once plants are subjected to arsenic, the form of the OJIP transient has undergone few alterations. The fluorescence liberation from this part is checked by response taking place in the donor part of PSII, while the I–P term matches to fluorescence quenching by oxidized plastoquinone (Abbaspoor et al. 2006). Whatever abiotic stress disrupts the organization and role of evolving OECs alters the quantity of oxygen evolution, hence boosting the fluorescence liberation due to the extinction phase J–I (Panda et al. 2006). In contrast, the transient increase from O to J is a result of the marked photochemical reduction from QA to QA- (Prakash et al. 2003).

In addition, *T. gallica* and *A. indicum* have the competence to waste excess energy which is considered as a particular phenomenon of halophytes that beat augmentation, inordinate potency reduction, the main source of ROS, thus escaping photo-inhibition of photosynthesis (Duarte et al. 2014). This indicates that most of the excitation energy dissipates as a non-photochemical mechanism, likely adaptive protection of the internal mechanism of the photosynthetic apparatus (Duarte et al. 2014).

Metals can induce an alteration of the photochemical process of photosystems II and affect the quantity of photosynthetic electron transport and the production of ATP and NADPH. PSII is usually assessed as the most delicate point for the poisonous action of metals (Jiang et al. 2008). To maintain the optimal functioning of photosystems and limit the formation of ROS, the plant uses regulatory mechanisms to manage the excess light energy absorbed by photosynthetic pigments (Niyogi et al. 1997). Stressed plants dissipate more energy to overcome excessive ion buildup and reduction in potency and prevent photodestruction of the photosynthetic system (Duarte et al. 2014).

The dissipation of this overabundance of energy in photosystems II in the kind of heat or in non-altering chemical responses (Lu et al. 2003) represents a coping mechanism for dealing with stress, rather a negative consequence of it. On the other hand, it is established that under normal physiological conditions, photoinhibition is controlled by the dissipation of excess energy in non-photochemical form (Krause and Jahns 2007).

7.5 Conclusion

The determination of heavy metals' effects on the process of photosynthesis, as a vital process for the plant, is an indispensable way for making a trustworthy approach to pick appropriate plants for phytoremediation and for the selection of As-supporting plants.

Overall, our research demonstrates a big tolerance of both halophytes to As contamination, revealing *A. indicum* as a proper plant for phyto-attenuation of As-contaminated soils like salt marshes or sebkha through a high As contamination in salty medium and the improvement in chlorophyll biosynthesis and photosynthesis apparatus as documented in this study. The data obtained in this chapter allow us to distinguish the tolerance pathway adapted by *T. galica* in stress conditions and can be productive to explain the approach adopted by the halophyte toward oxidative stress caused by arsenic.

The current results also have important implications for agronomic, as these two halophytic plants are known for their virtues in traditional medicine and even as human and animal food sources in the case of *A. indicum*.

References

Abbaspoor, M., Teicher, H.B., and Streibig, J.C. (2006). The effect of root-absorbed PSII inhibitors on Kautsky curve parameters in sugar beet. *Weed Res.* 46: 226–235.

Ali, B., Qian, P., Jin, R. et al. (2014). Physiological and ultra-structural changes in *Brassica napus* seedlings induced by cadmium stress. *Biol. Plant.* 58: 131–138.

Anjum, N.A., Duarte, B., Caçador, I. et al. (2016). Biophysical and biochemical markers of metal/metalloid-impacts in salt marsh halophytes and their implications. *Front. Environ. Sci.* 4: 1–13.

Arena, C., Figlioli, F., Sorrentino, M.C. et al. (2017). Ultrastructural, protein and photosynthetic alterations induced by Pb and Cd in *Cynara cardunculus* L., and its potential for phytoremediation. *Ecotox. Environ. Safe* 145: 83–89.

Busch, F., Hunter, N.P.A., and Ensminger, I. (2009). Biochemical constrains limit the potential of the photochemical reflectance index as a predictor of effective quantum efficiency of photosynthesis during the winter-spring transition in Jack pine seedlings. *Funct. Plant Biol.* 36: 1016–1026.

Cabrita, M.T., Duarte, B., Cesário, R. et al. (2019). Mercury mobility and effects in the salt-marsh plant *Halimione portulacoides*: uptake, transport, and toxicity and tolerance mechanisms. *Sci. Total Environ.* 650: 111–120. https://doi.org/10.1016/j.scitotenv.2018.08.335.

Caçador, I., Caetano, M., Duarte, B., and Vale, C. (2009). Stock and losses of trace metals from salt marsh plants. *Mar. Environ. Res.* 67: 75–82.

Dave, R., Singh, P.K., Tripathi, P. et al. (2013). Arsenite tolerance is related to proportional thiolic metabolite synthesis in rice (*Oryza sativa* L.). *Arch. Environ. Contam. Toxicol.* 64: 235–242.

Demmig-Adams, B. and Adams, W.W. (1996). The role of xanthophyll cycle carotenoids in the protection of photosynthesis. *Trends Plant Sci.* 1: 21–26.

Duarte, B. and Cacador, I. (2021). Iberian halophytes as agroecological solutions for degraded lands and biosaline agriculture. *Sustain. For.* 13: 1005. https://doi.org/10.3390/su13021005.

Duarte, B., Sleimi, N., and Caçador, I. (2014). Biophysical and biochemical constraints imposed by salt stress: learning from halophytes. *Front. Plant Sci.* 5: 1–10.

Duman, F., Ozturk, F., and Aydin, Z. (2010). Biological responses of duckweed (*Lemna minor* L.) exposed to the inorganic arsenic species As (III) and As (V): effects of concentration and duration of exposure. *Ecotoxicology* 19: 983–993.

Figlioli, F., Sorrentino, M.C., Memoli1, V. et al. (2019). Overall plant responses to Cd and Pb metal stress in maize: growth pattern, ultrastructure, and photosynthetic activity. *Environ. Sci. Pollut. Res.* 26 (2): 1781–1790. https://doi.org/10.1007/s11356-018-3743-y.

Finnegan, P.M. and Chen, W. (2012). Arsenic toxicity: the effects on plant metabolism. *Front. Physiol.* 3 (182): 1–18. https://doi.org/10.3389/fphys.2012.00182.

Flora, S.J.S. (2011). Arsenic-induced oxidative stress and its reversibility. *Free Radic. Biol. Med.* 51: 257–281.

Gaikwad, D.J., Pal, A., and Mondal, S. (2020). Effect of arsenic toxicity on chlorophyll content and antioxidative enzymes activities in rice seedlings. *Int. J. Environ. Clim. Change* 10 (12): 411–421. https://doi.org/10.9734/IJECC/2020/v10i1230318.

Gangwar, S., Singh, V.P., Tripathi, D.K. et al. (2014). Plant responses to metal stress: the emerging role of plant growth hormones in toxicity alleviation. In: *Emerging Technologies and Management of Crop Stress Tolerance* (ed. P. Ahmad and S. Rasool), 215–248. San Diego, CA: Academic Press https://doi.org/10.1016/B978-0-12-800875-1.00010-7.

González, P.S., Talano, M.A., Oller, A.L.W. et al. (2014). Update on mechanisms involved in arsenic and chromium accumulation, translocation and homeostasis in plants. In: *Heavy Metal Remediation* (ed. D.K. Gupta and S. Chatterjee), 45–72. New York: Nova Science Publishers ISBN: 978-1-63321-568-9.

Guo, H., Chen, H., Hong, C. et al. (2017). Exogenous malic acid alleviates cadmium toxicity in *Miscanthus sacchariflorus* through enhancing photosynthetic capacity and restraining ROS accumulation. *Ecotoxicol. Environ. Saf.* 141: 119–128.

Guo, H., Hong, C., Chen, X. et al. (2016). Different growth and physiological responses to cadmium of the three *Miscanthus* species. *PLoS One* 11 (4): e0153475.

Gusman, G.S., Oliveira, J.A., Farnese, F.S., and Cambraia, J. (2013). Arsenate and arsenite: the toxic effects on photosynthesis and growth of lettuce plants. *Acta Physiol. Plant.* 35: 1201–1209.

Hajiboland, R. (2014). Reactive oxygen species and photosynthesis. In: *Oxidative Damage to Plants* (ed. P. Ahmad), 1–63. San Diego, CA: Academic Press https://doi.org/10.1016/B978-0-12-799963-0.00001-0.

Hasanuzzaman, M., Nahar, K., Alam, M.M. et al. (2014). Potential use of halophytes to remediate saline soils. *Biomed. Res Int.* 2014 (589341): 1–12.

Heber, U., Bilger, W., and Shuvalov, V.A. (2006). Thermal energy dissipation in reaction centres and in the antenna of photosystem II protects desiccated poikilohydric mosses against photo-oxidation. *J. Exp. Bot.* 57: 2993–3006.

Hendrickson, L., Bail, M.C., Osmond, C.B. et al. (2003). Assessment of photoprotection mechanisms of grapevines at low temperature. *Funct. Plant Biol.* 30: 631–642.

Jiang, H.X., Chen, L.S., Zheng, J.G. et al. (2008). Aluminum induced effects on photosystem II photochemistry in citrus leaves assessed by the chlorophyll a fluorescence transient. *Tree Physiol.* 28: 1863–1871.

Karimi, N., Shayesteh, L.S., Ghasmpour, H., and Alavi, M. (2013). Effects of arsenic on growth, photosynthetic activity, and accumulation in two new hyperaccumulating populations of *Isatis cappadocica* desv. *J. Plant Growth Regul.* 32: 823–830.

Khalid, S., Shahid, M., Niazi, N.K. et al. (2016). Arsenic behaviour in soil-plant system: biogeochemical reactions and chemical speciation influences. In: *Enhancing Cleanup of Environmental Pollutants|| Non Biological Approaches* (ed. N. Anjum, S.S. Gill and N. Tuteja), 97–140. Cham: Springer.

Khan, M., Daud, M.K., Basharat, A. et al. (2016). Alleviation of lead-induced physiological, metabolic, andultramorphological changes in leaves of upland cotton through glutathione. *Environ. Sci. Pollut. Res.* 23: 8431–8440. https://doi.org/10.1007/s11356-015-5959-4.

Khan, M., Ahmad, R., Khan, M.D. et al. (2018). Trace elements in abiotic stress tolerance. In: *Plant Nutrients and Abiotic Stress Tolerance* (ed. M. Hasanuzzaman, M. Fujita, H. Oku, et al.), 137–151. Singapore: Springer https://doi.org/10.1007/978-981-10-9044-8_5.

Khan, M., Nawaz, N., Ali, I. et al. (2019). Regulation of photosynthesis under metal stress. In: *Photosynthesis, Productivity and Environmental Stress* (ed. P. Ahmad, M.A. Ahanger, M.N. Alyemeni and P. Alam), 95–105. Hoboken, NJ: Wiley https://doi.org/10.1002/9781119501800.ch6.

Kofroňová, M., Hrdinová, A., Mašková, P. et al. (2020). Multi-component antioxidative system and robust carbohydrate status, the essence of plant arsenic tolerance. *Antioxidants* 9: 283. https://doi.org/10.3390/antiox9040283.

Krause, G.H. and Jahns, P. (2007). Non-photochemical energy dissipation determined by chlorophyll fluorescence quenching: characterization and function. In: *Plant Cell Monographs*, vol. 19 (ed. G.C. Papageorgiou and Govindjee), 463–495. Dordrecht: Springer https://doi.org/10.1007/978-1-4020-3218-9_18.

Lambrev, P.H., Miloslavina, Y., Jahns, P., and Holzwartha, A.R. (2012). On the relationship between non-photochemical quenching and photoprotection of photosystem II. *Biochim. Biophys. Acta* 1817 (5): 760–769.

Li, N., Wang, J., and Song, W.Y. (2016). Arsenic uptake and translocation in plants. *Plant Cell Physiol.* 57: 4–13.

Liu, X., Wu, H., Ji, C. et al. (2013). An integrated proteomic and metabolomic study on the chronic effects of mercury in *Suaeda salsa* under an environmentally relevant salinity. *PLoS One* 8 (5): e6404.

Lu, C., Qiu, N., Lu, Q., and Marcelis, L.F.M. (2003). Photoinhibition and the xanthophyll cycle are not enhanced in the salt-acclimated halophyte *Artemisia anethifolia*. *Plant Physiol.* 118: 532–537.

Malik, J.A., Goel, S., Sandhir, R., and Nayyar, H. (2011). Uptake and distribution of arsenic in chickpea: effects on seed yield and seed composition. *Commun. Soil Sci. Plant Anal.* 42: 1728–1738. https://doi.org/10.1080/00103624.2011.584593.

Mecwan Neha, V., Yagnik, B.N., and Solanki Hitesh, A. (2018). Bioremediation of arsenic contaminated soil and its impacts on plants and human health. *Int. J. Res. Advent Technol* 6 (5): 351–358.

Meharg, A.A. and Hartley-Whitaker, J. (2002). Arsenic uptake and metabolism in arsenic resistant and nonresistant plant species. *New Phytol.* 154: 29–44.

Mei, L., Daud, M.K., Ullah, N. et al. (2015). Pretreatment with salicylic acid and ascorbic acid significantly mitigate oxidative stress induced by copper in cotton genotypes. *Environ. Sci. Pollut. Res.* 22: 9922–9931. https://doi.org/10.1007/s11356-015-4075-9.

Mishra, R.K., Kumar, J., Srivastava, P.K. et al. (2017). PSII photochemistry, oxidative damage and anti-oxidative enzymes in arsenate-stressed *Oryza sativa* L. seedlings. *Chem. Ecol.* 33: 34–50. https://doi.org/10.1080/02757540.2016.1265516.

Moronsito, T., Caffari, S., Dall'Osto, L., and Bassi, R. (2003). Mechanistic aspect of the xanthophyll dynarnics in higher plant thylakoids. *Physiol. Plant.* 119: 347–354.

Müller, P., Li, X.-P., and Niyogi, K.K. (2001). Non-photochemical quenching. A response to excess light energy. *Plant Physiol.* 125: 1558–1566.

Naeem, M., Aftab, T., Ansari, A.A. et al. (2019). Arsenic exposure modulates physiological attributes and artemisinin biosynthesis in *Artemisia annua* L. *Int. J. Herb. Med.* 7: 19–26.

Niyogi, K.K., Bjorkman, O., and Grossman, A.R. (1997). The role of specific xanthophylls in photoprotection. *Proc. Natl Acad. Sci. USA* 94: 14162–14167.

Otero, S., Nunez-Olivera, E., Martinez-Abaigar, J. et al. (2006). Effects of cadmium and enhanced UV radiation on the physiology and the concentration of UV-asborbing compounds of the aquatic liverwort *Jungermannia exsertifolia* subsp. *Cordifolia*. *Photochem. Photobiol. Sci.* 5: 760–769.

Panda, D., Rao, D.N., Sharma, S.G. et al. (2006). Submergence effects on rice genotypes during seedling stage: probing of submergence driven changes of photosystem II by chlorophyll a fluorescence induction O-J-I-P transient. *Photosynthetica* 44: 69–75.

Patra, M., Bhowmik, N., Bandopadhyay, B., and Sharma, A. (2004). Comparison of mercury, lead and arsenic with respect to genotoxic effects on plant systems and the development of genetic tolerance. *Environ. Exp. Bot.* 52: 199–223.

Prakash, J.S.S., Srivastava, A., Strasser, R.J., and Mohanty, P. (2003). Senescence-induced alternation in the photosystem II functions of *Cucumis sativus* cotyledons: probing of senescence driven alternation of photosystem II by chlorophyll a fluorescence induction O-J-I-P transients. *Indian J. Biochem. Biophys.* 40: 160–168.

Rai, R., Agrawal, M., and Agrawal, S. (2016). Impact of heavy metals on physiological processes of plants: with special reference to photosynthetic system. In: *Plant Responses to Xenobiotics* (ed. A. Singh, S. Prasad and R. Singh), 127–140. Singapore: Springer https://doi.org/10.1007/978-981-10-2860-1.

Reboredo, F.H., Pelica, J., Lidon, F.C. et al. (2021). The tolerance of *Eucalyptus globulus* to soil contamination with arsenic. *Plan. Theory* 10: 627. https://doi.org/10.3390/plants10040627.

Reeves, R.D. and Baker, A.J. (2000). Metal-accumulating plants. In: *Phytoremediation of Toxic Metals: Using Plants to Clean up the Environment* (ed. I. Raskin and B.D. Finsley), 193–229. New York: Wiley.

Santos, D., Duarte, B., and Caçador, I. (2015). Biochemical and Photochemical feedbacks of acute Cd toxicity in *Juncus acutus* seedlings: the role of non-functional. *Estuar. Coast. Shelf Sci.* 167: 228–239.

Sghaier, D.B., Duarte, B., Bankaji, I. et al. (2015). Growth, chlorophyll fluorescence and mineral nutrition in the halophyte *Tamarix gallica* cultivated in combined stress conditions: arsenic and NaCl. *J. Photochem. Photobiol. B* 149: 204–214. https://doi.org/10.1016/j.jphotobiol.2015.06.003.

Sghaier, D.B., Pedro, S., Diniz, M.S. et al. (2016). Tissue localization and distribution of As and Al in the halophyte *Tamarix gallica* under controlled conditions. *Front. Mar. Sci.* 3: 274.

Sghaier, D.B., Pedro, S., Duarte, B. et al. (2019). Arsenic tolerance mechanism in halophyte plant: the case of *Tamarix gallica*. In: *Halophytes and Climate Change: Adaptive Mechanisms and Potential Uses* (ed. M. Hasanuzzaman, S. Shabala and M. Fujita), 255–265. CABI https://doi.org/10.1079/9781786394330.0255.

Sghaier, D.B., Pedro, S., Duarte, B. et al. (2020). Arsenic accumulation, compartmentation, and complexation in arthrocnemum indicum. In: *Plant Ecophysiology and Adaptation Under Climate Change: Mechanisms and Perspectives I* (ed. M. Hasanuzzaman). Singapore: Springer https://doi.org/10.1007/978-981-15-2156-0_25.

Sharma, P. and Dubey, R.S. (2005). Lead toxicity in plants. *Braz. J. Plant Physiol.* 17 (1): 35–52.

Sharma, A., Kumar, V., Shahzad, B. et al. (2020). Photosynthetic response of plants under different abiotic stresses: a review. *J. Plant Growth Regul.* 39: 509–531. https://doi.org/10.1007/s00344-019-10018-x.

Singh, H.P., Mahajan, P., Kaur, S. et al. (2013). Chromium toxicity and tolerance in plants. *Environ. Chem. Lett.* 11 (3): 229–254.

Stoeva, N., Berova, M., and Zlatev, Z. (2003). Physiological response of maize to arsenic contamination. *Biol. Plant.* 47: 449–452.

Van Der Ent, A., Baker, A.J.M., Reeves, R.D. et al. (2013). Hyperaccumulators of metal and metalloid trace elements: facts and fiction. *Plant Soil* 362: 319–334.

Vincent, D. and Zivy, M. (2007). Plant proteome responses to abiotic stress. In: *Plant Proteomics* (ed. J. Šamaj and J.J. Thelen), 346–364. Berlin, Heidelberg, Germany: Springer-Verlag.

Zhang, X., Li, M., Yang, H. et al. (2018). Physiological responses of *Suaeda glauca* and *Arabidopsis thaliana* in phytoremediation of heavy metals. *J. Environ. Manag.* 223: 132–139.

Zhao, F.J., Ma, J.F., Meharg, A.A., and McGrath, S.P. (2009). Arsenic uptake and metabolism in plants. *New Phytol.* 181: 777–794.

Zou, J.H., Yu, K.L., Zhang, Z.G. et al. (2009). Antioxidant response system and chlorophyll fluorescence in chromium (VI) – treated *Zea mays* L. seedlings. *Acta Biol. Cracov. Ser. Bot.* 51: 23–33.

8

Genomic and Transcriptional Regulation During Arsenic Stress

Madhu Tiwari[1], *Maria Kidwai*[1], *Neelam Gautam*[1,2], *and Debasis Chakrabarty*[1,2]

[1] *Biotechnology and Molecular Biology Division, CSIR-National Botanical Research Institute, Lucknow, Uttar Pradesh, India*
[2] *Academy of Scientific and Innovative Research (AcSIR), Ghaziabad, Uttar Pradesh, India*

8.1 Introduction

Arsenic (As) is a toxic metalloid with high pervasiveness in the environment (Podgorski and Berg 2020). In 2019, the Agency of Toxic Substance and Disease Registry (ATSDR) (US Department of Health and Human Services) prioritized As on the topmost position due to its toxic effect on human health (https://www.atsdr.cdc.gov/spl/index.html). Worldwide, 108 countries are affected by groundwater As contamination in which southeast Asian countries are most affected (Shaji et al. 2020). Among all forms of As, the inorganic As, arsenate (AsV), and arsenite (AsIII), are highly toxic to humans and plants (Tripathi et al. 2007). Long-term exposure to As can cause cancer, skin lesions, cardiovascular disease, and diabetes in humans (Jomova et al. 2011).

Natural as well as anthropogenic activities are involved in As contamination in the environment. Arsenic-contaminated groundwater adversely affects the huge human population (approximately 230 million people) and is mainly an outcome of anthropogenic activities (Shaji et al. 2020). In soil, both forms of inorganic As, i.e. AsV and AsIII, predominate depending on environmental conditions. Under anaerobic conditions, AsV predominates over AsIII. However, under aerobic conditions like in aquatic systems or paddy fields, AsV is oxidized to AsIII and, thus, AsIII is the dominating species (Lim et al. 2014).

Plants are exposed to As through irrigation with As-contaminated groundwater (Huq et al. 2006). Although As is a nonessential metalloid and has no physiological role in a living organism (Khalid et al. 2017), plants are reported to uptake As from soil or water (Abbas et al. 2018). Heikens (2006) reported that the consumption of As-contaminated food is the main source of human exposure to As (Heikens 2006). Plant uptake both forms of As via absorption through root in terrestrial plants and through submerged leaves in aquatic plants (Abbas et al. 2018). The mechanism through which AsV and AsIII are primarily taken up in plants is completely different. Being a phosphate analogue, AsV is taken up by

Arsenic in Plants: Uptake, Consequences and Remediation Techniques, First Edition.
Edited by Prabhat Kumar Srivastava, Rachana Singh, Parul Parihar, and Sheo Mohan Prasad.

the phosphate transporter (Meharg and Macnair 1990; Zhao et al. 2009). In contrast, AsIII is mainly taken up by the aquaporin channels (Bienert et al. 2008; Chen et al. 2017b). Both AsV and AsIII interfere with metabolic pathways in plants. AsV, being a phosphate analogue, replaces the phosphate group in adenosine triphosphate (ATP) and disturbs the ATP-driven processes in the cell (Sherwood et al. 2011). Also, As severely affects the redox state of the cell and causes oxidative stress in plants (Sharma 2012). Plants showed various compromised physiological parameters like reduced chlorophyll synthesis, photosystem damage, and reduced photosynthetic activities on exposure with As (Abbas et al. 2018). Huhmann et al. 2017 showed a huge loss in rice productivity since the last 25 years in Bangladesh (approximately 7–26% $year^{-1}$) because of the use of As-polluted water during irrigation. Thus, As contamination in plants ultimately leads to low productivity which is critical to serving such a huge population in the world. Additionally, As loading into the vascular bundle and its further upward translocation enhances the As accumulation in grains/edible parts of plants. The subsequent intake of these As-contaminated food products serves as a source of the contaminated food chain in the ecosystem.

Keeping in view the critical climatic conditions, Muehe et al. 2019 estimated the huge loss and contamination in rice grains in near future. On exposure to As, plants reorganize the genetic and transcriptional regulations of various stress-responsive genes in order to resist the As toxicity (Chakrabarty et al. 2009; Fu et al. 2014; Hwang et al. 2017). To understand the coping capabilities and strategies in plants during As toxicity, the study of differential regulation of As-responsive gene, transcription factors (TFs), and transporters will provide a good strategy to cope with As toxicity. By manipulating the As-responsive genes, one can develop a transgenic plant with higher yield and lesser As accumulation in edible plant parts. In this chapter, we summarized the *in silico, in vitro,* and *in vivo* studies of functionally characterized genes that responded against As toxicity in various plants.

8.2 Study of Differentially Regulated Genes During Arsenic Stress in Plants

Stress-triggered genes are broadly divided into two categories – one is regulatory while the other is functional genes. Regulatory genes are comprised of TFs such as myeloblastosis (MYB), Auxin response factors (ARF), WRKY, MYC, etc., which regulate the other genes by binding to the specific binding sites at the promoter of the target genes. Whereas the functional genes encode for the metabolites, antioxidants, transporters, etc., that are directly involved in stress response. Arsenic being a heavy toxic metalloid disturbs the plant's metabolic and physiological pathways. Therefore, understanding the plant responses under As toxicity is a major concern for improving crop productivity. Previously, genomics studies have shed light on the genetic basis of As tolerance in plants. They have identified the genes involved in the As uptake, transport, and detoxification. Genome-wide analysis (GWA) of various As-treated plants had been done which demonstrated immense As-responsive genes present in the genome. Transcriptome analysis of plants under As stress has also identified various defense-related genes. Now, several technologies have been in use to study the transcriptome data such as next-generation sequencing (NGS) and RNA sequencing (RNA-Seq) (Lowe et al. 2017). Such technologies have been helpful to identify specific genes in different plants like *Arabidopsis*, rice, and maize that are involved in plant defense (Chakrabarty et al. 2009; Ghosh et al. 2017; Shukla et al. 2018). Earlier, the mechanism of As-induced stress and tolerance in two contrasting genotypes of *Brassica juncea*, viz. TPM-1 (tolerant) and TM-4 (sensitive) were investigated by Srivastava et al. 2009. Comparative transcriptional profiling was performed on both the genotypes. In TPM-1, upregulation of genes encoding for candidates of hormones' (auxin and jasmonate) biosynthesis pathways and sulfate transporters was observed, whereas downregulation of ethylene (ET) and cytokinin biosynthesis genes was observed. Notably, this downregulation of genes only occurred when the plant was exposed to As for a short duration of time but longer exposure results in upregulation of these genes. The observation suggested that an early perception downstream signaling pathways confer As tolerance in TPM-1. To unravel the major genes and signaling pathways involved in AsV stress in *B. juncea*, Srivastava et al. 2015 carried out the transcriptome profiling of *B. juncea* under AsV stress. Total 1285 genes showed modulated expression pattern that mainly belongs to kinases and hormonal (auxin, jasmonate, abscisic acid, and ET) signaling pathways. Also, a significant expression of nitrogen, sulfur, and lipid metabolism genes along with photosynthesis was observed.

Earlier, Norton et al. 2008 conducted transcriptome analysis in roots of rice seedlings of two different varieties that were AsV-tolerant (Bala) and sensitive (Azucena). There was at least twofold upregulation in a total of 576 probe sets and 622 probe sets were downregulated. Most of the differentially expressed genes (DEGs) belong to the stress proteins, transporters, and TFs. The upregulation of glutathione biosynthesis, metabolism and transporter genes indicates the relation between glutathione conjugation of AsV that can be considered as a detoxification response during As stress.

The comparative transcriptome analysis of rice under AsV and AsIII treatment was also conducted by other researchers. They demonstrated that there were DEGs under both the treatments in rice seedlings. Most of the genes were defense-related, transporters, heat shock proteins, metallothioneins, and regulatory genes. However, there were some sets of genes that were uniquely upregulated in either AsV or AsIII response. For instance, cytochrome P450 was upregulated in root only under AsV toxicity and glutaredoxin (Grxs) was upregulated in shoots only under AsIII toxicity (Chakrabarty et al. 2009). Further, the changes in transcript level and signaling pathways were also analyzed in rice roots under AsV treatment. Transcriptome analysis showed induced expression of various genes encoding heat shock factors, TFs, components of signaling pathways such as protein kinase, mitogen-activated protein kinase (MAPK) and calcium-dependent protein kinase (CDPK). The genes involved in the cell cycle and secondary cell wall biogenesis were reported to be repressed. Also, the expression of GARP-G2-like and C3H TFs were found to be modulated under AsV stress (Huang et al. 2012).

Arsenic accumulation and tolerance vary with genotypic variation in plants that are regulated by various environmental and genetic factors. To unravel the reason behind As-induced differential response in different genotypes, Rai et al. 2015 performed comparative transcriptome analysis in roots of high- and low-As-accumulating (HARG and LARG) rice genotypes under As toxicity. It was found that there was a genotype-dependent As tolerance and accumulation in rice genotypes which results in the operation of different regulating pathways. Also, the presence of unique Cis-motifs at DEGs suggested that the differential As accumulation in HARG and LARG rice genotypes is determined genetically and not regulated at the transcription level. Similarly, Shukla et al. 2018 investigated the difference in molecular and genetic responses of two *Arabidopsis* accessions, viz. As-tolerant (Col-0) and As-sensitive (Slavi-1). Comparative GWA of both the accessions showed that 120 DEGs were common in both the accessions, whereas 168 and 548 genes were differentially expressed in Col-0 and Slavi-1, respectively. Among these DEGs, some were associated with the antioxidant defense system, transport, TFs, and plant development.

Hwang et al. 2017 generated an As-tolerant type 1 (ATT1) rice mutant and compared the transcriptome data of wild type (WT) and ATT1 rice mutant to identify and characterize the As-responsive genes. Diverse genes related to mitochondrial transport, photosynthesis, and lipid biosynthesis were downregulated under As stress in both. However, some sulfate transporters were upregulated under AsV stress. To investigate the molecular mechanism behind As response in chickpea, whole genome-wide expression was performed by Yadav et al. 2019. It was observed that total 490 genes were upregulated and 645 genes were downregulated under As stress and the pathways which were affected were phenylpropanoid biosynthesis, indole alkaloid biosynthesis, pentose, and glucuronate interconversions.

The transcriptional response of *Arabidopsis* to AsV stress was studied by whole-genome oligonucleotide microarrays. The data suggested that the genes involved in stress response such as antioxidant-related genes were highly upregulated (Abercrombie et al. 2008). To understand the As response in *Arabidopsis* accessions, Col-0 and Ws-2, genome-wide gene expression was compared between both the genotypes. From the analysis, it was observed that the genes related to oxidative stress response such as heat shock protein, aquaporins, and transporters were common in both, but the ET-signaling-pathway genes were only upregulated in Col-0 As-tolerant accession (Fu et al. 2014).

Previously, GWA mapping of 349 genetically diverse *Arabidopsis* accessions was performed to identify the genetic loci which control the As accumulation in leaves. The study showed the presence of arsenate reductase (ArsC), which reduces AsV to AsIII. Thus, they named it high arsenic content 1 (HAC1). The presence of HAC1 in the pericycle limits loading of As into the xylem (Chao et al. 2014). Recently, in *Arabidopsis*, As response due to genetic variation was studied by Peña-Garcia et al. 2021 using a genome-wide association study (GWAS). They identified a genomic region associated with AsV response and revealed it as an arsenic stress-related F-box (ASRF) protein gene which was involved in AsV tolerance in *Arabidopsis*. Transcriptome analysis of the wild-type and *asrf* mutant in *Arabidopsis*, revealed an altered phosphate signaling and homeostasis in *asrf* mutant under AsV stress. This demonstrated that ASRF is involved in As tolerance in *Arabidopsis*.

Illumina-based transcriptomic profiling was performed under As stress on *Panax notoginseng*, a famous herbal medicine. By comparative analysis between As-treated (HAs) and non-treated (CK) plants, 1725 DEGs were identified with 763 upregulated and 962 downregulated genes. Total 20 genes were selected for quantitative real time PCR (qRT-PCR). Among them, seven genes coding for ginsenoside and flavonoid biosynthesis pathways were downregulated in HAs and oxidative stress-related genes (GSTs), TFs (HSF and MYBs), and molecular chaperones were upregulated. Genes encoding ATP-binding cassette (ABC) transporters, Multi-antimicrobial extrusion protein (MATE) and ET metabolism and As accumulation transporter (natural resistance-associated macrophage proteins) were differentially expressed (Liu et al. 2016). Arsenic effect on flavonoid content and metabolic pathway were analyzed in main roots of *Panax notoginseng* using RNA-Seq.

A distinct pattern of As-responsive genes was identified in As treated plant, which were also involved in secondary metabolism. Also, genes of major enzymes related to the flavonoid biosynthesis pathway were identified with a key gene chalcone synthase (DcCHS2) (Zu et al. 2018a). Similarly, the effect of As on saponin biosynthesis was studied using RNA-Seq. Transcriptomes of the main root of *Panax notoginseng* at the early flowering stage with different As concentrations were used for comparison using RNA-Seq. The result manifested the expression of key saponin biosynthesis pathway genes such as cytochrome P450 (CYP71D444 and CYP73A100) and glycosyltransferase gene (UGTPg25) depicting that As regulates transcript expression in saponin biosynthesis (Zu et al. 2018b).

Purple willow is a phytoremediator but the genetic mechanism, involved in As tolerance, metabolism, and phytoextraction, remains obscure. Therefore, to unravel the potential behind As tolerance in purple willow, As-responsive genetic factors were studied from their transcriptomic response to As stress. Different As-treated root, stem, and leaf transcriptomes were compared using RNA-Seq. DEGs were observed which code for a protein involved in As uptake via the root, its sequestration into the vacuole, and its further translocation. Hence, the results gave an insight into the genetic capability of purple willow for being a model species of As phytoremediator (Yanitch et al. 2017). Recently, to study the molecular response of upland rice (*Oryza sativa* indica) toward As stress, transcriptome analysis of AsIII and AsV-treated upland rice seedlings was performed. Results of data analysis showed that 2893 and 1844 genes were responsive to AsIII and AsV treatment, respectively, where 915 genes were shared in both treatments. Major changes were observed in genes belonging to transcriptional regulation, hormone signaling, and redox transporters. Genes for TF such as WRKY and AP2-EREBP were downregulated, whereas NAC was upregulated. Similarly, genes encoding superoxide dismutase (SOD), peroxidase, and catalase were also downregulated (Di et al. 2021).

The novel mechanism of gene regulation involves microRNAs (miRNAs), which regulate the expression of genes necessary for plant growth, development, and in various stress responses. To examine the role of miRNA in As stress, microarray profiling of miRNA was performed in *B. juncea*. It was observed that the predicted targets of miRNA belong to development (e.g. miR156 and miR169), hormonal biosynthesis/function (e.g. miR319 and miR167) and in sulfur uptake, transport, and in assimilation (miR395, miR838, and miR854) (Srivastava et al. 2013). High-throughput sequencing technology was used to study the transcriptome of rice under different times and doses of AsIII treatment. Also, the study focussed on metabolic regulation and their interaction with miRNA. Genes involved in jasmonic acid (JA) biosynthesis and signaling, heavy metal transport, and lipid metabolism were found in similar responses to AsIII stress. In addition, 36 new AsIII-responsive miRNA were identified, out of which 14 were involved in the regulation of genes of transporters, signaling, and metabolism (Yu et al. 2012).

In addition, microarray hybridization was carried out in two contrasting rice accessions to identify the miRNA which is responsive under AsIII and AsV toxicity. It was identified that the expressions of 14 miRNA were differentially regulated in AsIII and AsV stresses. Among these, various miRNA families like miR396, miR399, miR408, miR528, miR1861, miR2102, and miR2907 were found to be significantly upregulated. In contrast, some miRNA families like miR164, miR171, miR395, miR529, miR820, miR1432, and miR1846 were downregulated. The predicted targets of these miRNA families proved their important role in As stress response (Sharma et al. 2015). Ghosh et al. 2017 investigated the AsV-responsive miRNA in maize leaf, and from deep sequencing, they demonstrated 22 up- and 33 downregulated miRNA. From the gene ontology (GO) analysis, they found that the predicted targets of these miRNA were the genes, which relate to various biological processes such as metabolic (miR159b-5p.3, zma_460, zma_468, and zma_469) and hormonal signaling pathways (miR167d, miR159e.2, and miR319b-3p), plant growth and development (miR156s, miR166e-3p, miR166m, and miR319b-3p), and in reactive oxygen species (ROS) generation (miR167d). In one study, the genetic basis of As tolerance in rice was investigated by Murugaiyan et al. 2019, who performed quantitative trait locus (QTL) mapping in the early stage of rice seedling. From the analysis, some novel QTLs were identified associated with As tolerance and accumulation in which twenty-five genes showed transcriptional regulation for As toxicity (Murugaiyan et al. 2019).

8.3 Genetic Study of Arsenic-Responsive Genes in Plants

Plant roots absorb water and mineral from the soil for their growth and development. Various transporters are present on the root membrane that are directly involved in the absorption of minerals or water from the soil (Palmer et al. 2014; Sasaki et al. 2016). Along with several minerals, As is also present in the soil. Arsenate (AsV) is the predominant As species present in the soil, while arsenite (AsIII) is the dominant species in paddy fields (Lim et al. 2014). Both As species are toxic for plant growth and development (Finnegan and Chen 2012). Arsenic uptake and its subsequent translocation inside the plant utilize various plant transporters that are mainly used for nutrient uptake and translocation. Also, various detoxification genes are reported to confer tolerance against As toxicity in plants. Here we will discuss various AsV and AsIII transporters as well as the detoxification genes that mitigate the As toxicity in plants.

8.3.1 Genetic Study of Transporters Involved in Arsenic Uptake and Translocation

Both As species, i.e. AsV and AsIII are chemically different and utilize different plant root transporters to enter the plant cell. Till now, several transporters related to As uptake, translocation, and accumulation have been studied in detail.

8.3.1.1 Transporters Involved in Arsenate Uptake in Plants

AsV is a chemical analogue of phosphate (Tawfik and Viola 2011), and thus plants become very sensitive to AsV as it can enter plant cells by various phosphate transporters (Zhao et al. 2009). Interestingly, various phosphate transporters have been identified after treating plants with AsV (Catarecha et al. 2007; Chen et al. 2011; Wang et al. 2016). In *Arabidopsis*, *AtPHT1;1* and *AtPHT1;4* have been recognized as AsV transporters as their mutants showed enhanced AsV tolerance compared to wild-type plants (Shin et al. 2004). The *AtPHT1;1* higher expressing lines of *Arabidopsis* showed higher AsV uptake (LeBlanc et al. 2013), while *pht1;1* mutant *Arabidopsis* plant showed lesser AsV uptake (Catarecha et al. 2007). The *Arabidopsis* plant with high expressing *AtPHT1* or *AtPHT7* family under the regulation of 35S promoter showed enhanced AsV uptake (LeBlanc et al. 2013). Interestingly, Catarecha et al. 2007 has identified the AsV-tolerant mutant *Arabidopsis* plant named *pht1;1–3*. On close genetic observation of these mutants, it was identified that *AtPH1;1* possesses a semidominant allele (*pht1;1–3*) that affects the AsV uptake in these mutants. One study had also revealed the role of phosphate transporter facilitator (PHF) protein during AsV toxicity. PHF proteins facilitate the trafficking of phosphate transporter proteins on the plasma membrane. A *phf1* mutant plant had higher AsV tolerance than the wild-type plant, suggesting its function as an AsV transporter in a plant (González et al. 2005).

Under the Pi starvation condition, the *Arabidopsis* seedling was reported to activate different phosphate transporters that maintain the intracellular Pi acquisition. The two of these transporters named *Pht1;9* and *Pht1;8* have been reported for Pi acquisition as well as AsV uptake (Remy et al. 2012). Another phosphate transporter, *Pht1;5* from *Arabidopsis* has been reported to remobilize the Pi/AsV in plants. However, these transporters have no significant role in Pi acquisition. By using a reverse genetic approach, the *Pht1;5* transporter was found to be involved in source to sink mobilization of AsV (Nagarajan et al. 2011). Further, a phosphate transporter in *Arabidopsis*, *AtPht1;7*, which had higher transcript abundance in flower tissue rather than root or shoot tissues of the plant was also found to be responsive under AsV stress. When *Pht1;7* was expressed under the control of a constitutive promoter, the higher *Pht1;7* expressing lines of *Arabidopsis* showed increased AsV uptake as well as translocation (LeBlanc et al. 2013).

A vacuolar phosphate transporter in *Arabidopsis*, *AtVPT1*, has been recognized to sequester Pi in vacuole under high intracellular Pi concentration. In comparison to the wild-type plant, the high *VPT1* expressing lines of *Arabidopsis* showed higher As content (Luan et al. 2018). In contrast, the mutant *VPT1* lines showed lower As accumulation. However, whether *VPT1* is directly linked with AsV vacuolar sequestration is still unraveled (Luan et al. 2018). There are also several transporters reported in rice that are directly linked with AsV uptake and accumulation. *OsPT* (phosphate transporters in rice) expression analysis in rice seedlings showed three *OsPTs*, namely *OsPT2*, *OsPT4*, and *OsPT8*, which had a higher transcript abundance under phosphate starvation. However, *OsPT1*, in contrast, was not upregulated under the phosphate-limiting condition (Kamiya et al. 2013). It was identified that under the Pi-limiting condition and the AsV-rich condition, *OsPT2*, *OsPT4*, and *OsPT8* get suppressed. The observation suggests the probable impact of AsV on Pi signaling in rice. *OsPT1* overexpressing lines showed a higher As accumulation and translocation to the upper plant part (Kamiya et al. 2013). Similarly, another phosphate transporter, *OsPHT1;8* was found to be involved in Pi uptake (Jia et al. 2011), also supposed to serve during AsV uptake. Wu et al. 2011 has shown a higher accumulation of As when *OsPHT1;8* was overexpressed in rice (Wu et al. 2011). In addition, the rice *OsPHT1;8* mutant accumulated lesser As in grain compared to wild-type plant and thus found to be more tolerant against As toxicity (Wang et al. 2016). Similarly, a rice mutant of *OsPHT1;4* showed a drastic decrease, approximately 44%, of As accumulation in grains suggesting its role in As uptake and translocation (Cao et al. 2017; Ye et al. 2017). Further, the high-expressing *OsPHT1;4* rice lines were more sensitive to As toxicity and also showed a higher As content in rice (Ye et al. 2017).

The AsV transporters have also been studied in As hyperaccumulator plant, *Pteris vittata*. In *Pteris*, *PvPHT1;3* phosphate transporter showed a higher AsV uptake (DiTusa et al. 2016).

PvPHT1;3 is mainly expressed in the gametophyte of the *Pteris* plant and its expression is responsive under Pi-deficient and AsV-supplemented conditions. AsV showed a comparatively higher affinity with *PvPHT1;3* than *AtPHT1:5* (DiTusa et al. 2016). *PvPHT1;4*, a phosphate transporter, localized on fronds and roots of *Pteris*, showed induced expression in roots under Pi-deficient and AsV-supplemented conditions, suggesting its probable role in AsV uptake (Sun et al. 2019). Cao et al. (2019), by heterologous expression of *PvPHT1;3* in *Nicotiana*, found the increased As absorption and accumulation (Cao et al. 2019). Also, the heterologous expression of *PvPHT1;3* in yeast showed AsV transport activity (Sun et al. 2019).

8.3.1.2 Transporters for AsIII Uptake in Plants

In contrast to AsV, AsIII is present as a neutral molecule in the environment (Zhao et al. 2010). AsIII is a dithiol and affects the enzyme/protein activity by binding to its cysteine residue (Tawfik and Viola 2011). AsIII uptake has been done by aquaporin channels which are well known for transporting water and other molecules in plants. In plants, nodulin-26-like intrinsic proteins (*NIPs*) act as aquaporins and AsIII transport through NIPs is evolutionary conserved. Also, inside the plant, AsIII is transported through some plant intrinsic proteins like tonoplast intrinsic proteins (*TIPs*) and plasma membrane intrinsic proteins (*PIPs*) (Bienert and Jahn 2010).

In plants, different *NIPs* have been recognized to function as an AsIII transporter, for instance, *AtNIP5;1* and *AtNIP6:1* in *Arabidopsis*, *LjNIP5;1* and *LjNIP6;1* in lotus, *OsNIP2;1* and *OsNIP3;2* in rice (Bienert et al. 2008). On that note, AsIII transporters/channels are bidirectional, i.e. involved in AsIII uptake as well as efflux to the external medium. The *OsNIP3;2* is localized on the plasma membrane of lateral roots of rice. The expression pattern analysis reveals that the *OsNIP3;2* channel gets suppressed under low AsIII concentration. However, under high AsIII concentration, the *OsNIP3;2* channel get upregulated. Overall, *OsNIP3;2* channel uptakes AsIII through lateral roots in rice under high AsIII concentration (Chen et al. 2017b). Further, another aquaporin group in rice, *OsNIP1;1* and *OsNIP3;3*, was reported to have higher permeability for AsIII. Overexpression of both genes does not alter the AsIII uptake but showed a marked decrease in root to shoot translocation of AsIII. The overexpressing lines of *OsNIP1;1* and *OsNIP3;3* in rice were also reported to accumulate lesser As in the grains. The observation suggested that both genes restrict the xylem loading of As by channelizing a leak-out route for AsIII efflux (Sun et al. 2018). Additionally, the *PIPs* aquaporins also reported for arsenite uptake. Some *PIPs* from rice (*OsPIP2;4*, *OsPIP2;6*, and *OsPIP2;7*) have been heterologously expressed in *Xenopus laevis* oocyte that becomes resistant to AsIII toxicity (Mosa et al. 2012). The PIPs (*OsPIP2;4*, *OsPIP2;6*, and *OsPIP2;7*) overexpressing lines of *Arabidopsis* showed enhanced tolerance against AsIII toxicity with the addition of relatively increased biomass accumulation (Mosa et al. 2012).

8.3.1.3 Genes Involved in Intracellular AsV to AsIII Conversion in Plants

To study the major intracellular As species in plants, the xylem sap of AsV-treated plant was analyzed for As speciation. The xylem sap mainly possesses AsIII and not AsV (Su et al. 2010; Zhao et al. 2009), which concludes that although AsV uptake occurs in a plant, AsV is reduced in AsIII in the root cells (Dhankher et al. 2006; Xu et al. 2007). Dhankher et al. (2002) has performed the simultaneous heterologous expression of two bacterial genes in *Arabidopsis*, one of the bacterial genes encodes for ArsC and the other encodes for gamma-glutamylcysteine synthetase (gamma-ECS). Both genes, ArsC and gamma-ECS were cloned under the regulation of two different constitutive promoters, namely rubisco SRS1p small subunit and ACT2pt actin promoter, respectively. The overexpressing ArsC and gamma-ECS transgenic lines of *Arabidopsis* showed higher As tolerance. These overexpressing lines accumulated approximately threefolds more As than the wild-type plants and, thus, build up more tolerance toward As (Dhankher et al. 2002). In *Arabidopsis*, *AtHAC1* (or *AtATQ1*) functions as AsV reductase and reduces AsV to AsIII in the root cell (Chao et al. 2014; Sánchez-Bermejo et al. 2014). Two genes, *OsHAC1;1* and *OsHAC1;2* are predominantly expressed in the root and possess AsV reductase activity in rice plants (Shi et al. 2016). The overexpressing *OsHAC1;1* and *OsHAC1;2* lines of rice had lesser As accumulation and showed enhanced As tolerance (Shi et al. 2016). While knocked-out lines of *hac1;1* and *hac1;2* showed lesser AsV reduction in roots and more As accumulation in roots and shoots than wild-type plants. Recently, *Arabidopsis thaliana* AsV reductase, *HAC1* has also been characterized in different root tissues of *Arabidopsis* and it was identified that the expression of *HAC1* is more effective in the epidermis, cortex, and endodermis (outer tissues of the root) to confer AsIII efflux and AsV tolerance. However, it is less efficient in the stele and thus limits the As accumulation in shoots (Fischer et al. 2021). Arsenite, in addition to the main transported As species in plants, is also reported to be effluxed into the external environment by the plants. One of the AsV reductase genes in rice, i.e. *OsHAC4* has also been reported to possess tolerance against AsV toxicity. The high expressing *OsHAC4* lines of rice showed increased AsV to AsIII reduction, as well as increased AsIII efflux to the external medium. Overall, *OsHAC4* overexpressing lines showed enhanced As tolerance by accumulating lesser As in the plant (Xu et al. 2017). Also, overexpression of *OsHAC1;1* and *OsHAC1;2* in rice significantly increased the reduction of AsV to AsIII followed by its exit to the external medium (Shi et al. 2016). In addition, the knockout *hac1* lines in both *Arabidopsis* and rice showed decreased AsIII efflux, which gets accumulated in the shoot, confirming the role of *HAC1* as AsV reductase (Chao et al. 2014; Sánchez-Bermejo et al. 2014; Shi et al. 2016; Xu et al. 2017). In *P. vittata*, *PvACR3* and *PvACR3;1* have been identified as AsIII transporters. These transporters are homologs of yeast *ACR3* AsIII efflux transporter (Indriolo et al. 2010). The heterologous expression of *PvACR3* in *Arabidopsis* showed its localization on the plasma membrane and increased translocation of AsIII toward shoot from root (Indriolo et al. 2010; Wang et al. 2018). Both observations suggested that *PvACR3* worked as an efflux transporter and was responsible for the xylem loading of As (Wang et al. 2018). However, *PvACR3;1* was found to be localized on the vacuolar membrane when heterologously expressed in *A. thaliana* and *Nicotiana* (Chen et al. 2017). *PvACR3;1* is supposed to involve As accumulation rather than root to shoot translocation in plants (Chen et al. 2017a). Duan et al. 2012 overexpressed the *ACR3* gene (from *Saccharomyces cerevisiae*) under the control of (CaMV) 35S promoter in rice. The transgenic rice, expressing *ACR3* gene, showed higher AsIII efflux and lesser As accumulation in grains of rice. Overexpressing lines showed a 30% reduced concentration of As in roots and shoots than in the wild-type plants.

Like ArsC, Grxs also have an AsV reductase activity and play an important role in AsV to AsIII reduction (Sundaram et al. 2008, 2009). Heterologous expression of one of the Grx from *Pteris vitta* "*PvGrx5*" in *Arabidopsis* showed enhanced As tolerance by improving the As reduction and regulating the As level in cells (Sundaram et al. 2009). Two other rice Grxs, namely *OsGrx_C7* and *OsGrx_C2.1* were also showed improved As tolerance by dropping the intracellular As accumulation in transgenic *Arabidopsis* plants (Verma et al. 2016a). More recently, Kumar et al. (2020) identified chickpea Grxs, CaGrx to have a potential role in As stress by accumulating lesser As content and improving various biochemical and physiological parameters in *Arabidopsis* under As stress.

8.3.1.4 Transporters for As Translocation

In rice, silicon transporters are involved in the translocation of AsIII. *Lsi1* (silicon influx transporter) is mainly involved in AsIII uptake. *Lsi2* (silicon efflux transporter), on the other hand, is involved in AsIII translocation in plants. Mutant *Lsi2* lines of rice showed decreased transport of AsIII in the xylem. As a result, decreased As accumulation has been observed in shoot and grain (Ma et al. 2008). Interestingly, one study also recognizes the role of rice nodes in As storage. The *Lsi2* expression profile analysis showed its higher expression not only in the root but also at nodes when *Lsi2* mutants were treated with AsIII. The higher As accumulation was observed at nodes and flag leaf but not in grain (Chen et al. 2015). Another As transporter, *NIP2;2* or *Lsi6* is involved in AsIII transport from shoot to upper plant part (Yamaji and Ma 2009). Wang et al. 2017 reported that R2R3 MYB TF *OsARM1* (ARSENITE-RESPONSIVE MYB1) are involved in the regulation of As-associated transporter genes *OsLsi1*, *OsLsi2*, and *OsLsi6* in rice. Thus, *OsARM1* regulated the AsIII uptake from the soil as well as root to the aboveground part translocation. In comparison to the wild-type plants, *OsLsi1*, *OsLsi2*, and *OsLsi6* AsIII transporters had higher expression in the *OsARM1* knockout lines but lower expressions of these genes were found in *OsARM1* overexpressing lines. Thus, the knockout lines of *OsARM1* showed enhanced As tolerance while the overexpressed *OsARM1* become more sensitive to AsIII stress by regulating As-associated transporter genes. In addition, one of the *ABCC* genes in rice, *OsABCC7* was found to be downregulated under AsIII stress (Yu et al. 2012). It was later observed that *OsABBC7* is localized on the plasma membrane of root stelar tissues and is involved in translocation of As in a long-distance manner from root to other aerial parts of the plant (Tang et al. 2019).

Additionally, it was identified that when plants were supplemented with AsV, AsIII was the only As species observed in the external medium (Liu et al. 2010; Xu et al. 2007; Zhao et al. 2010). The observation further confirms that AsIII efflux transporters should be present on the root cells. The process takes place when As is present in the cell in abundance and becomes a burden to the cell. In rice, *OsLsi1* (aquaporin channel) is involved in AsIII influx as well as efflux. The *Lsi1* mutants showed approximately only 20% of AsIII efflux to the external medium. This observation suggested that *Lsi1* is not the main transporter involved in AsIII efflux to the external medium but it is a bidirectional transporter (Zhao et al. 2010). As discussed earlier, *OsLsi2*, in contrast, is mainly present in lateral roots on exodermis and endodermis root cells. The observation suggested that *OsLsi2* is although an efflux transporter but the direction of AsIII translocation is not toward the external medium but the vascular bundle for root to shoot translocation of AsIII. In *Arabidopsis*, an auxin efflux transporter, *AtPIN2* was reported to be localized on root cap and recently suggested to be involved in AsIII efflux in

Arabidopsis (Abas et al. 2006; Rahman et al. 2010; Ashraf et al. 2019). However, the key efflux transporters of AsIII in plants have not been identified yet.

8.3.1.5 Genetic Study of As Detoxification Genes in Plants

Arsenic entrance to plants activates various detoxification genes that are involved in conferring tolerance to plants. Numerous studies have been conducted to observe the differential gene responses toward As stress in hydroponically grown plants such as rice via quantitative real-time polymerase chain reaction (qRT-PCR) and semiquantitative RT-PCR. For example, Gupta and Ahmad 2014 compared the AsV-induced gene response in two rice cultivars, viz. PB1 and IR-64. Genes such as glutathione reductase (GR), glutathione synthetase (GS), and metallothionine-2 (MT-2) were taken into the study as they are an important candidate for detoxification of metalloids. The data showed an upregulation of all the genes in both the genotypes but a much higher expression was observed in tolerant variety PBI, due to which it accumulated low As.

Gasic and Korban (2007) expressed a phytochelatin (PC)-encoding gene of *Arabidopsis AtPCS1 for* the invention of As accumulation and tolerance in *B. juncea*. The transgenic line, expressing *AtPCS1* under the regulation of native promoter showed significantly higher tolerance in Indian mustard to As stress compared with wild-type plants. But it was incapable to increase the As accumulation potential in Indian mustard plants. Further, the overexpression of other PCs in plants showed better tolerance against As, which might also influence the As accumulation (Liu et al. 2010; Pal and Rai 2010). PC from *Ceratophyllum demersum CdPCS1* was expressed in a phytochelatin synthase (PCS)-deficient mutant of *Arabidopsis*, i.e. *cad1–3*. *CdPCS1* complements the effect of *cad1–3* mutant of *Arabidopsis* and restored the As tolerance in *cad1–3* mutant. Overexpression of *CdPCS1* in tobacco showed enhanced As accumulation without affecting the plant growth and providing the tolerance against As (Shukla et al. 2012). Similarly, the overexpression of *CdPCS1* in rice showed increased As detoxification by accumulating more As in root and lesser As content in rice grains. Hayashi et al. 2017 reported that higher expression of *OsPCS1* in rice resulted in low As content in rice grains. Other PC biosynthesis genes have also been reported to confer tolerance against As. Interestingly, a bacterial γ-ECS gene expressing transgenic lines of *Arabidopsis* showed significantly higher γ-ECS, glutathione (GSH), and phytochelatin (PC) accumulation as well as higher tolerance in comparison to the wild-type plants under As toxicity (Li et al. 2004). LeBlanc et al. (2011) developed a bacterial γ-ECS gene expressing genetically engineered eastern cottonwood. These transgenic lines showed an enhanced level of thiols and γ-ECS activity and showed tolerance toward AsV stress. Furthermore, in a transgenic plant, γ-ECS-expressing roots accumulated significantly higher As compared with the control roots while significantly lesser As accumulation was observed in transgenic lines compared with wild-type lines. By overexpressing *AtPCs1* under the strong constitutive *Arabidopsis* actin regulatory sequence in *Arabidopsis* showed higher tolerance by accumulating 20–100 times more biomass when treated with 250 and 300 μM AsV than wild-type plants (Li et al. 2004). Wojas et al. 2010 aimed to develop the transgenic *Nicotiana tabacum* plants by simultaneously expressing *AtPCS1* and *CePCS* genes to decipher their role in As tolerance. The expression analysis of these genes in transgenic plants showed that both the genes *AtPCS1* and *CePCS* have higher tolerance against As stress (*CePCS* plants most tolerant) and accumulate higher As content in roots and leaves. For the first time, Wojas et al. (2010) reported that the *PCS* response against As stress in transgenic plants depends on As concentration which is present in the external medium. The transgenic plants of *AtPCS1* and *CePCS* gene showed higher As accumulation in the roots and leaves at the lower concentration of As about 50 mM. In contrast, the reverse result was found in transgenic plants at a higher concentration of As about 200 mM (toxic concentration) due to several conflicts in the thiol metabolism of *PCS* transgenic plants. So, the transcript levels were decreased in the roots and leaves of transgenic plants compared with wild-type plants.

In plants, different enzymes of the glutathione pathway showed a significant role and had the potential for phytoremediation (Li et al. 2004; Gasic and Korban 2007). However, only the dual-gene *GSH1*- and *AsPCS1*-transformed transgenic lines showed increased As tolerance and accumulation (Guo et al. 2008). Overexpression of *AsPCS1* and *GSH1* in *Arabidopsis* had an increased As tolerance as well as a higher As accumulation (Guo et al. 2008). In plants, different enzymes of the glutathione pathway play a crucial role in phytoremediation (Li et al. 2004; Gasic and Korban 2007; Guo et al. 2008). Interestingly, the transgenic lines of these genes are not always successful in all the studied plants with respect to As tolerance and accumulation, except for dual-gene transformations with *GSH1* and *AsPCS1* (Guo et al. 2008). The greater As detoxification capability was found in the transgenic lines which are overexpressed with cysteine-, GSH-, and PC biosynthesis-regulating genes (Tripathi et al. 2007). Navaza et al. 2006 proposed that overexpressed gamma-glutamylcysteine synthetase and GS-encoding genes in *Brassica juncea* had a higher accumulation of As.

Shukla et al. (2013) investigated that for enhancing the metal-accumulation ability of plants, the potential candidates are the use of synthetic PCs (*ECs*). Three synthetic *PCs* were expressed in wild-type *Arabidopsis* (Col-0 background) and a PC-deficient mutant (*cad1–3*) of *Arabidopsis*. The *cad1–3* mutant showed better tolerance under As stress. Compared to the wild-type, EC expressing *Arabidopsis* plants showed the enhanced the accumulation of heavy metal(loid)s. Shukla et al. (2013) also suggested that both ECs and PCs are not functionally different with respect to conferring As tolerance in plants.

It has been reported that intracellular AsIII forms stable complexes with thiol-containing metal-chelating PCs and GSH. Two OsPCS genes, *OsPCS1* and *OsPCS2* were found to have a role in As accumulation. When the *OsPCS1* and *OsPCS2* genes were downregulated with RNA-mediated gene silencing in seeds in rice, by using the seed-specific promoter, approximately 35% decrease in As content was observed in seeds compared to WT seeds (Das et al. 2017). Two vacuolar transporters, named ABC transporters, *AtABCC1* and *AtABCC2* were found to be linked with vacuolar sequestration of PC–As complexes. The double knockout, *abcc1* and *abcc2*, *Arabidopsis* plants were highly sensitive against arsenic toxicity (Song et al. 2010). An *AtABCC1* ortholog of rice, *OsABCC1* gets upregulated under high As concentration. However, under low As concentration, *OsABCC1* was not found to be upregulated. Expression pattern analysis of *OsABCC* revealed its expression in many organs including root cells. Under As toxicity, *OsABCC1* sequesters more As in the companion cells of phloem. As a result, the As translocation toward grain is checked (Song et al. 2010). The observation was further confirmed by analysis of knockout *OsABCC1* transgenic rice lines. The knockout lines, unlike wild-type, were more sensitive for As toxicity (Song et al. 2010).

Like PCs, metallothioneins are also bound with metals ions and form metal complexes that get sequestered into vacuoles by transporters. Metallothionein gene from *Arabidopsis*, *AtMT2b* was expressed in *N. tabacum*. The transgenic lines showed a significant reduction of As content in the plant roots but more As accumulation was found in shoots while no change was observed in total As uptake by the plant (Grispen et al. 2009). Another cysteine-rich protein *OsGSTL2*, overexpressed in *A. thaliana* also has been shown tolerance toward different heavy metals including As and other abiotic stresses (Kumar et al. 2013).

To detoxify the As in plants, another strategy is to convert inorganic As (Asi) into organic As (Aso) and volatilize As (Ye et al. 2012). For instance, a gene encodes, the enzyme AsIII S-adenosylmethionine methyltransferase (*ArsM*) identified from the bacterium *Rhodopseudomonas palustris* is responsible for As biomethylation (Qin et al. 2006; Chen et al. 2014). *in vivo*, As methylation is unknown in the higher plant system (Bentley and Chasteen 2002; Gebel 2002; Qin et al. 2006). Meng et al. 2011 reported the *ArsM*, overexpressing lines in rice showed lesser As accumulation compared to As-treated control plants due to the Asi conversion to a volatile As form. Monomethyl AsV and dimethyl AsV, which are methylated forms of As, were detected in various transgenic plant parts. Verma et al. (2016b) isolated an As methyltransferase gene (*WaarsM*) from a fungus, *Westerdykella aurantiaca* and overexpressed in *Arabidopsis*. The transgenic line of *Arabidopsis* accumulated the lesser As as compared to the wild-type plants due to volatilization of As by *WaarsM* and showed enhanced As tolerance. Verma et al. (2016) further studied the effect of overexpression of *WaarsM* in rice and found a higher As tolerance, which must be due to higher As volatilization by *WaarsM*. Recently, for the first time, Kidwai et al. 2019 demonstrated the role of Rice class III peroxidase toward the As tolerance. Overexpressing the rice peroxidase (class III) (OsPRX38) in *Arabidopsis* resulted in a higher lignin formation in the apoplastic region of the root and thus checked the entry of As inside the plant. Similarly, Tiwari et al. 2022 reported that *OsGSTU5* confer tolerance by accumulating more As in root compared to shoot in rice.

8.4 Concluding Remarks and Future Prospects

Reducing the toxicity effect of As on plants as well as developing transgenic crops with a safer As content is one of the important tasks for public health. To understand the genetic basis of As tolerance and sensitivity in plants, a number of *in silico* data analyses such as transcriptome, microarray data, and genome-wide association analyses have been performed. Various defense-regulated genes, TFs, and antioxidants were found to be differentially regulated during As toxicity in various plants. Various plant transporters have also been recognized in plant roots that uniquely transport either AsV or AsIII from external medium to inside the plant. AsIII either transported toward the upper plant part or effluxed to the external medium with the help of various transporters. Inside the plant, As mainly conjugates with PC or Glutathione and sequester the As into the vacuoles by utilizing vacuolar transporters. Metallothionine is also important to detoxify the As toxicity in plants. Some of the microbial genes have also been identified to convert inorganic As into organic volatile forms that evaporate the As to the external environment. Overall, by utilizing various approaches, plants try to cope under As toxicity. However, various genes related to As stress in plants remain elusive such as AsIII efflux transporters that effluxed AsIII to the external medium. Other detoxification genes that are differentially expressed in various transcriptome analysis have not been identified yet. Identification of alleles of As-related detoxification genes (transporters and defense-related genes) would be a good approach for marker-assisted breeding in crop plants to increase the As tolerance in plants.

Acknowledgments

The authors would like to acknowledge the Director CSIR-NBRI for the facilities to carry out this work. This manuscript bears the CSIR-NBRI communication number CSIR-NBRI_MS/2022/05/01.

References

Abas, L., Benjamins, R., Malenica, N. et al. (2006). Intracellular trafficking and proteolysis of the *Arabidopsis* auxin-efflux facilitator PIN2 are involved in root gravitropism. *Nature Cell Biology* 8: 249–256.

Abbas, G., Murtaza, B., Bibi, I. et al. (2018). Arsenic uptake, toxicity, detoxification, and speciation in plants: physiological, biochemical, and molecular aspects. *International Journal of Environmental Research and Public Health* 15: 59.

Abercrombie, J.M., Halfhill, M.D., Ranjan, P. et al. (2008). Transcriptional responses of *Arabidopsis* thaliana plants to As (V) stress. *BMC Plant Biology* 8: 1–15.

Ashraf, M.A., Umetsu, K., Ponomarenko, O. et al. (2019). PIN FORMED 2 facilitates the transport of Arsenite in *Arabidopsis thaliana*. *BioRxiv* 710160. https://doi.org/10.1101/710160.

Bentley, R. and Chasteen, T.G. (2002). Microbial methylation of metalloids: arsenic, antimony, and bismuth. *Microbiology and Molecular Biology Reviews* 66: 250–271.

Bienert, G.P. and Jahn, T.P. (2010). Major intrinsic proteins and arsenic transport in plants: new players and their potential role. In: *MIPs and Their Role in the Exchange of Metalloids*, vol. 679 (ed. T.P. Jahn and G.P. Bienert), 111–125. New York: Springer.

Bienert, G.P., Thorsen, M., Schüssler, M.D. et al. (2008). A subgroup of plant aquaporins facilitate the bi-directional diffusion of $As(OH)_3$ and $Sb(OH)_3$ across membranes. *BMC Biology* 6: 1–15.

Cao, Y., Sun, D., Ai, H. et al. (2017). Knocking out OsPT4 gene decreases arsenate uptake by rice plants and inorganic arsenic accumulation in rice grains. *Environmental Science & Technology* 51: 12131–12138.

Cao, Y., Feng, H., Sun, D. et al. (2019). Heterologous expression of *Pteris vittata* phosphate transporter PvPht1; 3 enhances arsenic translocation to and accumulation in tobacco shoots. *Environmental Science & Technology* 53: 10636–10644.

Catarecha, P., Segura, M.D., Franco-Zorrilla, J.M. et al. (2007). A mutant of the *Arabidopsis* phosphate transporter PHT1; 1 displays enhanced arsenic accumulation. *The Plant Cell* 19: 1123–1133.

Chakrabarty, D., Trivedi, P.K., Misra, P. et al. (2009). Comparative transcriptome analysis of arsenate and arsenite stresses in rice seedlings. *Chemosphere* 74: 688–702.

Chao, D.Y., Chen, Y., Chen, J. et al. (2014). Genome-wide association mapping identifies a new arsenate reductase enzyme critical for limiting arsenic accumulation in plants. *PLoS Biology* 12: e1002009.

Chen, J., Liu, Y., Ni, J. et al. (2011). OsPHF1 regulates the plasma membrane localization of low-and high-affinity inorganic phosphate transporters and determines inorganic phosphate uptake and translocation in rice. *Plant Physiology* 157: 269–278.

Chen, J., Sun, G.X., Wang, X.X. et al. (2014). Volatilization of arsenic from polluted soil by *Pseudomonas putida* engineered for expression of the arsM arsenic (III) *S*-adenosine methyltransferase gene. *Environmental Science & Technology* 48: 10337–10344.

Chen, Y., Moore, K.L., Miller, A.J. et al. (2015). The role of nodes in arsenic storage and distribution in rice. *Journal of Experimental Botany* 66: 3717–3724.

Chen, Y., Hua, C.Y., Jia, M.R. et al. (2017a). Heterologous expression of *Pteris vittata* arsenite antiporter PvACR3; 1 reduces arsenic accumulation in plant shoots. *Environmental Science & Technology* 51: 10387–10395.

Chen, Y., Sun, S.K., Tang, Z. et al. (2017b). The Nodulin 26-like intrinsic membrane protein OsNIP3; 2 is involved in arsenite uptake by lateral roots in rice. *Journal of Experimental Botany* 68: 3007–3016.

Das, N., Bhattacharya, S., Bhattacharyya, S. et al. (2017). Identification of alternatively spliced transcripts of rice phytochelatin synthase 2 gene OsPCS2 involved in mitigation of cadmium and arsenic stresses. *Plant Molecular Biology* 94: 167–183.

Dhankher, O.P., Li, Y., Rosen, B.P. et al. (2002). Engineering tolerance and hyperaccumulation of arsenic in plants by combining arsenate reductase and γ-glutamylcysteine synthetase expression. *Nature Biotechnology* 20: 1140–1145.

Dhankher, O.P., Rosen, B.P., McKinney, E.C. et al. (2006). Hyperaccumulation of arsenic in the shoots of *Arabidopsis* silenced for arsenate reductase (ACR2). *Proceedings of the National Academy of Sciences of the United States of America* 103: 5413–5418.

Di, X., Zheng, F., Norton, G.J. et al. (2021). Physiological responses and transcriptome analyses of upland rice following exposure to arsenite and arsenate. *Environmental and Experimental Botany* 183: 104366.

DiTusa, S.F., Fontenot, E.B., Wallace, R.W. et al. (2016). A member of the Phosphate transporter 1 (Pht1) family from the arsenic-hyperaccumulating fern *Pteris vittata* is a high-affinity arsenate transporter. *New Phytologist* 209: 762–772.

Duan, G., Kamiya, T., Ishikawa, S. et al. (2012). Expressing ScACR3 in rice enhanced arsenite efflux and reduced arsenic accumulation in rice grains. *Plant and Cell Physiology* 53: 154–163.

Finnegan, P. and Chen, W. (2012). Arsenic toxicity: the effects on plant metabolism. *Frontiers in Physiology* 3: 182.

Fischer, S., Sánchez-Bermejo, E., Xu, X. et al. (2021). Targeted expression of the arsenate reductase HAC1 identifies cell type specificity of arsenic metabolism and transport in plant roots. *Journal of Experimental Botany* 72: 415–425.

Fu, S.F., Chen, P.Y., Nguyen, Q.T.T. et al. (2014). Transcriptome profiling of genes and pathways associated with arsenic toxicity and tolerance in *Arabidopsis*. *BMC Plant Biology* 14: 1–16.

Gasic, K. and Korban, S.S. (2007). Transgenic Indian mustard (*Brassica juncea*) plants expressing an *Arabidopsis* phytochelatin synthase (AtPCS1) exhibit enhanced As and Cd tolerance. *Plant Molecular Biology* 64: 361–369.

Gebel, T.W. (2002). Arsenic methylation is a process of detoxification through accelerated excretion. *International Journal of Hygiene and Environmental Health* 205: 505–508.

Ghosh, S., Singh, K., Shaw, A.K. et al. (2017). Insights into the miRNA-mediated response of maize leaf to arsenate stress. *Environmental and Experimental Botany* 137: 96–109.

González, E., Solano, R., Rubio, V. et al. (2005). PHOSPHATE TRANSPORTER TRAFFIC FACILITATOR1 is a plant-specific SEC12-related protein that enables the endoplasmic reticulum exit of a high-affinity phosphate transporter in *Arabidopsis*. *The Plant Cell* 17: 3500–3512.

Grispen, V.M., Irtelli, B., Hakvoort, H.W. et al. (2009). Expression of the *Arabidopsis* metallothionein 2b enhances arsenite sensitivity and root to shoot translocation in tobacco. *Environmental and Experimental Botany* 66: 69–73.

Guo, J., Dai, X., Xu, W., and Ma, M. (2008). Overexpressing GSH1 and AsPCS1 simultaneously increases the tolerance and accumulation of cadmium and arsenic in *Arabidopsis* thaliana. *Chemosphere* 72: 1020–1026.

Gupta, M. and Ahmad, M.A. (2014). Arsenate induced differential response in rice genotypes. *Ecotoxicology and Environmental Safety* 107: 46–54.

Hayashi, S., Kuramata, M., and Abe, T. (2017). Phytochelatin synthase Os PCS 1 plays a crucial role in reducing arsenic levels in rice grains. *The Plant Journal* 91: 840–848.

Heikens, A. (2006). Arsenic contamination of irrigation water, soil and crops in Bangladesh: risk implications for sustainable agriculture and food safety in Asia. Rap Publication (FAO).

Huang, T.L., Nguyen, Q.T.T., Fu, S.F. et al. (2012). Transcriptomic changes and signalling pathways induced by arsenic stress in rice roots. *Plant Molecular Biology* 80: 587–608.

Huhmann, B.L., Harvey, C.F., Uddin, A. et al. (2017). Field study of rice yield diminished by soil arsenic in Bangladesh. *Environmental Science & Technology* 51: 1553–11560.

Huq, S.I., Joardar, J.C., Parvin, S. et al. (2006). Arsenic contamination in food-chain: transfer of arsenic into food materials through groundwater irrigation. *Journal of Health, Population, and Nutrition* 24: 305.

Hwang, S.G., Chapagain, S., Lee, J.W. et al. (2017). Genome-wide transcriptome profiling of genes associated with arsenate toxicity in an arsenic-tolerant rice mutant. *Plant Physiology and Biochemistry* 120: 40–51.

Indriolo, E., Na, G., Ellis, D. et al. (2010). A vacuolar arsenite transporter necessary for arsenic tolerance in the arsenic hyperaccumulating fern *Pteris vittata* is missing in flowering plants. *The Plant Cell* 22: 2045–2057.

Jia, H., Ren, H., Gu, M. et al. (2011). The phosphate transporter gene OsPht1; 8 is involved in phosphate homeostasis in rice. *Plant Physiology* 156: 1164–1175.

Jomova, K., Jenisova, Z., Feszterova, M. et al. (2011). Arsenic: toxicity, oxidative stress and human disease. *Journal of Applied Toxicology* 31: 95–107.

Kamiya, T., Islam, R., Duan, G. et al. (2013). Phosphate deficiency signaling pathway is a target of arsenate and phosphate transporter OsPT1 is involved in As accumulation in shoots of rice. *Soil Science and Plant Nutrition* 59: 580–590.

Khalid, S., Shahid, M., Niazi, N.K. et al. (2017). Arsenic behaviour in soil-plant system: biogeochemical reactions and chemical speciation influences. In: *Enhancing Cleanup of Environmental Pollutants* (ed. N. Anjum, S. Gill, N. Tuteja, et al.), 97–140. Cham: Springer.

Kidwai, M., Dhar, Y.V., Gautam, N. et al. (2019). *Oryza sativa* class III peroxidase (OsPRX38) overexpression in *Arabidopsis thaliana* reduces arsenic accumulation due to apoplastic lignification. *Journal of Hazardous Materials* 362: 383–393.

Kumar, A., Dubey, A.K., Kumar, V. et al. (2020). Over-expression of chickpea glutaredoxin (CaGrx) provides tolerance to heavy metals by reducing metal accumulation and improved physiological and antioxidant defence system. *Ecotoxicology and Environmental Safety* 192: 110252.

Kumar, S., Asif, M.H., Chakrabarty, D. et al. (2013). Differential expression of rice lambda class GST gene family members during plant growth, development, and in response to stress conditions. *Plant Molecular Biology Reporter* 31: 569–580.

LeBlanc, M.S., Lima, A., Montello, P., and Kim, T. (2011). Enhanced arsenic tolerance of transgenic eastern cottonwood plants expressing gamma-glutamylcysteine synthetase. *International Journal of Phytoremediation* 13: 657–673.

LeBlanc, M.S., McKinney, E.C., Meagher, R.B. et al. (2013). Hijacking membrane transporters for arsenic phytoextraction. *Journal of Biotechnology* 163: 1–9.

Li, Y., Dhankher, O.P., Carreira, L. et al. (2004). Overexpression of phytochelatin synthase in *Arabidopsis* leads to enhanced arsenic tolerance and cadmium hypersensitivity. *Plant and Cell Physiology* 45: 1787–1797.

Lim, K.T., Shukor, M.Y., and Wasoh, H. (2014). Physical, chemical, and biological methods for the removal of arsenic compounds. *BioMed Research International* 2014: 503784.

Liu, W.J., Wood, B.A., Raab, A. et al. (2010). Complexation of arsenite with phytochelatins reduces arsenite efflux and translocation from roots to shoots in *Arabidopsis*. *Plant Physiology* 152: 2211–2221.

Liu, Y., Mi, Y., Zhang, J. et al. (2016). Illumina-based transcriptomic profiling of Panax notoginseng in response to arsenic stress. *Botanical Studies* 57: 1–11.

Lowe, R., Shirley, N., Bleackley, M. et al. (2017). Transcriptomics technologies. *PLoS Computational Biology* 13: 1005457.

Luan, M., Liu, J., Liu, Y. et al. (2018). Vacuolar phosphate transporter 1 (VPT1) affects arsenate tolerance by regulating phosphate homeostasis in *Arabidopsis*. *Plant and Cell Physiology* 59: 1345–1352.

Ma, J.F., Yamaji, N., Mitani, N. et al. (2008). Transporters of arsenite in rice and their role in arsenic accumulation in rice grain. *Proceedings of the National Academy of Sciences of the United States of America* 105: 9931–9935.

Meharg, A.A. and Macnair, M.R. (1990). An altered phosphate uptake system in arsenate-tolerant *Holcus lanatus* L. *New Phytologist* 116: 29–35.

Meng, X.Y., Qin, J., Wang, L.H. et al. (2011). Arsenic biotransformation and volatilization in transgenic rice. *New Phytologist* 191: 49–56.

Mosa, K.A., Kumar, K., Chhikara, S. et al. (2012). Members of rice plasma membrane intrinsic proteins subfamily are involved in arsenite permeability and tolerance in plants. *Transgenic Research* 21: 1265–1277.

Muehe, E.M., Wang, T., Kerl, C.F. et al. (2019). Rice production threatened by coupled stresses of climate and soil arsenic. *Nature Communications* 10: 1–10.

Murugaiyan, V., Ali, J., Mahender, A. et al. (2019). Mapping of genomic regions associated with arsenic toxicity stress in a backcross breeding populations of rice (*Oryza sativa* L.). *Rice* 12: 1–14.

Nagarajan, V.K., Jain, A., Poling, M.D. et al. (2011). *Arabidopsis* Pht1; 5 mobilizes phosphate between source and sink organs and influences the interaction between phosphate homeostasis and ethylene signaling. *Plant Physiology* 156: 1149–1163.

Navaza, A.P., Montes-Bayón, M., LeDuc, D.L. et al. (2006). Study of phytochelatins and other related thiols as complexing biomolecules of As and Cd in wild type and genetically modified *Brassica juncea* plants. *Journal of Mass Spectrometry* 41: 323–331.

Norton, G.J., Lou-Hing, D.E., Meharg, A.A. et al. (2008). Rice–arsenate interactions in hydroponics: whole genome transcriptional analysis. *Journal of Experimental Botany* 59: 2267–2276.

Pal, R. and Rai, J.P.N. (2010). Phytochelatins: peptides involved in heavy metal detoxification. *Applied Biochemistry and Biotechnology* 160: 945–963.

Palmer, N.A., Saathoff, A.J., Waters, B.M. et al. (2014). Global changes in mineral transporters in tetraploid switchgrasses (*Panicum virgatum* L.). *Frontiers in Plant Science* 4: 549.

Peña-Garcia, Y., Shinde, S., Natarajan, P. et al. (2021). Arsenic Stress-Related F-Box (ASRF) gene regulates arsenic stress tolerance in *Arabidopsis thaliana*. *Journal of Hazardous Materials* 407: 124831.

Podgorski, J. and Berg, M. (2020). Global threat of arsenic in groundwater. *Science* 368: 845–850.

Qin, J., Rosen, B.P., Zhang, Y. et al. (2006). Arsenic detoxification and evolution of trimethylarsine gas by a microbial arsenite *S*-adenosylmethionine methyltransferase. *Proceedings of the National Academy of Sciences of the United States of America* 103: 2075–2080.

Rahman, A., Takahashi, M., Shibasaki, K. et al. (2010). Gravitropism of *Arabidopsis thaliana* roots requires the polarization of PIN2 toward the root tip in meristematic cortical cells. *The Plant Cell* 22: 1762–1776.

Rai, A., Bhardwaj, A., Misra, P. et al. (2015). Comparative transcriptional profiling of contrasting rice genotypes shows expression differences during arsenic stress. *The Plant Genome* 8: https://doi.org/10.3835/plantgenome2014.09.0054.

Remy, E., Cabrito, T.R., Batista, R.A. et al. (2012). The Pht1; 9 and Pht1; 8 transporters mediate inorganic phosphate acquisition by the *Arabidopsis thaliana* root during phosphorus starvation. *New Phytologist* 195: 356–371.

Sánchez-Bermejo, E., Castrillo, G., Del Llano, B. et al. (2014). Natural variation in arsenate tolerance identifies an arsenate reductase in *Arabidopsis thaliana*. *Nature Communications* 5: 1–9.

Sasaki, A., Yamaji, N., and Ma, J.F. (2016). Transporters involved in mineral nutrient uptake in rice. *Journal of Experimental Botany* 67: 3645–3653.

Shaji, E., Santosh, M., Sarath, K.V. et al. (2020). Arsenic contamination of groundwater: a global synopsis with focus on the Indian Peninsula. *Geoscience Frontiers*.

Sharma, I. (2012). Arsenic induced oxidative stress in plants. *Biologia* 67: 447–453.

Sharma, D., Tiwari, M., Lakhwani, D. et al. (2015). Differential expression of microRNAs by arsenate and arsenite stress in natural accessions of rice. *Metallomics* 7: 174–187.

Sherwood, C.L., Lantz, R.C., Burgess, J.L. et al. (2011). Arsenic alters ATP-dependent Ca^{2+} signaling in human airway epithelial cell wound response. *Toxicological Sciences* 121: 191–206.

Shi, S., Wang, T., Chen, Z. et al. (2016). OsHAC1; 1 and OsHAC1; 2 function as arsenate reductases and regulate arsenic accumulation. *Plant Physiology* 172: 1708–1719.

Shin, H., Shin, H.S., Dewbre, G.R. et al. (2004). Phosphate transport in *Arabidopsis*: Pht1; 1 and Pht1; 4 play a major role in phosphate acquisition from both low-and high-phosphate environments. *The Plant Journal* 39: 629–642.

Shukla, D., Kesari, R., Mishra, S. et al. (2012). Expression of phytochelatin synthase from aquatic macrophyte *Ceratophyllum demersum* L. enhances cadmium and arsenic accumulation in tobacco. *Plant Cell Reports* 31: 1687–1699.

Shukla, D., Tiwari, M., Tripathi, R.D. et al. (2013). Synthetic phytochelatins complement a phytochelatin-deficient *Arabidopsis* mutant and enhance the accumulation of heavy metal (loid) s. *Biochemical and Biophysical Research Communications* 434: 664–669.

Shukla, T., Khare, R., Kumar, S. et al. (2018). Differential transcriptome modulation leads to variation in arsenic stress response in *Arabidopsis thaliana* accessions. *Journal of Hazardous Materials* 351: 1–10.

Song, W.Y., Park, J., Mendoza-Cózatl, D.G. et al. (2010). Arsenic tolerance in *Arabidopsis* is mediated by two ABCC-type phytochelatin transporters. *Proceedings of the National Academy of Sciences of the United States of America* 107: 21187–21192.

Srivastava, S., Srivastava, A.K., Suprasanna, P. et al. (2009). Comparative biochemical and transcriptional profiling of two contrasting varieties of *Brassica juncea* L. in response to arsenic exposure reveals mechanisms of stress perception and tolerance. *Journal of Experimental Botany* 60: 3419–3431.

Srivastava, S., Srivastava, A.K., Suprasanna, P. et al. (2013). Identification and profiling of arsenic stress-induced microRNAs in *Brassica juncea*. *Journal of Experimental Botany* 64: 303–315.

Srivastava, S., Srivastava, A.K., Sablok, G. et al. (2015). Transcriptomics profiling of Indian mustard (*Brassica juncea*) under arsenate stress identifies key candidate genes and regulatory pathways. *Frontiers in Plant Science* 6: 646.

Su, Y.H., McGrath, S.P., and Zhao, F.J. (2010). Rice is more efficient in arsenite uptake and translocation than wheat and barley. *Plant and Soil* 328: 27–34.

Sun, S.K., Chen, Y., Che, J. et al. (2018). Decreasing arsenic accumulation in rice by overexpressing OsNIP 1; 1 and Os NIP 3; 3 through disrupting arsenite radial transport in roots. *New Phytologist* 219: 641–653.

Sun, D., Feng, H., Li, X. et al. (2019). Expression of new *Pteris vittata* phosphate transporter PvPht1; 4 reduces arsenic translocation from the roots to shoots in tobacco plants. *Environmental Science & Technology* 54: 1045–1053.

Sundaram, S., Rathinasabapathi, B., Ma, L.Q. et al. (2008). An arsenate-activated glutaredoxin from the arsenic hyperaccumulator fern *Pteris vittata* L. regulates intracellular arsenite. *Journal of Biological Chemistry* 283: 6095–6101.

Sundaram, S., Wu, S., Ma, L.Q., and Rathinasabapathi, B. (2009). Expression of a *Pteris vittata* glutaredoxin PvGRX5 in transgenic *Arabidopsis thaliana* increases plant arsenic tolerance and decreases arsenic accumulation in the leaves. *Plant, Cell & Environment* 32: 851–858.

Tang, Z., Chen, Y., Miller, A.J. et al. (2019). The C-type ATP-binding cassette transporter OsABCC7 is involved in the root-to-shoot translocation of arsenic in rice. *Plant and Cell Physiology* 60: 1525–1535.

Tawfik, D.S. and Viola, R.E. (2011). Arsenate replacing phosphate: alternative life chemistries and ion promiscuity. *Biochemistry* 50: 1128–1134.

Tiwari, M., Kidwai, M., Dutta, P. et al. (2022). A tau class glutathione-S-transferase (OsGSTU5) confers tolerance against arsenic toxicity in rice by accumulating more arsenic in root. *Journal of Hazardous Materials* 426: 128100.

Tripathi, R.D., Srivastava, S., and Mishra, S. (2007). Arsenic hazards: strategies for tolerance and remediation by plants. *Trends in Biotechnology* 25: 158–165.

Verma, P.K., Verma, S., Pande, V. et al. (2016a). Overexpression of rice glutaredoxin OsGrx_C7 and OsGrx_C2. 1 reduces intracellular arsenic accumulation and increases tolerance in *Arabidopsis thaliana*. *Frontiers in Plant Science* 7: 740.

Verma, S., Verma, P.K., Pande, V. et al. (2016b). Transgenic *Arabidopsis thaliana* expressing fungal arsenic methyltransferase gene (WaarsM) showed enhanced arsenic tolerance via volatilization. *Environmental and Experimental Botany* 132: 113–120.

Wang, P., Zhang, W., Mao, C. et al. (2016). The role of OsPT8 in arsenate uptake and varietal difference in arsenate tolerance in rice. *Journal of Experimental Botany* 67: 6051–6059.

Wang, F.Z., Chen, M.X., Yu, L.J. et al. (2017). OsARM1, an R2R3 MYB transcription factor, is involved in regulation of the response to arsenic stress in rice. *Frontiers in Plant Science* 8: 1868.

Wang, C., Na, G., Bermejo, E.S. et al. (2018). Dissecting the components controlling root-to-shoot arsenic translocation in *Arabidopsis thaliana*. *New Phytologist* 217: 206–218.

Wojas, S., Clemens, S., SkŁodowska, A. et al. (2010). Arsenic response of AtPCS1-and CePCS-expressing plants–effects of external As (V) concentration on As-accumulation pattern and NPT metabolism. *Journal of Plant Physiology* 167: 169–175.

Wu, Z., Ren, H., McGrath, S.P. et al. (2011). Investigating the contribution of the phosphate transport pathway to arsenic accumulation in rice. *Plant Physiology* 157: 498–508.

Xu, X.Y., McGrath, S.P., and Zhao, F.J. (2007). Rapid reduction of arsenate in the medium mediated by plant roots. *New Phytologist* 176: 590–599.

Xu, J., Shi, S., Wang, L. et al. (2017). OsHAC4 is critical for arsenate tolerance and regulates arsenic accumulation in rice. *New Phytologist* 215: 1090–1101.

Yadav, B.S., Singh, S., Srivastava, S. et al. (2019). Whole transcriptome expression profiling and biological network analysis of chickpea during heavy metal stress. *Journal of Plant Biochemistry and Biotechnology* 28: 45–352.

Yamaji, N. and Ma, J.F. (2009). A transporter at the node responsible for intervascular transfer of silicon in rice. *The Plant Cell* 21: 2878–2883.

Yanitch, A., Brereton, N.J., Gonzalez, E. et al. (2017). Transcriptomic response of purple willow (*Salix purpurea*) to arsenic stress. *Frontiers in Plant Science* 8: 1115.

Ye, J., Rensing, C., Rosen, B.P. et al. (2012). Arsenic biomethylation by photosynthetic organisms. *Trends in Plant Science* 17: 155–162.

Ye, Y., Li, P., Xu, T. et al. (2017). OsPT4 contributes to arsenate uptake and transport in rice. *Frontiers in Plant Science* 8: 2197.

Yu, L.J., Luo, Y.F., Liao, B. et al. (2012). Comparative transcriptome analysis of transporters, phytohormone and lipid metabolism pathways in response to arsenic stress in rice (*Oryza sativa*). *New Phytologist* 195: 97–112.

Zhao, F.J., Ma, J.F., Meharg, A.A. et al. (2009). Arsenic uptake and metabolism in plants. *New Phytologist* 181: 777–794.

Zhao, F.J., Ago, Y., Mitani, N. et al. (2010). The role of the rice aquaporin Lsi1 in arsenite efflux from roots. *New Phytologist* 186: 392–399.

Zu, Y., Cheng, S., Li, Z. et al. (2018a). Effects of arsenic on flavonoids accumulation and metobolism pathway in main roots of *Panax notoginseng*. *International Journal of Agriculture and Biology* 6: 1375–1381.

Zu, Y., Li, Z., Mei, X. et al. (2018b). Transcriptome analysis of main roots of *Panax notoginseng* identifies genes involved in saponin biosynthesis under arsenic stress. *Plant Gene* 16: 1–7.

9

Proteomic Regulation During Arsenic Stress

Naina Marwa[1,2], *Sunil Kumar Gupta*[3], *Gauri Saxena*[2], *Vivek Pandey*[1], *and Nandita Singh*[1]

[1] *Plant Ecology and Climate Change Science Division, CSIR-National Botanical Research Institute, Lucknow, Uttar Pradesh, India*
[2] *Department of Botany, University of Lucknow, Lucknow, Uttar Pradesh, India*
[3] *CAS Key Laboratory of Tropical Forest Ecology, Xishuangbanna Tropical Botanical Garden, Chinese Academy of Sciences, Menglun, Yunnan, People's Republic of China*

9.1 Introduction

Proteomics is an intensive study of proteins that involves techniques to recognize proteins synthesized complementing the expressed genome. In contrast to genomics, proteomics is a best approach to determine distinctively expressed proteins, as they display the certain activity in relation to metabolic reactions and regulatory pathways and thus facilitating in designing targeted investigations (Sacheti et al. 2014). The attempts to study the diverse protein expression due to external environmental stress are gaining researcher's interest over the years.

There are generally two types of arsenic phytotoxicity, i.e. arsenate "As(V)" and arsenite "As(III)." Many of the previous studies already illustrated that As(III) is relatively more toxic than As(V) to plants; comparative transcriptomic (Chakrabarty et al. 2009), and proteomic (Ahsan et al. 2008, 2009; Duquesnoy et al. 2009) investigations reported that relatively more genes and/or proteins were differentially expressed under As(V) exposure than As(III). Shilev et al. (2007) reported an increase in protein expressions under environmental stress conditions, such as toxic pollutants, heat stress, and nutrient deficiencies. Weiss et al. (2009) and Parvatiyar et al. (2005) have used molecular methodologies to observe the effect of As stress in *Herminiimonas asoxydans* and *Pseudomonas aeruginosa*. However, Srivastava et al. (2012) demonstrated the As(V) resistance mechanism in *Staphylococcus* sp. using the more advanced methods. The As metabolism and resistance act differentially in different bacteria; detoxification of As and its speciation affects the bacterial biogeochemical cycle under As toxicity (Singh et al. 2016a). Proteins involved in carbon sequestration, pathogen-responsive (PR) proteins, and energy metabolism-related proteins have significantly changed their expression level under As stress.

Arsenic in Plants: Uptake, Consequences and Remediation Techniques, First Edition.
Edited by Prabhat Kumar Srivastava, Rachana Singh, Parul Parihar, and Sheo Mohan Prasad.

9.1.1 Proteins in Antioxidative Defense Strategies

In plants, intercellular biochemical and molecular responses come into action due to metal toxicity, and numerous self-guarding approaches are involved in this process. The metal triggered signals are detected and transported using signal pathways that ultimately stimulate the launch of transcription factors altering the expression of involved genes, aiming for the plant's survivability. Plants' potential to endure severe environmental conditions depends on their potential to identify stress triggered signals and their translation into appropriate reactions (Sarwat et al. 2013). Initiation of stress-associated genes is recognized to include definite signal transduction pathways such as mitogen-activated protein kinase (MAPK) cascade, Ca-calmodulin system, and reactive oxygen species (ROS) signaling.

Kärenlampi et al. (2000) described that plants under heavy metal stress stimulate the production of free radicals, which disrupts plant's foremost intercellular biomolecules such as proteins, lipids, and DNA. However, plants are naturally adapted with different enzymatic or non-enzymatic antioxidant systems, signaling pathways, and metabolites against the effects of ROS and subsequent oxidative stress (Piotrowska et al. 2009; Ahammed et al. 2013). Gupta et al. (2009) described that the plant's defense system includes enzymes like superoxide dismutase (SOD), catalase (CAT), guaiacol peroxidase (GPX), and ascorbate peroxidase (APX) and other antioxidant compounds such as glutathione (GSH), carotenoids, etc.

SOD, the frontline plant's defense, facilitates to convert superoxide radicals to hydrogen peroxide (H_2O_2), which is subsequently reduced to H_2O and O_2 either by APX in ascorbate–glutathione cycle or by GPX and CAT in the cytoplasm and other cellular compartments (Gupta et al. 2009). The antioxidant defense system is governed by another enzyme, glutathione reductase (GR), which regulates a high GSH/GSSG (glutathione disulfide) ratio, essential for safeguarding against oxidative damage (Srivastava et al. 2012).

Maksymiec (2007) suggested that signaling molecules and protein expressions under abiotic stress involve the intricate signal transduction system that initiates by detecting the heavy metal resulting in stimulation of metal-responsive genes by transcription, which results in the reduction of lethal effects on the plant. So, the protein kinases are important to understand how plants regulate the arsenic (As) stress. Upon extracellular signals, cells within upregulate major kinases, like tyrosine kinase-like (TKL), receptor-like cytoplasmic kinases (RLCKs), CT10 regulator of the kinase (CRK), and wall-associated kinases (WAK), leading to the launch of signaling cascades.

Previously, Giri et al. (2011) and Tanaka et al. (2012) have already discussed the role of TKL under abiotic stress. Numerous arsenic-receptive kinases belonged to the TKL group when gene expressions were related to As stress through microarray study. According to Becraft (2002), TKLs are serine–threonine kinases, which function as signaling regulators to turn "on" and "off" other proteins. They are regarded as TKL as their sequence resembles tyrosine kinases. Maximum TKLs activated in plants under As stress belong to receptor-like kinases (RLK) (a subfamily of kinases belonging to the TKL group). Huang et al. (2012) stated the enhanced protein expression of 15 members of diverse RLCKs related to the signal transduction in rice when treated with As. RLCKs have been observed to moderately related functions in various biotic and abiotic stresses (Becraft 2002).

In plants, two other proteins, CRK and WAK, involved in intercellular signaling have also been investigated to be upregulated under As stress (Wagner and Kohorn 2001; Huang et al. 2012). CRK is connector proteins that bind to the tyrosine-phosphorylated proteins and transfer the signals to cytoplasmic proteins through phosphorylation. WAK protein family comprises cell wall receptors, having one domain adhered to the cell wall while the other serine/threonine-protein kinase domain to the cytoplasm, enabling it to directly transmit signals (He et al. 1999).

MAPKs characterize an evolutionary preserved signaling device in eukaryotes (Ichimura et al. 2002). MAPKs are serine/threonine/tyrosine definite signaling proteins (Pearson et al. 2001). Different signaling pathways use MAPKs to monitor the diversity of cellular activities in response to extracellular stimulus. MAPKs are part of a phosphorylation cascade composed of three sequentially activated protein kinases: MAPKKK, MAPK kinase (MAPKK), and MAPK (Widmann et al. 1999). Heavy metals are among the many environmental abiotic stress stimuli involved in the activation of MAPK signaling cascades (Agrawal et al. 2002; Yeh et al. 2004). Previous studies have reported the upregulation of MAPKKK and MAPK mRNA in plants under metal stress (Fu et al. 2002; Kim et al. 2003). The triggering of MAPKs in heavy metal stress initiates through NADPH (nicotinamide adenine dinucleotide phosphate) oxidases, ROS, CDPK (calcium-dependent protein kinase), phosphatidylinositol 3-kinase (PI3K), and other MAPKs (Wang et al. 1998; Yeh et al. 2007).

Huang et al. (2012) established a relation between As treatment and MAPK proteins regulation in rice plants. The study also revealed the upregulation of this signaling cascade – 11 MAPKKKs, one MAPK, and 10 PP2C genes. In roots, MAPKs synthesis intensified within few minutes of As exposure. This major upsurge was detected in OsMPK3 and OsMPK4 levels in particular (Huang et al. 2012). Similarly, Rao et al. (2011) also observed the upregulation of MAPK signaling constituents, namely, OsMPK3, OsMPK4, and OsMKK4, in As stress settings. MAPKs play a chief role in ROS regulation, which is the key source of heavy metal toxicity (Pitzschke and Hirt 2009). Gupta et al. (2009) described the stimulation of MAPKs under As stress, which participates in defense mechanism, ultimately resulting in increased incorporation of sulfate.

9.2 Molecular Chaperones in Response to Arsenic Stress

It is well documented that As accumulation results in toxic cellular reactions in plants (Marwa et al. 2019, 2020). Being the structural analogue, As(V) possesses a high affinity for phosphate transporters (Meharg and Macnair 1992). Bun-Ya et al. (1991) identified high-affinity phosphate transporter (PHO84) in *Saccharomyces cerevisiae*. In plants, the high-affinity phosphate transporter family (PHT1) was first observed in *Arabidopsis thaliana*. Ever since the detection of PHT1 transporters, numerous studies have described their homologs in diverse plant species (Rausch and Bucher 2002). Phosphate transporters, PHT1;1 and PHT1;4, contribute to phosphate uptake in *A. thaliana* at low and high phosphate concentration. Mutant *A. thaliana* having a semi-dominant allele for Pi transporter PHT1;1 displayed a comparatively double increase in As accumulation into wild-type plants, besides the overexpression of the same transporter in wild-type plants boosted As accumulation (Catarecha et al. 2007).

However, some plants are inherently tolerant to substantial levels of As and accumulate high concentrations of the metalloid in their system. Ma et al. (2001) first reported *Pteris vittata*, a fern, as an arsenic hyperaccumulator whose fronds accumulated the contaminated inorganic forms of As at higher amounts. In *P. vittata*, PvPht1;3 and PvPht1;5 revealed an analogous affinity for phosphate, but PvPht1;3 indicated a higher affinity for As(V). The expression of PvPht1;3 was induced by a low level of Pi and As(V), which additionally improved the amount of As(V) uptake (DiTusa et al. 2016). Out of various As species, Arsenite (As(III)) is considered extremely toxic for both plants and animals by binding to the sulfhydryl-containing enzyme and protein resulting into interference of cellular functions (Singh et al. 2016b). Numerous transporters that are involved in As(III) uptake have been reported and summarized in Table 9.1.

Falandysz and Borovička (2013) described, *Sarcosphaera coronaria*, an ectomycorrhizal ascomycetes fungus, which could accumulate high concentrations of methylarsonic acid in its fruit bodies. The ACR3 arsenite effluxer, membrane protein, in yeast offers As resistance to its heterologous expression in *A. thaliana* (Ali et al. 2012). Two genes similar to the

Table 9.1 A complete list of transporters involved in arsenite (As(III)) uptake in plants.

Transporters involved in As(III) uptake	
ABC-type transporters	
ABC transporters coupled with arsenic compound resistance 3 (ACR3) facilitates the vacuolar As(III) sequestration either directly or by complex formation	Indriolo et al. (2010)
OsABCC1; engaged in As purification and reduction of rice grains	Song et al. (2014)
Aquaporins	
Nodulin-26-like intrinsic proteins (NIP) subfamily members (NIP1;1, NIP1;2, NIP5;1, NIP6;1, and NIP7;1) of transport As(III) in *A. thaliana*	Bienert et al. (2008); Isayenkov and Maathuis (2008); Kamiya et al. (2009); Ma et al. (2008)
OsNIP3;2, an intrinsic membrane protein, was reported to be involved in As(III) uptake in rice	Chen et al. (2017)
Tonoplast intrinsic protein (TIP) (localized in tonoplast and endosomal membranes) PvTIP4;1, observed to transport As(III) in *P. vittata*	He et al. (2016)
Inositol transporters (INTs) (localized on both vacuolar membrane and plasma membrane) AtINT2 and AtINT4 in *Saccharomyces cerevisiae* studied for long-distance transport of As(III) and postulated to function in phloem loading of As(III)	Duan et al. (2015)
Silicon transporters	
Lsi1, an influx transporter that aids in As(III) uptake; Lsi2, the efflux transporter and is involved in the transport of As(III) to xylem	Ma et al. (2008)
Lsi1 was also found to uptake methylated-As species in rice	Li et al. (2009)

yeast's arsenite effluxer, ACR3, have been isolated from *P. vittata*, PvACR3 and PvACR3;1. Out of these, PvACR3 has a specific role in As tolerance in the plants (Indriolo et al. 2010). As^{5+} gets reduced to As^{3+} intercellularly when taken up by roots. PvACR3 induces the As^{3+} efflux for vacuolar sequestration. In angiosperms, the ACR3 gene is absent; interestingly, when PvACR3 is expressed in transgenic *A. thaliana*, its As tolerance efficacy increases. Moreover, Chen et al. (2013) detected the germination of PvACR3 transgenic seeds of *A. thaliana* at very high levels of As that are potentially toxic for wild-type plants. Paddy cultivated in soils where *P. vittata* was previously grown has been observed to sequester comparatively reduced As in straw and grains (Ye et al. 2011).

The two studied rice varieties, Indica rice varieties, were relatively more As tolerant than japonica ones (Dasgupta et al. 2004). In their experiments, they detected an As tolerant gene AsTol on chromosome 6 between markers RZ516 and RG213. Mosa et al. (2012) improved As^{3+} tolerance in *Arabidopsis* through overexpressing plasma membrane intrinsic proteins from rice, OsPIP2;4, OsPIP2;6, and OsPIP2;7. These rice transporters might play a crucial role in As tolerance. Mutant *Arabidopsis* plants for auxin transporters have been observed to be more sensitive to As^{3+} stress. Auxin applied exogenously increases tolerance against As^{3+} in such mutants (Krishnamurthy and Rathinasabapathi, 2013).

9.3 Participation of Protein in CO_2 Assimilation and Photosynthetic Activity

Enough studies indicate severe damage to photosynthetic apparatus and CO_2 assimilation (Duquesnoy et al. 2009). Proteins such as oxygen-evolving enhancer protein, RuBisCO small and large subunits, RuBisCO activase, and ATP synthase involved in the Calvin or Krebs cycle are mostly shown differential expression during As(V) stress (Duquesnoy et al. 2009). As(V) stress lowered the CO_2 assimilation due to decreased activity of RuBisCO activity (Ahsan et al. 2010). Fructose bisphosphate aldolase (FBA) is a vital protein and plays an essential role in glycolysis and the Calvin cycle. The primary function of FBA is to reversibly catalyze the conversion of fructose 1,6-bisphosphate (FBP) to glyceraldehyde 3-phosphate (GAP) and dihydroxyacetone 3-phosphate (DHAP). As(V) impaired the phosphorylation reactions by replacing Pi from biomolecules (Finnegan and Chen 2012). As(V) esters are much less stable than Pi esters and quickly hydrolyze under normal physiological conditions, resulting in uncoupling of phosphorylation (Baer et al. 1981; Finnegan and Chen 2012). Many reports suggest that As stress leads to decreased FBA in plants and impairs the glycolysis, pentose phosphate pathway (PPP), and turnover rate of the Calvin cycle (Fridlyand and Scheibe 1999; Nakahara et al. 2003; Pandey et al. 2012). Fructose 1,6-bisphosphatase (FBPase) is also an important enzyme that participates in the glycolysis and Krebs cycle. During As stress, glyceraldehyde 3-phosphate and fructose-1,6-bisphosphate may be converted to glucose 6-phosphate for re-entry into the PPP for nicotinamide adenine dinucleotide phosphate (NADPH) productions. Furthermore, FBA can participate in the regeneration of Ribulose biphosphate (RUBP) (Fridlyand and Scheibe 1999). As accumulation is known to reduce photosynthetic efficiency by disturbing PSII protein precursors situated in the chloroplast thylakoid membrane (Maksymiec 1997). PII protein precursors were found to be overexpressed with As(III) treatments in *Agrostis tenuis* As(V) (Duquesnoy et al. 2009). This overexpression could be considered as a metabolic disturbance in response to As(III) stress but not as a specific tolerant action. These results reveal that As adversely affects CO_2 assimilation and decreases photosynthetic activity in As affected zones.

9.4 Pathogen-Responsive Proteins (PR) in Response to Arsenic Stress

Pathogen-responsive (PR) proteins were initially reported in tobacco plants that were infected by tobacco mosaic virus (TMV) (Van Loon and Van Kammen 1970; Bol et al. 1990). Initially, PR just consisted of five major classes, i.e. PR1, PR2, PR3, PR4, and PR5. These all classes were first reported in tobacco plants, identified through biochemical and molecular approaches (Bol et al. 1990). Although scientists have sequenced the genome of many model plants species, such as *A. thaliana* (The *Arabidopsis* Genome Initiative 2000), moss *Physicomitrella patens* (Rensing et al. 2008), rice (International Rice Genome Sequencing Project, 2005), and other important crops (Phytozome 2015), the challenges for successfully designating their protein product for plant biologists are always there. One enzyme, L-amino acid amidase, generally involved in proline biosynthesis in plants found to be more abundant in *Bacillus aryabhattai* under As stress (Singh et al. 2016a). A high proline level shows coping against high As stress. PR proteins are important to comprehend the quantitative and qualitative adverse responses to the As stress in plants.

9.5 Participation of Proteins in Energy Metabolism

The arsenic role to influence energy-related proteins, leading to overall impairment of plant metabolism, was observed in rice (Ahsan et al. 2010) and *Anabaena* sp. (Pandey et al. 2012). Transketolase (TK) is an amphibolic enzyme universally required for PPP for NADPH production and Calvin cycle. Kaiser (1979) reported As to be abundant maybe due to $H_2 0_2$ accumulation during As stress and high competition with TK catalyzed reaction by As (Gorbach et al. 1981). Phosphoglycerate kinase (PGK) is a monomeric two-domain enzyme involved in the Calvin cycle and energy-conserving phase of glycolysis. PGK's primary function is to generate ATP and increase the respiration rate under stressful conditions. During stress conditions energy requirement for plants is always high, depicting overexpression of energy metabolism-related proteins. Hence, several reports show an increase in the abundance of PGK protein during As stress. Some energy-related proteins, for instance, malate dehydrogenase (MDH) ATP synthase (α, β, and γ subunits), also showed differential expression upon treatment with As(V). Bona et al. (2010) reported differentially expressed two proteins in fronds of *P. vittata* related to bioenergetics metabolism viz. MDH and ATP synthase beta subunit. As much as regard the direct influence of arsenic on ATP synthase expression, Requejo and Tena (2005) showed its more abundant in maize roots treated with arsenic. ATP-dependent protease is crucial in controlling short-lived regulatory proteins that are generally involved in maintaining abnormal or damaged proteins by their removal. Suzuki et al. (1997) found that chaperon and peptidase activity proteases are dependent on ATP-dependent proteases. Clp proteases (ClpP, plastome-encoded proteolytic subunit, and ClpC, nuclear-encoded ATPase subunit) are two important enzymes in chloroplast development, and housekeeping gets affected under As stress (Shanklin et al. 1995; Crafts-Brandner et al. 1996; Weave et al. 1999). These results about energy-metabolism-related proteins show their involvement in coping As stress and regulating normal homeostasis in the plant life cycle.

9.6 Possible Pan-interactomics

Over the years, the intensive study of intercellular arsenic biology has been well reported. The challenge is in incorporating proteomic, transcriptomic, and metabolomic data to provide a broad representation of the metal's contribution in molecular, cellular, and organismal processes. This will fundamentally include connecting lower order experimental data to higher order omics studies (Figure 9.1).

The terminology "interactome" states the comprehensive molecular interaction involving all possible physical exchanges between cellular macromolecules (DNA, RNA, proteins, lipids, ions, metabolites, etc.) and the significances of such interactions. The concept of pan-interactomics comprises DNA-protein, protein-RNA, protein–protein interactions (PPIs), protein-ions, and various probable interfaces between biomolecules (Srivastava et al. 2020). This mapping technique functions as an instrument to detect the cellular links in various physiological conditions creating a database for the problem. This intensive study generates a large amount of data that is further processed by bioinformatic tools. The processed data can be used to predict unexplored genes, or pathways, or mutations. Conversely, the computational investigation requires a cross-examination through experiments or vice versa.

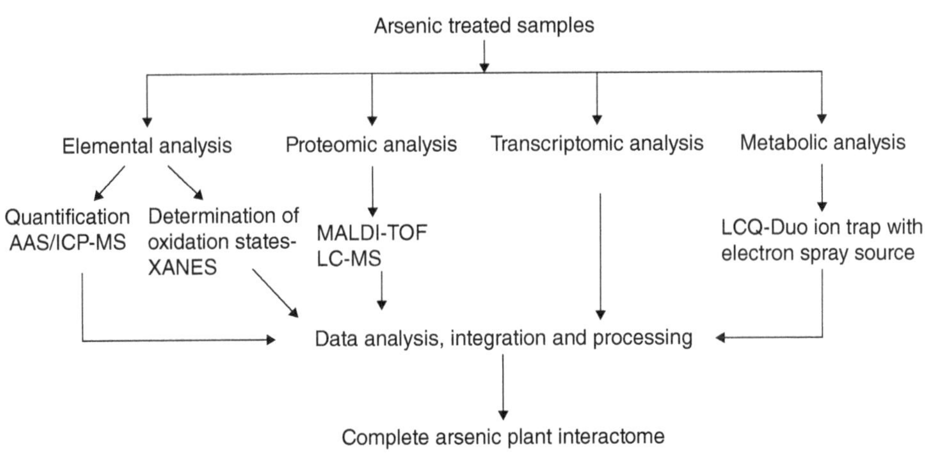

Figure 9.1 Diagrammatic illustration of a future planned methodology for any suggested experiment in which data obtained from the different analysis is integrated to form a complete plant-As interactome.

9.7 Conclusion

Due to the adverse effects of arsenic on the ecosystem, it has become a global concern over the years. A structural analogue to phosphate enters the plant system through phosphate transporters and eventually becomes part of the food chain. In plants, prolonged arsenic exposure has evolved various signaling molecules like proteins, hormones, ROS, etc., which enable them to survive and activate the defense system. Deciphering the complex arsenic signaling mechanisms of plants will provide an understanding of the altered functioning of the plants. Employing advanced computational tools provides information about unexplored pathways that can be exploited to develop arsenic-resistant plant cultivars or reduce or minimize arsenic toxicity.

References

Agrawal, G.K., Rakwal, R., and Iwahashi, H. (2002). Isolation of novel rice (*Oryza sativa* L.) multiple stress responsive MAP kinase gene, OsMSRMK2, whose mRNA accumulates rapidly in response to environmental cues. *Biochemical and Biophysical Research Communications* 294 (5): 1009–1016.

Ahammed, G.J., Choudhary, S.P., Chen, S. et al. (2013). Role of brassinosteroids in alleviation of phenanthrene–cadmium co-contamination-induced photosynthetic inhibition and oxidative stress in tomato. *Journal of Experimental Botany* 64 (1): 199–213.

Ahsan, N., Lee, D.G., Alam, I. et al. (2008). Comparative proteomic study of arsenic-induced differentially expressed proteins in rice roots reveals glutathione plays a central role during As stress. *Proteomics* 8: 3561–3576.

Ahsan, N., Lee, D.G., Kim, K.H. et al. (2010). Analysis of arsenic stress-induced differentially expressed proteins in rice leaves by two-dimensional gel electrophoresis coupled with mass spectrometry. *Chemosphere* 78 (3): 224–231.

Ahsan, N., Renaut, J., and Komatsu, S. (2009). Recent developments in the application of proteomics to the analysis of plant responses to heavy metals. *Proteomics* 9: 2602–2621.

Ali, W., Isner, J.C., Isayenkov, S.V. et al. (2012). Heterologous expression of the yeast arsenite efflux system ACR3 improves *Arabidopsis thaliana* tolerance to arsenic stress. *New Phytologist* 194 (3): 716–723.

Baer, C.D., Edwards, J.O., and Rieger, P.H. (1981). Kinetics of the hydrolysis of arsenate(V) triesters. *Inorganic Chemistry* 20: 905–907.

Becraft, P.W. (2002). Receptor kinase signaling in plant development. *Annual Review of Cell and Developmental Biology* 18 (1): 163–192.

Bienert, G.P., Thorsen, M., Schüssler, M.D. et al. (2008). A subgroup of plant aquaporins facilitate the bi-directional diffusion of $As(OH)_3$ and $Sb(OH)_3$ across membranes. *BMC Biology* 6 (1): 26.

Bol, B.J.C., Linthorst, J.F., and Cornelissen, H.J.M. (1990). Plant pathogenesis-related proteins induced by virus infection. *Annual Review of Phytopathology* 28: 113–138.

Bona, E., Cattaneo, C., Cesaro, P. et al. (2010). Proteomic analysis of *Pteris vittata* fronds: two arbuscular mycorrhizal fungi differentially modulate protein expression under arsenic contamination. *Proteomics* 10: 3811–3834.

Bun-Ya, M., Nishimura, M., Harashima, S., and Oshima, Y.A.S.U.J.I. (1991). The PHO84 gene of Saccharomyces cerevisiae encodes an inorganic phosphate transporter. *Molecular and Cellular Biology* 11 (6): 3229–3238.

Catarecha, P., Segura, M.D., Franco-Zorrilla, J.M. et al. (2007). A mutant of the Arabidopsis phosphate transporter PHT1; 1 displays enhanced arsenic accumulation. *The Plant Cell* 19 (3): 1123–1133.

Chakrabarty, D., Trivedi, P.K., Misra, P. et al. (2009). Comparative transcriptome analysis of arsenate and arsenite stresses in rice seedlings. *Chemosphere* 74: 688–702.

Chen, Y., Sun, S.K., Tang, Z. et al. (2017). The Nodulin 26-like intrinsic membrane protein OsNIP3; 2 is involved in arsenite uptake by lateral roots in rice. *Journal of Experimental Botany* 68 (11): 3007–3016.

Chen, Y., Xu, W., Shen, H. et al. (2013). Engineering arsenic tolerance and hyperaccumulation in plants for phytoremediation by a PvACR3 transgenic approach. *Environmental Science & Technology* 47 (16): 9355–9362.

Crafts-Brandner, S.J., Klein, R.R., Klein, P. et al. (1996). Coordination of protein and mRNA abundances of stromal enzymes and mRNA abundances of the Clp protease subunits during senescence of *Phaseolus vulgaris* (L.) leaves. *Planta* 200: 312–318.

Dasgupta, T., Hossain, S.A., Meharg, A.A., and Price, A.H. (2004). An arsenate tolerance gene on chromosome 6 of rice. *New Phytologist* 45–49.

DiTusa, S.F., Fontenot, E.B., Wallace, R.W. et al. (2016). A member of the Phosphate transporter 1 (Pht1) family from the arsenic-hyperaccumulating fern *Pteris vittata* is a high-affinity arsenate transporter. *New Phytologist* 209 (2): 762–772.

Duan, G.L., Hu, Y., Schneider, S. et al. (2015). Inositol transporters AtINT2 and AtINT4 regulate arsenic accumulation in *Arabidopsis* seeds. *Nature Plants* 2 (1): 1–6.

Duquesnoy, I., Goupil, P., Nadaud, I. et al. (2009). Identification of Agrostis tenuis leaf proteins in response to As(V) and As(III) induced stress using a proteomics approach. *Plant Science* 176: 206–213.

Falandysz, J. and Borovička, J. (2013). Macro and trace mineral constituents and radionuclides in mushrooms: health benefits and risks. *Applied Microbiology and Biotechnology* 97 (2): 477–501.

Finnegan, P.M. and Chen, W. (2012). Arsenic toxicity: the effects on plant metabolism. *Frontiers in Physiology* 3: 182.

Fridlyand, L.E. and Scheibe, R. (1999). Regulation of the Calvin cycle for CO_2 fixation as an example for general control mechanism in metabolic cycles. *Biosystems* 51: 79–93.

Fu, S.F., Chou, W.C., Huang, D.D., and Huang, H.J. (2002). Transcriptional regulation of a rice [*Oryza sativa*] mitogen-activated protein kinase gene, OsMAPK4, in response to environmental stresses. *Plant and Cell Physiology (Japan)* 43 (8): 958–963.

Giri, J., Vij, S., Dansana, P.K., and Tyagi, A.K. (2011). Rice A20/AN1 zinc-finger containing stress-associated proteins (SAP1/11) and a receptor-like cytoplasmic kinase (OsRLCK253) interact via A20 zinc-finger and confer abiotic stress tolerance in transgenic *Arabidopsis* plants. *New Phytologist* 191 (3): 721–732.

Gorbach, Z.V., Maglysh, S.S., Kubyshin, V.L., and Ostrovskii, I.M. (1981). Purification and properties of rat liver transketolase. *Biokhimiia* 46: 1963–1969.

Gupta, M., Sharma, P., Sarin, N.B., and Sinha, A.K. (2009). Differential response of arsenic stress in two varieties of *Brassica juncea* L. *Chemosphere* 74 (9): 1201–1208.

He, Z., Yan, H., Chen, Y. et al. (2016). An aquaporin Pv TIP 4; 1 from Pteris vittata may mediate arsenite uptake. *New Phytologist* 209 (2): 746–761.

He, Z.H., Cheeseman, I., He, D., and Kohorn, B.D. (1999). A cluster of five cell wall-associated receptor kinase genes, Wak1–5, are expressed in specific organs of *Arabidopsis*. *Plant Molecular Biology* 39 (6): 1189–1196.

Huang, T.L., Nguyen, Q.T.T., Fu, S.F. et al. (2012). Transcriptomic changes and signalling pathways induced by arsenic stress in rice roots. *Plant Molecular Biology* 80 (6): 587–608.

Ichimura, K., Shinozaki, K., Tena, G. et al. (2002). Mitogen-activated protein kinase cascades in plants: a new nomenclature. *Trends in Plant Science* 7 (7): 301–308.

Indriolo, E., Na, G., Ellis, D. et al. (2010). A vacuolar arsenite transporter necessary for arsenic tolerance in the arsenic hyperaccumulating fern *Pteris vittata* is missing in flowering plants. *The Plant Cell* 22 (6): 2045–2057.

International Rice Genome Sequencing Project (2005). The map-based sequence of the rice genome. *Nature* 436: 793–800.

Isayenkov, S.V. and Maathuis, F.J. (2008). The *Arabidopsis thaliana* aquaglyceroporin AtNIP7; 1 is a pathway for arsenite uptake. *FEBS Letters* 582 (11): 1625–1628.

Kaiser, W. (1979). Reversible inhibition of the Calvin cycle and activation of oxidative pentose phosphate cycle in isolated intact chloroplasts by hydrogen peroxide. *Planta* 145: 377–382.

Kamiya, T., Tanaka, M., Mitani, N. et al. (2009). NIP1; 1, an aquaporin homolog, determines the arsenite sensitivity of *Arabidopsis thaliana*. *Journal of Biological Chemistry* 284 (4): 2114–2120.

Kärenlampi, S., Schat, H., Vangronsveld, J. et al. (2000). Genetic engineering in the improvement of plants for phytoremediation of metal polluted soils. *Environmental Pollution* 107 (2): 225–231.

Kim, C.Y., Liu, Y., Thorne, E.T. et al. (2003). Activation of a stress-responsive mitogen-activated protein kinase cascade induces the biosynthesis of ethylene in plants. *The Plant Cell* 15 (11): 2707–2718.

Krishnamurthy, A. and Rathinasabapathi, B. (2013). Auxin and its transport play a role in plant tolerance to arsenite-induced oxidative stress in *Arabidopsis thaliana*. *Plant, Cell & Environment* 36 (10): 1838–1849.

Li, R.Y., Ago, Y., Liu, W.J. et al. (2009). The rice aquaporin Lsi1 mediates uptake of methylated arsenic species. *Plant Physiology* 150 (4): 2071–2080.

Ma, J.F., Yamaji, N., Mitani, N. et al. (2008). Transporters of arsenite in rice and their role in arsenic accumulation in rice grain. *Proceedings of the National Academy of Sciences of the United States of America* 105 (29): 9931–9935.

Ma, L.Q., Komar, K.M., Tu, C. et al. (2001). A fern that hyperaccumulates arsenic. *Nature* 409 (6820): 579–579.

Maksymiec, W. (1997). Effect of copper on cellular processes in higher plants. *Photosynthetica* 34: 321–342.

Maksymiec, W. (2007). Signalling responses in plants to heavy metal stress. *Acta Physiologiae Plantarum* 29 (3): 177–187.

Marwa, N., Mishra, N., Singh, N. et al. (2020). Effect of rhizospheric inoculation of isolated arsenic (As) tolerant strains on growth, As-uptake and bacterial communities in association with Adiantum capillus-veneris. *Ecotoxicology and Environmental Safety* 196: 110498.

Marwa, N., Singh, N., Srivastava, S. et al. (2019). Characterizing the hypertolerance potential of two indigenous bacterial strains (*Bacillus flexus* and *Acinetobacter junii*) and their efficacy in arsenic bioremediation. *Journal of Applied Microbiology* 126 (4): 1117–1127.

Meharg, A.A. and Macnair, M.R. (1992). Suppression of the high affinity phosphate uptake system: a mechanism of arsenate tolerance in *Holcus lanatus* L. *Journal of Experimental Botany* 43 (4): 519–524.

Mosa, K.A., Kumar, K., Chhikara, S. et al. (2012). Members of rice plasma membrane intrinsic proteins subfamily are involved in arsenite permeability and tolerance in plants. *Transgenic Research* 21 (6): 1265–1277.

Nakahara, K., Yamamoto, H., Miyake, C., and Yokota, A. (2003). Purification and characterization of class-I and class-II fructose-1,6-bisphosphate aldolase from the cyanobacterium *Synechocystis* sp. PCC 6803. *Plant & Cell Physiology* 44: 326–333.

Pandey, S., Rai, R., and Rai, L.C. (2012). Proteomics combines morphological, physiological and biochemical attributes to unravel the survival strategy of *Anabaena* sp. PCC7120 under arsenic stress. *Journal of Proteomics* 75: 921–937.

Parvatiyar, K., Alsabbagh, E.M., Ochsner, U.A. et al. (2005). Global analysis of cellular factors and responses involved in *Pseudomonas aeruginosa* resistance to arsenite. *Journal of Bacteriology* 187 (14): 4853–4864.

Pearson, G., Robinson, F., Beers Gibson, T. et al. (2001). Mitogen-activated protein (MAP) kinase pathways: regulation and physiological functions. *Endocrine Reviews* 22 (2): 153–183.

Phytozome (2015). http://www.phytozome.com (displayed 16 January 2014).

Piotrowska, A., Bajguz, A., Godlewska-Żyłkiewicz, B. et al. (2009). Jasmonic acid as modulator of lead toxicity in aquatic plant *Wolffia arrhiza* (Lemnaceae). *Environmental and Experimental Botany* 66 (3): 507–513.

Pitzschke, A. and Hirt, H. (2009). Disentangling the complexity of mitogen-activated protein kinases and reactive oxygen species signaling. *Plant Physiology* 149 (2): 606–615.

Rao, K.P., Vani, G., Kumar, K. et al. (2011). Arsenic stress activates MAP kinase in rice roots and leaves. *Archives of Biochemistry and Biophysics* 506 (1): 73–82.

Rausch, C. and Bucher, M. (2002). Molecular mechanisms of phosphate transport in plants. *Planta* 216 (1): 23–37.

Rensing, S.A., Lang, D., Zimmer, A.D. et al. (2008). The Physcomitrella genome reveals evolutionary insights into the conquest of land by plants. *Science* 319: 64–69. https://doi.org/10.1126/science.1150646.

Requejo, R. and Tena, M. (2005). Proteome analysis of maize roots reveals that oxidative stress is a main contributing factor to plant arsenic toxicity. *Phytochemistry* 66: 1519–1528.

Sacheti, P., Patil, R., Dube, A. et al. (2014). Proteomics of arsenic stress in the gram-positive organism *Exiguobacterium* sp. PS NCIM 5463. *Applied Microbiology and Biotechnology* 98 (15): 6761–6773.

Sarwat, M., Ahmad, A., Abdin, M.Z., and Ibrahim, M.M. (ed.) (2013). *Stress Signaling in Plants: Genomics and Proteomics Perspective*, vol. 2. New York: Springer.

Shanklin, J., Dewitt, N.D., and Flanagan, J.M. (1995). The stroma of higher plant plastids contains ClpP and ClpC, functional homologs of *Escherichia coli* ClpP and ClpA: an archetypal two-component ATP-dependent protease. *Plant Cell* 7: 1713–1722.

Shilev, S., Lopez, A.F., Prieto, M.S., and Puebla, E.D.S. (2007). Induced protein profile changes in arsenate tolerant and sensitive *Pseudomonas fluorescens* strains. *Journal of Environmental Engineering and Landscape Management* 15: 221–226.

Singh, N., Gupta, S.K., Marwa, N. et al. (2016a). Arsenic mediated modifications in Bacillus aryabhattai and their biotechnological applications for arsenic bioremediation. *Chemosphere* 164: 524–534.

Singh, N., Marwa, N., Mishra, J. et al. (2016b). Brevundimonas diminuta mediated alleviation of arsenic toxicity and plant growth promotion in *Oryza sativa* L. *Ecotoxicology and Environmental Safety* 125: 25–34.

Song, W.Y., Yamaki, T., Yamaji, N. et al. (2014). A rice ABC transporter, OsABCC1, reduces arsenic accumulation in the grain. *Proceedings of the National Academy of Sciences of the United States of America* 111 (44): 15699–15704.

Srivastava, G.P., Yadav, N., Yadav, B.N. et al. (2020). Pan-interactomics and its applications. In: *Pan-Genomics: Applications, Challenges, and Future Prospects* (ed. D. Barh, S. Soares, S. Tiwari and V. Azevedo), 397–435. London: Academic Press ISBN 9780128170762, https://doi.org/10.1016/B978-0-12-817076-2.00021-4. (https://www.sciencedirect.com/science/article/pii/B9780128170762000214).

Srivastava, S., Verma, P.C., Singh, A. et al. (2012). Isolation and characterization of *Staphylococcus* sp. strain NBRIEAG-8 from as contaminated site of West Bengal. *Applied Microbiology and Biotechnology* 95: 1275–1291.

Suzuki, C.K., Rep, M., Maarten van Dijl, J. et al. (1997). ATP dependent proteases that also chaperone protein biogenesis. *Trends in Biochemical Sciences* 22: 118–123.

Tanaka, H., Osakabe, Y., Katsura, S. et al. (2012). Abiotic stress-inducible receptor-like kinases negatively control ABA signaling in *Arabidopsis*. *The Plant Journal* 70 (4): 599–613.

The *Arabidopsis* Genome Initiative (2000). Analysis of the genome sequence of the flowering plant *Arabidopsis thaliana*. *Nature* 408: 796–815.

Van Loon, L.C. and Van Kammen, A. (1970). Polyacrylamide disc electrophoresis of the soluble leaf proteins from *Nicotiana tabacum* var. Samsun and Samsun NN II. Changes in protein constitution after infection with tobacco mosaic virus. *Virology* 40: 199.

Wagner, T.A. and Kohorn, B.D. (2001). Wall-associated kinases are expressed throughout plant development and are required for cell expansion. *The Plant Cell* 13 (2): 303–318.

Wang, X., Martindale, J.L., Liu, Y., and Holbrook, N.J. (1998). The cellular response to oxidative stress: influences of mitogen-activated protein kinase signalling pathways on cell survival. *Biochemical Journal* 333 (2): 291–300.

Weave, L.M., Froehlich, J.E., and Amasino, R.M. (1999). Chloroplast-targeted ERD1 protein declines but its mRNA increases during senescence in *Arabidopsis*. *Plant Physiology* 119: 1209–1216.

Weiss, S., Carapito, C., Cleiss, J. et al. (2009). Enhanced structural and functional genome elucidation of the arsenite oxidizing strain *Herminiimonas arsenicoxydans* by proteomics data. *Biochimie* 91: 192–203.

Widmann, C., Gibson, S., Jarpe, M.B., and Johnson, G.L. (1999). Mitogen-activated protein kinase: conservation of a three-kinase module from yeast to human. *Physiological Reviews* 79 (1): 143–180.

Ye, W.L., Khan, M.A., McGrath, S.P., and Zhao, F.J. (2011). Phytoremediation of arsenic contaminated paddy soils with *Pteris vittata* markedly reduces arsenic uptake by rice. *Environmental Pollution* 159 (12): 3739–3743.

Yeh, C.M., Chien, P.S., and Huang, H.J. (2007). Distinct signalling pathways for induction of MAP kinase activities by cadmium and copper in rice roots. *Journal of Experimental Botany* 58 (3): 659–671.

Yeh, C.M., Hsiao, L.J., and Huang, H.J. (2004). Cadmium activates a mitogen-activated protein kinase gene and MBP kinases in rice. *Plant and Cell Physiology* 45 (9): 1306–1312.

10

Metabolomic Regulation During the Arsenic Stress

Pooja Sharma[1], Anuj Kumar Tiwari[2], Neeraj Kumar Dubey[3], Charu Chaturvedi[4], Amit Prakash Raghuvanshi[4], and Surendra Pratap Singh[4]

[1] *Department of Environmental Microbiology, School for Environmental Sciences, Babasaheb Bhimrao Ambedkar Central University, Lucknow, Uttar Pradesh, India*
[2] *Department of Botany, Bhavan's Mehta Mahavidyalaya, Kaushambi, Uttar Pradesh, India*
[3] *Department of Botany, Rashtriya Snatkottar Mahavidyalaya, Jaunpur, Uttar Pradesh, India*
[4] *Plant Molecular Biology Laboratory, Department of Botany, Dayanand Anglo-Vedic (PG) College, Chhatrapati Shahu Ji Maharaj University, Kanpur, Uttar Pradesh, India*

10.1 Introduction

Chemical, physical, and microbiological processes combine to generate As mobilization in the subsurface, and several concepts are offered to elucidate the mechanism of As organization (Mondal et al. 2006; Drewniak et al. 2012). International Agency of Research and Cancer (IARC) classified two inorganic forms of arsenic, i.e. As(V) and As(III), as Class 1 carcinogens to humans (IARC 2004). Because of a mixture of anaerobic rising circumstances and certain plant physiological properties, rice poses a particular issue for the As accumulation in food chain. Rice provides roughly half of the inorganic arsenic (iAs) intake for people in South Asia showing high iAs in their drinking water (Meharg et al. 2009; Tripathi et al. 2012a, 2012b). Furthermore, in the United States, overuse of methylated arsenical insecticides has resulted in As buildup in soil (Williams et al. 2005). Furthermore, As anxiety has an impact on the yield of crop plants that accumulate a high level of As. The understanding of As absorption and metabolism has been discussed in this chapter based on previous and new research. Additionally, toxic and resistant procedures are examined to understand better the impacted biological processes and responsible compounds within the plants.

Arsenic is a worldwide pollutant, and iAs is the most common form that causes a variety of biological problems. Arsenic toxicity has been linked to a variety of disorders, including cardiovascular disease, respiratory disease, cancer, kidney disease, neuropsychological impairment, and diabetes (Bräuner et al. 2014; Bailey et al. 2016; Kuo et al. 2017). According to World Health Organization (WHO) recommendation on arsenic-polluted drinkable water containing arsenic levels $>10\,g\,l^{-1}$, arsenic-polluted groundwater with arsenic levels

Arsenic in Plants: Uptake, Consequences and Remediation Techniques, First Edition.
Edited by Prabhat Kumar Srivastava, Rachana Singh, Parul Parihar, and Sheo Mohan Prasad.

$>10\,g\,l^{-1}$ may impact more than 200 million people globally (Edition 2011; Wu et al. 2021). Furthermore, those who eat rice as a staple meal in China, Southeast Asia, Taiwan, Bengal Delta, and South America may be at risk for arsenic-related disorders (Islam et al. 2016). The Agency for Toxic Substances and Disease Registry (ATSDR) has placed arsenic at the top of its priority list of dangerous materials. Arsenic, being less hazardous than cadmium, mercury, and lead, is the most often found element and has the greatest possibility of human contamination (Li et al. 2011; Pierce et al. 2011; Tchounwou et al. 2012). Moreover, arsenic invades the organs as a toxic nonessential element by impersonating vital nutrients and minerals, including phosphate, glucose, and glycerol, and blocking their transport systems (Bräuner et al. 2014; Tuulaikhuu et al. 2015). Arsenic causes toxicity in these processes through a variety of mechanisms, including oxidative stress, altering *S*-adenosylmethionine metabolism, binding to sulfhydryl groups, and intrusive with oxidative phosphorylation (Sun et al. 2016). Arsenic exposure has been linked to mitochondrial dysfunction, endocrinological problems, poor lipid metabolism, and DNA oxidative damage (Abhyankar et al. 2012; Khan et al. 2020).

Understanding the real-world toxicology of the environment is necessary to decrease the environmental hazard features that affect human health. Male infertility induced by arsenic has been linked to changes in profiles of hormonal, metabolic abnormalities linked to spermatogenesis and induces oxidative stress (Renu et al. 2018). Infertility-associated metabolic markers, we hypothesized that energy expenditure in spermatogenesis, antioxidant defense, like oxidative stress, fatty acid oxidation in mitochondria, and regulation of hormone in sperm genesis, may all play a crucial role in sterility (Shen et al. 2013; Zhang et al. 2014). Arsenic-mediated energy metabolism problems have been identified as one of the key contributing causes for poor scientific consequences in epidemiological investigations (Pan et al. 2013; Laine et al. 2015; Shapiro et al. 2015). Investigations in laboratory animals have shown that arsenic accumulates in the prostate gland, testicular tissue, epididymis, and seminal vesicle after prolonged exposure to arsenic-contaminated drinking water (Chiou et al. 2008; Chen et al. 2020). According to current research, arsenic-induced testicular damage can be mediated worldwide through the "pituitary-testicular axis" (Jana et al. 2006; Huang et al. 2016). The design of the metabolome systems added a better understanding of arsenic in male reproduction by illuminating the black box of conventional epidemiology. Several obstacles must be overcome to create appropriate molecular interactions. Integrating omics data from epidemiology studies like the genome, proteome, transcriptome, and metabolome allows linking network in a systems approach. "The important middle metabolites, counting those linked with male reproductive disorders in arsenic exposure-induced metabolic disorders and those produced in response to arsenic exposure in male reproductive disorder-related metabolic processes, must be identified before the building of the arsenic-induced male reproductive links." Any perturbation association of metabolic routes that involves ecological hazard requires attention as phenotypic alterations are included in the human metabolome. Using cross-talk with arsenic-stimulated metabolic indicators through the disrupted system, metabolic indicators to indicate/reflect phenotypic changes may be able to bond the hole between arsenic exposure and male imitation. This chapter aimed to outline the risk of arsenic exposure on plants and provide a technique for metabolomic management under arsenic stress. The gap areas in the current knowledge and future possibilities are summarized in this chapter.

10.2 Arsenic Uptake/Translocation in Plants

It seems critical for different species of plants to comprehend and control the arsenic threat from polluted soils (Table 10.1). In a reducing environment, ferric oxyhydroxides release the adsorbed arsenic, resulting in increase in the arsenic accessibility to the plants. There are significant varietal differences in the degree of Fe-plaque production and sequestration (Tripathi et al. 2014). In rice, higher sequestering poses gradually decreasing As translocation to aboveground tissues from root iron plaque (Liu et al. 2004). Rice varieties featuring larger porosities showed greater rates of radial oxygen decline and were better at restricting As transfer to open-pit tissues leading to the generation of further Fe-plaques (Wu et al. 2012). However, considerations other than the amount of iron plaque in the rhizosphere, such as ionic species, plant type, rhizosphere pH, their strengths, or the occurrence of other elements, may influence As uptake. Understanding the processes of As absorption, accumulating, and absorption in crop plants can help us understand the harmful impacts of As on plants (Panda et al. 2010). The majority of the As absorbed accumulates in the roots and, to a lesser extent, in the plant's aboveground parts (Garg and Singla 2011).

Inorganic arsenic comes in a variety of forms in the environment, and the rates of intake, transportation, and uptake in various tissues differ greatly. The capture and transfer of different rice plant species are controlled by several carriers. As(V) is a chemical equivalent of phosphate. As(V) and phosphate compete to accumulate via the root-mediated method transport. Many phosphate transporters (PHTs) involved in As(V) absorption in the *Arabidopsis thaliana* have already been identified, including AtPHT1-1, AtPHT1-4, AtPHT1-5, AtPHT1-8, and AtPHT1-9 (Shin et al. 2004; Catarecha et al. 2007; Remy et al. 2012; Fontenot et al. 2015; Cao et al. 2017). Pht1;9 deficiency improved As(V) acceptance, but overexpression enhanced Pi starvation and As(V) activity (Remy et al. 2012). Categorized a phosphate "transporter 1 (Pht1) family member, PvPht1, from *Pteris vittata*" showed a high As(V) affinity and promoted high As(V) absorption and buildup (Di Tusa et al. 2016). There are 13 OsPT genes in the rice PHT gene family, spanning "OsPT1 to

Table 10.1 Different forms of As compounds in the environment.

Sr. No.	Compound	Formula
1.	Arsenocholine	$(CH_3)As{+}CH_2\ CH_2OH$, AC
2.	Trimethylarsine	$(CH_3)_3As$
3.	Dimethylarsine	$(CH_3)_2AsH$
4.	Dimethylarsenous acid	$(CH_3)_2AsOH$, DMAIII
5.	Monomethylarsonic acid	$CH_3AsO(OH)_2$, MMAV
6.	Arsenobetaine	$(CH_3)_3As{+}CH_2COO^-$, AB
7.	Methylarsine	CH_3AsH_2
8.	Dimethylarsinic acid	$(CH_3)_2AsO(OH)$, DMAV
9.	Trimethylarsinic oxide	$(CH3)_3AsO$, TMAO

Source: Mishra et al. (2021)/Springer Nature.

OsPT13" (Paszkowski et al. 2002). "Out of these, OsPT1, OsPT4, and OsPT8 functions have been investigated" (Kamiya et al. 2013; Wang et al. 2016; Cao et al. 2017; Ye et al. 2017). Increased expression of these phosphate genes has been found to promote As(V) absorption, whereas knockdown leads to As(V) accumulation and limited absorption. Ye et al. (2017) recently revealed that OsPT4 is essential for As(V) absorption in their functional characterization of the gene. Within around cells the As(V) is either reduced to As(III) or loaded onto xylem vessels using phosphate conveyors like OsHAC1;1, OsHARC1;2, or OsHARC4 (Mendoza-Cózatl et al. 2011; Wu et al. 2011; Shi et al. 2016; Xu et al. 2017).

Contrary to As(V), As(III) enters the root epidermal cells via aquaporin channels (Li et al. 2009). Several nodulin-like intrinsic proteins' (NIPs) genetic material participating in As(III) accumulate has been labeled in "*Arabidopsis thaliana*, namely AtNIP1-1, AtNIP1-2, AtNIP5-1, AtNIP3-1, AtNIP6-1, AtNIP7-1" (Mizutani et al. 2006; Isayenkov and Maathuis 2008; Tanaka et al. 2008; Kamiya and Fujiwara 2009; Pang et al. 2010; Xu et al. 2015). In rice, OsNIP2;1 (Lsi1) facilitates the uptake of As(III), whereas OsNIP2;2 (Lsi2) mediates As(III) efflux from root cells to the xylem in rice plant (Ma et al. 2008; Ma and Yamaji 2015). Similarly, Lsi1 and Lsi2 proteins are found on the plasma membranes of exodermal and endodermal cells, while Lsi1 is located on the cell's distal side, and Lsi2 is found on the cell's proximal side (Ma and Yamaji 2006; Ma et al. 2007). The reversible activity of the Lsi1 protein channel allows some As(III) to be released into the rhizosphere (Zhao et al. 2010a). "OsNIP2;1 and OsNIP2;2 transporters, other NIPs including OsNIP1;1, OsNIP3;1, and OsNIP3;2 have been reported to play a role in As(III) efflux" (Bienert et al. 2008; Ma et al. 2008; Sun et al. 2018). Small basic intrinsic proteins (SIPs), plasma membrane intrinsic proteins (PIPs), uncategorized intrinsic proteins (XIPs), tonoplast intrinsic proteins (TIPs), and natural resistance-associated macrophage proteins (OsNRAMP1) are among the aquaporin subfamilies implicated in As(III) trafficking (Mosa et al. 2012; Tiwari et al. 2014; Maurel et al. 2015). In *Arabidopsis*, the expression of rice transporter OsNRAMP1 resulted in elevated As(III) tolerance and the improved As(III) tolerance. When compared with wild-type, transgenic plants have higher accumulation in their shoots system. The role of this transporter in xylem loading of As(III) from roots to shoots has been hypothesizing (Tiwari et al. 2014). When rice "OsPIP2;4, OsPIP2;6, and OsPIP2;7" transporters were overexpressed in *Arabidopsis*, the mutants showed greater As(III) tolerance and biomass accumulation relative to wild-type plants. The transgenic plants demonstrated active inflow and outflow of As in the roots after a brief exposure to As(III). After long-term exposure to As(III), there was no appreciable buildup of As in root as well as in the shoot tissues (Mosa et al. 2012). In *Arabidopsis*, transgenic plants expressing PvTIP4;1, which is a *Pteris vittata* TIP class carrier blamable for the As(III) uptake, displayed higher As(III) sensitivity and higher arsenic when linked to wild type (He et al. 2016).

10.3 Arsenic Removal Efficiency in Plants

Plants can detoxify As by using a variety of detoxifying mechanisms. For preserving the free As concentration at a low, the first phase involves the reduction of As and the creation of metal-binding peptides like phytochelatins (PCs) and metallothioneins (MTs) to create complexes with As followed by their vacuolar sequestration (Tripathi et al. 2012). The second phase is to switch-on the antioxidant defense cascade to prevent As-mediated oxidative damage (Zhao et al. 2010b). Plants used a variety of methods to fight the stress conditions due to toxic metalloids like As. Plants produce metal-binding peptides; GSH, MTs, and PCs actively concentrate As within the vacuoles and reduce its availability to plants (Verbruggen et al. 2009). Decontamination processes are also crucial in transforming harmful forms of As into less toxic As forms and their efflux from cell via particular binding protein (Dixit et al. 2016; Singh et al. 2017). The synthesis of suitable solutes or osmolytes, like glycinebetaine, coincides with these mechanisms. Proline and mannitol also minimize the As-induced toxicity (Matysik et al. 2002; Mishra and Dubey 2006; Garg and Singla 2011).

10.4 Toxicity of Arsenic on Plants Metabolism

Arsenic is an incredibly toxic metal (loid) that is a powerful non-threshold class-one carcinogen in human beings (Smith et al. 2002; Li et al. 2017). The main potential causes of As in the ecosystem are anthropogenic and industrial activities (Patel et al. 2005; Zhao et al. 2009). Countries like the United States, India, Bangladesh, France, China, Vietnam, Brazil, Nicaragua, and others have recorded high As concentrations (Patel et al. 2005; McClintock et al. 2012; Singh et al. 2015). Arsenic penetrates the tissues via primary route of eating food with high levels of As buildup. As toxicity affects a variety of severe and persistent health problems in humans, involving lung, kidney, liver, and bladder malignancies, as well as negative effects on the cardiac, neurology, hematologic, and cardiovascular systems (Sahoo and Kim 2013; Jaishankar et al. 2014). Arsenic phytotoxicity was already linked to plant yield loss due to the deleterious impact on health, growth, and metabolic functions (Li et al. 2006; Singh et al. 2015).

Plants are susceptible to As, and they have a range of negative consequences, including a reduction in nodulation in the legume (Lafuente et al. 2010, 2015) (Figure 10.1). Arsenomics is a recently defined method for analyzing As-induced changes in a plant's metabolome,

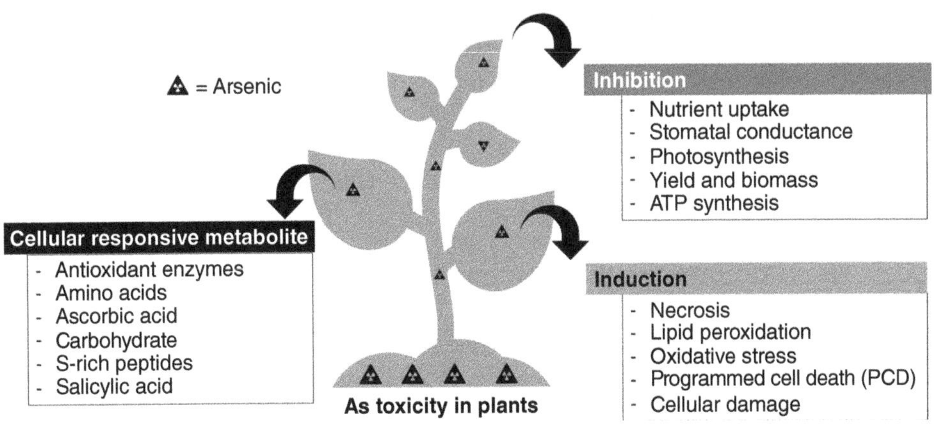

Figure 10.1 Effect of arsenic on plant metabolism: the As includes and inhibits the various metabolic activities and underlying pathways and processes and generates cellular responses during arsenic stress.

proteome, and transcriptome (Tripathi et al. 2012). Exogenous use of several potential compounds in agricultural plants modulates arsenic absorption and metabolism, including toxicity responses. Arsenic is a natural source and widespread hazardous metalloid that remains in the environment, frequently combined with sulfur and other metals. It has a wide range of effects on many life forms, making it a global problem. As contamination has spread through China, the United States, Southeast Asia, and Europe (Srivastava et al. 2016). Furthermore, various human actions, such as the widespread use of As-built substances, herbicides, insecticides, and fertilizers, are to be blamed for the elevated levels of As in the ecosystem (Sharma et al. 2014). The As content in groundwater and drinking water has increased due to the aforementioned natural and anthropogenic processes and is now in the range of $10–50\,g\,l^{-1}$, which is harmful to numerous living organisms (WHO 2001).

As is taken in a pentavalent form in the plant system and is promptly relegated to "trivalent form by arsenate reductase (ArsR) or reduced glutathione." NIPs and aquaglyceroporins allow As(III) to enter plant cells and "thus reacts with the thiol (-SH) groups of macromolecule and disrupt the protein structure and damages the membrane structure leading to aberration in cellular processes" (Ma et al. 2008; Zhao et al. 2010). Furthermore, As(V) is absorbed by plants via PHTs, which replace inorganic phosphate (iP) in the phosphorylation process, generating energy currency ATP (DiTusa et al. 2016). The toxicity of arsenic has not caused a substantial health risk, but it has created a severe food safety hazard, especially in Southeast Asia where rice is the main sugar source. Many agronomic measures and traditional breeding procedures have been used to reduce the buildup of As in plants. Consequently, the application and long-term viability of these approaches in the management of arsenic toxicity are limiting. Provided the severity of the crisis and the people impacted by arsenite poisoning, new ways for producing rice with low arsenic absorption are desperately required. Current advances in nanobiotechnology are two potentially viable techniques that have the opportunities to alleviate this challenge.

Aside from that As metabolism showed by various research that translocation and uptake in plants are energy-related procedures. However, the processes for As uptake, change, and purification in plants are unknown and require more research. Identifying the cellular metabolic pathways that are harmed by As is an area where there is still a lot to understand. This chapter examines the application of metabolomic knowledge in the relief of arsenic toxicity, the obstacles connected with these technologies, and the information openings that require additional investigation to produce rice with low arsenic accumulation or insignificant arsenic attention for improved nutrition protection.

10.5 Metabolome Regulation and Plants Tolerance

The very first step in As metabolism is the reduction of As(V) to As(III). Enzymatic and nonenzymatic reductions are also possible. As(V) reduced to As(III) during nonenzymatic process involves the oxidation of two reduced glutathione (GSH) molecules to generate oxidized GSH, which are quickly recycled into two GSH molecules by GSH reductase. Therefore, for the efficient reduction of As(V), this reaction is rather sluggish. The arsenate reductase reduces As(V) into As(III) directly through enzymatic processes (Mukhopadhyay et al. 2000; Mukhopadhyay and Rosen 2002; Foyer et al. 2011). Employing genome-wide association (GWA) mapping, researchers discovered a gene called high arsenic content1 (HAC1) in *Arabidopsis* that can convert As(V) into As(III) and allow As(III) efflux to return back to the soil, which limit the As(III) uptake into the root and transfer to the aerial shoot (Chao et al. 2014). HAC1 loss resulted in a decreased As(III) efflux but elevated As(III) translocation in plant shoots system. Different arsenate reductases in rice, such as OsHAC1;1, OsHAC1;2, and OsHAC4, have been identified and functionally characterized (Shi et al. 2016; Xu et al. 2017). Arsenite efflux from the root back to the soil is known to be facilitated by these proteins. In rice, their overexpression resulted in enhanced arsenite outflow into the outer middle and low arsenic accumulation. In contrast, knockout mutation resulted in decreased As(III) outflow and increased As buildup in the stems and seed. Because not all As(III) is effluxed out of the cell, the remaining As(III) is additionally processed by constructing facilities with sulfhydryl-rich particles for the plants to react to the improved As(III) levels. GSH, a precursor of PC biosynthesis and a restricting component for elevated phytochelatins (PCS) production, and this one performs a significant role in arsenic stress in rice (Guo et al. 2008). The PC synthases catalyze formation of PC from glutathione, forming conjugates with As and then transported into the cell vacuoles via transporters ABCC1/ABCC2. Two PCS genes, OsPCS1 and OsPCS2, are found to catalyze PC production in rice (Yamazaki et al. 2018). Additionally, "glutathione S-transferases (GSTs) enzymes catalyze the conjugation of AsIII with GSH to produce the AsIII-GSH complex" using GSH as a substrate (Kumar and Trivedi 2018). Polyphenols act as antioxidants and lower the activity of reactive oxygen species (ROS)-producing enzymes, allowing for As-stress tolerance. We sought to compile recent developments on methods of As accumulation; biochemical, toxicity assessment of physiological, and molecular aspects exerted by As; and the functions of several molecules in decreasing As impacts in yield. Exogenous treatment of nitric oxide, salicylic acid, potassium, and phosphate, as well as proline and polyphenols, protects plants from As-mediated damage. Arsenic contamination, often known as arsenic poisoning, is a major problem that causes plants to suffer from a variety of physiological and biochemical reactions. Increasing population has caused higher levels of As in the environment, which reaches agricultural areas via polluted water irrigation, growth, and decreases harvest efficiency, impacting all living forms. As toxicity generates oxidative stress, which destroys membranes and changes macromolecule structure, leading to cell death.

10.6 Concluding Remarks

Arsenic is a very hazardous metalloid that is observed in abundance in the environment. Arsenic is a very healthy stimulator of cellular ROS in plants, whether directly by converting arsenate to arsenite or indirectly by attaching to the thiol groups of antioxidants and inactivating enzymes. Plants suffer from a variety of metabolic problems as a consequence of arsenic-mediated oxidative stress. As a strategy against As, the modification of cellular thiol compounds for protection against ROS-induced damage has been utilized. To resist this stress, plants have a well-developed antioxidant system such as nonenzymatic and enzymatic activity content that regulates the ROS.

Acknowledgments

We are grateful to the Science and Engineering Research Board (SERB), a statutory body under the Department of Science and Technology (DST), Government of India, for the Start-up Research Grant (SRG/2019/001940).

References

Abhyankar, L.N., Jones, M.R., Guallar, E., and Navas-Acien, A. (2012). Arsenic exposure and hypertension: a systematic review. *Environmental Health Perspectives* 120 (4): 494–500.

Bailey, K.A., Smith, A.H., Tokar, E.J. et al. (2016). Mechanisms underlying latent disease risk associated with early-life arsenic exposure: current research trends and scientific gaps. *Environmental Health Perspectives* 124 (2): 170–175.

Bienert, G.P., Thorsen, M., Schüssler, M.D. et al. (2008). A subgroup of plant aquaporins facilitate the bi-directional diffusion of $As(OH)_3$ and $Sb(OH)_3$ across membranes. *BMC Biology* 6 (1): 1–15.

Bräuner, E.V., Nordsborg, R.B., Andersen, Z.J. et al. (2014). Long-term exposure to low-level arsenic in drinking water and diabetes incidence: a prospective study of the diet, cancer and health cohort. *Environmental Health Perspectives* 122 (10): 1059–1065.

Cao, Y., Sun, D., Ai, H. et al. (2017). Knocking out OsPT4 gene decreases arsenate uptake by rice plants and inorganic arsenic accumulation in rice grains. *Environmental Science & Technology* 51 (21): 12131–12138.

Catarecha, P., Segura, M.D., Franco-Zorrilla, J.M. et al. (2007). A mutant of the Arabidopsis phosphate transporter PHT1; 1 displays enhanced arsenic accumulation. *The Plant Cell* 19 (3): 1123–1133.

Chao, D.Y., Chen, Y., Chen, J. et al. (2014). Genome-wide association mapping identifies a new arsenate reductase enzyme critical for limiting arsenic accumulation in plants. *PLoS Biology* 12 (12): e1002009.

Chen, H., Liu, G., Qiao, N. et al. (2020). Toxic effects of arsenic trioxide on spermatogonia are associated with oxidative stress, mitochondrial dysfunction, autophagy and metabolomic alterations. *Ecotoxicology and Environmental Safety* 190: 110063.

Chiou, T.J., Chu, S.T., Tzeng, W.F. et al. (2008). Arsenic trioxide impairs spermatogenesis via reducing gene expression levels in testosterone synthesis pathway. *Chemical Research in Toxicology* 21 (8): 1562–1569.

DiTusa, S.F., Fontenot, E.B., Wallace, R.W. et al. (2016). A member of the Phosphate transporter 1 (Pht1) family from the arsenic-hyperaccumulating fern *Pteris vittata* is a high-affinity arsenate transporter. *New Phytologist* 209 (2): 762–772.

Dixit, G., Singh, A.P., Kumar, A. et al. (2016). Reduced arsenic accumulation in rice (*Oryza sativa* L.) shoot involves sulfur mediated improved thiol metabolism, antioxidant system and altered arsenic transporters. *Plant Physiology and Biochemistry* 99: 86–96.

Drewniak, L., Maryan, N., Lewandowski, W. et al. (2012). The contribution of microbial mats to the arsenic geochemistry of an ancient gold mine. *Environmental Pollution* 162: 190–201.

Edition, F. (2011). Guidelines for drinking-water quality. *WHO Chronicle* 38 (4): 104–108.

Fontenot, E.B., Ditusa, S.F., Kato, N. et al. (2015). Increased phosphate transport of A rabidopsis thaliana P ht1; 1 by site-directed mutagenesis of tyrosine 312 may be attributed to the disruption of homomeric interactions. *Plant, Cell & Environment* 38 (10): 2012–2022.

Foyer, C.H., Noctor, G., and Hodges, M. (2011). Respiration and nitrogen assimilation: targeting mitochondria-associated metabolism as a means to enhance nitrogen use efficiency. *Journal of Experimental Botany* 62 (4): 1467–1482.

Garg, N. and Singla, P. (2011). Arsenic toxicity in crop plants: physiological effects and tolerance mechanisms. *Environmental Chemistry Letters* 9 (3): 303–321.

Guo, J., Dai, X., Xu, W., and Ma, M. (2008). Overexpressing GSH1 and AsPCS1 simultaneously increases the tolerance and accumulation of cadmium and arsenic in *Arabidopsis thaliana*. *Chemosphere* 72 (7): 1020–1026.

He, Z., Yan, H., Chen, Y. et al. (2016). An aquaporin Pv TIP 4; 1 from *Pteris vittata* may mediate arsenite uptake. *New Phytologist* 209 (2): 746–761.

Huang, Q., Luo, L., Alamdar, A. et al. (2016). Integrated proteomics and metabolomics analysis of rat testis: mechanism of arsenic-induced male reproductive toxicity. *Scientific Reports* 6 (1): 1–12.

International Agency of Research and Cancer (2004). *IARC Monographs on the Evaluation of Carcinogenic Risks to Humans. Vol.* 84: *Some Drinking-Water Disinfectants and Contaminants, Including Arsenic.* Vienna: IARC.

Isayenkov, S.V. and Maathuis, F.J. (2008). The *Arabidopsis thaliana* aquaglyceroporin AtNIP7; 1 is a pathway for arsenite uptake. *FEBS Letters* 582 (11): 1625–1628.

Islam, S., Rahman, M.M., Islam, M.R., and Naidu, R. (2016). Arsenic accumulation in rice: consequences of rice genotypes and management practices to reduce human health risk. *Environment International* 96: 139–155.

Jaishankar, M., Tseten, T., Anbalagan, N. et al. (2014). Toxicity, mechanism and health effects of some heavy metals. *Interdisciplinary Toxicology* 7 (2): 60.

Jana, K., Jana, S., and Samanta, P.K. (2006). Effects of chronic exposure to sodium arsenite on hypothalamo-pituitary-testicular activities in adult rats: possible an estrogenic mode of action. *Reproductive Biology and Endocrinology* 4 (1): 1–13.

Kamiya, T. and Fujiwara, T. (2009). Arabidopsis NIP1; 1 transports antimonite and determines antimonite sensitivity. *Plant and Cell Physiology* 50 (11): 1977–1981.

Kamiya, T., Islam, R., Duan, G. et al. (2013). Phosphate deficiency signaling pathway is a target of arsenate and phosphate transporter OsPT1 is involved in As accumulation in shoots of rice. *Soil Science and Plant Nutrition* 59 (4): 580–590.

Khan, K.M., Chakraborty, R., Bundschuh, J. et al. (2020). Health effects of arsenic exposure in Latin America: an overview of the past eight years of research. *Science of the Total Environment* 710: 136071.

Kumar, S. and Trivedi, P.K. (2018). Glutathione *S*-transferases: role in combating abiotic stresses including arsenic detoxification in plants. *Frontiers in Plant Science* 9: 751.

Kuo, C.C., Moon, K.A., Wang, S.L. et al. (2017). The association of arsenic metabolism with cancer, cardiovascular disease, and diabetes: a systematic review of the epidemiological evidence. *Environmental Health Perspectives* 125 (8): 087001.

Lafuente, A., Pajuelo, E., Caviedes, M.A., and Rodríguez-Llorente, I.D. (2010). Reduced nodulation in alfalfa induced by arsenic correlates with altered expression of early nodulins. *Journal of Plant Physiology* 167 (4): 286–291.

Lafuente, A., Pérez-Palacios, P., Doukkali, B. et al. (2015). Unraveling the effect of arsenic on the model M edicago–E nsifer interaction: a transcriptomic meta-analysis. *New Phytologist* 205 (1): 255–272.

Laine, J.E., Bailey, K.A., Rubio-Andrade, M. et al. (2015). Maternal arsenic exposure, arsenic methylation efficiency, and birth outcomes in the Biomarkers of Exposure to ARsenic (BEAR) pregnancy cohort in Mexico. *Environmental Health Perspectives* 123 (2): 186–192.

Li, G., Sun, G.X., Williams, P.N. et al. (2011). Inorganic arsenic in Chinese food and its cancer risk. *Environment International* 37 (7): 1219–1225.

Li, R.Y., Ago, Y., Liu, W.J. et al. (2009). The rice aquaporin Lsi1 mediates uptake of methylated arsenic species. *Plant Physiology* 150 (4): 2071–2080.

Li, W.X., Chen, T.B., Huang, Z.C. et al. (2006). Effect of arsenic on chloroplast ultrastructure and calcium distribution in arsenic hyperaccumulator *Pteris vittata* L. *Chemosphere* 62 (5): 803–809.

Li, Z., Zhang, D., Xiong, X. et al. (2017). A potent Cas9-derived gene activator for plant and mammalian cells. *Nature Plants* 3 (12): 930–936.

Liu, W.J., Zhu, Y.G., Smith, F.A., and Smith, S.E. (2004). Do iron plaque and genotypes affect arsenate uptake and translocation by rice seedlings (*Oryza sativa* L.) grown in solution culture? *Journal of Experimental Botany* 55 (403): 1707–1713.

Ma, J.F. and Yamaji, N. (2006). Silicon uptake and accumulation in higher plants. *Trends in Plant Science* 11 (8): 392–397.

Ma, J.F. and Yamaji, N. (2015). A cooperative system of silicon transport in plants. *Trends in Plant Science* 20 (7): 435–442.

Ma, J.F., Yamaji, N., Mitani, N. et al. (2008). Transporters of arsenite in rice and their role in arsenic accumulation in rice grain. *Proceedings of the National Academy of Sciences of the Unites States of America* 105 (29): 9931–9935.

Ma, J.F., Yamaji, N., Tamai, K., and Mitani, N. (2007). Genotypic difference in silicon uptake and expression of silicon transporter genes in rice. *Plant Physiology* 145 (3): 919–924.

Matysik, J., Alia, Bhalu, B., and Mohanty, P. (2002). Molecular mechanisms of quenching of reactive oxygen species by proline under stress in plants. *Current Science* 525–532.

Maurel, C., Boursiac, Y., Luu, D.T. et al. (2015). Aquaporins in plants. *Physiological Reviews* 95 (4): 1321–1358.

McClintock, T.R., Chen, Y., Bundschuh, J. et al. (2012). Arsenic exposure in Latin America: biomarkers, risk assessments and related health effects. *Science of the Total Environment* 429: 76–91.

Meharg, A.A., Williams, P.N., Adomako, E. et al. (2009). Geographical variation in total and inorganic arsenic content of polished (white) rice. *Environmental Science & Technology* 43 (5): 1612–1617.

Mendoza-Cózatl, D.G., Jobe, T.O., Hauser, F., and Schroeder, J.I. (2011). Long-distance transport, vacuolar sequestration, tolerance, and transcriptional responses induced by cadmium and arsenic. *Current Opinion in Plant Biology* 14 (5): 554–562.

Mishra, R.K., Tiwari, S., Patel, A., and Prasad, S.M. (2021). Arsenic contamination, speciation, toxicity and defense strategies in plants. *Brazilian Journal of Botany* 1–10.

Mishra, S. and Dubey, R.S. (2006). Inhibition of ribonuclease and protease activities in arsenic exposed rice seedlings: role of proline as enzyme protectant. *Journal of Plant Physiology* 163 (9): 927–936.

Mizutani, M., Watanabe, S., Nakagawa, T., and Maeshima, M. (2006). Aquaporin NIP2; 1 is mainly localized to the ER membrane and shows root-specific accumulation in Arabidopsis thaliana. *Plant and Cell Physiology* 47 (10): 1420–1426.

Mondal, P., Majumder, C.B., and Mohanty, B. (2006). Laboratory based approaches for arsenic remediation from contaminated water: recent developments. *Journal of Hazardous Materials* 137 (1): 464–479.

Mosa, K.A., Kumar, K., Chhikara, S. et al. (2012). Members of rice plasma membrane intrinsic proteins subfamily are involved in arsenite permeability and tolerance in plants. *Transgenic Research* 21 (6): 1265–1277.

Mukhopadhyay, C.K., Mazumder, B., and Fox, P.L. (2000). Role of hypoxia-inducible factor-1 in transcriptional activation of ceruloplasmin by iron deficiency. *Journal of Biological Chemistry* 275 (28): 21048–21054.

Mukhopadhyay, R. and Rosen, B.P. (2002). Arsenate reductases in prokaryotes and eukaryotes. *Environmental Health Perspectives* 110: 745–748.

Pan, W.C., Seow, W.J., Kile, M.L. et al. (2013). Association of low to moderate levels of arsenic exposure with risk of type 2 diabetes in Bangladesh. *American Journal of Epidemiology* 178 (10): 1563–1570.

Panda, S.K., Upadhyay, R.K., and Nath, S. (2010). Arsenic stress in plants. *Journal of Agronomy and Crop Science* 196 (3): 161–174.

Pang, Y., Li, L., Ren, F. et al. (2010). Overexpression of the tonoplast aquaporin AtTIP5; 1 conferred tolerance to boron toxicity in *Arabidopsis*. *Journal of Genetics and Genomics* 37 (6): 389–397.

Paszkowski, U., Kroken, S., Roux, C., and Briggs, S.P. (2002). Rice phosphate transporters include an evolutionarily divergent gene specifically activated in arbuscular mycorrhizal symbiosis. *Proceedings of the National Academy of Sciences of the United States of America* 99 (20): 13324–13329.

Patel, K.S., Shrivas, K., Brandt, R. et al. (2005). Arsenic contamination in water, soil, sediment and rice of central India. *Environmental Geochemistry and Health* 27 (2): 131–145.

Pierce, B.L., Argos, M., Chen, Y. et al. (2011). Arsenic exposure, dietary patterns, and skin lesion risk in Bangladesh: a prospective study. *American Journal of Epidemiology* 173 (3): 345–354.

Remy, E., Cabrito, T.R., Batista, R.A. et al. (2012). The Pht1; 9 and Pht1; 8 transporters mediate inorganic phosphate acquisition by the *Arabidopsis thaliana* root during phosphorus starvation. *New Phytologist* 195 (2): 356–371.

Renu, K., Madhyastha, H., Madhyastha, R. et al. (2018). Review on molecular and biochemical insights of arsenic-mediated male reproductive toxicity. *Life Sciences* 212: 37–58.

Sahoo, P.K. and Kim, K. (2013). A review of the arsenic concentration in paddy rice from the perspective of geoscience. *Geosciences Journal* 17 (1): 107–122.

Shapiro, G.D., Dodds, L., Arbuckle, T.E. et al. (2015). Exposure to phthalates, bisphenol A and metals in pregnancy and the association with impaired glucose tolerance and gestational diabetes mellitus: the MIREC study. *Environment International* 83: 63–71.

Sharma, A.K., Tjell, J.C., Sloth, J.J., and Holm, P.E. (2014). Review of arsenic contamination, exposure through water and food and low cost mitigation options for rural areas. *Applied Geochemistry* 41: 11–33.

Shen, H., Xu, W., Zhang, J. et al. (2013). Urinary metabolic biomarkers link oxidative stress indicators associated with general arsenic exposure to male infertility in a Han Chinese population. *Environmental Science & Technology* 47 (15): 8843–8851.

Shi, S., Wang, T., Chen, Z. et al. (2016). OsHAC1; 1 and OsHAC1; 2 function as arsenate reductases and regulate arsenic accumulation. *Plant Physiology* 172 (3): 1708–1719.

Shin, H., Shin, H.S., Dewbre, G.R., and Harrison, M.J. (2004). Phosphate transport in Arabidopsis: Pht1; 1 and Pht1; 4 play a major role in phosphate acquisition from both low-and high-phosphate environments. *The Plant Journal* 39 (4): 629–642.

Singh, P.K., Indoliya, Y., Chauhan, A.S. et al. (2017). Nitric oxide mediated transcriptional modulation enhances plant adaptive responses to arsenic stress. *Scientific Reports* 7 (1): 1–13.

Singh, V.P., Singh, S., Kumar, J., and Prasad, S.M. (2015). Investigating the roles of ascorbate-glutathione cycle and thiol metabolism in arsenate tolerance in ridged Luffa seedlings. *Protoplasma* 252 (5): 1217–1229.

Smith, A.H., Lopipero, P.A., Bates, M.N., and Steinmaus, C.M. (2002). Arsenic epidemiology and drinking water standards. *Science* 296 (5576): 2145–2146.

Srivastava, S., Upadhyay, M.K., Tripathi, R.D., and Dhankher, O.P. (2016). Arsenic transport, metabolism and toxicity in plants. *International Journal of Plant and Environment* 2 (1 and 2): 17–28.

Sun, H.J., Xiang, P., Luo, J. et al. (2016). Mechanisms of arsenic disruption on gonadal, adrenal and thyroid endocrine systems in humans: a review. *Environment International* 95: 61–68.

Sun, S.K., Chen, Y., Che, J. et al. (2018). Decreasing arsenic accumulation in rice by overexpressing Os NIP 1; 1 and Os NIP 3; 3 through disrupting arsenite radial transport in roots. *New Phytologist* 219 (2): 641–653.

Tanaka, M., Wallace, I.S., Takano, J. et al. (2008). NIP6; 1 is a boric acid channel for preferential transport of boron to growing shoot tissues in *Arabidopsis*. *The Plant Cell* 20 (10): 2860–2875.

Tchounwou, P.B., Yedjou, C.G., Patlolla, A.K., and Sutton, D.J. (2012). Heavy metal toxicity and the environment. *Molecular, Clinical and Environmental Toxicology* 101: 133–164.

Tiwari, M., Sharma, D., Dwivedi, S. et al. (2014). Expression in A rabidopsis and cellular localization reveal involvement of rice NRAMP, OsNRAMP 1, in arsenic transport and tolerance. *Plant, Cell & Environment* 37 (1): 140–152.

Tripathi, P., Mishra, A., Dwivedi, S. et al. (2012a). Differential response of oxidative stress and thiol metabolism in contrasting rice genotypes for arsenic tolerance. *Ecotoxicology and Environmental Safety* 79: 189–198.

Tripathi, R.D., Tripathi, P., Dwivedi, S. et al. (2012b). Arsenomics: omics of arsenic metabolism in plants. *Frontiers in Physiology* 3: 275.

Tripathi, R.D., Tripathi, P., Dwivedi, S. et al. (2014). Roles for root iron plaque in sequestration and uptake of heavy metals and metalloids in aquatic and wetland plants. *Metallomics* 6 (10): 1789–1800.

Tuulaikhuu, B.A., Romaní, A.M., and Guasch, H. (2015). Arsenic toxicity effects on microbial communities and nutrient cycling in indoor experimental channels mimicking a fluvial system. *Aquatic Toxicology* 166: 72–82.

Verbruggen, N., Hermans, C., and Schat, H. (2009). Mechanisms to cope with arsenic or cadmium excess in plants. *Current Opinion in Plant Biology* 12 (3): 364–372.

Wang, P., Zhang, W., Mao, C. et al. (2016). The role of OsPT8 in arsenate uptake and varietal difference in arsenate tolerance in rice. *Journal of Experimental Botany* 67 (21): 6051–6059.

WHO (World Health Organization) (2001). *IPCS Environmental Health Criteria* 224 *Arsenic and Arsenic Compounds*. Geneva: International Programme on Chemical Safety, World Health Organization.

Williams, P.N., Price, A.H., Raab, A. et al. (2005). Variation in arsenic speciation and concentration in paddy rice related to dietary exposure. *Environmental Science & Technology* 39 (15): 5531–5540.

Wu, C., Ye, Z., Li, H. et al. (2012). Do radial oxygen loss and external aeration affect iron plaque formation and arsenic accumulation and speciation in rice? *Journal of Experimental Botany* 63 (8): 2961–2970.

Wu, Y., Ding, R., Zhang, X. et al. (2021). Meet-in-metabolite analysis: a novel strategy to identify connections between arsenic exposure and male infertility. *Environment International* 147: 106360.

Wu, Z., Ren, H., McGrath, S.P. et al. (2011). Investigating the contribution of the phosphate transport pathway to arsenic accumulation in rice. *Plant Physiology* 157 (1): 498–508.

Xu, J., Shi, S., Wang, L. et al. (2017). OsHAC4 is critical for arsenate tolerance and regulates arsenic accumulation in rice. *New Phytologist* 215 (3): 1090–1101.

Xu, W., Dai, W., Yan, H. et al. (2015). Arabidopsis NIP3; 1 plays an important role in arsenic uptake and root-to-shoot translocation under arsenite stress conditions. *Molecular Plant* 8 (5): 722–733.

Yamazaki, S., Ueda, Y., Mukai, A. et al. (2018). Rice phytochelatin synthases Os PCS 1 and Os PCS 2 make different contributions to cadmium and arsenic tolerance. *Plant Direct* 2 (1): e00034.

Ye, Y., Li, P., Xu, T. et al. (2017). OsPT4 contributes to arsenate uptake and transport in rice. *Frontiers in Plant Science* 8: 2197.

Zhang, J., Huang, Z., Chen, M. et al. (2014). Urinary metabolome identifies signatures of oligozoospermic infertile men. *Fertility and Sterility* 102 (1): 44–53.

Zhao, F.J., Ago, Y., Mitani, N. et al. (2010a). The role of the rice aquaporin Lsi1 in arsenite efflux from roots. *New Phytologist* 186 (2): 392–399.

Zhao, F.J., Ma, J.F., Meharg, A.A., and McGrath, S.P. (2009). Arsenic uptake and metabolism in plants. *New Phytologist* 181 (4): 777–794.

Zhao, F.J., McGrath, S.P., and Meharg, A.A. (2010b). Arsenic as a food chain contaminant: mechanisms of plant uptake and metabolism and mitigation strategies. *Annual Review of Plant Biology* 61: 535–559.

11

Role of Phytohormones in Regulating Arsenic-Induced Toxicity in Plants

Ummey Aymen[1], Marya Khan[1], Rachana Singh[2], Parul Parihar[2,3], and Neha Pandey[4]

[1] *Department of Botany, Lovely Professional University, Phagwara, Punjab, India*
[2] *Department of Botany, University of Allahabad, Prayagraj, Uttar Pradesh, India*
[3] *Department of Bioscience and Biotechnology, Banasthali Vidyapith, Rajasthan, India*
[4] *Department of Botany, CMP PG College, Prayagraj, Uttar Pradesh, India*

11.1 Arsenic and Its Source

Arsenic is an element that is neither grouped under metals nor nonmetals but is a metalloid having physical and chemical characteristics between the two groups. Naturally, arsenic exists within our environment in all possible ecosystems (Mandal et al. 2002; Bhattacharya et al. 2007; Hare et al. 2019) but its concentration is in trace amounts, which do not have a negative impact on the living ecosystems and the primary sources are found to be in the form of igneous and sedimentary rocks which later on get accumulated within the soil and the plant life during soil formation and nutrient absorption by plants respectively. However, anthropogenic activities (chromated copper arsenate [CCA]-treated wood being the largest anthropogenic arsenic contributor) have caused arsenic to accumulate within ecosystems in toxic amounts due to which it is now being characterized as a carcinogen (Jang et al. 2016; Cardoso et al. 2018; Zhou and Xi 2018) and was listed among the top five toxic elements in the US CERCLA (Comprehensive Environment Response, Compensation, and Liability) Act of hazardous substances (Pfeifer et al. 2002; Rosen and Liu 2009). Arsenic exists in four oxidation states (−3, 0, +3, +5) with varying toxicity levels and the inorganic form of +3 oxidation state having the highest toxicity (Webb 1963; Wang and Mulligan 2006; Srivastava et al. 2011; Abbas et al. 2018). These toxic states would hardly create pollution earlier because their overall concentration would add up to be very low but as of now the sources adding arsenic to our environment has drastically increased since mineral extractions and mining procedures have increased quite a lot and arsenic is associated with a variety of elements and compounds like gold, lead, copper, iron hydroxides and sulfides (Mandal et al. 2002) and their presence in various industrial products like pesticides, insecticides, paint products (Webb 1963; Gavhane et al. 2021) in addition to the natural sources like volcanic eruptions, weathering etc. all lead to arsenic toxicity (Xu et al. 2018).

Arsenic in Plants: Uptake, Consequences and Remediation Techniques, First Edition.
Edited by Prabhat Kumar Srivastava, Rachana Singh, Parul Parihar, and Sheo Mohan Prasad.

Arsenic toxicity is caused due to the reason that arsenic, being a trace element should be present in the ecosystems only in trace amounts but its overexploitation has led to its accumulation higher than the normal levels that are set by the World Health Organization (WHO) and the main reason for this is the nonbiodegradable nature of arsenic due to which it keeps on cycling, mobilizing and affecting each part of the ecosystem, be it atmosphere, lithosphere, hydrosphere or biosphere. The mobility of arsenic depends on a number of factors like source, oxidation state, speciation, pH, redox potential etc. (Nriagu and Azcue 1990; Pfeifer et al. 2002; Langdon et al. 2003; Kumarathilaka et al. 2018). Taking the case of the lithosphere, arsenic is found in the immobilized state in the earth's crust along with various important metals and nonmetals and activities like mining, smelting, and volcanoes are responsible for arsenic immobilization in the atmosphere either in the form of gas (As_2SO_3) or dust particles (Jang et al. 2016). Arsenic immobilization also occurs from the lithosphere to the hydrosphere by weathering of oceanic rocks and the presence of submarine volcanoes. The oxidation states of arsenic in the lithosphere are mainly +3 and +5 along with an organic form, which is present in the least quantities (Ascar et al. 2008; Majumdar et al. 2019). Moreover, because of the factors mentioned earlier (Bhattacharya et al. 2002), their concentrations keep on changing by getting precipitated or being free (Polizzotto et al. 2006; Jang et al. 2016); being better trapped by the soil or just leached (Ronzan et al. 2019); oxidized or reduced both biologically and chemically in soil and water (Rhine et al. 2006; Zhang and Selim 2008); soil mineralogy like some ions (chlorides, nitrates, and sulfates) having no impact while ions like phosphates inhibiting arsenic mobility (Livesey and Huang 1981; Sisr et al. 2007); lower pH conditions favoring arsenic bonding with aluminum and iron while higher pH conditions favoring arsenic and calcium bonding. Ferric hydroxide is considered one of the important factors to determine arsenic immobility in the soil and is directly proportional to the latter's concentration (Jang et al. 2016) and also arsenic has great affinity for iron oxides and ions like phosphates displace arsenic from iron oxides and, under anerobic situations, get mobilized into the soil or water like groundwater (Livesey and Huang 1981).

Taking the case of the hydrosphere wherein not only the oceanic sedimentary and igneous rocks are responsible for introducing arsenic into the water bodies (arsenites and arsenates being the most common) but also the soil that contains arsenic (due to natural or anthropogenic activities), upon erosion, introduces it into the water bodies (Ng et al. 2004). Although arsenic use has been banned for farming practices, being nonbiodegradable, its concentrations are still found in the soil and water bodies.

Hydrosphere and lithosphere are not aloof but we know that all these ecosystems are connected and thus when the arsenic-polluted water is used for irrigating soils, the lithosphere also becomes toxic and volatilization of arsenic compounds in the soil due to chemical or microbial action keeps the cycle in a continuous state. A direct source of arsenic to both hydrosphere and lithosphere can occur due to geothermal water sources (Villanueva-Estrada et al. 2013).

For the atmosphere, mining, smelting, and volcanoes introduce arsenic in the form of gases and dust particles that get suspended and the more the anthropogenic activities like factories dealing with arsenic extractions, the more polluted is the atmosphere due to arsenic toxicity (Matschullat 2000).

Finally, the biosphere component is the worst hit by arsenic and arsenic accumulation and cycling in this component occurs through all other interconnected components, that is atmosphere through arsenic dust which gets inhaled by living organisms, lithosphere through the vegetables, and crops that grow from arsenic-contaminated soils, and, lastly, the most common, that is hydrosphere which is the main cause of toxicity in living beings, be that in the form of contaminated water or contaminated aquatic life (Jang et al. 2016).

11.2 Uptake and Transport of Arsenic Within Plants

Food is obtained either directly or indirectly from plants and due to increased arsenic accumulation in the various ecosystems of the environment, arsenic eventually finds its way into plants as well and, hence, all other consumers as well. It has been observed that arsenic accumulation within rice grains occurs and since rice is a staple food for most parts of the world, especially Asia, thus, to prevent toxicity, it is essential to understand the mechanism of arsenic accumulation within the plants (Williams et al. 2007; Zhu et al. 2008). In fact, some plants have the ability to hyperaccumulate these elements within them and are aptly called hyperaccumulators and could be used as bioremediation techniques to decontaminate soils and grow arsenic-free healthy food plants.

Since arsenic is present in different oxidation states in different environments, taking the case of arsenates, which is the form in which arsenic is predominant in oxygen-containing environments, but its availability is not too much due to its affinity for iron oxides and hydroxides (Wenzel et al. 2002; Ma et al. 2019). Intake of this arsenate by the plants occurs through the phosphate transporters, which obviously have more affinity toward phosphates than arsenates and the gradient is operated by means of protons (Asher and Reay 1979; Ullrich-Eberius et al. 1989; Meharg et al. 1994; Lindsay and Maathuis 2017; Zvobgo et al. 2018). These phosphate transporters mostly belong to the phosphate transporter 1 (Pht1) family and over a hundred types get expressed in the root cells (Rausch and Bucher 2002; Bucher 2007). This has been demonstrated by a double mutant of *Arabidopsis thaliana* (Pht1Δ4Δ being defective in phosphate transporters, Pht1;1 and Pht1;4) that proved quite resistant to arsenic stress rather than wild-type plants (Shin et al. 2004). Similar tolerance was showed by *A. thaliana* mutant for phosphate transporter traffic facilitator 1 (PHF1) (González et al. 2005). Various such studies were carried out and proved similar results indicating that reducing arsenate uptake leads to arsenic tolerance or decreased arsenic accumulation within plants.

Rice is grown in reducing environments, flooded or semi-flooded, and arsenite is the dominant form of arsenic found in such conditions (Marin et al. 1993; Inskeep et al. 2002; Takahashi et al. 2004; Xu et al. 2008). Moreover, this state has more bioavailability than arsenate (Xu et al. 2008) and the transportation mechanism within the plants occurs through some aquaglyceroporins (Bhattacharjee and Rosen 2007), which came to the limelight because of microbial world studies like that of *Escherichia coli* and yeast wherein glycerol facilitator (GlpF) and glycerol channel protein (Fps1p), respectively, proved to uptake arsenite as well. Moreover, in the case of yeast, hexose permease genes were mutated and that led to reduced arsenic uptake proving to be responsible for the arsenite uptake mechanism and also when glucose was being added, this led to competitive inhibition with arsenic and reduced arsenic uptake as well (Liu et al. 2004). This competitive inhibition mechanism was further demonstrated by reduced arsenite uptake upon glycerol and antimonite use while phosphate use had no effect (Meharg and Jardine 2003). Studies have been carried out independently proving the same results that nodulin26-like intrinsic proteins (NIPs), which are plant aquaporins are also responsible for prime arsenite uptake into the plant vascular system as was proved by Bienert et al. (2008) by introducing these in yeast from *A. thaliana* (AtNIP5;1 and AtNIP6;1), rice (OsNIP2;1 and OsNIP3;2), and *Lotus japonicus* (LjNIP5;1 and LjNIP6;1) and upon dwelling into more NIP observations, it was observed that AtNIP7;1 was much better arsenite transporter than the ones stated earlier (AtNIP5;1 and AtNIP6;1).

Ma et al. (2008) conducted similar studies for rice plants and found Lsi1 transporter of plants (or OsNIP2;1) when expressed in yeasts and *Xenopus laevis* oocytes showed increased arsenite uptake and the transporter is present on the outer/distal side of both exodermal and endodermal cells. Mutation of the gene expressing the protein led to arsenic-resistant comparative to wild-type plants and the same transporter is a silicon transporter and is quite obvious by studying the structures of arsenite and silicic acid (Ma et al. 2006). Three other NIPs were also taken into account (OsNIP1;1, OsNIP2;2 [also named Lsi6], and OsNIP3;1) but did not enhance arsenic uptake to a wide observable difference.

Aquaporins are the broad group of protein channels, which are subdivided into four subtypes and NIPs, are one of the four subdivisions (others being Plasma membrane intrinsic proteins (PIPs), Tonoplast intrinsic proteins (TIPs), and Small basic intrinsic proteins (SIPs) (Chaumont et al. 2005; Maurel et al. 2008) and among this subdivision, some are little permeable to water and showed better permeability to large uncharged molecules like glycerol and arsenites and hence being called aquaglyceroporins (Wallace et al. 2006; Bienert et al. 2008; Isayenkov and Maathuis 2008; Ma et al. 2008) while others show little permeability even toward these molecules like glycerol (Ma et al. 2006; Mitani et al. 2008). This specificity depends upon the pore constriction differences, either NPA boxes or aromatic/arginine selectivity filters (Wallace et al. 2006; Maurel et al. 2008). The latter group of NIPs are further divided into three groups on the basis of the size of the pore that they form, NIP I, II, and III with NIP III having the largest pore but arsenite transport was not affected much and was almost similarly transported through all of these three types while each of the group is otherwise specific to water and other larger solutes (Wallace et al. 2006; Mitani et al. 2008).

Among the NIPs, Lsi1 already discussed earlier has a counterpart on the opposite side, that is, toward the inner/proximal end of exodermal and endodermal cells called Lsi2, where Lsi1 played a role in the influx or uptake of arsenites, Lsi2 is responsible for efflux of these arsenites toward the xylem and is primarily responsible for Si transportation (Ma et al. 2007). Mutations in the genes responsible for the expression of Lsi2 transporter led to a prominent decrease in arsenic uptake and the xylem sap studied had also very concentrations of arsenic relative to the wild type (Ma et al. 2008). However, when Si is added to normal plants, it causes competitive inhibition leading to lower arsenic transport and eventually lower arsenic levels within the plant but when the transporter (Lsi2) is mutated, the ability to transport either arsenite or silicate is negatively affected and the silicate addition has no effect over the arsenic uptake by the plant. At the same time, when we observe the phenomenon in the case of rice plants, the uptake of silicate and arsenite is very efficient and transportation occurs in a balanced manner for both the elements and proving the reason for rice being an efficient accumulator of arsenic. Lsi2 is therefore the actual reason for less arsenic accumulation within the plants because *lsi1* and *lsi2* mutants were compared with wild types and it was *lsi2* mutant that showed the least arsenic levels. This also proved that the reason behind lower arsenic levels upon silicate exposure in the case of arsenates being the substratum (arsenates and silicates have different transporters) is due to the fact that arsenates get converted to arsenites within the reducing root environment and further not being able to pass the proximal exodermis and endodermis of the *lsi2* mutant (Guo et al. 2005, 2007). We can exploit this property of Lsi2 transporter by using Si fertilizers, which will cause competitive inhibition of arsenic to enter the plants and hence prevent their accumulation (Ma and Yamaji 2006).

11.3 Mechanism of Arsenic Efflux by Plant Roots

It is the arsenic efflux toward the xylem and some leakage through the external medium that gives us an idea about whether the plant is a hyperaccumulator or simply takes up arsenic within itself being arsenic stressed. Efflux of arsenate toward the external medium has not been demonstrated but is based on the observations and the pathway that arsenates follow being similar to phosphates and since phosphate efflux toward the external medium has been observed (Xu et al. 2007). Thus, the arsenate efflux or leakage toward the external medium has been observed and occurs in much higher concentrations in arsenic-sensitive plants while the leakage is quite less in the case of hyperaccumulators (Mimura 1999). Arsenate is quite stable in aerobic conditions but as soon as arsenate enters the root cells, it gets converted to arsenites and these studies were carried out in tomatoes and rice (Xu et al. 2007). So, the efflux to the external medium is not just comprised of arsenates but arsenites as well and the reason is competitive inhibition of phosphates, which inhibits both intakes of arsenates and thereafter conversion of arsenates to arsenites and their efflux toward the exterior medium and the observations were recorded in tomato root cells, which had arsenate accumulations already and when these cells were grown in As-free media, both arsenate and arsenite efflux into the medium was recorded (Xu et al. 2007). Arsenite efflux has been observed in a wide range of plants (*A. thaliana*, *Holcus lanatus*, *Triticum aestivum*, *Hordeum vulgare*, *Zea mays*), with the least efflux in hyperaccumulator plants (*Pteris vittate*), which do most of the arsenic accumulation and its sequestration and this phenomenon has put a doubt on a number of hydroponic studies that took into account only the arsenic efflux and not the accumulation, although the efflux amount was much higher (about three times) than the arsenic accumulated, calculated after 24 hours of exposure (Zhao et al. 2009).

Just as arsenates get reduced in the root cells, similarly arsenites after efflux into the outer environment get oxidized by the aerobic environment as well as the microbes and this redox cycle of arsenates and arsenites in the rhizosphere and the root cells keep on cycling according to the environmental conditions (Oscarson et al. 1981; Macur et al. 2004). The conversion of arsenites to arsenates was proved when arsenite measurement was done near rhizospheres by means of rhizoboxes in the case of sunflower and maize roots (Ultra et al. 2007b; Ultra Jr et al. 2007a; Vetterlein et al. 2007) and the microbes in the vicinity convert them to arsenates for which they are specialized and hence leading to arsenic detoxification (Bhattacharjee and Rosen 2007).

Arsenic efflux process is also energy-dependent by means of proton gradient as was observed by arsenite inhibition by means of proton decoupler, CCCP, protonophore carbonylcyanide *m*-chlorophenylhydrazone and the same requirement was observed in case of microbes that are involved with reduction of arsenates to arsenites and in contrary to plants, which possess some anion channels, possess protein complexes like ArB or ArAB (in *E. coli*) (Dey et al. 1994) or efflux carrier protein, Acr3p (in yeast and fungi) (Wysocki et al. 1997) coupled with proton channels.

As we already know that aquaporins are responsible for arsenic uptake, but some of these can be bidirectional as well and hence help in arsenic efflux as well and are not dependent on the energy of the plant but on diffusion phenomenon, that is, arsenic concentration (Mitani et al. 2008). Among these aquaglyceroporins, many NIPs as mentioned earlier for uptake are also responsible for arsenic efflux and hence proving arsenic tolerance, and this arsenic resistance has been proved for many microbes but is not as clear in the case of plants. For instance, AtNIP5;1, AtNIP6;1, OsNIP2;1, OsNIP3;1, and LjNIP6;1 genes in yeast caused arsenite efflux and hence arsenic tolerance reported by Bienert et al. (2008). Similarly, in the yeast acr3Δ mutant, AtNIP7;1 gene expression also led to arsenic tolerance due to the arsenic efflux (Isayenkov and Maathuis 2008). A study performed by Logoteta et al. (2008) in *H. lanatus* ecotypes, one being resistant to arsenic stress and the other sensitive and he observed that the influx and efflux for both the ecotypes were proportional with the resistant one accumulating lesser arsenic than the nonresistant one, which brought a conclusion that efflux of arsenic does not imply sensitivity or resistance to the metalloid but it definitely does help to remove out the excess arsenic inside the plant whenever excess concentrations need to be disposed off, creating a passageway for the plant to reduce the arsenic concentration burden and could be enhanced by biotechnological tools to serve a much useful and active purpose in arsenic efflux or sequestration.

11.4 Impact of Arsenic on Metabolism and its Toxicity in Plants

Arsenic is toxic even in little quantities and the most common states, arsenates and arsenites are also toxic with the latter form being the most toxic and the form that actually is present within the plant and has been observed to have a receding influence on both root and shoot growth by having a physiological effect and hence also influencing various essential developmental processes like reproduction leading to a loss in crop yield and harvest (Garg and Singla 2011; Singh et al. 2018, 2020). Today, the research on arsenic toxicity studies the biomolecular impact of arsenic and in a plant, it is the root that first fights the arsenic stress and tries to suppress it but it depends upon the plant type and various plants have shown to translocate arsenic into different harvesting plant organs and tissues like leaves mostly and plant-type adaptation determines whether the plant can withstand the stress or is sensitive to it. Arsenic exposure stimulates various biochemical pathways within cells like on just entering the root cells, phosphate transporters are disturbed, which lead to extra electrolyte leaking (Singh et al. 2006), oxidative stress gets generated due to cell membrane damage and lipid peroxidation giving rise to a chain of reactions and parallelly antioxidant defences coming to rescue like ascorbate, glutathione (GSH) and phytochelatins (PCs) (Schmöger et al. 2000; Li et al. 2004; Geng et al. 2006; Singh et al. 2006; Khan et al. 2009) especially in the root cells and anthocyanins in the leaf portions (Catarecha et al. 2007) and since the plant is under a lot of stress, water is required in excess and transpiration rate decreases (Stoeva and Bineva 2003). After such observations, arsenic was used as a stimulant in very little quantities only and hence led to a positive impact on growth (Woolson et al. 1971; Carbonell-Barrachina et al. 1997, 1998a; Miteva 2002; Garg and Singla 2011) and it was ensured that it was arsenic that directly led to the positive impact on metabolic growth rate by taking the observations under microbe-free environments (Chen et al. 2010). On the other hand, when concentrations are toxic, plant metabolism can be disrupted to such an extent that death can occur but some plants have adapted to such conditions and have brought such changes in their metabolism that even toxic arsenic concentrations do not disrupt their normal physiological mechanisms by efflux either toward the outer environment or toward shoot organs like leaves without causing any damage to these organs (Ma et al. 2001; Visoottiviseth et al. 2002; Ellis et al. 2006; Pickering et al. 2006; Zhao et al. 2009; Chen et al. 2021) and it is these hyperaccumulator plants that can enhance our research phytoremediation on a practical level by employing such crops in arsenic-contaminated lands and making arsenic-sensitive plants resistant by modifying them according to hyperaccumulators.

Toxicity in plants is not caused by arsenates and arsenites only but their conversion to organic forms by means of methylation are more toxic (Marin et al. 1992; Styblo et al. 1997, 2000; Lin et al. 1999; Quaghebeur and Rengel 2003; Schwerdtle et al. 2003; Raab et al. 2005, 2007a; Xu et al. 2007; Naranmandura et al. 2011) and give rise to MMA^{V} (monomethylarsonic acid), DMA^{V} (dimethylarsinic acid), and $TMAO^{V}$ (trimethylarsine oxide) by methylation of arsenates (Abedin et al. 2002) and MMA^{III}, DMA^{III}, TMA^{III} by methylation of arsenites and are volatile (Mestrot et al. 2011). The involvement of microbes was also tested by performing arsenic stress response in axenic conditions where the methylated forms were not produced (Lomax et al. 2012). Similar to arsenites, methylated forms (MMA and DMA) are also transported through aquaporin channels due to their similarity with arsenites (Li et al. 2009) but their intake is comparatively slower (Abedin et al. 2002; Raab et al. 2007b; Abbas and Meharg 2008) and mobility within the plant comparatively faster (Marin et al. 1992; Burló et al. 1999; Raab et al. 2007a; Li et al. 2009; Carey et al. 2010, 2011). Although their influx systems are quite similar to basic forms, their efflux systems are yet complicated to comprehend and Lsi transporters have not been found to involve here.

A generalized view of the uptake and translocation of different phytotoxic forms of arsenic was known, with arsenites being taken up more readily and DMA^{V} least readily and later their translocation within the plant showed the opposite trend but different plants showed varied results based on the soil type and the plant type for phytotoxic response to these forms of arsenic. For this, studies were carried out hydroponically (Raab et al. 2007a, 2007b) so that bioavailability of arsenic forms was similar for all plants and did not become a differentiating factor (Marin et al. 1992; Carbonell-Barrachina et al. 1998a; Abbas and Meharg 2008). Even in hydroponic studies, phytotoxic responses varied like in the case of two *Spartina* species, As^{V} (almost similar to As^{III}) was the most phytotoxic (Carbonell-Barrachina et al. 1998b) while for rice and maize plants, DMA^{V} showed most phytotoxicity (Marin et al. 1992; Abbas and Meharg 2008).

The toxicological studies have one more factor that is limiting, which is the arsenic extract being obtained during the research. Various studies have shown that arsenic calculations obtained were not accurate because of inefficient collection of total arsenic content from the plants as in a sunflower study, 15–20% of the arsenic content could not be included (Raab et al. 2005), 75% of the arsenic in beetroot could not be extracted due to high arsenic toxicity (Száková et al. 2010). All these disparities and miscalculations ask for a cellular and biochemical analysis of the arsenic effects and efficient biotechnological tools to study them. Moreover, interlinking these pathways with other stress stimulating pathways like antioxidant activation and apoptotic mechanisms, which are equally important for a clear picture of the arsenic impact on plants and the mechanism that can prevent such adverse effects.

11.5 Phytohormones, Their Role and Interaction with Heavy Metals

Phytohormones are the chemical messengers that perform vital growth and development processes within a plant. Within these chemical messengers also, the primary ones are totally responsible for growth processes while some specialized hormones have adapted to tolerate stress conditions. These phytohormones have been analyzed to target ROS and antioxidant systems and regulate their balance in such a way that stress conditions are relieved (Sytar et al. 2013; Choudhury et al. 2017; Lajayer et al. 2017), which has paved the way for artificial biosynthesis of these phytohormones and their exogenous applications have proved effective as the natural ones and led to stress tolerance (Sytar et al. 2019). These stress-tolerating plants are the key to cleansing our environment from heavy metals like arsenic without getting negatively affected since these plants detoxify arsenic or immobilize the arsenic forms in such a way that they become unavailable to their forms of life and hence harmless. Present studies focus on the mechanisms that these plants adapt to render these heavy metals harmless and employ them in metal-contaminated lands and aim to develop more such plants through genetic engineering techniques, which is a cost-effective as well as an eco-friendly approach toward metal decontamination.

Plant hormonal signaling is diverse and complex and where a single hormone may be stimulating a number of pathways, at the same time, many hormones could be required for a single signal generation and (Gray 2004) since phytohormones are the connecting links that mediate epigenetic and genetic responses, we need to understand these complex mechanisms operating by which we will be able to better understand the plant types suitable for specific environmental contaminations and what exogenous applications are required. Some common and well-known phytohormones are auxins (IAAs), cytokinins (CKs), abscisic acid (ABA), ethylene (ET), gibberellins (GAs), jasmonic acid (JA), salicylic acid (SA), brassinosteroids (BRs), strigolactones (SLs), etc. (Sytar et al. 2019). Auxin is the main growth hormone that is responsible for the length of a plant body and yet along with the role of elongation, it signals ascorbate-GSH cycle operation and metal phytoextraction and hence its presence in excess concentrations during stress alleviates metal stress conditions as well (Kazan 2013). Arsenic stress has proved to increase IAA levels along with miRNA expression which improved growth (Srivastava et al. 2013) and the signaling pathways of IAA, ET, and ROS cooperate to decrease the stress levels (Camacho-Cristóbal et al. 2015). Cytokinin is one more biomass increasing phytohormone specific for shoot activity but has also additional roles to increase the antioxidant activity when needed like that of auxin along with a function for increasing the transpiration rate. One of the toxic effects of arsenic stress is that it reduces transpiration, which would eventually lead to water stress and hence cytokinin can come to the rescue to relieve such water-stressed conditions. Cytokinin concentration usually decreases during metal stress and has been found to be antagonistic to ABA (Pospíšilová 2003a) and it is their cross talk and the cross-talk with other signaling pathways that determines the outcome for the plant cells (Ha et al. 2012). Similarly, GA is also a plant elongating hormone with its point of action around the nodal portions (GA_1) (Sponsel and Hedden 2010) but its signaling has the main role to play in the nitrogen cycle where it reduces the nitric oxide accumulation and regulates the activities of enzymes involved. Besides these functions, it plays a role in the transport and translocation of metals within the plant and hence prevents metal toxicity (Maggio et al. 2010). Their concentrations have shown an increase upon metal stress but high metal concentrations lead to their decrease in concentration (Atici et al. 2005). Abscisic acid is the first hormone that comes to mind when plant stress is being talked about and in case of metal stress as well, its concentration rises (Hollenbach et al. 1997; Bücker-Neto et al. 2017) and a cascade of signaling begins (O'Brien and Benkova 2013) since it is responsible for the regulation of about 10% protein-encoding genes (Nemhauser et al. 2006; Wani et al. 2016). Salicylic acid can prove very essential in terms of arsenic stress tolerance due to their roles in terms of ion uptake and transport and could prove vital for influx and efflux regulations as they are involved with electrolyte leakages and improving mineral nutrition. Moreover, they play a role in reducing the levels of ROS and reversing the lipid peroxidation (Rivas-San and Plasencia 2011). SA synchronously with JA and ET enhances the stress response (Jia et al. 2013). ET enzymes also increased during metal stress (Skottke et al. 2011; Bücker-Neto et al. 2017) and ET concentration depended upon the type of metal stress and the concentration, wherein higher concentrations led to ET decrease as well but comparably lower metal concentration also increased ET concentrations (Gora and Clijsters 1989). BRs further enhance the functions of auxin by cell division and elongation and are also involved in the regulation of stomatal and vascular differentiation, which could be useful when the plant is stressed due to arsenic accumulation. Parallelly, the antioxidant defense system is also stimulated by signaling (Cao et al. 2005). As many as 70 BRs have been found in plants (Mandava 1988; Vardhini 2014) and most of them have shown the potential to fight metal stress like 24-Epibrassinolide (EBL) (Shahzad et al. 2018). Jasmonic acid is one of the key phytohormones, which acts during stress conditions and its exogenous applications have resulted in various stress tolerance mechanisms in a number of research studies. They shift the plant energy from growth mode to enhancing physiological processes like increased antioxidant biosynthesis and various other secondary metabolites. SLs is quite a recently discovered phytohormone and has also stress resistance functions, be it in the case of drought, salinity, nutrient deficiency or here in case of heavy metals. They provide disease resistance to the plant and improve its various architectural modifications in addition to the phytoremediation process. Thus, studies have come to conclude that phytohormones in spite of playing essential roles in growth and development also enable a plant to better adjust to the environment and mould the physiology according to the various stress conditions prevailing (Nishiyama et al. 2011; Colebrook et al. 2014; Xu et al. 2016; Bücker-Neto et al. 2017). Further, upon understanding the vital roles of phytohormones, exogenous applications done proved to be safe as well as effective for the plant as well as the environment (Zhu et al. 2012, 2013; Agami and Mohamed 2013; Masood et al. 2016) and similar studies in association with heavy metals are diverse and in progress (Bücker-Neto et al. 2017).

11.6 Mechanism of Detoxification of Heavy Metals with Special Emphasis on Arsenic by Phytohormones

Phytohormones are the plant's major signaling molecules and when conditions of stress like arsenic arise, they are bound to signal other pathways to get activated. Although auxins, cytokinins, and GAs are responsible for overall growth and development regulation, the newer class of phytohormones is specifically specialized to tolerate stressful conditions and regulate the antioxidant mechanism of the plant and these include ABAs, JAs, SAs, SLs, ETs, etc. (Zhang et al. 2018). One of the well-known and essential elements required for phytohormone signaling is the SKP1-CULLIN-F-BOX complex (SCF), which is responsible for proteasome degradation and is vital for activating and reducing proteins that are required during stressful conditions created during arsenic stress (Guilfoyle et al. 2015). It is not a single phytohormone regulation but crosstalk between ABAs and SLs (Van Ha et al. 2014) and then between SLs and IAA (Domagalska and Leyser 2011; Shinohara et al. 2013) and various other interconnected pathways whose net output results in targeting of the F-box of SCF and proceeding the signal transduction pathway (Guilfoyle et al. 2015).

We take the case of jasmonates which are cyclopentanone compounds and whose two forms, Jasmonic acid and Methyl Jasmonate have been studied the most due to their mitigative effects against stressful conditions (Chen et al. 2014; Okada et al. 2015). These phytohormones have showed genetic regulation for various plant processes like activation of plant defense pathways, cell communication, regulation of growth and development, decrease in malondialdehyde content, and ROS species like degradation of OH and H_2O_2 etc., during metal stress conditions (Soares et al. 2010; Wasternack and Hause 2013; Hanaka et al. 2015; Farooq et al. 2016). The other way by which ROS species are tolerated is that jasmonates protect and increase the pigment biosynthesis, which also protects the cells from ROS-damaging impact (Poonam et al. 2013). Metal stress levels are also controlled by stomatal closure regulation so that metal accumulation within the plant is prevented due to the stoppage of transpiration (Hossain et al. 2011; Chen et al. 2014). Exogenous applications have proved efficient in increasing antioxidant levels and the mechanisms are not yet clear but they somehow manage to stimulate endogenous phytohormone levels, which eventually ameliorate heavy metals (Maksymiec and Krupa 2002; Chen et al. 2014). Moreover, PCs and metallothioneins (MTs), which also play vital roles in heavy metal resistance have been found to increase due to exogenous jasmonate applications (Chen et al. 2014). Among the heavy metals like arsenic and cadmium-contaminated soils, it was observed that arsenic gets accumulated to much higher levels within plant roots like for rice roots and jasmonates like jasmonic acid, methyl jasmonate, and OPDA mitigated the toxic effects but yet the plants were negatively impacted. Thus, more research is needed and the mitigation effects need to be studied and techniques employed to increase their impact (Ronzan et al. 2019).

BRs are phytosteroidal hormones (Bhardwaj et al. 2014) that have been found to play active roles in many developmental processes (Rajewska et al. 2016; Que et al. 2017) and have been found to improve horticulture by reducing the stress conditions that they face. Their signaling role involves regulation of those genes that cause cell differentiation and cell expansion and hence contributing to the growth and also other genes that control metabolism as it has been recorded to regulate as many as a hundred genes (Fariduddin et al. 2014; Bhardwaj et al. 2014; Wei and Li 2016). Their concentrations have been found to directly impact signaling pathways like antioxidant activity, PC biosynthesis, and reducing ROS and lipid peroxidation (Rajewska et al. 2016). Various studies link BRs to the amelioration of both biotic and abiotic stresses (Vardhini et al. 2016). They regulate the biosynthesis of such proteins (like MTs and PCs) and enzymes that do the metal binding and hence lead to metal compartmentalization (Rajewska et al. 2016; Sharma et al. 2018; Yu et al. 2018). pH changes are also performed by BRs which lead to metal biosorption (Bajguz 2002). But there are studies where BRs alone could not improve arsenic stress like in one of the hydroponic studies, it was observed that upon arsenic and cadmium stress induction, BR application decreased rice growth but when the phytohormone was supplemented with iron plaque, metal toxicity improved to a great extent. Thus, along with the nature of metal stress, concentration and supplementations should be properly tested for better yields, stress resistance, and growth processes (Xu et al. 2018). SLs are mainly associated with rhizospheric processes and their interaction with microbes (Foo and Reid 2013; Koltai and Kapulnik 2013; Kapulnik and Koltai 2014; Van Zeijl et al. 2015), and are known for their role against nutrient and biotic stress (Marzec et al. 2013; Marzec and Muszynska 2015; Sun et al. 2016). This happens by their signaling mechanism (Marzec 2016a; Saeed et al. 2017) and their little concentrations are found in the plant (Dun et al. 2009; Xie et al. 2010) and more than 1000 SLs exist in nature but only 19 have yet been known (Zwanenburg and Pospíšil 2013; Ćavar et al. 2015). Metal stresses lead to shoot branching at times and this phytohormone negates the effect by preventing this branching while at the same time, deflecting this biomass growth toward root and root environment development for better nutrient absorption and excess metal efflux (Gilroy and Jones 2000; Koltai and Kapulnik 2011). Their effects have been studied in a number of plants like rice, *Arabidopsis*, pea, *Petunia*, and it was observed that SLs act in a coordinative manner with other phytohormones like auxins and their links with antioxidants like GSH is also quite pronounced in dealing with stress conditions (Xie et al. 2010; Marquez-Garcia et al. 2014). SL-deficient plants showed higher arsenic accumulation and much oxidative stress as in one of the studies conducted over rice, it was found that in wild plants, which contained SL concentrations in a good amount did the Vacuolar sequestration of arsenic and maintained the antioxidant system in a balanced manner while SL-deficient plants showed high malondialdehyde levels, quite high oxidative stress, and high arsenic accumulation (Mostofa et al. 2021).

These studies suggest that phytohormones produce such signals in plants upon arsenic and other metal contamination detection which activate various interlinked pathways that act in a coordinative manner to relieve the stress conditions by compartmentalization and detoxification of the metals. However, various factors govern the metal detoxification process like plant genotype, soil environment, nature of metal, the concentration of the metal, presence of nutrients, and other unknown factors, which need to be understood and taken into account for proper application of phytohormones against metal stress.

11.7 Exogenous Application of Phytohormones over Detoxification

Phytohormones have proved to be efficient in tolerating metal stress and activating such signaling pathways, which strengthen the defense system of the plant and, as already stated, arsenic toxicity leads to a lot of negative impacts on developmental processes, and oxidative stress conditions are created whose mitigation is handled by interlinking phytohormonal signaling and other crosslinks. The recent research aims to use exogenous phytohormonal treatment to ameliorate excess metal stress which is otherwise fatal to the plant under normal conditions. This technique is cost-effective as well as eco-friendly and the plant acts in a manner similar to its natural phytohormones (Bücker-Neto et al. 2017). Cytokinin levels already increase upon metal stress induction but along with an exogenous application, their signaling enhances biosynthesizing of more stress-resistant secondary metabolites like phenolics and proline (Pospíšilová 2003b; Ergün and Öncel 2012). Similar is the case for ABA and GAs when metal stress of Zn and Pb is induced but the response is quite opposite upon Cd stress induction (Fuhrer 1982; Ahmad et al. 2015) and hence lead to the conclusion that different stress conditions involve different signaling mechanisms for tolerance. Exogenous auxin, GA and citric acid application led to an increase in biomass of the plant in various metal-contaminated soils (Aderholt et al. 2017). Cell division is a characteristic property of cytokinin induction whose exogenous application has also shown increased biomass accumulation and better photosynthetic rate (Tassi et al. 2008; Cassina et al. 2011). Exogenous BRs have been seen to show enhanced antioxidant activity, decrease free radicals, and along with increasing defense responses, also enhance auxin activity leading to overall growth and development (catalase, peroxidase, superoxide dismutase, and GSH reductase enzymes) (Choudhary et al. 2011; Rady 2011; Ramakrishna and Rao 2012). Exogenous salicylic acid also increased antioxidant machinery and pigment systems, which protect the plant as well as enhance photosynthesis and mineral nutrition (Shi and Zhu 2008; Belkhadi et al. 2010; Kazemi et al. 2010; Xu et al. 2015a, 2015b). Exogenous jasmonates along with endogenous level increases during various metal-induced stress have shown stress alleviation by controlling transpiration and reducing photosynthetic damage and enhancing gene expression of stress proteins like type-2 MT (Chen et al. 2014; Per et al. 2018). Exogenous ET has proved quite effective for horticultural crops that are under various metal stress-contaminated soils (Agarwal et al. 2012). Exogenous SLs like GR24 has proved quite efficient in plant growth and development under metal stress and has shown enhanced plant pigment biosynthesis (Marzec 2016b). Thus, the exogenous application has proved to enhance the endogenous activity of different phytohormones and no major difference was observed between the activity of the two but it is the dosage that matters and creates varied changes in the plant and metal detoxification (Piotrowska-Niczyporuk and Bajguz 2014).

11.8 Conclusion

Heavy metals are required by plants in trace quantities but due to anthropogenic activities, their concentrations are drastically increasing and plants are trying to cope with such toxic environments and it is the plants that could be the key factors for detoxifying and immobilizing these heavy metals like arsenic from the environment. Otherwise, heavy metals like arsenic will contaminate not only our soils but also each ecosystem as their cycle passes through each one and has entered the biosphere through complex food webs and the hydrosphere being the largest source of contamination for it. Arsenic toxicity (along with other metal toxicities) is a global issue and each country is taking measures to combat the problem but employing plants for the same is one of the popular solutions due to its nature being eco-friendly as well as cost-effective. Within the plants, it is the phytohormones that operate at a level to bring epigenetic and genetic levels to a balance and make the plants adaptable to such situations where they thrive. Most plants have been able to mitigate these effects but a very small number of plants and researches are being carried out to introduce such hyperaccumulation and detoxification characteristics, especially in plants like rice and various horticultural plants so that the means of food as well as income are flourished. The challenges imposed are that each plant according to its genotype accumulates arsenic differently and a uniform system is not yet developed to categorize their concentrations and feasibility. Thus, more studies need to be devoted to analyze what factors affect them, which conditions are suitable for what phytohormones, and will exogenous applications prove successful for all plant types and for this, long-term field and economical studies of phytoremediation need to be performed for successful application.

References

Abbas, G., Murtaza, B., Bibi, I. et al. (2018). Arsenic uptake, toxicity, detoxification, and speciation in plants: physiological, biochemical, and molecular aspects. *International Journal of Environmental Research and Public Health* 15 (1): 59.

Abbas, M.H. and Meharg, A.A. (2008). Arsenate, arsenite and dimethyl arsinic acid (DMA) uptake and tolerance in maize (*Zea mays* L.). *Plant and Soil* 304 (1): 277–289.

Abedin, M.J., Feldmann, J., and Meharg, A.A. (2002). Uptake kinetics of arsenic species in rice plants. *Plant Physiology* 128 (3): 1120–1128.

Aderholt, M., Vogelien, D.L., Koether, M., and Greipsson, S. (2017). Phytoextraction of contaminated urban soils by *Panicum virgatum* L. enhanced with application of a plant growth regulator (BAP) and citric acid. *Chemosphere* 175: 85–96.

Agami, R.A. and Mohamed, G.F. (2013). Exogenous treatment with indole-3-acetic acid and salicylic acid alleviates cadmium toxicity in wheat seedlings. *Ecotoxicology and Environmental Safety* 94: 164–171.

Agarwal, G., Choudhary, D., Singh, V.P., and Arora, A. (2012). Role of ethylene receptors during senescence and ripening in horticultural crops. *Plant Signaling & Behavior* 7 (7): 827–846.

Ahmad, A., Hadi, F., and Ali, N. (2015). Effective phytoextraction of cadmium (Cd) with increasing concentration of total phenolics and free proline in *Cannabis sativa* (L.) plant under various treatments of fertilizers, plant growth regulators and sodium salt. *International Journal of Phytoremediation* 17 (1): 56–65.

Ascar, L., Ahumada, I., and Richter, P. (2008). Effect of biosolid incorporation on arsenic distribution in Mollisol soils in central Chile. *Chemosphere* 70 (7): 1211–1217.

Asher, C.J. and Reay, P.F. (1979). Arsenic uptake by barley seedlings. *Functional Plant Biology* 6 (4): 459–466.

Atici, Ö., Ağar, G., and Battal, P. (2005). Changes in phytohormone contents in chickpea seeds germinating under lead or zinc stress. *Biologia Plantarum* 49 (2): 215–222.

Bajguz, A. (2002). Brassinosteroids and lead as stimulators of phytochelatins synthesis in *Chlorella vulgaris*. *Journal of Plant Physiology* 159 (3): 321–324.

Belkhadi, A., Hediji, H., Abbes, Z. et al. (2010). Effects of exogenous salicylic acid pre-treatment on cadmium toxicity and leaf lipid content in *Linum usitatissimum* L. *Ecotoxicology and Environmental Safety* 73 (5): 1004–1011.

Bhardwaj, R., Sharma, I., Kapoor, D. et al. (2014). Brassinosteroids: improving crop productivity and abiotic stress tolerance. In: *Physiological Mechanisms and Adaptation Strategies in PlantsUunder Changing Eenvironment* (ed. P. Ahmad and M. Wani), 161–187. New York: Springer https://doi.org/10.1007/978-1-4614-8600-8_6.

Bhattacharjee, H. and Rosen, B.P. (2007). Arsenic metabolism in prokaryotic and eukaryotic microbes. In: *Molecular Microbiology of Heavy Metals* (ed. D.H. Nies and S. Silver), 371–406. Berlin, Heidelberg: Springer https://doi.org/10.1007/7171_2006_086.

Bhattacharya, P., Frisbie, S.H., Smith, E. et al. (2002). Arsenic in the environment: a global perspective. In: *Handbook of Heavy Metals in the Environment* (ed. B. Sarker), 147–215. New York: Marcell Dekker Inc.

Bhattacharya, P., Welch, A.H., Stollenwerk, K.G. et al. (2007). Arsenic in the environment: biology and chemistry. *Science of the Total Environment* 379 (2–3): 109–120.

Bienert, G.P., Thorsen, M., Schüssler, M.D. et al. (2008). A subgroup of plant aquaporins facilitate the bi-directional diffusion of $As(OH)_3$ and $Sb(OH)_3$ across membranes. *BMC Biology* 6 (1): 1–15.

Bucher, M. (2007). Functional biology of plant phosphate uptake at root and mycorrhiza interfaces. *The New Phytologist* 173 (1): 11–26.

Bücker-Neto, L., Paiva, A.L.S., Machado, R.D. et al. (2017). Interactions between plant hormones and heavy metals responses. *Genetics and Molecular Biology* 40 (1): 373–386.

Burlo, F., Guijarro, I., Carbonell-Barrachina, A.A. et al. (1999). Arsenic species: effects on and accumulation by tomato plants. *Journal of Agricultural and Food Chemistry* 47 (3): 1247–1253.

Camacho-Cristóbal, J.J., Martín-Rejano, E.M., Herrera-Rodríguez, M.B. et al. (2015). Boron deficiency inhibits root cell elongation via an ethylene/auxin/ROS-dependent pathway in *Arabidopsis* seedlings. *Journal of Experimental Botany* 66 (13): 3831–3840.

Cao, S., Xu, Q., Cao, Y. et al. (2005). Loss-of-function mutations in DET2 gene lead to an enhanced resistance to oxidative stress in *Arabidopsis*. *Physiologia Plantarum* 123 (1): 57–66.

Carbonell-Barrachina, A.A., Aarabi, M.A., DeLaune, R.D. et al. (1998a). The influence of arsenic chemical form and concentration on *Spartina patens* and *Spartina alterniflora* growth and tissue arsenic concentration. *Plant and Soil* 198 (1): 33–43.

Carbonell-Barrachina, A.A., Burló, F., and Mataix, J. (1998b). Response of bean micronutrient nutrition to arsenic and salinity. *Journal of Plant Nutrition* 21 (6): 1287–1299.

Carbonell-Barrachina, A.A., Burló, F., Burgos-Hernandez, A. et al. (1997). The influence of arsenite concentration on arsenic accumulation in tomato and bean plants. *Scientia Horticulturae* 71 (3–4): 167–176.

Cardoso, A.P.F., Al-Eryani, L., and States, J.C. (2018). Arsenic-induced carcinogenesis: the impact of miRNA dysregulation. *Toxicological Sciences* 165 (2): 284–290.

Carey, A.M., Norton, G.J., Deacon, C. et al. (2011). Phloem transport of arsenic species from flag leaf to grain during grain filling. *The New Phytologist* 192 (1): 87–98.

Carey, A.M., Scheckel, K.G., Lombi, E. et al. (2010). Grain unloading of arsenic species in rice. *Plant Physiology* 152 (1): 309–319.

Cassina, L., Tassi, E., Morelli, E. et al. (2011). Exogenous cytokinin treatments of an Ni hyper-accumulator, *Alyssum murale*, grown in a serpentine soil: implications for phytoextraction. *International Journal of Phytoremediation* 13 (sup1): 90–101.

Catarecha, P., Segura, M.D., Franco-Zorrilla, J.M. et al. (2007). A mutant of the *Arabidopsis* phosphate transporter PHT1; 1 displays enhanced arsenic accumulation. *Plant Cell* 19 (3): 1123–1133.

Ćavar, S., Zwanenburg, B., and Tarkowski, P. (2015). Strigolactones: occurrence, structure, and biological activity in the rhizosphere. *Phytochemistry Reviews* 14 (4): 691–711.

Chaumont, F., Moshelion, M., and Daniels, M.J. (2005). Regulation of plant aquaporin activity. *Biology of the Cell* 97 (10): 749–764.

Chen, J., Yan, Z., and Li, X. (2014). Effect of methyl jasmonate on cadmium uptake and antioxidative capacity in *Kandelia obovata* seedlings under cadmium stress. *Ecotoxicology and Environmental Safety* 104: 349–356.

Chen, J.X., Cao, Y., Yan, X. et al. (2021). Novel PvACR3; 2 and PvACR3; 3 genes from arsenic-hyperaccumulator *Pteris vittata* and their roles in manipulating plant arsenic accumulation. *Journal of Hazardous Materials* 415: 125647.

Chen, W., Chi, Y., Taylor, N.L. et al. (2010). Disruption of ptLPD1 or ptLPD2, genes that encode isoforms of the plastidial lipoamide dehydrogenase, confers arsenate hypersensitivity in *Arabidopsis*. *Plant Physiology* 153 (3): 1385–1397.

Choudhary, S.P., Kanwar, M., Bhardwaj, R. et al. (2011). Epibrassinolide ameliorates Cr (VI) stress via influencing the levels of indole-3-acetic acid, abscisic acid, polyamines and antioxidant system of radish seedlings. *Chemosphere* 84 (5): 592–600.

Choudhury, F.K., Rivero, R.M., Blumwald, E., and Mittler, R. (2017). Reactive oxygen species, abiotic stress and stress combination. *The Plant Journal* 90 (5): 856–867.

Colebrook, E.H., Thomas, S.G., Phillips, A.L., and Hedden, P. (2014). The role of gibberellin signalling in plant responses to abiotic stress. *The Journal of Experimental Biology* 217 (1): 67–75.

Dey, S., Dou, D., and Rosen, B.P. (1994). ATP-dependent arsenite transport in everted membrane vesicles of *Escherichia coli*. *The Journal of Biological Chemistry* 269 (41): 25442–25446.

Domagalska, M.A. and Leyser, O. (2011). Signal integration in the control of shoot branching. *Nature Reviews Molecular Cell Biology* 12 (4): 211–221.

Dun, E.A., Brewer, P.B., and Beveridge, C.A. (2009). Strigolactones: discovery of the elusive shoot branching hormone. *Trends in Plant Science* 14 (7): 364–372.

Ellis, D.R., Gumaelius, L., Indriolo, E. et al. (2006). A novel arsenate reductase from the arsenic hyperaccumulating fern *Pteris vittata*. *Plant Physiology* 141 (4): 1544–1554.

Ergün, N. and Öncel, I. (2012). Effects of some heavy metals and heavy metal hormone interactions on wheat (*Triticum aestivum* L. cv. Gun 91) seedlings. *African Journal of Agricultural Research* 7 (10): 1518–1523.

Fariduddin, Q., Yusuf, M., Ahmad, I., and Ahmad, A. (2014). Brassinosteroids and their role in response of plants to abiotic stresses. *Biologia Plantarum* 58 (1): 9–17.

Farooq, M.A., Gill, R.A., Islam, F. et al. (2016). Methyl jasmonate regulates antioxidant defense and suppresses arsenic uptake in *Brassica napus* L. *Frontiers in Plant Science* 7: 468.

Foo, E. and Reid, J.B. (2013). Strigolactones: new physiological roles for an ancient signal. *Journal of Plant Growth Regulation* 32 (2): 429–442.

Fuhrer, J. (1982). Early effects of excess cadmium uptake in *Phaseolus vulgaris*. *Plant, Cell and Environment* 5 (4): 263–270.

Garg, N. and Singla, P. (2011). Arsenic toxicity in crop plants: physiological effects and tolerance mechanisms. *Environmental Chemistry Letters* 9 (3): 303–321.

Gavhane, S.K., Sapkale, J.B., Susware, N.K., and Sapkale, S.J. (2021). Impact of heavy metals in riverine and estuarine environment: a review. *Research Journal of Chemistry and Environment* 25 (5): 226–233.

Geng, C.N., Zhu, Y.G., Hu, Y. et al. (2006). Arsenate causes differential acute toxicity to two P-deprived genotypes of rice seedlings (*Oryza sativa* L.). *Plant and Soil* 279 (1): 297–306.

Gilroy, S. and Jones, D.L. (2000). Through form to function: root hair development and nutrient uptake. *Trends in Plant Science* 5 (2): 56–60.

González, E., Solano, R., Rubio, V. et al. (2005). PHOSPHATE TRANSPORTER TRAFFIC FACILITATOR1 is a plant-specific SEC12-related protein that enables the endoplasmic reticulum exit of a high-affinity phosphate transporter in *Arabidopsis*. *Plant Cell* 17 (12): 3500–3512.

Gora, L. and Clijsters, H. (1989). Effect of copper and zinc on the ethylene metabolism in *Phaseolus vulgaris* L. In: *Biochemical and Physiological Aspects of Ethylene Production in Lower and Higher Plants*, vol. *26* (ed. H. Clijsters, M. De Proft, R. Marcelle and M. Van Poucke), 219–228. Dordrecht: Springer https://doi.org/10.1007/978-94-009-1271-7_25.

Gray, W.M. (2004). Hormonal regulation of plant growth and development. *PLoS Biology* 2 (9): e311.

Guilfoyle, T., Hagen, G., Larrieu, A., and Vernoux, T. (2015). Comparison of plant hormone signalling systems. *Essays in Biochemistry* 58: 165–181.

Guo, W., Hou, Y.L., Wang, S.G., and Zhu, Y.G. (2005). Effect of silicate on the growth and arsenate uptake by rice (*Oryza sativa* L.) seedlings in solution culture. *Plant and Soil* 272 (1): 173–181.

Guo, W., Zhu, Y.G., Liu, W.J. et al. (2007). Is the effect of silicon on rice uptake of arsenate (As^V) related to internal silicon concentrations, iron plaque and phosphate nutrition? *Environmental Pollution* 148 (1): 251–257.

Ha, S., Vankova, R., Yamaguchi-Shinozaki, K. et al. (2012). Cytokinins: metabolism and function in plant adaptation to environmental stresses. *Trends in Plant Science* 17 (3): 172–179.

Hanaka, A., Maksymiec, W., and Bednarek, W. (2015). The effect of methyl jasmonate on selected physiological parameters of copper-treated *Phaseolus coccineus* plants. *Plant Growth Regulation* 77 (2): 167–177.

Hare, V., Chowdhary, P., Kumar, B. et al. (2019). Arsenic toxicity and its remediation strategies for fighting the environmental threat. In: *Emerging and Eco-Friendly Approaches for Waste Management* (ed. R.N. Bharagava and P. Chowdhary), 143–170. Singapore: Springer.

Hollenbach, B., Schreiber, L., Hartung, W., and Dietz, K.J. (1997). Cadmium leads to stimulated expression of the lipid transfer protein genes in barley: implications for the involvement of lipid transfer proteins in wax assembly. *Planta* 203 (1): 9–19.

Hossain, M.A., Munemasa, S., Uraji, M. et al. (2011). Involvement of endogenous abscisic acid in methyl jasmonate-induced stomatal closure in *Arabidopsis*. *Plant Physiology* 156 (1): 430–438.

Inskeep, W.P., McDermott, T.R., and Fendorf, S. (2002). Arsenic (V)/(III) cycling in soils and natural waters: chemical and microbiological processes. *Environmental Chemistry of Arsenic* 183–215.

Isayenkov, S.V. and Maathuis, F.J. (2008). The *Arabidopsis thaliana* aquaglyceroporin AtNIP7; 1 is a pathway for arsenite uptake. *FEBS Letters* 582 (11): 1625–1628.

Jang, Y.C., Somanna, Y., and Kim, H. (2016). Source, distribution, toxicity and remediation of arsenic in the environment – a review. *International Journal of Applied Environment Sciences* 11 (2): 559–581.

Jia, C., Zhang, L., Liu, L. et al. (2013). Multiple phytohormone signalling pathways modulate susceptibility of tomato plants to *Alternaria alternata* f. sp. *lycopersici*. *Journal of Experimental Botany* 64 (2): 637–650.

Kapulnik, Y. and Koltai, H. (2014). Strigolactone involvement in root development, response to abiotic stress, and interactions with the biotic soil environment. *Plant Physiology* 166 (2): 560–569.

Kazan, K. (2013). Auxin and the integration of environmental signals into plant root development. *Annals of Botany* 112 (9): 1655–1665.

Kazemi, N., Khavari-Nejad, R.A., Fahimi, H. et al. (2010). Effects of exogenous salicylic acid and nitric oxide on lipid peroxidation and antioxidant enzyme activities in leaves of *Brassica napus* L. under nickel stress. *Scientia Horticulturae* 126 (3): 402–407.

Khan, I., Ahmad, A., and Iqbal, M. (2009). Modulation of antioxidant defence system for arsenic detoxification in Indian mustard. *Ecotoxicology and Environmental Safety* 72 (2): 626–634.

Koltai, H. and Kapulnik, Y. (2011). Strigolactones as mediators of plant growth responses to environmental conditions. *Plant Signaling & Behavior* 6 (1): 37–41.

Koltai, H. and Kapulnik, Y. (2013). Unveiling signaling events in root responses to strigolactone. *Molecular Plant* 6 (3): 589–591.

Kumarathilaka, P., Seneweera, S., Meharg, A., and Bundschuh, J. (2018). Arsenic accumulation in rice (*Oryza sativa* L.) is influenced by environment and genetic factors. *Science of the Total Environment* 642: 485–496.

Lajayer, B.A., Ghorbanpour, M., and Nikabadi, S. (2017). Heavy metals in contaminated environment: destiny of secondary metabolite biosynthesis, oxidative status and phytoextraction in medicinal plants. *Ecotoxicology and Environmental Safety* 145: 377–390.

Langdon, C.J., Piearce, T.G., Meharg, A.A., and Semple, K.T. (2003). Interactions between earthworms and arsenic in the soil environment: a review. *Environmental Pollution* 124 (3): 361–373.

Li, R.Y., Ago, Y., Liu, W.J. et al. (2009). The rice aquaporin Lsi1 mediates uptake of methylated arsenic species. *Plant Physiology* 150 (4): 2071–2080.

Li, Y., Dhankher, O.P., Carreira, L. et al. (2004). Overexpression of phytochelatin synthase in *Arabidopsis* leads to enhanced arsenic tolerance and cadmium hypersensitivity. *Plant and Cell Physiology* 45 (12): 1787–1797.

Lin, S., Cullen, W.R., and Thomas, D.J. (1999). Methylarsenicals and arsinothiols are potent inhibitors of mouse liver thioredoxin reductase. *Chemical Research in Toxicology* 12 (10): 924–930.

Lindsay, E.R. and Maathuis, F.J. (2017). New molecular mechanisms to reduce arsenic in crops. *Trends in Plant Science* 22 (12): 1016–1026.

Liu, Z., Boles, E., and Rosen, B.P. (2004). Arsenic trioxide uptake by hexose permeases in *Saccharomyces cerevisiae*. *The Journal of Biological Chemistry* 279 (17): 17312–17318.

Livesey, N.T. and Huang, P.M. (1981). Adsorption of arsenate by soils and its relation to selected chemical properties and anions. *Soil Science* 131 (2): 88–94.

Logoteta, B., Xu, X.Y., Macnair, M.R., McGrath, S.P. and Zhao, F.J., 2008. Arsenic tolerance mechanisms in *Holcus lanatus*: is arsenite efflux involved. In *20th New Phytologist Symposium. Arsenic: Unravelling Its Metabolism and Speciation in Plants*, Aberdeen, Scotland, UK (26–27 June 2008). Programme, Abstracts and Participants, p. 37. New Phytologist Trust.

Lomax, C., Liu, W.J., Wu, L. et al. (2012). Methylated arsenic species in plants originate from soil microorganisms. *The New Phytologist* 193 (3): 665–672.

Ma, J., Lei, M., Weng, L. et al. (2019). Fractions and colloidal distribution of arsenic associated with iron oxide minerals in lead-zinc mine-contaminated soils: comparison of tailings and smelter pollution. *Chemosphere* 227: 614–623.

Ma, J.F. and Yamaji, N. (2006). Silicon uptake and accumulation in higher plants. *Trends in Plant Science* 11 (8): 392–397.

Ma, J.F., Tamai, K., Yamaji, N. et al. (2006). A silicon transporter in rice. *Nature* 440 (7084): 688–691.

Ma, J.F., Yamaji, N., Mitani, N. et al. (2007). An efflux transporter of silicon in rice. *Nature* 448 (7150): 209–212.

Ma, J.F., Yamaji, N., Mitani, N. et al. (2008). Transporters of arsenite in rice and their role in arsenic accumulation in rice grain. *Proceedings of the National Academy of Sciences of the United States of America* 105 (29): 9931–9935.

Ma, L.Q., Komar, K.M., Tu, C. et al. (2001). A fern that hyperaccumulates arsenic. *Nature* 409 (6820): 579–579.

Macur, R.E., Jackson, C.R., Botero, L.M. et al. (2004). Bacterial populations associated with the oxidation and reduction of arsenic in an unsaturated soil. *Environmental Science & Technology* 38 (1): 104–111.

Maggio, A., Barbieri, G., Raimondi, G., and De Pascale, S. (2010). Contrasting effects of GA_3 treatments on tomato plants exposed to increasing salinity. *Journal of Plant Growth Regulation* 29 (1): 63–72.

Majumdar, A., Upadhyay, M.K., Kumar, J.S. et al. (2019). Ultra-structure alteration via enhanced silicon uptake in arsenic stressed rice cultivars under intermittent irrigation practices in Bengal delta basin. *Ecotoxicology and Environmental Safety* 180: 770–779.

Maksymiec, W. and Krupa, Z. (2002). Jasmonic acid and heavy metals in *Arabidopsis* plants-a similar physiological response to both stressors? *Journal of Plant Physiology* 159 (5): 509–515.

Mandal, B.K. and Suzuki, K.T. (2002). Arsenic round the world: a review. *Talanta* 58 (1): 201–235.

Mandava, N.B. (1988). Plant growth-promoting brassinosteroids. *Annual Review of Plant Physiology and Plant Molecular Biology* 39 (1): 23–52.

Marin, A.R., Masscheleyn, P.H., and Patrick, W.H. (1992). The influence of chemical form and concentration of arsenic on rice growth and tissue arsenic concentration. *Plant and Soil* 139 (2): 175–183.

Marin, A.R., Masscheleyn, P.H., and Patrick, W.H. (1993). Soil redox-pH stability of arsenic species and its influence on arsenic uptake by rice. *Plant and Soil* 152 (2): 245–253.

Marquez-Garcia, B., Njo, M., Beeckman, T.O.M. et al. (2014). A new role for glutathione in the regulation of root architecture linked to strigolactones. *Plant, Cell and Environment* 37 (2): 488–498.

Marzec, M. and Muszynska, A. (2015). In silico analysis of the genes encoding proteins that are involved in the biosynthesis of the RMS/MAX/D pathway revealed new roles of strigolactones in plants. *International Journal of Molecular Sciences* 16 (4): 6757–6782.

Marzec, M. (2016a). Perception and signaling of strigolactones. *Frontiers in Plant Science* 7: 1260.

Marzec, M. (2016b). Strigolactones as part of the plant defence system. *Trends in Plant Science* 21 (11): 900–903.

Marzec, M., Muszynska, A., and Gruszka, D. (2013). The role of strigolactones in nutrient-stress responses in plants. *International Journal of Molecular Sciences* 14 (5): 9286–9304.

Masood, A., Khan, M.I.R., Fatma, M. et al. (2016). Involvement of ethylene in gibberellic acid-induced sulfur assimilation, photosynthetic responses, and alleviation of cadmium stress in mustard. *Plant Physiology and Biochemistry* 104: 1–10.

Matschullat, J. (2000). Arsenic in the geosphere – a review. *Science of the Total Environment* 249 (1–3): 297–312.

Maurel, C., Verdoucq, L., Luu, D.T., and Santoni, V. (2008). Plant aquaporins: membrane channels with multiple integrated functions. *Annual Review of Plant Biology* 59: 595–624.

Meharg, A.A. and Jardine, L. (2003). Arsenite transport into paddy rice (*Oryza sativa*) roots. *The New Phytologist* 157: 39–44.

Meharg, A.A., Naylor, J., and Macnair, M.R. (1994). Phosphorus nutrition of arsenate-tolerant and nontolerant phenotypes of velvetgrass. *Journal of Environmental Quality* 23 (2): 234–238. American Society of Agronomy, Crop Science Society of America, and Soil Science Society of America.

Mestrot, A., Feldmann, J., Krupp, E.M. et al. (2011). Field fluxes and speciation of arsines emanating from soils. *Environmental Science & Technology* 45 (5): 1798–1804.

Mimura, T. (1999). Regulation of phosphate transport and homeostasis in plant cells. *International Review of Cytology* 191: 149–200.

Mitani, N., Yamaji, N., and Ma, J.F. (2008). Characterization of substrate specificity of a rice silicon transporter, Lsi1. *Pflügers Archiv - European Journal of Physiology* 456 (4): 679–686.

Miteva, E. (2002). Accumulation and effect of arsenic on tomatoes. *Communications in Soil Science and Plant Analysis* 33 (11–12): 1917–1926.

Mostofa, M.G., Rahman, M.M., Nguyen, K.H. et al. (2021). Strigolactones regulate arsenate uptake, vacuolar-sequestration and antioxidant defense responses to resist arsenic toxicity in rice roots. *Journal of Hazardous Materials* 415: 125589.

Naranmandura, H., Xu, S., Sawata, T. et al. (2011). Mitochondria are the main target organelle for trivalent monomethylarsonous acid (MMAIII)-induced cytotoxicity. *Chemical Research in Toxicology* 24 (7): 1094–1103.

Nemhauser, J.L., Hong, F., and Chory, J. (2006). Different plant hormones regulate similar processes through largely nonoverlapping transcriptional responses. *Cell* 126 (3): 467–475.

Ng, K.S., Ujang, Z., and Le-Clech, P. (2004). Arsenic removal technologies for drinking water treatment. *Reviews in Environmental Science and Biotechnology* 3 (1): 43–53.

Nishiyama, R., Watanabe, Y., Fujita, Y. et al. (2011). Analysis of cytokinin mutants and regulation of cytokinin metabolic genes reveals important regulatory roles of cytokinins in drought, salt and abscisic acid responses, and abscisic acid biosynthesis. *The Plant Cell* 23 (6): 2169–2183.

Nriagu, J.O. and Azcue, J.M. (1990). Environmental sources of arsenic in food. *Advanced Journal of Environment Science and Technology* 23: 103–127.

O'Brien, J.A. and Benková, E. (2013). Cytokinin cross-talking during biotic and abiotic stress responses. *Frontiers in Plant Science* 4: 451.

Okada, K., Abe, H., and Arimura, G.I. (2015). Jasmonates induce both defense responses and communication in monocotyledonous and dicotyledonous plants. *Plant and Cell Physiology* 56 (1): 16–27.

Oscarson, D.W., Huang, P.M., Defosse, C., and Herbillon, A. (1981). Oxidative power of Mn(IV) and Fe(III) oxides with respect to As(III) in terrestrial and aquatic environments. *Nature* 291 (5810): 50–51.

Per, T.S., Khan, M.I.R., Anjum, N.A. et al. (2018). Jasmonates in plants under abiotic stresses: crosstalk with other phytohormones matters. *Environmental and Experimental Botany* 145: 104–120.

Pfeifer, H.R., Beatrizotti, G., Berthoud, J. et al. (2002). Natural arsenic-contamination of surface and ground waters in Southern Switzerland (Ticino). *Bulletin for Applied Geology* 7 (1): 81–103.

Pickering, I.J., Gumaelius, L., Harris, H.H. et al. (2006). Localizing the biochemical transformations of arsenate in a hyperaccumulating fern. *Environmental Science & Technology* 40 (16): 5010–5014.

Piotrowska-Niczyporuk, A. and Bajguz, A. (2014). The effect of natural and synthetic auxins on the growth, metabolite content and antioxidant response of green alga *Chlorella vulgaris* (Trebouxiophyceae). *Plant Growth Regulation* 73 (1): 57–66.

Polizzotto, M.L., Harvey, C.F., Li, G. et al. (2006). Solid-phases and desorption processes of arsenic within Bangladesh sediments. *Chemical Geology* 228 (1–3): 97–111.

Poonam, S., Kaur, H., and Geetika, S. (2013). Effect of jasmonic acid on photosynthetic pigments and stress markers in *Cajanus cajan* (L.) Millsp. seedlings under copper stress. *American Journal of Plant Sciences* 4: 817–823.

Pospíšilová, J. (2003a). Interaction of cytokinins and abscisic acid during regulation of stomatal opening in bean leaves. *Photosynthetica* 41 (1): 49–56.

Pospíšilová, J. (2003b). Participation of phytohormones in the stomatal regulation of gas exchange during water stress. *Biologia Plantarum* 46 (4): 491–506.

Quaghebeur, M. and Rengel, Z. (2003). The distribution of arsenate and arsenite in shoots and roots of *Holcus lanatus* is influenced by arsenic tolerance and arsenate and phosphate supply. *Plant Physiology* 132 (3): 1600–1609.

Que, F., Wang, G.L., Xu, Z.S. et al. (2017). Transcriptional regulation of brassinosteroid accumulation during carrot development and the potential role of brassinosteroids in petiole elongation. *Frontiers in Plant Science* 8: 1356.

Raab, A., Schat, H., Meharg, A.A., and Feldmann, J. (2005). Uptake, translocation and transformation of arsenate and arsenite in sunflower (*Helianthus annuus*): formation of arsenic–phytochelatin complexes during exposure to high arsenic concentrations. *The New Phytologist* 168 (3): 551–558.

Raab, A., Williams, P.N., Meharg, A., and Feldmann, J. (2007a). Uptake and translocation of inorganic and methylated arsenic species by plants. *Environment and Chemistry* 4 (3): 197–203.

Raab, A., Wright, S.H., Jaspars, M. et al. (2007b). Pentavalent arsenic can bind to biomolecules. *Angewandte Chemie* 119 (15): 2648–2651.

Rady, M.M. (2011). Effect of 24-epibrassinolide on growth, yield, antioxidant system and cadmium content of bean (*Phaseolus vulgaris* L.) plants under salinity and cadmium stress. *Scientia Horticulturae* 129 (2): 232–237.

Rajewska, I., Talarek, M., and Bajguz, A. (2016). Brassinosteroids and response of plants to heavy metals action. *Frontiers in Plant Science* 7: 629.

Ramakrishna, B. and Rao, S.S.R. (2012). 24-Epibrassinolide alleviated zinc-induced oxidative stress in radish (*Raphanus sativus* L.) seedlings by enhancing antioxidative system. *Plant Growth Regulation* 68 (2): 249–259.

Rausch, C. and Bucher, M. (2002). Molecular mechanisms of phosphate transport in plants. *Planta* 216 (1): 23–37.

Rhine, E.D., Phelps, C.D., and Young, L.Y. (2006). Anaerobic arsenite oxidation by novel denitrifying isolates. *Environmental Microbiology* 8 (5): 899–908.

Rivas-San Vicente, M. and Plasencia, J. (2011). Salicylic acid beyond defence: its role in plant growth and development. *Journal of Experimental Botany* 62 (10): 3321–3338.

Ronzan, M., Piacentini, D., Fattorini, L. et al. (2019). Auxin-jasmonate crosstalk in *Oryza sativa* L. root system formation after cadmium and/or arsenic exposure. *Environmental and Experimental Botany* 165: 59–69.

Rosen, B.P. and Liu, Z. (2009). Transport pathways for arsenic and selenium: a minireview. *Environment International* 35 (3): 512–515.

Saeed, W., Naseem, S., and Ali, Z. (2017). Strigolactones biosynthesis and their role in abiotic stress resilience in plants: a critical review. *Frontiers in Plant Science* 8: 1487.

Schmöger, M.E., Oven, M., and Grill, E. (2000). Detoxification of arsenic by phytochelatins in plants. *Plant Physiology* 122 (3): 793–802.

Schwerdtle, T., Walter, I., and Hartwig, A. (2003). Arsenite and its biomethylated metabolites interfere with the formation and repair of stable BPDE-induced DNA adducts in human cells and impair XPAzf and Fpg. *DNA Repair* 2 (12): 1449–1463.

Shahzad, B., Tanveer, M., Che, Z. et al. (2018). Role of 24-epibrassinolide (EBL) in mediating heavy metal and pesticide induced oxidative stress in plants: a review. *Ecotoxicology and Environmental Safety* 147: 935–944.

Sharma, I., Sharma, A., Pati, P., and Bhardwaj, R. (2018). Brassinosteroids reciprocates heavy metals induced oxidative stress in radish by regulating the expression of key antioxidant enzyme genes. *Brazilian Archives of Biology and Technology* 61.

Shi, Q. and Zhu, Z. (2008). Effects of exogenous salicylic acid on manganese toxicity, element contents and antioxidative system in cucumber. *Environmental and Experimental Botany* 63 (1–3): 317–326.

Shin, H., Shin, H.S., Dewbre, G.R., and Harrison, M.J. (2004). Phosphate transport in *Arabidopsis*: Pht1; 1 and Pht1; 4 play a major role in phosphate acquisition from both low-and high-phosphate environments. *The Plant Journal* 39 (4): 629–642.

Shinohara, N., Taylor, C., and Leyser, O. (2013). Strigolactone can promote or inhibit shoot branching by triggering rapid depletion of the auxin efflux protein PIN1 from the plasma membrane. *PLoS Biology* 11 (1): e1001474.

Singh, N., Ma, L.Q., Srivastava, M., and Rathinasabapathi, B. (2006). Metabolic adaptations to arsenic-induced oxidative stress in *Pteris vittata* L. and *Pteris ensiformis* L. *Plant Science* 170 (2): 274–282.

Singh, R., Parihar, P., and Prasad, S.M. (2018). Sulphur and calcium simultaneously regulates photosynthetic performance and nitrogen metabolism status in As-challenged *Brassica juncea* L. seedlings. *Frontiers in Plant Science* 9: 772.

Singh, R., Parihar, P., and Prasad, S.M. (2020). Interplay of calcium and nitric oxide in improvement of growth and arsenic-induced toxicity in mustard seedlings. *Scientific Reports* 10 (1): 1–12.

Sisr, L., Mihaljevič, M., Ettler, V. et al. (2007). Effect of application of phosphate and organic manure-based fertilizers on arsenic transformation in soil columns. *Environmental Monitoring and Assessment* 135 (1): 465–473.

Skottke, K.R., Yoon, G.M., Kieber, J.J., and DeLong, A. (2011). Protein phosphatase 2A controls ethylene biosynthesis by differentially regulating the turnover of ACC synthase isoforms. *PLoS Genetics* 7 (4): e1001370.

Soares, A.M.D.S., Souza, T.F.D., Jacinto, T., and Machado, O.L.T. (2010). Effect of methyl jasmonate on antioxidative enzyme activities and on the contents of ROS and H_2O_2 in *Ricinus communis* leaves. *Brazilian Journal of Plant Physiology* 22 (3): 151–158.

Sponsel, V.M. and Hedden, P. (2010). Gibberellin biosynthesis and inactivation. In: *Plant Hormones* (ed. P.J. Davies), 63–94. Dordrecht: Springer.

Srivastava, P.K., Vaish, A., Dwivedi, S. et al. (2011). Biological removal of arsenic pollution by soil fungi. *Science of the Total Environment* 409 (12): 2430–2442.

Srivastava, S., Srivastava, A.K., Suprasanna, P., and D'souza, S.F. (2013). Identification and profiling of arsenic stress-induced microRNAs in *Brassica juncea*. *Journal of Experimental Botany* 64 (1): 303–315.

Stoeva, N. and Bineva, T. (2003). Oxidative changes and photosynthesis in oat plants grown in As-contaminated soil. *Bulgarian Journal of Plant Physiology* 29 (1–2): 87–95.

Styblo, M., Del Razo, L.M., Vega, L. et al. (2000). Comparative toxicity of trivalent and pentavalent inorganic and methylated arsenicals in rat and human cells. *Archives of Toxicology* 74 (6): 289–299.

Styblo, M., Serves, S.V., Cullen, W.R., and Thomas, D.J. (1997). Comparative inhibition of yeast glutathione reductase by arsenicals and arsenothiols. *Chemical Research in Toxicology* 10 (1): 27–33.

Sun, H., Bi, Y., Tao, J. et al. (2016). Strigolactones are required for nitric oxide to induce root elongation in response to nitrogen and phosphate deficiencies in rice. *Plant, Cell and Environment* 39 (7): 1473–1484.

Sytar, O., Kumar, A., Latowski, D. et al. (2013). Heavy metal-induced oxidative damage, defense reactions, and detoxification mechanisms in plants. *Acta Physiologiae Plantarum* 35 (4): 985–999.

Sytar, O., Kumari, P., Yadav, S. et al. (2019). Phytohormone priming: regulator for heavy metal stress in plants. *Journal of Plant Growth Regulation* 38 (2): 739–752.

Száková, J., Havlík, J., Valterová, B. et al. (2010). The contents of risk elements, arsenic speciation, and possible interactions of elements and betalains in beetroot (*Beta vulgaris*, L.) growing in contaminated soil. *Open Life Sciences* 5 (5): 692–701.

Takahashi, Y., Minamikawa, R., Hattori, K.H. et al. (2004). Arsenic behavior in paddy fields during the cycle of flooded and non-flooded periods. *Environmental Science & Technology* 38 (4): 1038–1044.

Tassi, E., Pouget, J., Petruzzelli, G., and Barbafieri, M. (2008). The effects of exogenous plant growth regulators in the phytoextraction of heavy metals. *Chemosphere* 71 (1): 66–73.

Ullrich-Eberius, C.I., Sanz, A., and Novacky, A.J. (1989). Evaluation of arsenate-and vanadate-associated changes of electrical membrane potential and phosphate transport in *Lemna gibba* G1. *Journal of Experimental Botany* 119–128.

Ultra, V.U. Jr., Tanaka, S., Sakurai, K., and Iwasaki, K. (2007a). Arbuscular mycorrhizal fungus (*Glomus aggregatum*) influences biotransformation of arsenic in the rhizosphere of sunflower (*Helianthus annuus* L.). *Soil Science and Plant Nutrition* 53 (4): 499–508.

Ultra, V.U., Tanaka, S., Sakurai, K., and Iwasaki, K. (2007b). Effects of arbuscular mycorrhiza and phosphorus application on arsenic toxicity in sunflower (*Helianthus annuus* L.) and on the transformation of arsenic in the rhizosphere. *Plant and Soil* 290 (1): 29–41.

Van Ha, C., Leyva-González, M.A., Osakabe, Y. et al. (2014). Positive regulatory role of strigolactone in plant responses to drought and salt stress. *Proceedings of the National Academy of Sciences of the United States of America* 111 (2): 851–856.

Van Zeijl, A., Liu, W., Xiao, T.T. et al. (2015). The strigolactone biosynthesis gene DWARF27 is co-opted in *Rhizobium* symbiosis. *BMC Plant Biology* 15 (1): 1–15.

Vardhini, B.V. (2014). Brassinosteroids' role for amino acids, peptides and amines modulation in stressed plants – a review. In: *Plant Adaptation to Environmental Change: Significance of Amino Acids and Their Derivatives* (ed. N.A. Anjum, S.S. Gill and R. Gill), 300–316. CAB International.

Vardhini, B.V. (2016). Brassinosteroids are potential ameliorators of heavy metal stresses in plants. In: *Plant Metal Interaction* (ed. P. Ahmad), 209–237. Amsterdam: Elsevier.

Vetterlein, D., Szegedi, K., Ackermann, J. et al. (2007). Competitive mobilization of phosphate and arsenate associated with goethite by root activity. *Journal of Environmental Quality* 36 (6): 1811–1820.

Villanueva-Estrada, R.E., Prol-Ledesma, R.M., Rodríguez-Díaz, A.A. et al. (2013). Arsenic in hot springs of Bahía Concepción, Baja California Peninsula, México. *Chemical Geology* 348: 27–36.

Visoottiviseth, P., Francesconi, K., and Sridokchan, W. (2002). The potential of Thai indigenous plant species for the phytoremediation of arsenic contaminated land. *Environmental Pollution* 118 (3): 453–461.

Wallace, I.S., Choi, W.G., and Roberts, D.M. (2006). The structure, function and regulation of the nodulin 26-like intrinsic protein family of plant aquaglyceroporins. *Biochimica et Biophysica Acta (BBA) – Biomembranes* 1758 (8): 1165–1175.

Wang, S. and Mulligan, C.N. (2006). Occurrence of arsenic contamination in Canada: sources, behavior and distribution. *Science of the Total Environment* 366 (2–3): 701–721.

Wani, S.H., Kumar, V., Shriram, V., and Sah, S.K. (2016). Phytohormones and their metabolic engineering for abiotic stress tolerance in crop plants. *The Crop Journal* 4 (3): 162–176.

Wasternack, C. and Hause, B. (2013). Jasmonates: biosynthesis, perception, signal transduction and action in plant stress response, growth and development. An update to the 2007 review in Annals of Botany. *Annals of Botany* 111 (6): 1021–1058.

Webb, J.L. (1963). *Enzyme and Metabolic Inhibitors*. New York: Academic Press.

Wei, Z. and Li, J. (2016). Brassinosteroids regulate root growth, development, and symbiosis. *Molecular Plant* 9 (1): 86–100.

Wenzel, W.W., Brandstetter, A., Wutte, H. et al. (2002). Arsenic in field-collected soil solutions and extracts of contaminated soils and its implication to soil standards. *Journal of Plant Nutrition and Soil Science* 165 (2): 221–228.

Williams, P.N., Villada, A., Deacon, C. et al. (2007). Greatly enhanced arsenic shoot assimilation in rice leads to elevated grain levels compared to wheat and barley. *Environmental Science & Technology* 41 (19): 6854–6859.

Woolson, E.A., Axley, J.H., and Kearney, P.C. (1971). The chemistry and phytotoxicity of arsenic in soils: I. Contaminated field soils. *Soil Science Society of America Journal* 35 (6): 938–943.

Wysocki, R., Bobrowicz, P., and Ułaszewski, S. (1997). The *Saccharomyces cerevisiae* ACR3 gene encodes a putative membrane protein involved in arsenite transport. *The Journal of Biological Chemistry* 272 (48): 30061–30066.

Xie, X., Yoneyama, K., and Yoneyama, K. (2010). The strigolactone story. *Annual Review of Phytopathology* 48: 93–117.

Xu, B., Yu, J.Y., Xie, T. et al. (2018). Brassinosteroids and iron plaque affect arsenic and cadmium uptake by rice seedlings grown in hydroponic solution. *Biologia Plantarum* 62 (2): 362–368.

Xu, L.L., Fan, Z.Y., Dong, Y.J. et al. (2015a). Effects of exogenous salicylic acid and nitric oxide on physiological characteristics of two peanut cultivars under cadmium stress. *Biologia Plantarum* 59 (1): 171–182.

Xu, X.Y., McGrath, S.P., and Zhao, F.J. (2007). Rapid reduction of arsenate in the medium mediated by plant roots. *The New Phytologist* 176 (3): 590–599.

Xu, X.Y., McGrath, S.P., Meharg, A.A., and Zhao, F.J. (2008). Growing rice aerobically markedly decreases arsenic accumulation. *Environmental Science & Technology* 42 (15): 5574–5579.

Xu, Y.X., Mao, J., Chen, W. et al. (2016). Identification and expression profiling of the auxin response factors (ARFs) in the tea plant (*Camellia sinensis* (L.) O. Kuntze) under various abiotic stresses. *Plant Physiology and Biochemistry* 98: 46–56.

Xu, Z., Jiang, Y., and Zhou, G. (2015b). Response and adaptation of photosynthesis, respiration, and antioxidant systems to elevated CO_2 with environmental stress in plants. *Frontiers in Plant Science* 6: 701.

Yu, M.H., Zhao, Z.Z., and He, J.X. (2018). Brassinosteroid signaling in plant–microbe interactions. *International Journal of Molecular Sciences* 19 (12): 4091.

Zhang, H. and Selim, H.M. (2008). Reaction and transport of arsenic in soils: equilibrium and kinetic modeling. *Advances in Agronomy* 98: 45–115.

Zhang, S., Deng, Y.Z., and Zhang, L.H. (2018). Phytohormones: the chemical language in *Magnaporthe oryzae*-rice pathosystem. *Mycology* 9 (3): 233–237.

Zhao, F.J., Ma, J.F., Meharg, A.A., and McGrath, S.P. (2009). Arsenic uptake and metabolism in plants. *The New Phytologist* 181 (4): 777–794.

Zhou, Q. and Xi, S. (2018). A review on arsenic carcinogenesis: epidemiology, metabolism, genotoxicity and epigenetic changes. *Regulatory Toxicology and Pharmacology* 99: 78–88.

Zhu, X.F., Jiang, T., Wang, Z.W. et al. (2012). Gibberellic acid alleviates cadmium toxicity by reducing nitric oxide accumulation and expression of IRT1 in *Arabidopsis thaliana*. *Journal of Hazardous Materials* 239: 302–307.

Zhu, X.F., Wang, Z.W., Dong, F. et al. (2013). Exogenous auxin alleviates cadmium toxicity in *Arabidopsis thaliana* by stimulating synthesis of hemicellulose 1 and increasing the cadmium fixation capacity of root cell walls. *Journal of Hazardous Materials* 263: 398–403.

Zhu, Y.G., Williams, P.N., and Meharg, A.A. (2008). Exposure to inorganic arsenic from rice: a global health issue? *Environmental Pollution* 154 (2): 169–171.

Zvobgo, G., LwalabaWaLwalaba, J., Sagonda, T. et al. (2018). Phosphate alleviates arsenate toxicity by altering expression of phosphate transporters in the tolerant barley genotypes. *Ecotoxicology and Environmental Safety* 147: 832–839.

Zwanenburg, B. and Pospíšil, T. (2013). Structure and activity of strigolactones: new plant hormones with a rich future. *Molecular Plant* 6 (1): 38–62.

12

Influence of Some Chemicals in Mitigating Arsenic-Induced Toxicity in Plants

Palin Sil and Asok K. Biswas

Plant Physiology and Biochemistry Laboratory, Centre for Advanced Study, Department of Botany, University of Calcutta, Kolkata, West Bengal, India

12.1 Introduction

Extensive anthropogenic activities and/or natural processes have exacerbated environmental pollution causing a persistent setback to the agroecosystems. Heavy metal pollution has become a severe environmental hazard owing to the undesirable effects of these substances on the ecosystem. “Heavy metal” refers to the group of metals and metalloids having an atomic density greater than $4 \pm 1\,\mathrm{g\,cm^{-3}}$ (Hawkes 1997). Though few metals in trace amounts are necessary for various metabolic activities of plants, heavy metal uptake beyond the permissible level is phytotoxic and leads to substantial heavy metal accumulation in the plant body that impedes metabolic processes by displacing or disturbing the protein structure, disrupts the functionality of essential biomolecules, and adversely affects cellular integrity (Nagajyoti et al. 2010). This has severe consequences on crop productivity and lowers grain quality by inducing deleterious effects on several physiological processes from seed germination to growth and yield.

Arsenic belongs to Group VA of the periodic table having an atomic density of $5.7\,\mathrm{g\,cm^{-3}}$. The acclaimed environmental contaminant has been classified as a non-threshold Group I carcinogen by the US Environmental Protection Agency (EPA) and International Agency for Research on Cancer (IARC) (Rosas-Castor et al. 2014; Niazi et al. 2018). Its electro negativity is not sufficient to confer it a metallic character and is therefore, considered a metalloid. Among the wide spectrum of pollutants, this potentially toxic element comprises roughly $5\,\mathrm{mg\,kg^{-1}}$ of earth’s crust and exists in the soil at concentrations varying from 0.1 to $40\,\mathrm{mg\,kg^{-1}}$ with an average of $2\,\mathrm{mg\,kg^{-1}}$ in igneous and sedimentary rocks (Mandal and Suzuki 2002). More than 200 arsenic-containing minerals exist naturally, of which arsenates, arsenites, arsenides, sulfides, sulfosalts, silicates, oxides, elemental arsenic, and metal alloys are common. Entry of the carcinogenic metalloid into the aquatic and terrestrial environments takes place through natural as well as anthropogenic sources (Hettick et al. 2015). Natural phenomena of arsenic entry include weathering, biological, and

Arsenic in Plants: Uptake, Consequences and Remediation Techniques, First Edition.
Edited by Prabhat Kumar Srivastava, Rachana Singh, Parul Parihar, and Sheo Mohan Prasad.

volcanic activities. Oxidation of arsenopyrite bedrock, reductive dissolution of ferric-oxyhydroxides, carbon, and microbial reduction are several theories that have been put forward to elucidate the mechanism and natural causes of arsenic leaching (Dey et al. 2014). Additionally, mining, smelting, milling, combustion of fossil fuels, municipal sludge of land, coal fly ash, additives to livestock feed, and use of arsenic-containing pesticides in agriculture also cause arsenic contamination. According to the World Health Organization (WHO) and the US EPA, the permissible threshold of arsenic in drinking water is $10\,\mu g\,l^{-1}$. Nevertheless, several countries like Argentina, Canada, Chile, Ghana, Vietnam, Laos, Mexico, the United States, China, and the Indian subcontinent contain arsenic in drinking water that exceeds this threshold (Basu et al. 2014). In India, arsenic contamination in groundwater is predominant in regions that include the alluvial terrain (areas of Assam, Manipur, Uttar Pradesh, Bihar, Punjab, Haryana, Jharkhand, and West Bengal) and the hard-rock terrain (areas of Karnataka and Chhattisgarh). The Bengal basin comprising riverine deposits of Bengal Delta Plains (BDP) from the three rivers – Ganges, Meghna, and Brahmaputra – is regarded as the most intensely contaminated geological region in the world where arsenic contents in water have been documented to up to $3200\,\mu g\,l^{-1}$ against the safe limit of $10\,\mu g\,l^{-1}$ (McCarty et al. 2011). Concentrations of arsenic in the Indo-Gangetic basin, areas of Uttar Pradesh, Chhattisgarh, Bihar, Assam and deltaic regions of West Bengal range from 3.34 to $105\,mg\,kg^{-1}$ (Mallick et al. 2011). Adjoining contaminated areas with similar geologic settings in the Indian subcontinent include regions of Bangladesh, Nepal, Pakistan, and Myanmar. Arsenic pollution in both soil and water has become a serious environmental crisis in cultivable lands where the metalloid concentrates in consumable portions of plants and ultimately enters the human body (Farooq et al. 2016b).

Arsenic level in non-contaminated soil fluctuates from 0.1 to $40\,mg\,kg^{-1}$ with an average of $7.2\,mg\,kg^{-1}$ but symptoms of arsenic-induced phytotoxicity are manifested beyond $20\,mg\,kg^{-1}$(Srivastava et al. 2013). Predominant categories of this environmental contaminant include arsenate [As (V)], arsenite [As (III)], monomethylarsonic acid (MMA), dimethylarsenic acid (DMA), trimethylarsine oxide (TMAO), arsenobetaine (AsB), arsenocholine (AsC), arsenosugars (AsS), arsenolipids, etc. (Singh et al. 2015a). Physiological and electrophysiological studies have demonstrated similar uptake pathways for both As (V) and phosphate (Pi) in higher plants as As (V) and Pi are structurally analogous but the transporters have a higher affinity for Pi in comparison to As (V) and involve cotransport of Pi or As (V) and protons (Ullrich-Eberius et al. 1989; Meharg and Mc Nair 1994). However, members of the nodulin 26-like intrinsic protein (NIPs) or aquaporins mediate the influx of arsenite faster than arsenate through plant roots (Li et al. 2014; Mukhopadhyay et al. 2014). Uptake of methylated As species, viz. MMA and DMA, occurs with lesser efficiency in comparison to arsenate or arsenite (Zheng et al. 2013).

Phytotoxicity of arsenic is evident from morphological to molecular level and depends on the availability of arsenic in soil and the chemical species present in the plant (Figure 12.1). A decrease in germination and seedling growth under arsenic toxicity has been documented in *Oryza sativa*, *Cicer arietinum*, and *Spinacea oleracea* (Abedin et al. 2002; Choudhury et al. 2011; Malik et al. 2011; Srivastava and Sharma 2013b). Arsenic accumulation in plants predominantly occurs in roots but to a lesser extent in the aboveground parts upsets plant metabolism, reduces growth, hinders crucial biochemical as well as metabolic

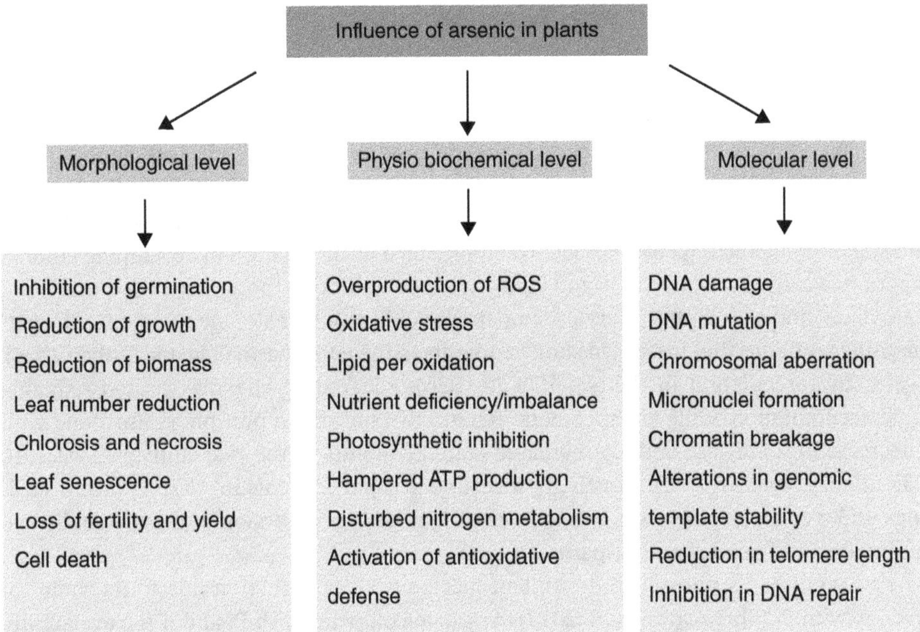

Figure 12.1 Schematic representation of arsenic-induced effects in plants.

processes, and compromises plant reproductive capacity through losses in productivity eventually resulting in plant death (Finnegan and Chen 2012). Additionally, nutrient uptake, absorption, and transport are impeded (Sil et al. 2019b). Transformation of As (V) to As (III) inside the cell generates reactive oxygen species (ROS) that results in oxidative stress and increases malondialdehyde (MDA) as well as H_2O_2 levels leading to damage in cellular membranes. The high affinity of As (III) for the sulfhydryl groups of enzymes disrupts protein structure and affects vital processes that deplete cellular glutathione levels making the plants susceptible to ROS-induced damages (Heidari 2017). The generation of oxidative stress upregulates the antioxidant defense mechanisms that include non enzymatic and enzymatic antioxidants. The non enzymatic antioxidants, viz. carotenoids, ascorbic acid, tocopherol, and glutathione along with the antioxidative enzymes, particularly superoxide dismutase (SOD), catalase (CAT), ascorbate peroxidase (APX), and glutathione peroxidase (GPx), have a decisive function in H_2O_2 degradation (Singh et al. 2016). Reduction of arsenate to arsenite by arsenate reductase in a glutathione-dependent reaction followed by complexation of As (III) with phytochelatins (PCs) as As (III)–PC complexes in vacuoles constitutes the strategy for detoxification of arsenic (Li et al. 2015).

Reduced chlorophyll synthesis or increased chlorophyll destruction under arsenic toxic conditions disorganizes the sub-microstructure of the chloroplast membrane system that drastically affects photosynthesis as well as transpiration (Nagajyoti et al. 2010; Farnese et al. 2014; Sil et al. 2019b). This hinders the activity of ribulose-1, 5-bisphosphate carboxylase/oxygenase (RuBisCO) and eventually reduces CO_2 fixation. A significant decrease in pigment contents under high arsenic levels has been attributed to affect carbon dioxide fixation in *Zea mays* and *Trifolium pratense* seedlings (Emamverdian et al. 2015;

Hasanuzzaman et al. 2017). Lettuce plants exposed to As (V) and As (III) had lower net photosynthesis rate, stomatal conductance, and internal CO_2 concentration (Gusman et al. 2013). An increase in minimal fluorescence (F_0) and a decrease in maximum variable chlorophyll fluorescence (F_v) under arsenic exposure were reported in maize. The $F_v : F_0$ and $F_v : F_m$ ratios decreased considerably that reduced the photochemical efficiency of PSII (Stoeva et al. 2003).

Arsenic stress also affects carbohydrate metabolism. Enhancement in activities of sugar and starch-metabolizing enzymes led to soluble sugar accumulation in developing embryo axes of rice cultivars, Malviya-36 and Pant-12, but inhibition in the activities of α-amylase-, β-amylase-, and starch phosphorylase impaired mobilization of storage reserves to the germinating seedlings that increased starch contents in the endosperm (Jha and Dubey 2005). Again, an upregulation in the activities of sucrose-degrading enzymes caused reducing sugar accumulation while enhancement in the activity of starch phosphorylase along with a decrease in α-amylase activity increased starch contents in the rice cultivar MTU 1010 (Choudhury et al. 2011). Contrastingly, a decline in sugar contents in *Albizia procera* seedlings under arsenate stress was due to lowered synthesis or diversion of the metabolites to other pathways (Pandey and Tripathi 2011).

Arsenate disrupts phosphate-dependent facets of ATP production due to its chemical analogy with Pi (Shrivastava et al. 2015). Arsenate competes with Pi and if the availability of arsenate exceeds that of Pi, arsenate is transported to the cell where it forms unstable ADP-As (V) adducts. This sets up an ineffectual cycle of uncoupling oxidative phosphorylation and photophosphorylation that disrupts energy flow in cells (Tawfik and Viola 2011; Shrivastava et al. 2015). As (V) also displaces phosphate from crucial biomolecules while As (III) binds with sulfhydryl groups of respiratory cycle enzymes and impairs tissue respiration as well as energy production in cells (Ullrich-Eberius et al. 1989). The binding of As (III) to the lipoic acid moiety of pyruvate dehydrogenase (PDH) and oxoglutarate dehydrogenase (OGDH) complexes inhibits the activity of these enzymes that deplete the mitochondrial NADH pool and produces ROS to generate oxidative stress (Shen et al. 2013; Sil et al. 2018). Such inhibition can block the TCA cycle and, ultimately, decrease ATP production resulting in cellular damage and death.

Arsenate stress also hampers the activities of enzymes associated with nitrogen and ammonium assimilatory pathways that perturb nitrogen metabolism. Reductions in levels of nitrate, nitrite, and amino acid occurred along with ammonia accumulation in wheat seedlings (Sil et al. 2019b). Additionally, inhibitions of nitrate reductase (NR) and nitrite reductase (NR and NiR) activities affected the total protein contents that reduced growth in rice and wheat (Singh et al. 2009a; Sil et al. 2019b).

Arsenic contamination, therefore, lowers crop quality as well as productivity. Consumption of cereals and vegetables grown in arsenic-polluted soils irrigated with arsenic-polluted water forms a route of arsenic exposure and has led to the incidence of arsenic-related disorders in more than 200 million people worldwide that include about 38 million people residing in West Bengal (India) and Bangladesh (Flora 2011). Accumulation of arsenic is alarming as toxicological concerns of the metalloid are a menace to human health owing to its adverse effects on food safety and demand urgent solutions (Mishra et al. 2014). This necessitates the need for immediate strategies to tackle the emerging issues of arsenic stress on crop plants. Reducing arsenic uptake in crops would be an efficient approach to mitigate the perils of diet-based arsenic poisoning. Limiting the amount of arsenic entry into the plant and directing its deposition to tissues having no dietary value can possibly reduce the ingression of the toxicant into the food chain. The way out in the prevailing crisis is environmental remediation or production of arsenic-tolerant crops. Different technologies for *in situ* remediations exist, which include immobilization, phytostabilization, phytoextraction, physical separation, extraction, etc. These traditional methods are, however, time-consuming, expensive and may have negative impacts on ecosystems (McGrath et al. 2001). Soil remediation via physical and chemical techniques appears to be unfeasible due to the wide expanse of pollution. However, remediation utilizing biological systems appears to be attainable. In the present scenario, therefore, quick, inexpensive, and environment-friendly techniques seem to be the imperative requisite. A reasonable option would be to utilize chemicals that would curtail the uptake of arsenic and lower arsenic toxicity. Exogenous supplies of proline (Singh et al. 2015b), phytohormones (Chandrakar et al. 2016a; Kaya et al. 2020), NO (Singh et al. 2016), phosphate (Choudhury et al. 2011; Saha et al. 2017b), selenate (Pandey and Gupta 2018), silicate (Sil and Biswas 2020), and calcium (Singh et al. 2020) have been documented to protect crop plants against arsenic. These chemicals diminish arsenic toxicity by limiting arsenic uptake and control membrane damage. Utilization of these substances alters the expression of stress-related genes to boost up the antioxidative defense that quenches metal-incited ROS assembly in conjunction with lipid peroxidation and facilitates growth (Abbas et al. 2018). Considering the outcomes of As pollution on plant productivity worldwide, it becomes necessary to minimize the dilemma. This article compiles aspects of the possible ameliorative effects of certain exogenous supplies under As stress. The protective role of these chemicals would help combat the effects of arsenic in plants and improve productivity.

12.2 Role of Phosphorus

Phosphorus, the essential nutrient is an important component in the cellular metabolism of plants. It modulates gene transcription and enzyme activity; consequently, its storage, assimilation as well as metabolism are significant in the growth and development of plants. The energy-rich phosphate bonds have a vital role in the energy transfer of numerous biological reactions (Pessarakli et al. 2015). Belonging to group VA of the periodic table, phosphorus has a similar electronic configuration and chemical properties as arsenic and competes for the same uptake pathway (Ullrich-Eberius et al. 1989; Meharg and MacNair 1994; Meharg and Hartley-Whitaker 2002). Being the analogue of Pi, arsenate competes and replaces Pi in ATP forming unstable ADP–As complexes inside the cytoplasm. This uncouples ATP synthesis from electron transport and hampers cellular ATP production, which disrupts energy flow in cells and retards growth (Shrivastava et al. 2015). However, according to Meharg (2005), high cellular phosphorus status in plants downregulates the arsenate/Pi plasmalemma transporters and competes to replace arsenate during uptake as both ions use similar uptake channels for transport, thereby reducing the effects of arsenic toxicity.

Studies have elucidated the beneficial effects of phosphate administration under arsenic stress in several plant species (Pigna et al. 2009; Choudhury et al. 2011; Srivastava and Sharma 2014; Saha et al. 2017b; Zvobgo et al. 2019). Phosphate administration significantly modulated the arsenate-induced phytotoxic effects on growth in rice, black gram, and barley cultivars along with an improvement in water contents to an appreciable extent (Choudhury et al. 2011; Srivastava and Sharma 2014; Saha et al. 2017b; Zvobgo et al. 2019). The anatomical damages to the root tissues were repaired by phosphate application as well. Again, Pigna et al. (2009) reported that phosphate treatment in irrigation water containing arsenic reinstated the As-induced inhibition of growth in wheat (*Triticum aestivum*) by reducing As uptake. Enhanced availability of plant-available Pi in rhizosphere due to phosphate treatments increased the cytoplasmic phosphate concentration that eventually downregulated the arsenate–phosphate plasma membrane transporter and lowered arsenic uptake. Addition of phosphate under $100\,\mu mol\,l^{-1}$ As also reduced As uptake and significantly altered the antioxidative defense mechanisms as well as the relative expression of genes in two barley (*Hordeum vulgare*) genotypes, viz. As-tolerant ZDB160 and As-sensitive ZDB475 (Zvobgo et al. 2019). Additionally, phosphate addition significantly suppressed arsenate uptake in FSMY and ZY90 genotypes of tobacco (*Nicotiana tabacum*) differing in arsenate tolerance. Greater phosphate uptake reduced the influx of arsenate, and improved nutrient uptake in plants (Zvobgo et al. 2015). Suppressed arsenate accumulation due to phosphate treatments altered cellular pH, decreased the activity of arsenate reductase, and limited conversion of arsenate to arsenite, thereby lowering ROS production. Enhanced accrual of oxidative stress markers, viz. proline, H_2O_2, and MDA under arsenate stress also declined with phosphate treatments along with differential alterations in the antioxidative enzyme activities, viz. CAT, SOD, and APX. Damages to the photosynthetic machinery could be repaired with phosphate nutrition in the arsenate-challenged plants (Choudhury et al. 2011; Srivastava and Sharma 2014; Zvobgo et al. 2015; Niazi et al. 2017; Saha et al. 2017b). The pigment contents along with Hill activity improved with Pi addition in the arsenate-contaminated growth medium that enhanced plant biomass. Reducing and nonreducing sugar contents decreased and starch contents were increased in the arsenate-challenged rice cultivars with phosphate treatments (Choudhury et al. 2011). The activities of sugar-metabolizing enzymes were also positively modulated with phosphate to differential extents. The activities of the TCA cycle enzymes increased considerably and the accumulation of TCA cycle intermediates were lowered with phosphate treatments (Saha et al. 2017a). The ascorbate–glutathione cycle was also altered along with the activities of glutathione-metabolizing enzymes in arsenate-treated rice cultivars with phosphate treatments. The joint application of phosphate along with arsenate decreased glutathione and ascorbate content and partially helped overcome the toxic effects of arsenic (Saha et al. 2017b). A decrease in glutathione reductase (GR) and GPX activities along with an escalation in the activity of glutathione-*S* transferase (GST) decreased oxidative stress in the rice cultivars that restored normal growth. The joint application of phosphate with arsenate also reduced arsenate uptake that lowered oxidative stress leading to a decline in PC production. Contrastingly, phosphate treatments upregulated the activity of arsenate reductase in wheat that subsequently reduced arsenate, increased arsenite levels, and sequestered it to vacuole as As (III)–PC complexes forming a crucial mechanism of As tolerance (Pigna et al. 2009). Phosphate nutrition, therefore, increased Pi uptake and downregulated the arsenate/Pi transporters resulting in high cellular Pi levels and improved arsenate tolerance. Phosphate treatments further reduced As-imposed ROS that decreased lipid peroxidation and, ultimately, lowered tissue damage. Application of phosphate lowered oxidative stress and improved plant growth in comparison to As treatments. Thus, enhanced P accumulation with phosphate application improved arsenate tolerance and helped combat the effects of arsenic stress in plants. Phosphorus fertilization is, therefore, competent to trim down the outcomes of arsenic toxicity without enhancing arsenic levels in the shoot.

12.3 Role of Nitric Oxide

The protective role of the gaseous signaling molecule nitric oxide (NO) against arsenic toxicity has been well investigated (Singh et al. 2009a; Hasanuzzaman and Fujita 2013; Singh et al. 2016). Exogenous NO donors attenuate oxidative stress by mediating the reduction of ROS in plants and confer tolerance by decreasing metal accumulation. Sodium nitroprusside (SNP, an NO donor) treatment (100 μM) to arsenic exposed hydroponically grown rice seedlings' improved growth and chlorophyll contents, abated oxidative stress by limiting arsenic accumulation in root as well as shoot (Singh et al. 2016). Exogenous SNP treatments enhanced the antioxidant defense and glyoxalase system in wheat that rendered tolerance to As-induced injuries (Hasanuzzaman and Fujita 2013). SNP supplementation also enhanced the relative water content (RWC), chlorophyll, proline, acsorbate (ASA), and GSH/GSSG ratio. The activities of CAT, GPX, GR, MDHAR (monodehydroascorbate reductase), and DHAR (dehydroascorbate reductase) were also increased, which helped in the effective regeneration of ASA in wheat seedlings subjected to As stress. Substantive reduction in MDA and H_2O_2 levels was noted as NO has been documented to quench ROS and modulate redox regulation by limiting oxidative injury and cell death. SNP treatments further enhanced the activities of glyoxalase enzymes, viz. glyoxalase I (Gly I) and (glyoxalase II) Gly II under all As regimes indicating efficient detoxification of methyl glyoxal (Hasanuzzaman and Fujita 2013). NO supplementation under arsenate exposure in rice also decreased the expression levels of OsLsi1 and OsLsi2 genes that subsequently reduced translocation of As to shoot (Singh et al. 2016). Downregulation of OsLsi1 and OsLsi2 along with activation of ABC transporters not only reduced As accumulation in the root but also improved vacuolar sequestration of As (III)–PC complexes and lessened As transport to the shoot. Decreased As accumulation in shoot indicated lesser accumulation of the toxicant in grains, thereby lowering As toxicity in the food chain. NO-mediated antioxidant action was further justified by the reduction in ROS profile along with H_2O_2 and MDA while the endogenous NO contents were enhanced. The decline in SOD activity along with increments in APX, CAT, and GPX activities are the NO-regulated mechanisms of antioxidative defense in wheat and rice (Sun et al. 2014; Singh et al. 2016). The profound impact of exogenous NO on the expression levels of OsYSL2, OsFRDL1, OsIRT1, and OsIRO2 genes associated with iron transport reverted iron deficiency induced by As (V) in rice. As-induced reduction in GSH contents and GSH/GSSG ratio were resumed by NO supplementation while the level of PCs declined. The activity of NR in As (V)-exposed and NO-supplemented plants was diminished, which increased the nitrite level. Enhanced NR activity possibly indicated a higher synthesis of antioxidant NO that was consumed during neutralization of free radicals produced under As (V) excess.

12.4 Role of Hydrogen Sulfide

The novel gaseous signal molecule hydrogen sulfide (H_2S) at low cellular concentration promotes plant growth and development. It also influences numerous physiological processes that involve seed germination, root organogenesis, stomatal movement, photosynthesis, and fruit ripening, and regulates the advent of senescence (Corpas and Palma 2020). H_2S is a part of the plant sulfur metabolism that regulates redox interactions and protects the stress-sensitive proteins from ROS-mediated oxidation by stimulating the persulfidation of cysteine (Cys) residues. H_2S, a potent inducer of osmolytes and antioxidants, also interacts with other secondary messengers like Ca^{2+}, H_2O_2, and NO to stimulate plant growth. However, the beneficial effects of H_2S are evident only at optimum concentrations. Information garnered through research indicates that the application of H_2S using different donors fortifies plants against diverse stress responses that include drought, salinity, hypoxia, heavy metal toxicity, and extremes of temperatures principally by inducing the antioxidative defense (Banerjee et al. 2018). The application of exogenous H_2S enhances the antioxidant machinery at both gene and protein levels by triggering the accumulation of non enzymatic and enzymatic antioxidants to quench ROS. The potential priming agent, sodium hydrosulfide (NaHS), is considered an H_2S donor since it elicits endogenous H_2S accrual. Hydroponic experiments with pea seedlings subjected to As (V) stress revealed improvement in growth, photosynthesis, and nitrogen content, with NaHS treatment that coordinated with enhancement in H_2S and NO contents. ROS-mediated damages to macromolecules, viz. lipids, proteins and membranes were counterbalanced with NaHS application as well. Lower MDA accumulation in NaHS-supplemented plants repaired damages to the cellular membranes while redox states of glutathione and ascorbate were restored. NaHS supplementation escalated the activities of enzymes connected with the ascorbate-glutathione cycle (AsA-GSH cycle) and was competent to alleviate the As-mediated injuries by maintaining a high endogenous Cys level that possibly formed PCs and scavenged arsenic (Singh et al. 2015c).

12.5 Role of Calcium

Calcium has vital functions in mediating diverse aspects of growth and metabolism in plants. It is considered to be the second messenger in the signal transduction network under both nonstressed as well as stressed conditions (Singh et al. 2020). Numerous investigations have elucidated the involvement of Ca^{2+} in regulating the activities of enzymes associated with photosynthetic electron transport, Calvin cycle, and antioxidative defense (Ahmad et al. 2015; Singh et al. 2018a). The essential macronutrient functions in preserving the integrity of cell wall and cell membranes by activating the plasma membrane ATPase that pumps back nutrients lost under stress (Palta 2000). Calcium also activates calmodulin to control metabolic activities and improves plant growth under environmental stress (Marschner 1995). The function of exogenous Ca in mitigating stresses imposed by metals has been extensively investigated (Alam et al. 2006; Ahmad et al. 2015; Singh et al. 2018b; Singh et al. 2020). Exposure to calcium reduced uptake of arsenic and modulated the antioxidative as well as glyoxalase systems in arsenate-stressed rice seedlings

(Rahman et al. 2015). Hydroponically grown rice seedlings (cv. BRRI dhan29) subjected to sodium arsenate treatment (0.5–1 mM concentrations) showed suppressed growth along with chlorosis. Enhanced As accumulation resulted in overproduction of ROS that disrupted the antioxidant as well as glyoxalase systems. Exogenous supply of Ca (as 10 mM calcium chloride) improved tolerance of the seedlings to As stress and restored plant growth by decreasing arsenic uptake as well as accumulation. Calcium supplementation also reduced ROS production by enhancing the antioxidant defense and glyoxalase systems. Enhancement in AsA contents and upregulation in the activities of MDHAR, DHAR, CAT, GPX, and SOD along with Gly I and Gly II enzymes effectively counteracted the As-induced negative impacts and improved growth. Calcium singly or in combination with other chemicals has also been demonstrated to provide cumulative effects in orchestrating plant responses against arsenate toxic conditions (Singh et al. 2018a, 2018b, 2020). Exogenous sulfur and calcium individually as well as in combination ameliorated the inhibitory effect of As by improving biomass accumulation, nutrient uptake, photosynthetic pigment contents, antioxidative defense in *Brassica juncea* seedlings and markedly reduced the negative impact of As (Singh et al. 2018a). Supplemental S and Ca and their combinatorial application further reduced $O_2^{\cdot -}$, H_2O_2, and MDA contents as well as electrolyte leakage and augmented the photosynthetic performance that improved protein and carbohydrate contents. The activities of nitrate-assimilating enzymes, viz. NR and NiR as well as the ammonia-assimilating enzymes, viz. glutamine synthetase (GS) and glutamate synthase (GOGAT) were favorably regulated to sustain nitrogen metabolism under As-challenged conditions (Singh et al. 2018b). Again, the interactive effect of calcium with NO revived the As-induced damages in phenotypic appearance in *Brassica* seedlings (Singh et al. 2020). Activities of antioxidant enzymes were triggered to counteract As-induced oxidative stress that declined the accumulation of ROS. Ca^{2+} and NO treatments also rescued the *Brassica* seedlings from the As-induced losses in photosynthetic pigments leading to the recovery in photosynthetic rate and restored structural attributes of PS II. As-induced losses in ascorbate and glutathione contents were improved and syntheses of thiol compounds were upregulated to maintain the cellular redox activity. Enhanced production of thiol compounds promoted peptides and proteins to chelate the metalloid.

12.6 Role of Proline

The proteinogenic amino acid proline is considered to be a compatible osmolyte with multiple functions in stress adaptation, recovery, and signaling. It is an osmoprotectant that scavenges ROS, stabilizes cellular proteins and membranes, and shields plants from the detrimental outcomes of stress (Ashraf and Foolad 2007). It also acts as a metal chelator and induces the production of PCs necessary for the detoxification of heavy metals. Furthermore, it activates diverse signaling processes in plants and regulates growth (Yang et al. 2009). Accumulation of endogenous proline under heavy metal stress in plants provides protection against ROS-induced damages and helps survive stressful conditions (Chandrakar et al. 2016b). Proline buildup in rice genotypes under arsenate and arsenite exposure helps counteract water-deficit stress (Dave et al. 2013). Proline has been documented to reduce As uptake by modifying cell wall structure and protecting cellular membranes from ROS-induced damages. It also directly quenches As-induced ROS generation by boosting up the antioxidative defense (Shahbaz et al. 2013). Expressions of stress-related genes are also modified by proline as well (Hayat et al. 2012). Proline application in *Solanum melongena* L. (eggplant) seedlings has been ascribed to effectively regulating the osmotic potential of cells that helped sustain growth under arsenic excess by reducing As accumulation (Singh et al. 2015d). Proline treatment mitigated the undesirable effects induced by As on photochemistry of photosystem II. The fluorescence parameters, viz. quantum yield of electron transport (φE0), yield of electron transport per trapped excitation (Ψ0), performance index of PSII (PIABS), and quantum yield of primary photochemistry (φP0) were stimulated in the proline-treated arsenate-stressed seedlings while the energy flux parameters including energy dissipation per active reaction center (DI0/RC), trapped energy per active reaction center (TR0/RC), electron transport per active reaction center (ET0/RC), absorption per active reaction center (ABS/RC), non-photochemical quenching (qP) and photochemical quenching were inhibited. The oxidative stress markers were significantly diminished while the activities of enzymatic antioxidants improved with proline treatments. Downregulation in proline dehydrogenase activity together with an upregulation in Δ1-pyrroline-5-carboxylate synthetase activity led to proline accumulation in the proline-treated seedlings that decreased accrual of arsenic, boosted up the anti-oxidative defense, and competently controlled oxidative stress resulting in better growth of eggplant seedlings under arsenate stress (Singh et al. 2015d).

12.7 Role of Phytohormones

Phytohormones are endogenous organic molecules having significant functions in growth and developmental processes and assist the plants to acclimate to diverse stresses. Plant hormones that include auxins (IAA), gibberellins (GA), cytokinins (CK), ethylene (ET), abscisic acid (ABA), brassinosteroids (BR), jasmonates (JA), and peptides in optimal concentrations mediate growth and reproduction by modulating enzyme activity, cell membrane permeability, and production of secondary metabolites (Wani et al. 2016). Exogenous supply of phytohormones decreases ROS levels along with lipid peroxidation, enhances anti-oxidative defense, stimulates cell division as well as photosynthesis and nitrogen metabolism, thereby improving plant growth (Sytar et al. 2019). These chemical messengers help plants sustain growth and development under abiotic as well as biotic stresses and can be safely administered to enhance tolerance under stress imposed by heavy metals (Huang et al. 2012; Srivastava et al. 2013; Singh et al. 2015a; Masood et al. 2016 Bücker-Neto et al. 2017).

Auxin stimulates plant development in response to light and gravity (Zhao 2010). Auxins also regulate plant metabolism under stress and disruption in its balance under stress has been connected with a reduction in growth (Fahad et al. 2015). A decrease in endogenous auxin contents under arsenic stress in *B. juncea* has been reported to decrease growth by altering the expression of approximately 69 microRNAs belonging to 18 plant miRNA families (Srivastava et al. 2013). Supplemental IAA under As stress, on the other hand, restored growth by amending the expression of miR854, miR319, and miR167 indicating the protective influence of IAA in As tolerance. According to Krishnamurthy and

Rathinasabapathi (2013), the sensitivity of auxin transporter mutants aux1, pin1, and pin2 to As (III) was greater in comparison to the wild type (WT) while auxin transport inhibitors lowered tolerance to arsenic due to high levels of H_2O_2. Supply of IAA developed As (III) tolerance of aux1 indicating the efficacy of auxin transport in tolerating As (III) stress.

The GAs are tetracyclic diterpenoid carboxylic acid compounds that promote physiological processes like flower initiation, fruit development, seed germination, leaf expansion, stem elongation, and assist plants to acclimatize in several stresses including those induced by heavy metals (HMs) (Sytar et al. 2019). HM stress reduces CK production and transport in plants. Exogenous application of CKs has been reported to stimulate antioxidant capacity and cell division in plants (Tassi et al. 2008). CK application also enhanced photosynthesis and influenced transpiration rate leading to an increase in plant biomass (Cassina et al. 2011; Piotrowska-Niczyporuk et al. 2012). Arsenate-imposed effects could be mitigated by pretreating mungbean seeds with GA3 (50 μM) and kinetin (10 μM) (Swarnakar 2017). Improvement in photosynthetic pigment contents in both the pretreated sets indicated the revival of damage to non cyclic electron transport by phytohormone treatments. Photolysis and Hill reaction were restored and the rate of electron flow from excited chlorophyll molecules to the electron acceptor dye DCPIP increased. Application of both GA3 and kinetin also revived sucrose metabolism to a considerable extent which promoted seedling growth.

SA, a phenolic substance and an endogenous natural signal molecule has been recognized to regulate both physiological and biochemical processes including growth, photosynthesis, nitrogen metabolism, and production of ethylene. The multifunctional growth regulator has a significant role in the defense responses of plants and is crucial in the alleviation of heavy metal-induced toxicity (Odjegba 2012; Mostofa and Fujita 2013). Research indicates that SA-mediated alleviation of oxidative damage is associated with an upregulation of antioxidative defense (Parkhey et al. 2014). SA-mediated accretion in H_2O_2 levels serves as a secondary messenger to heighten plant defense against stress (Guo et al. 2007). Exogenous SA (500 μM) remarkably boosted growth indices, germination percentage, activities of enzymatic antioxidants, and accumulation of proline and lowered As accumulation, ROS as well as LOX in As-stressed *Glycine max* seedlings (Chandrakar et al. 2016a). Both co-treatment and pretreatment of SA in conjunction with As (V) efficiently lowered the As (V) accumulation in rice (*Oryza sativa* L.), although co-application was more effective (Singh et al. 2015a). Application of SA (100 μM) restored As (V)-imposed plant growth and biomass production by declining translocation of As to shoots. Growth restoration by SA could be auxin-mediated or due to reduced As accumulation in the shoot. SA-mediated enhancement in endogenous NO increased the bioavailability of Fe and increased photosynthetic pigments that reverted As (V)-induced chlorosis in rice. Additionally, downregulation of OsLsi2 gene, under SA supplementation, reduced the uptake of As. SA supplements (1 mM) under As stress also reduced lipid peroxidation and increased chlorophyll and protein concentrations in wheat (Zengin 2015). SA supplementation effectively improved the endogenous NO contents in maize that reverted the effects of As (V) stress (Kaya et al. 2020). Application of SA (0.5 mM) + SNP (0.1 mM) jointly as foliar spray to the As (V)-treated maize plants lowered As content, boosted antioxidative defense along with photosynthetic traits, and was efficient in reverting the oxidative stress-related parameters to refurbish plant growth. Additionally, an upsurge in the enzyme activities of the ascorbate-glutathione

(AsA-GSH) cycle as well as glyoxalase cycle was noted. Application of NO scavenger, cPTIO blocked synthesis of NO and obliterated the beneficial effects of SA indicating the participation of NO in regulating tolerance (Kaya et al. 2020). Although SA treatment led to a slight accumulation of H_2O_2 but at higher concentrations of As (V), SA promoted the activities of antioxidant enzymes while H_2O_2 and MDA levels subsided. SA treatment also increased ascorbate levels as well as GSH/GSSG ratio to improve the cellular redox under As(V) stress in rice (Singh et al. 2015d). SA co-treatment with As further reduced PC accumulation with respect to the As (V)-challenged plants. The expression levels of Fe transporters (*OsNRAMP5* and *OsFRDL1*) hampered by As (V) treatments were restored by SA treatment that increased Fe accumulation and translocation in rice root and shoot.

Jasmonic acid (JA) and methyl jasmonate (MJ) are naturally occurring cyclopentanone compounds synthesized from linolenic acid that influence diverse physio-biochemical processes in plants. JAs are important in the signaling network of plants under several stresses (Yan et al. 2015). High concentrations of exogenous JS can inhibit growth, photosynthesis, and speed up senescence but its low concentration can increase plant resilience to environmental stresses by upregulating the antioxidant metabolism, osmolyte synthesis, and metabolite accumulation (Farooq et al. 2016a; Tam Ho et al. 2020). JAs alone or in coordination with other plant hormones has been known to counter stresses induced by cadmium, copper, lead, and arsenic (Yan et al. 2015; Hanaka et al. 2015; Bali et al. 2019; Mousavi et al. 2020).

The interactive effects of three doses of MJ (0, 0.1, and 1 μM) with two As levels (0 and 200 μM) reduced As uptake and caused significant recovery in growth as well as biomass in two As (V)-stressed *Brassica napus* cultivars; ZS 758 and Zheda 622 (Farooq et al. 2016a). Exogenous MJ improved the chlorophyll fluorescence yield (F_v/F_m) to sustain photosynthesis under stress. Decreased levels of ROS and lipid peroxidation due to MJ treatment lowered oxidative damage and protected the cell membrane. MJ-mediated enhancement in enzyme transcript levels triggered the antioxidative defense by altering transcription or translation of specific defense genes and scavenged ROS to attenuate the oxidative damage inducing tolerance to As stress in *B. napus* and *O. sativa* (Farooq et al. 2016a; Verma et al. 2020). The activities of enzymes related to the AsA–GSH cycle were enhanced along with improvement in cellular redox by increasing GR, DHAR, and MDHAR transcript levels. The activities of secondary metabolite-related enzymes, viz. cinnamyl alcohol dehydrogenase (CAD), polyphenol peroxidase (PPO), and phenylalanine ammonia-lyase (PAL) were further induced along with their relative mRNA levels to stimulate the phenylpropanoid pathway necessary for removal of ROS and minimizing oxidative stress in *B. napus* cultivars. Exogenous MJ also increased the expression of *LOX* that increased endogenous JA level and regulated inducible defensive genes necessary for mitigation of stress (Farooq et al. 2016b). Co-application of MJ along with As (III) altered the gene expressions in the downstream JA signaling pathway, As (III) uptake, translocation, and detoxification, to endure As (III) stress in *O. sativa* (Verma et al. 2020). MJ supplementation also improved biochemical and molecular traits as well as yield in two rice varieties (T. hashemi and Fajr) under arsenic toxicity (Mousavi et al., 2020). MJ elicited a decline in expression of arsenic transporter genes, *viz. Lsi1*, *Lsi2*, and *Lsi6* and upregulated the activities of antioxidant enzymes along with ASA–GSH cycle enzymes in the mentioned varieties that enhanced plant biomass as well as photosynthetic pigments, lowered As accumulation, and, consequently, improved growth. Supplemental MJ also reversed the As-induced reduction in Fe accumulation by enhancing the expression of Fe transporters under As toxicity. MJ application, therefore, improved the As-induced decrease in yield and yield components in the rice cultivars by improving the antioxidant enzymatic system to modulate ROS homeostasis and strengthened tolerance against As stress.

Phytohormone priming in As-contaminated areas can, therefore, definitely stimulate the antioxidative defense, ASA–GSH cycle, photosynthesis, nitrogen metabolism as well as cell division, which would consequently attenuate the detrimental effects of As stress and improve plant growth.

12.8 Role of Selenium

The essential micronutrient selenium has a significant role in antioxidative defense, uptake as well as accumulation of minerals, and regulates metabolism in animals and humans (Sharma et al. 2016). Being the catalytic center of different selenoproteins, viz. iodothyronine-deiodinase glutathione peroxidase (GSHPx) and thioredoxin reductase, it scavenges free radicals, protects against oxidative stresses and strengthens the immune system (Gupta and Gupta 2017). Additionally, selenium has immunity-boosting, anticancerous, and anti-ageing effects in humans. Selenium biofortification optimizes yield and quality by reducing heavy metal accumulation in crops, thereby providing health benefits to the consumers (Sharma et al. 2016). A number of studies have elucidated the effects of Se supplementation in hindering the absorption of heavy metals that comprise As, cadmium (Cd), and lead (Pb) (Malik et al. 2012; Ghosh et al. 2013; Wu et al. 2016; Sun et al. 2017). The importance of Se application on As uptake, antioxidative defense, and transcript dynamics in rice cultivars divulged antagonism between Se and As (Malik et al. 2012; Pandey et al. 2015a; Pandey and Gupta 2018). The interactive effects of As and Se altered nutrient transporters and thiol-related genes in genotypes of rice (*O. sativa* L.), Pusa Basmati1, and IR64 (Pandey and Gupta 2018). A combined strategy of As and Se (20 μM) administration repaired growth by reducing the accumulation of As and retrieved As-induced macro- and micronutrient deficiency. A decline in As content could be ascribed to the competition between Se and As as both bind with the amino acid, Cys, and utilize same transporter proteins for uptake. Histochemical localization data and accumulation patterns of ROS further revealed a decrease in superoxide radical ($O^{2^{\cdot}-}$), H_2O_2, ROS, and NO contents that lowered cell death in both varieties of rice indicating Se as a possible quencher of ROS (Pandey and Gupta 2018). Such response was connected with induction in the levels of nonprotein thiols (NPTs), GSH, and PCs along with improved activities of APX, CAT, and GPx indicating the effectiveness of Se supplementation in mitigating As-induced phytotoxicity in rice (Kumar et al. 2014). Additionally, induction in the activities of thiol-metabolizing enzymes, viz. γ-glutamylcysteine synthetase (γ-ECS), GST, and PC synthase (PCS) with Se amendments efficiently helped in the chelation and inactivation of ROS suggesting the protective role of Se in As detoxification (Kumar et al. 2014; Pandey and Gupta 2018). Upregulation in the expression pattern of genes linked with antioxidative defense and metalloid detoxification during As exposure were significantly downregulated under As + Se interaction strengthening the fact that Se improved plant growth by scavenging ROS. Transcript profiles of genes concerned with nutrient uptake and utilization were differentially induced and coordinated with the expression patterns of NPK-related genes having functions in nutrient availability and metabolism in rice cultivars (Pandey and Gupta 2018). Exogenous selenate appreciably repaired plant growth and nitrogen metabolism in As-challenged wheat seedlings (Ghosh et al. 2013). The antioxidative defense was positively modulated to augment growth while the activities of the crucial enzymes of nitrogen and ammonia assimilatory pathways were enhanced and there was less ammonia accumulation in the tissues. Thus, selenium supplementation at recommended ranges could prove to be a suitable enrichment technique to shield plants from the arsenate-imposed consequences owing to decreased uptake, translocation of arsenic or formation of nontoxic Se–metal complexes within the plants.

12.9 Role of Silicon

Silicon (Si), a tetravalent metalloid, is the most abundant metalloid in the soil. Although abundant, silicon always remains in combined form as oxides and silicates. Monosilicic acid (H_4SiO_4) is the bioavailable phase of Si in soil that exists at concentrations ranging from 0.1 to 0.6 mM (Ma and Takahashi 2002). Though silicon is not a prerequisite for plant growth, members of Poaceae and Cyperaceae require ample concentrations of this element for their growth and development. The International Plant Nutrition Institute, Georgia, USA has, therefore, considered silicon as a "beneficial substance" (Tubana et al. 2016). The benefits of silicon under unstressed condition is mostly benign. However, under environmental constraints, Si-induced external as well as internal mechanisms have been demonstrated to heighten resistance in several plants (Seal et al. 2018). Silicon-mediated mitigation of arsenic stress has been extensively investigated (Tripathi et al. 2013; Sanglard et al. 2014; Da Silva et al. 2015; Sil et al. 2018, 2019a, 2019b; Sil and Biswas 2020). Silicate amendments helped revive the arsenate-inflicted consequences in growth and biomass to appreciable extents indicating an ameliorative interaction of silicate on arsenate (Tripathi et al. 2015; Pontigo et al. 2017; Sil et al. 2018, 2019a, 2019b). Silicon-induced decline in abscisic acid as well as ethylene production along with improved cytokinin biosynthesis have been attributed to regenerate the hormonal imbalance in *O. sativa*, *Arabidopsis*, and *Sorghum* that improved growth (Kim et al. 2014; Markovich et al. 2017).

Arsenic and silicon are diagonally related in the periodic table and the chemical analogy between arsenite and silicic acid along with their similar uptake pathway influence their interaction. The structural resemblances between arsenite and silicic acid govern their entry to the root. Arsenite efflux to the xylem occurs via Si/arsenite efflux transporter Lsi 2. Silicon treatment has been reported to limit arsenite transport in rice by decreasing the expression of Lsi 2 (Fleck et al. 2013). Further deposition of silicon in the cell wall affords mechanical strength, whereas sodium metasilicate in the growth media elevates the rhizospheric pH that decreases the availability of heavy metals (Luyckx et al. 2017). Silicon-elicited immobilization of heavy metal in root limits translocation of toxic metal to the shoots while deposition of SiO_2 in the root apoplast and on the leaf surface obstructs the apoplastic flow of metallic ions (Lux et al. 2002; Shi et al. 2005a). Additionally, silicon co-precipitates metal–silicate complexes in the cell wall and aids in vacuolar compartmentalization of the metal (Neumann and Nieden 2001). Exudation of phenolic compounds in silicon-amended plants has been explicated to display metal-chelating abilities that assist in the detoxification process (Pontigo et al. 2017). Silicate amendments restricted the entry of arsenic in root and consequently lowered arsenic uptake in the shoot by *ex planta* and *in planta* mechanisms in rice and wheat (Sanglard et al., 2014; Sil et al. 2019a). A decrease in arsenic transport from root to shoot was accountable for the Si induced alleviation of arsenic toxicity (Sil and Biswas 2020). Increasing silicon concentration in the soil solution has been documented to lessen arsenic accumulation in rice due to competitive inhibition (Guo et al. 2009; Zhao et al. 2010). Silicon supply to different types of rice paddy soils trimmed down arsenic levels in rice grain, straw, husk, and flag leaf (Fleck et al. 2013). Arsenic uptake and translocation were lowered owing to the decrease in expression of Lsi 1 and Lsi 2 or because of competitive inhibition at Lsi 2. Such Si nutrition also restricted arsenic accumulation and significantly enhanced the straw biomass in six rice genotypes (Chuan et al. 2015).

Si fertilization further limited arsenate uptake, diminished lipid peroxidation, and oxidative stress in wheat. Escalation in polyamine synthesis with Si treatments also revived the ionic balance that boosted the antioxidative defense substantially (Sil et al. 2019b).

Silicon administration further controlled oxidative overload and refurbished cellular homeostasis by upregulating the antioxidative defense. Silicate co-treatments led to a substantive decline in the levels of proline, malondialdehyde as well as H_2O_2 and efficiently boosted defense by augmenting the antioxidant enzyme activities, viz. CAT, GR, SOD, catechol peroxidase (CPX), and ascorbic acid oxidase (AAO) in rice as well as wheat cultivars when exposed to arsenic stress (Tripathi et al. 2013; Sil et al. 2018, Sil and Biswas 2020). Silicon-induced intensification in ascorbate, Cys, and GSH contents also stimulated defense that counteracted reactive oxidative anions as well as lipid peroxidation to stabilize membrane integrity (Tripathi et al. 2013; Sil and Biswas 2020). Silicate treatments limited the entry of arsenic and declined the buildup of cytotoxic compounds as well as ROS that downregulated GST activity and decreased PC production (Sil and Biswas 2020).

Silicon supplementation could alleviate the undesirable outcomes by enhancing the photosynthetic efficiency in rice (Sanglard et al. 2014), maize (Da Silva et al. 2015), and wheat (Sil et al. 2019b) plants challenged with arsenic. Sanglard et al. (2016) reported that silicon efficiently reversed the negative effects of arsenic and distinctly alleviated arsenic-related impairments on the photosynthetic process in rice cultivar lsi1 mutant defective in Si uptake and its wild type (WT). Photosynthesis reverted to near control with silicon in time and the genotype-dependent approach in the mentioned rice cultivars was challenged with arsenite. Silicon also modified the arsenic-induced effects in maize by recuperating chlorophyll content together with photosynthetic efficiency. The alleviation was, however, dependent on an optimal dose of Si as doses higher than 1 m Ml^{-1} enhanced stress in maize (Da Silva et al. 2015). Incorporation of silicon in the nutrient media along with arsenate partially restored nutrient deprivation as Si-induced accumulation of polyamines hindered degradation of chlorophyll and reinstated ionic balance by retaining K and lowering the influx of Na. Supplemental silicate also enhanced Ca as well as Mg contents that revived chlorophyll promoting photosynthetic recovery, thus dampening the arsenate-induced effects that eventually enabled growth in wheat (Sil et al. 2019a). Silicon treatments restricted the uptake of arsenic by root and obstructed its translocation to shoot. Lesser availability of arsenic curbed H_2O_2 as well as MDA production improved the activities of photosynthetic enzymes, enhanced net photosynthetic rate, CO_2 demand, and stomatal conductance subsequently enhancing starch contents in wheat seedlings (Sil et al. 2019a). Additionally, deposition of silicon in cell walls has been explicated to enhance tissue resistance and sustain photosynthetic performance in plants (Ali et al. 2013).

Silicate application under arsenate stress improved the activities of respiratory enzymes in wheat seedlings. Supplemental silicate further decreased GABA accumulation by lowering the activities of GABA synthesizing enzymes (Sil et al. 2018). Induction in polyamine contents viz., putrescine, spermidine and spermine with silicate treatments helped to scavenge ROS and decreased H_2O_2 as well as lipid peroxidation (Sil et al. 2019b). Exogenous silicate also improved NO_3^- uptake and subsequently enhanced the activities of NR and NiR in arsenate stressed wheat seedlings (Sil et al. 2019b). Enhanced utilization of NH_4^+-N levels with silicate treatments lowered ammonia contents. Silicon supplementation also improved the activities of GS and GOGAT enzymes to restore the amino acid pool that refurbished growth.

Employment of silicon nanoparticles (Si-NPs) in recent times has encouraging outcomes in the agronomic sector with the potential to evolve highly productive superior varieties resistant to stressful conditions. Despite the availability of ample literature pertaining to the beneficial effects of silicon on several stresses, the nanoparticle-mediated alleviation of stress is still less explored. The mesoporous nature of Si-NPs makes them suitable nanocarriers to control agricultural damages during stresses and can be exploited to modernize the prevailing technology and devise sustainable strategies for improving crop yield under heavy metal toxicity (Cui et al. 2017; Rastogi et al. 2019). Furthermore, employment of Si-NPs may provide concrete solutions to numerous agriculture-related problems including drought, pathogenicity, and productivity. Hydroponic experiments in maize cultivar and its hybrid under arsenate stress revealed a reduction in arsenic accumulation along with oxidative stress markers and an enhancement in components of the ascorbate-glutathione cycle prominently in the Si-NP-fed seedlings than that of Si (Tripathi et al. 2016). Further, the levels of components of the ascorbate-glutathione cycle were higher in Si-NPs and arsenate combinations in comparison to Si, which subsequently counterbalanced ROS-mediated damage to macromolecules more effectively. The positive advantages of silicon fertilization by low-cost industrial by-products and natural silicates have the ability to alleviate stresses in an environment-friendly manner and can be a substitute for the rampant utilization of NPK fertilizers for re-establishing soil integrity and sustaining productivity. The use of silicon to enhance plant growth might provide a prospective solution to the impending crisis of arsenic pollution in agriculture in the near future.

12.10 Conclusion

The emerging crisis of arsenic contamination accounts for annual losses in worldwide agricultural productivity and calls out for immediate solutions. The inherent antioxidant defense in plants has the potential to minimize uncontrolled cellular oxidation under normal and stressful situations but supplies of chemicals counteract the undesired ROS generation by augmenting the defense cascade. Several scientific documentations on the

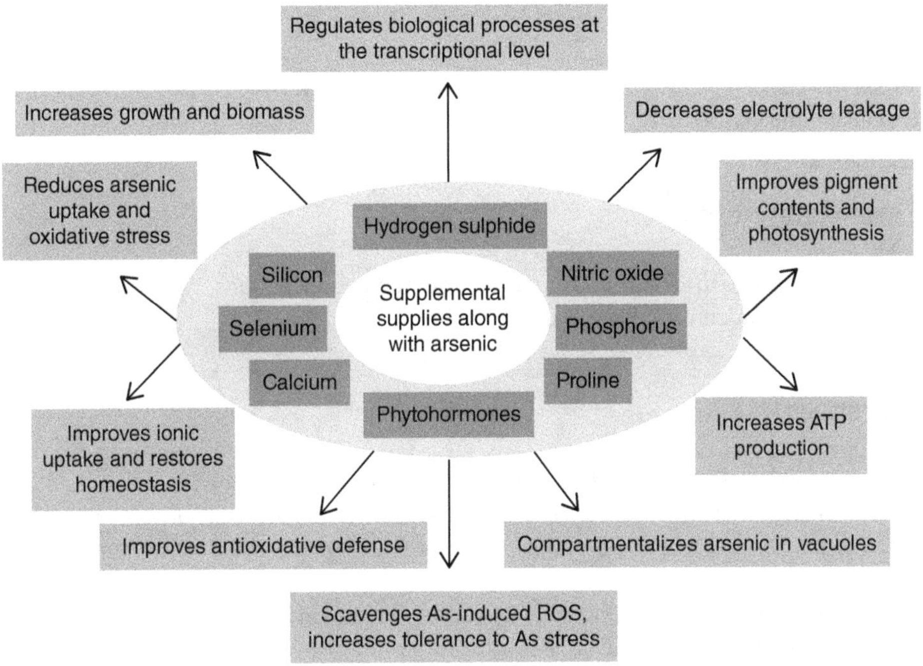

Figure 12.2 Schematic representation of amelioration of arsenic stress using different chemicals in plants.

benefits of such supplementation to a wide variety of crops have helped establish their importance as an agronomic practice in arsenate-contaminated areas and ensure plant productivity. These chemicals principally restrict the entry and limit the uptake of arsenic by modulating the biological processes at the transcriptional level and boosting the antioxidative defense (Figure 12.2). Reduced arsenic uptake also lowers the availability of arsenic in the aboveground parts including the grains. Upsurge in the enzymatic and non enzymatic antioxidative defense cascade equilibrates the accretion of ROS and revives the oxidative overload imposed by arsenic. Improvements in the contents of both micro as well as macronutrients with exogenous supplies of these compounds restore the ionic homeostasis that markedly improves growth. Additionally, these substances enhance the photosynthetic pigments to considerable extents that favorably affect the photosynthetic parameters improving productivity and yield. Joint treatments of such chemicals with arsenic lower the accumulation of stress markers that revive the respiratory cycle enzyme activities assisting to endure the arsenic-prompted alterations and restoring growth.

Thus, co-treatment of these chemicals individually with arsenate has the potential to trim down arsenic-induced negative consequences and may provide a low-cost, technologically simple and point-of-use solutions in arsenic-polluted soils. Despite diverse effects of these substances in plants, ample research is still required to attain a comprehensive knowledge regarding the consequences of these chemicals to counteract stressful environments. Understanding these processes would broaden the perspective of plant adaptation to adverse environmental circumstances that can be utilized to boost crop productivity in initiating cost-effective and environmentally benign agriculture.

Author Contributions

AKB conceived the idea and PS drafted the manuscript. Both AKB and PS actively participated in finalizing the manuscript.

Acknowledgments

We express gratitude to all scientists for their important contributions. We are also grateful for the support rendered by the Department Library.

References

Abbas, G., Murtaza, B., Bibi, I. et al. (2018). Arsenic uptake, toxicity, detoxification, and speciation in plants: physiological, biochemical, and molecular aspects. *International Journal of Environmental Research and Public Health* 15: 59. https://doi.org/10.3390/ijerph15010059.

Abedin, M.J., Feldmann, J., and Meharg, A.A. (2002). Uptake kinetics of arsenic species in rice plants. *Plant Physiology* 128 (3): 1120–1128.

Ahmad, P., Sarwat, M., Bhat, N.A. et al. (2015). Alleviation of cadmium toxicity in *Brassica juncea* L. (Czern. & Coss.) by calcium application involves various physiological and biochemical strategies. *PLoS One* 10: e0114571. https://doi.org/10.1371/journal.pone.0114571.

Alam, S., Kodama, R., Akiha, F. et al. (2006). Alleviation of manganese phytotoxicity in barley with calcium. *Journal of Plant Nutrition* 29 (1): 59–74.

Ali, S., Farooq, M.A., Yasmeneen, T. et al. (2013). The influence of silicon on barley growth, photosynthesis and ultra-structure under chromium stress. *Ecotoxicology and Environmental Safety* 89: 66–72. https://doi.org/10.1016/j.ecoenv.2012.11.015.

Ashraf, M. and Foolad, M.R. (2007). Roles of glycine betaine and proline in improving plant abiotic stress resistance. *Environmental and Experimental Botany* 59: 206–216. https://doi.org/10.1016/j.envexpbot.2005.12.006.

Bali, S., Jamwal, V.L., Kohli, S.K. et al. (2019). Jasmonic acid application triggers detoxification of lead (Pb) toxicity in tomato through the modifications of secondary metabolites and gene expression. *Chemosphere* 235: 734–748. https://doi.org/10.1016/j.chemosphere.2019.06.188.

Banerjee, A., Tripathi, D.K., and Roychoudhury, A. (2018). Hydrogen sulphide trapeze: environmental stress amelioration and phytohormone crosstalk. *Plant Physiology and Biochemistry* 132: 46–53. https://doi.org/10.1016/j.plaphy.2018.08.028.

Basu, A., Saha, D., Saha, R. et al. (2014). A review on sources, toxicity and remediation technologies for removing arsenic from drinking water. *Research on Chemical Intermediates* 40 (2): 447–485. http://link.springer.com/10.1007/s11164-012-1000-4.

Bücker-Neto, L., Sobral Paiva, A.L., Machado, R.D. et al. (2017). Interactions between plant hormones and heavy metals responses. *Genetics and Molecular Biology* 40: 1373–1386. https://doi.org/10.1590/1678-4685-GMB-2016-0087.

Cassina, L., Tassi, E., Morelli, E. et al. (2011). Exogenous cytokinin treatments of an NI Hyper-Accumulator, *Alyssum murale*, grown in a serpentine soil: implications for phytoextraction. *International Journal of Phytoremediation* 13: 90–101.

Chandrakar, V., Dubey, A., and Keshavkant, A. (2016a). Modulation of antioxidant enzymes by salicylic acid in arsenic exposed *Glycine max* L. *Journal of Soil Science and Plant Nutrition* 16: 3. https://doi.org/10.4067/S0718-95162016005000048.

Chandrakar, V., Naithani, S.C., and Keshavkant, S. (2016b). Arsenic-induced metabolic disturbances and their mitigation mechanisms in crop plants: a review. *Biologia* 71: 367–377.

Choudhury, B., Chowdhury, S., and Biswas, A.K. (2011). Regulation of growth and metabolism in rice (*Oryza sativa* L.) by arsenic and its possible reversal by phosphate. *Journal of Plant Interactions* 1: 15–24. https://doi.org/10.1080/17429140903487552.

Chuan, W., Qi, Z., Shengguo, X. et al. (2015). Effects of silicon (Si) on arsenic (As) accumulation and speciation in rice (*Oryza sativa* L.) genotypes with different radial oxygen loss (ROL). *Chemosphere* 138: 447–453. https://doi.org/10.1016/j.chemosphere.2015.06.081.

Corpas, F.J. and Palma, J.M. (2020). H_2S signaling in plants and applications in agriculture. *Journal of Advanced Research.* 24: 131–137. https://doi.org/10.1016/j.jare.2020.03.011.

Cui, J., Liu, T., Li, F. et al. (2017). Silica nanoparticles alleviate cadmium toxicity in rice cells: mechanisms and size effects. *Environmental Pollution* 228: 363–369. https://doi.org/10.1016/j.envpol.2017.05.014.

Da Silva, A.J., Nascimento, C.W., Gouveia-Neto, A.D.S., and Silva, E.A. Jr. (2015). Effects of silicon on alleviating arsenic toxicity in maize plants. *Revista Brasileira de Ciencia do Solo* 39: 289–296. https://doi.org/10.1590/01000683rbcs20150176.

Dave, R., Tripathi, R.D., Dwivedi, S. et al. (2013). Arsenate and arsenite exposure modulate antioxidants and amino acids in contrasting arsenic accumulating rice (*Oryza sativa* L.) genotypes. *Journal of Hazardous Materials* 262: 1123–1131.

Dey, T.K., Banerjee, P., Bakshi, M. et al. (2014). Groundwater arsenic contamination in West Bengal: current scenario, effects and probable ways of mitigation. *International Letters of Natural Sciences* 13: 45–58. http://dx.doi.org/10.18052/www.scipress.com/ILNS.13.45.

Emamverdian, A., Ding, Y., Mokhberdoran, F., and Xie, Y. (2015). Heavy metal stress and some mechanisms of plant defense response. *Scientific World Journal* 2015: 756120. https://doi.org/10.1155/2015/756120.

Fahad, S., Hussain, S., Bano, A. et al. (2015). Potential role of phytohormones and plant growth-promoting rhizobacteria in abiotic stresses: consequences for changing environment. *Environmental Science and Pollution Research* 22: 4907–4921.

Farnese, F.S., Alves, J.O., Gusman, G.S. et al. (2014). Effects of adding nitroprusside on arsenic stressed response of *Pistia stratiotes* L. under hydroponic conditions. *International Journal of Phytoremediation* 16: 123–137. https://doi.org/10.1080/15226514.2012.759532.

Farooq, M.A., Gill, R.A., Islam, F. et al. (2016a). Methyl jasmonate regulates antioxidant defense and suppresses arsenic uptake in *Brassica napus* L. *Frontiers in Plant Science* 7: 468.

Farooq, M.A., Islam, A., Ali, B. et al. (2016b). Arsenic toxicity in plants: cellular and molecular mechanisms of its transport and metabolism. *Environmental and Experimental Botany* 132: 42–52. https://doi.org/10.1016/j.envexpbot.2016.08.004.

Finnegan, P.M. and Chen, W. (2012). Arsenic toxicity: the effects on plant metabolism. *Frontiers in Physiology* 3 (182): 1–18. https://doi.org/10.3389/fphys.2012.00182.

Fleck, A.T., Mattusch, J., and Schenk, M.K. (2013). Silicon decreases the arsenic level in rice grain by limiting arsenite transport. *Journal of Plant Nutrition and Soil Science* 176: 785–794. https://doi.org/10.1002/jpln.201200440785.

Flora, S.J.S. (2011). Arsenic-induced oxidative stress and its reversibility. *Free Radical Biology and Medicine* 51: 257–281. https://doi.org/10.1016/j.freeradbiomed.2011.04.008.

Ghosh, S., Saha, J., and Biswas, A.K. (2013). Interactive Influence of arsenate and selenate on growth and nitrogen metabolism in wheat (*Triticum aestivum* L.) seedlings. *Acta Physiologiae Plantarum* 35: 1873–1885. https://doi.org/10.1007/s11738-013-1225-x.

Guo, B., Liang, Y.C., Zhu, Y.G., and Zhao, F.J. (2007). Role of salicylic acid in alleviating oxidative damage in rice roots (*Oryza sativa*) subjected to cadmium stress. *Environmental Polluion* 147: 743–749. https://doi.org/10.1016/j.envpol.2006.09.007.

Guo, W., Zhang, J., Teng, M., and Wang, L.H. (2009). Arsenic uptake is suppressed in a rice mutant defective in silicon uptake. *Journal of Plant Nutrition and Soil Science* 172: 867–874. https://doi.org/10.1002/jpln.200900060.

Gupta, M. and Gupta, S. (2017). An overview of selenium uptake, metabolism, and toxicity in plants. *Frontiers in Plant Science* 7: 2074. https://doi.org/10.3389/fpls.2016.02074.

Gusman, G.S., Oliveira, J.A., Farnese, F.S., and Cambraia, J. (2013). Arsenate and arsenite: the toxic effects on photosynthesis and growth of lettuce plants. *Acta Physiologia Plantarum* 35: 1201–1209. https://doi.org/10.1007/s11738012-1159-8.

Hanaka, A., Maksymiec, W., and Bednarek, W. (2015). The effect of methyl jasmonate on selected physiological parameters of copper-treated *Phaseolus coccineus* plants. *Plant Growth Regulation* 77: 167–177. https://doi.org/10.1007/s10725-015-0048-8.

Hasanuzzaman, M. and Fujita, M. (2013). Exogenous sodium nitroprusside alleviates arsenic-induced oxidative stress in wheat (*Triticum aestivum* L.) seedlings by enhancing antioxidant defense and glyoxalase system. *Ecotoxicology* 22: 584–596. https://doi.org/10.1007/s10646-013-1050-4.

Hasanuzzaman, M., Nahar, K., Rahman, A. et al. (2017). Actions of biological trace elements in plant abiotic stress tolerance. In: *Essential Plant Nutrients*, 213–274. Berlin, Germany: Springer.

Hawkes, J.S. (1997). Heavy metals. *Journal of Chemical Education* 74: 1369–1374.

Hayat, S., Hayat, Q., Alyemeni, M.N. et al. (2012). Role of proline under changing environments. *Plant Signaling and Behavior* 7 (11): 1456–1466. https://doi.org/10.4161/psb.21949.

Heidari, M. (2017). Plant responses to arsenic: uptake, metabolism and resistance to arsenic toxicity – a review. *Zemdirbyste Agriculture* 104 (4): 369–376. https://doi.org/10.13080/z-a.2017.104.047.

Hettick, B.E., Cañas-Carrell, J.E., French, A.D., and Klein, D.M. (2015). Arsenic: a review of the element's toxicity, plant interactions, and potential methods of remediation. *Journal of Agriculture and Food Chemistry* 63: 7097–7107. https://doi.org/10.1021/acs.jafc.5b02487.

Huang, T.L., Nguyen, Q.T.T., Fu, S.F. et al. (2012). Transcriptomic changes and signalling pathways induced by arsenic stress in rice roots. *Plant Molecular Biology* 80: 587–608.

Jha, A.B. and Dubey, R.S. (2005). Effect of arsenic on behaviour of enzymes of sugar metabolism in germinating rice seeds. *Acta Physiologia Plantarum* 27 (3B): 341–347. https://doi.org/10.1007/s11738-005-0010-x.

Kaya, C., Ashraf, M., Alyemeni Nasser, M. et al. (2020). Salicylic acid-induced nitric oxide enhances arsenic toxicity tolerance in maize plants by upregulating the ascorbate-glutathione cycle and glyoxalase system. *Journal of Hazardous Materials* 399: 123020. https://doi.org/10.1016/j.jhazmat.2020.123020.

Kim, Y.H., Khan, A.L., Kim, D.H. et al. (2014). Silicon mitigates heavy metal stress by regulating P-type heavy metal ATPases, *Oryza sativa* low silicon genes, and endogenous phytohormones. *BMC Plant Biology* 14: 13. http://www.biomedcentral.com/1471-2229/14/13.

Krishnamurthy, A. and Rathinasabapathi, B. (2013). Auxin and its transport play a role in plant tolerance to arsenite-induced oxidative stress in *Arabidopsis thaliana*. *Plant Cell Environment*. 36: 1838–1849. https://doi.org/10.1111/pce.12093.

Kumar, A., Singh, R.P., Singh, P.K. et al. (2014). Selenium ameliorates arsenic induced oxidative stress through modulation of antioxidant enzymes and thiols in rice (*Oryza sativa* L.). *Ecotoxicology* 23: 1153–1163.

Li, G.W., Santoni, V., and Maurel, C. (2014). Plant aquaporins: roles in plant physiology. *Biochimica et Biophysica Acta* 1840: 1574–1582. https://doi.org/10.1016/j.bbagen.2013.11.004.

Li, N., Wang, J., and Song, W.Y. (2015). Arsenic uptake and translocation in plants. *Plant and Cell Physiology* 57 (1): 1–10. https://doi.org/10.1093/pcp/pcv143.

Lux, A., Luxova, M., Hattori, T. et al. (2002). Silicification in sorghum (*Sorghum bicolor*) cultivars with different drought tolerance. *Physiologia Plantarum* 115: 87–92.

Luyckx, M., Hausman, J.F., Lutts, S., and Guerriero, G. (2017). Silicon and plants: current knowledge and technological perspectives. *Frontiers in Plant Science* 8: 411. https://doi.org/10.3389/fpls.2017.00411.

Ma, J.F. and Takahashi, E. (2002). *Soil, fertilizer, and plant silicon research in Japan*. Amsterdam, The Netherlands: Elsevier Science.

Malik, J.A., Goel, S., Kaur, N. et al. (2012). Selenium antagonises the toxic effects of arsenic on mungbean (*Phaseolus aureus* Roxb.) plants by restricting its uptake and enhancing the antioxidative and detoxification mechanisms. *Environmental and Experimental Botany* 77: 242–248.

Malik, J.A., Goel, S., Sandhir, R., and Nayyar, H. (2011). Uptake and distribution of arsenic in chickpea: effects on seed yield and seed composition. *Communications in Soil Science and Plant Analysis* 42 (14): 1728–1738. https://doi.org/10.1080/00103624.2011.584593.

Mallick, S., Sinam, G., and Sinha, S. (2011). Study on arsenate tolerant and sensitive cultivars of *Zea mays* L. differential detoxification mechanism and effect on nutrients status. *Ecotoxicology and Environmental Safety* 74 (5): 1316–1324. https://doi.org/10.1016/j.ecoenv.2011.02.012.

Mandal, K.B. and Suzuki, K.T. (2002). Arsenic round the world: a review. *Talanta* 58: 201–235.

Markovich, O., Steiner, E., Kouøil, S. et al. (2017). Silicon promotes cytokinin biosynthesis and delays senescence in *Arabidopsis* and *Sorghum*. *Plant Cell and Environment* https://doi.org/10.1111/pce.12913.

Marschner, H. (1995). *Mineral Nutrition in Higher Plants*. London: Academic Press.

Masood, A., Khan, M.I.R., Fatma, M. et al. (2016). Involvement of ethylene in gibberellic acid-induced sulfur assimilation, photosynthetic responses, and alleviation of cadmium stress in mustard. *Plant Physiology and Biochemistry* 104: 1–10.

McCarty, K.M., Hanh, H.T., and Kim, K.W. (2011). Arsenic geochemistry and human health in South East Asia. *Reviews on Environmental Health* 26 (1): 71–78. https://www.ncbi.nlm.nih.gov/pmc/articles/PMC3128386.

McGrath, S.P., Zhao, F.J., and Lombi, E. (2001). Plant and rhizosphere process involved in phytoremediation of metal-contaminated soils. *Plant and Soil* 232: 207–214. https://doi.org/10.1023/A:1010358708525.

Meharg, A.A. (2005). Mechanism of plant resistance to metal and metalloid ions and potential biotechnological applications. *Plant and Soil* 274: 163–174.

Meharg, A.A. and Hartley-Whitaker, J. (2002). Arsenic uptake and metabolism in arsenic resistant and non-resistant plant species. *New Phytologist* 154: 29–43. https://doi.org/10.1046/j.1469-8137.2002.00363.x.

Meharg, A.A. and Macnair, M.R. (1994). Relationship between plant phosphorus status and the kinetics of arsenate influx in clones of *Deschampsia cespitosa* (L.) beauv. that differ in their tolerance to arsenate. *Plant and Soil* 162: 99–106. https://doi.org/10.1007/BF01416094.

Mishra, B.K., Dubey, C.S., Shukla, D.P. et al. (2014). Concentration of arsenic by selected vegetables cultivated in the Yamuna flood plains (YFP) of Delhi, India. *Environmental Earth Sciences* 72 (9): 3281–3291. https://doi.org/10.1007/s12665-014-3232-7.

Mostofa, M.G. and Fujita, M. (2013). Salicylic acid alleviates copper toxicity in rice (*Oryza sativa* L.) seedlings by up-regulating antioxidative and glyoxalase systems. *Ecotoxicology.* 22: 959–973.

Mousavi, S.R., Niknejad, Y., Fallah, H., and Tari, D.B. (2020). Methyl jasmonate alleviates arsenic toxicity in rice. *Plant Cell Reports* 39: 1041–1060. https://doi.org/10.1007/s00299-020-02547-7.

Mukhopadhyay, R., Bhattacharjee, H., and Rosen, B.P. (2014). Aqua glyceroporins: generalized metalloid channels. *Biochimica et Biophysica Acta* 1840: 1583–1591. https://doi.org/10.1016/j.bbagen.2013.11.021.

Nagajyoti, P.C., Lee, K.D., and Sreekanth, T.V.M. (2010). Heavy metals, occurrence and toxicity for plants: a review. *Environmental Chemistry Letters* 8: 199–216. https://doi.org/10.1007/s10311-010-0297-8.

Neumann, D. and Nieden, U.Z. (2001). Silicon and heavy metal tolerance of higher plant. *Phytochemistry* 56: 685–692.

Niazi, N.K., Bibi, I., Fatimah, A. et al. (2017). Phosphate-assisted phytoremediation of arsenic by *Brassica napus* and *Brassica juncea*: morphological and physiological response. *International Journal of Phytoremediation* 19: 670–678.

Niazi, N.K., Bibi, I., Shahid, M. et al. (2018). Arsenic removal by perilla leaf biochar in aqueous solutions and groundwater: an integrated spectroscopic and microscopic examination. *Environmental Pollution* 232: 31–41. https://doi.org/10.1016/j.envpol.2017.09.051.

Odjegba, V.J. (2012). Exogenous salicylic acid alleviates arsenic toxicity in *Arabidopsis thaliana*. *Indian Journal of Innovations and Developments.* 7: 515–522.

Palta, J.P. (2000). Stress interactions at the cellular and membrane levels. *Horticultural Science* 25 (11): 1377–1381.

Pandey, C. and Gupta, M. (2018). Selenium amelioration of arsenic toxicity in rice shows genotypic variation: a transcriptomic and biochemical analysis. *Journal of Plant Physiology* 231: 168–181. https://doi.org/10.1016/j.jplph.2018.09.013.

Pandey, C., Raghuram, B., Sinha, A.K., and Gupta, M. (2015a). miRNA plays a role in the antagonistic effect of selenium on arsenic stress in rice seedlings. *Metallomics* 7: 857–866.

Pandey, P. and Tripathi, A.K. (2011). Effect of heavy metals on morphological and biochemical characteristics of *Albizia procera* (Roxb.) Benth. seedlings. *International Journal of Environmental Sciences* 1 (5): 1009–1018.

Parkhey, S., Tandan, M., and Keshavkant, S. (2014). Salycylic acid and acquisition of desiccation tolerance in *Pisum sativum*. *Seeds Biotechnology* 13: 217–225. https://doi.org/10.3923/biotech.2014.217.225.

Pessarakli, M., Haghighi, M., and Sheibanirad, A. (2015). Plant responses under environmental stress conditions. *Advances in Plants and Agriculture Research* 2 (6): 276–286. http://dx.doi.org/10.15406/apar.2015.02.00073.

Pigna, M., Cozzolino, V., Violante, A., and Meharg, A.A. (2009). Influence of phosphate on the arsenic uptake by Wheat (*Triticum durum* L.) irrigated with arsenic contaminated water at three different concentrations. *Water Air and Soil Pollution* 197: 371–380.

Piotrowska-Niczyporuk, A., Bajguz, A., Zambrzycka, E., and GodlewskaŻyłkiewicz, B. (2012). Phytohormones as regulators of HM biosorption and toxicity in green alga *Chlorella vulgaris* (Chlorophyceae). *Plant Physiology and Biochemistry* 52: 52–65.

Pontigo, S., Godoy, K., Jiménez, H. et al. (2017). Silicon-mediated alleviation of aluminum toxicity by modulation of Al/Si uptake and antioxidant performance in rye grass plants. *Frontiers in Plant Science* 8: 642. https://doi.org/10.3389/fpls.2017.00642.

Rahman, A., Mostofa, M.G., Alam, M.M. et al. (2015). Calcium mitigates arsenic toxicity in rice seedlings by reducing arsenic uptake and modulating the antioxidant defense and glyoxalase systems and stress markers. *BioMed Research International* https://doi.org/10.1155/2015/340812.

Rastogi, A., Tripathi, D.K.T., Yadav, S. et al. (2019). Application of silicon nanoparticles in agriculture. *3Biotech* 9: 90. https://doi.org/10.1007/s13205-019-1626-7.

Rosas-Castor, J., Guzmán-Mar, J., Hernández-Ramírez, A. et al. (2014). Arsenic accumulation in maize crop (*Zea mays*): a review. *Science of the Total Environment* 488: 176–187. https://doi.org/10.1016/j.scitotenv.2014.04.075.

Saha, J., Dutta, M., and Biswas, A.K. (2017a). Influence of arsenate and phosphate on the regulation of growth and TCA Cycle in the rice (*Oryza sativa* L.) cultivars IR64 and Nayanmani. *American Journal of Plant Sciences* 8: 1868–1887. https://doi.org/10.4236/ajps.2017.88127.

Saha, J., Majumder, B., Mumtaz, B., and Biswas, A.K. (2017b). Arsenic induced oxidative stress and thiol metabolism in two cultivars of rice and its possible reversal by phosphate. *Acta Physiologia Plantarum* 39: 263.

Sanglard, L.M.V.P., Detmann, K.C., Martins, S.C.V. et al. (2016). The role of silicon in metabolic acclimation of rice plants challenged with arsenic. *Environmental and Experimental Botany* 123: 22–36. https://doi.org/10.1016/j.envexpbot.2015.11.004.

Sanglard, L.M.V.P., Martins, S.C.V., Detmann, K.C. et al. (2014). Silicon nutrition alleviates the negative impacts of arsenic on the photosynthetic apparatus of rice leaves: an analysis of the key limitations of photosynthesis. *Physiologia Plantarum* 152: 355–366. https://doi.org/10.1111/ppl.12178.

Seal, P., Das, P., and Biswas, A.K. (2018). Versatile potentiality of silicon in mitigation of biotic and abiotic stresses in plants: a review. *American Journal of Plant Sciences* 9: 1433–1454. https://doi.org/10.4236/ajps.2018.97105.

Shahbaz, M., Mushtaq, Z., Andaz, F., and Masood, A. (2013). Does proline application ameliorate adverse effects of salt stress on growth, ions and photosynthetic ability of eggplant (*Solanum melongena* L.)? *Scientia Horticulturae* 164: 507–511.

Sharma, S., Kaur, N., Kaur, S., and Nayyar, H. (2016). Selenium as a nutrient in biostimulation and biofortification of cereals. *Indian Journal of Plant Physiology* 1–15.

Shen, S., Li, X.F., Cullen, W.R. et al. (2013). Arsenic binding to proteins. *Chemical Reviews* 113: 7769–7792. https://doi.org/10.1021/cr300015c.

Shi, X.H., Zhang, C.C., Wang, H., and Zhang, F.S. (2005a). Effect of Si on the distribution of Cd in rice seedlings. *Plant Soil.* 272: 53–60.

Shrivastava, A., Ghosh, D., Dash, A., and Bose, S. (2015). Arsenic contamination in soil and sediment in India: sources, effects, and remediation. *Current Pollution Reports* 1: 35–46. https://doi.org/10.1007/s40726-015-0004-2.

Sil, P. and Biswas, A.K. (2020). Silicon nutrition modulates arsenic-inflicted oxidative overload and thiol metabolism in wheat (*Triticum aestivum* L.) seedlings. *Environmental Science and Pollution Research* https://doi.org/10.1007/s11356-020-10369-z.

Sil, P., Das, P., and Biswas, A.K. (2018). Silicon induced mitigation of TCA cycle and GABA synthesis in arsenic stressed wheat (*Triticum aestivum* L.) seedlings. *South African Journal of Botany* 119: 340–352. https://doi.org/10.1016/ j.sajb.2018.09.035.

Sil, P., Das, P., and Biswas, A.K. (2019a). Impact of exogenous silicate amendments on nitrogen metabolism in wheat seedlings subjected to arsenate stress. *Silicon* 12: 535–545. https://doi.org/10.1007/s12633-019-00158-w.

Sil, P., Das, P., Biswas, S. et al. (2019b). Modulation of photosynthetic parameters, sugar metabolism, polyamine and ion contents by silicon amendments in wheat (*Triticum aestivum* L.) seedlings exposed to arsenic. *Environmental Science and Pollution Research* https://doi.org/ 10.1007/s11356-019-04896-7.

Singh, A.P., Dixit, G., Kumar, A. et al. (2016a). Nitric oxide alleviated arsenic toxicity by modulation of antioxidants and thiol metabolism in rice (*Oryza sativa* L.). *Frontiers in Plant Science* 6 (1272): https://doi.org/10.3389/fpls.2015.01272.

Singh, A.P., Dixit, G., Mishra, S. et al. (2015a). Salicylic acid modulates arsenic toxicity by reducing its root to shoot translocation in rice (*Oryza sativa* L.). *Frontiers in Plant Science* 6 (340): https://doi.org/10.3389/fpls.2015.00340.

Singh, H.P., Kaur, S., Batish, D.R. et al. (2009a). Nitric oxide alleviates arsenic toxicity by reducing oxidative damage in the roots of *Oryza sativa* (rice). *Nitric Oxide* 20: 289–297. https://doi.org/10.1016/j.niox.2009.02.004.

Singh, M., Singh, V.P., Dubey, G., and Prasad, S.M. (2015b). Exogenous proline application ameliorates toxic effects of arsenate in *Solanum melongena* L. seedlings. *Ecotoxicology and Environmental Safety* 117: 164–173. https://doi.org/10.1016/j.ecoenv.2015.03.021.

Singh, M., Singh, V.P., and Prasad, S.M. (2016b). Responses of photosynthesis, nitrogen and proline metabolism to salinity stress in *Solanum lycopersicum* under different levels of nitrogen supplementation. *Plant Physiology and Biochemistry* 109: 72–83. https://doi.org/10.1016/j.plaphy.2016.08.021.

Singh, N., Ma, L.Q., Vu, J.C., and Raj, A. (2009b). Effects of arsenic on nitrate metabolism in arsenic hyperaccumulating and non-hyperaccumulating ferns. *Environmental Pollution* 157: 2300–2305. https://doi.org/10.1016/j.envpol.2009.03.036.

Singh, R., Parihar, P., and Prasad, S.M. (2018a). Simultaneous exposure of sulphur and calcium hinder As toxicity: up-regulation of growth, mineral nutrients uptake and antioxidants system. *Ecotoxicology and Environmental Safety* 161: 318–331. https://doi.org/10.1016/j.ecoenv.2018.05.060.

Singh, R., Parihar, P., and Prasad, S.M. (2018b). Sulfur and calcium simultaneously regulate photosynthetic performance and nitrogen metabolism status in As-challenged *Brassica juncea* L. seedlings. *Frontiers in Plant Science* 19 (9): 772. https://doi.org/10.3389/fpls.2018.00772.

Singh, R., Parihar, P., and Prasad, S.M. (2020). Interplay of calcium and nitric oxide in improvement of growth and arsenic-induced toxicity in mustard seedlings. *Scientific Reports* 10 (1): 6900. https://doi.org/10.1038/s41598-020-62831-0.

Singh, R., Singh, S., Parihar, P. et al. (2015c). Arsenic contamination, consequences and remediation techniques: a review. *Ecotoxicology and Environmental Safety* 112: 247–270. https://doi.org/10.1016/j.ecoenv.2014.10.009.

Singh, V.P., Singh, S., Kumar, J., and Prasad, S.M. (2015d). Hydrogen sulfide alleviates toxic effects of arsenate in pea seedlings through up-regulation of the ascorbate–glutathione cycle: possible involvement of nitric oxide. *Journal of Plant Physiology* 181: 20–29. https://doi.org/10.1016/j.jplph.2015.03.015 J Plant Physiol.

Srivastava, S. and Sharma, Y.K. (2013a). Impact of arsenic toxicity on black gram and its amelioration using phosphate ISRN. *Toxicology* https://doi.org/10.1155/2013/340925.

Srivastava, S. and Sharma, Y.K. (2013b). Altered growth, photosynthetic machinery and induced oxidative stress in spinach in response to arsenic stress. *Journal of Plant Physiology and Pathology* 1: 4. https://doi.org/10.4172/2329-955X.1000113.

Srivastava, S. and Sharma, Y.K. (2014). Arsenic induced changes in growth and metabolism of black gram seedlings (*Vigna mungo* L.) and the role of phosphate as an ameliorating agent. *Environmental Processes* 1: 431–445. https://doi.org/10.1007/s40710-014-0035-5.

Srivastava, S., Srivastava, A.K., Suprasanna, P., and D'Souza, S.F. (2013). Identification and profiling of arsenic stress-induced microRNAs in *Brassica juncea*. *Journal of Experimental Botany* 64 (1): 303–315.

Stoeva, N., Berova, M., and Zlatev, Z. (2003). Physiological response of maize to arsenic contamination. *Biologia Plantarum* 47: 449–452. https://doi.org/10.1023/B:BIOP.0000023893.12939.48.

Sun, C., Lu, L., Liu, L. et al. (2014). Nitrate reductase–mediated early nitric oxide burst alleviates oxidative damage induced by aluminium through enhancement of antioxidant defences in roots of wheat (*Triticum aestivum*). *New Phytologist* 201: 1240–1250. https://doi.org/10.1111/nph.12597.

Sun, H., Wang, X., Wang, Y. et al. (2017). Alleviation of cadmium toxicity in cucumber (*Cucumis sativus*) seedlings by the application of selenium. *Spanish Journal of Agricultural Research* 14: 1105.

Swarnakar, A. (2017). Enhancement effect of gibberellic acid and kinetin on sucrose metabolism in mungbean seedlings under arsenate toxicity. *American Journal of Bioscience and Bioengineering* 5 (1): 50–55. https://doi.org/10.11648/j.bio.20170501.18.

Sytar, O., Kumari, P., Yadav, S. et al. (2019). Phytohormone priming: regulator for heavy metal stress in plants. *Journal of Plant Growth Regulation* 38: 739–752. https://doi.org/10.1007/s00344-018-9886-8.

Tam Ho, T., Murthy, H.N., and Park, S.Y. (2020). Methyl jasmonate induced oxidative stress and accumulation of secondary metabolites in plant cell and organ cultures. *International Journal of Molecular Sciences* 21: 716. https://doi.org/10.3390/ijms21030716.

Tassi, E., Pouget, J., Petruzzelli, G., and Barbafieri, M. (2008). The effects of exogenous plant growth regulators in the phytoextraction of HMs. *Chemosphere* 71 (1): 66–73.

Tawfik, D.S. and Viola, R.E. (2011). Arsenate replacing phosphate: alternative life chemistries and ion promiscuity. *Biochemistry* 50: 1128–1134. https://doi.org/10.1021/bi200002a.

Tripathi, D.K., Singh, S., Singh, V.P. et al. (2016). Silicon nanoparticles more efficiently alleviate arsenate toxicity than silicon in maize cultivar and hybrid differing in arsenate tolerance. *Frontiers in Environmental Science* 4: 46. https://doi.org/10.3389/fenvs.2016.00046.

Tripathi, D.K., Singh, V.P., Prasad, S.M. et al. (2015). Silicon-mediated alleviation of Cr (VI) toxicity in wheat seedlings as evidenced by chlorophyll florescence, laser induced breakdown spectroscopy and anatomical changes. *Ecotoxicology and Environmental Safety* 113: 133–144. https://doi.org/10.1016/j.ecoenv.2014.09.029.

Tripathi, P., Tripathi, R.D., Singh, R.P. et al. (2013). Silicon mediates arsenic tolerance in rice (*Oryza sativa* L.) through lowering of arsenic uptake and improved antioxidant defence system. *Ecological Engineering* 52: 96–103. https://doi.org/10.1016/j.ecoleng.2012.12.057.

Tubana, B., Babu, T., and Datnoff, L. (2016). A review of silicon in soils and plants and its role in US agriculture: history and future perspectives. *Soil Science* 181: 393–411. https://doi.org/10.1097/ss.0000000000000179.

Ullrich-Eberius, C., Sanz, A., and Novacky, A. (1989). Evaluation of arsenic and vandate-associated changes of electrical membrane potential and phosphate transport in *Lemna gibba* GL. *Journal of Experimental Botany* 40: 119–128. https://doi.org/10.1093/jxb/40.1.119.

Verma, G., Srivastava, D., Narayan, S. et al. (2020). Exogenous application of methyl jasmonate alleviates arsenic toxicity by modulating its uptake and translocation in rice (*Oryza sativa* L.). *Ecotoxicology and Environmental Safety* 15: 201–110735. https://doi.org/10.1016/j.ecoenv.2020.110735.

Wani, S.H., Kumar, V., Shriram, V., and Sahd, S.K. (2016). Phytohormones and their metabolic engineering for abiotic stress tolerance in crop plants. *The Crop Journal* 4: 162–176.

Wu, Z., Yin, X., Bañuelos, G.S. et al. (2016). Indications of selenium protection against cadmium and lead toxicity in oilseed rape (*Brassica napus* L.). *Frontiers in Plant Science* 7: 1875. https://doi.org/10.3389/fpls.2016.01875.

Yan, Z., Zhang, W., Chen, J., and Li, X. (2015). Methyl jasmonate alleviates cadmium toxicity in *Solanum nigrum* by regulating metal uptake and antioxidative capacity. *Biologia Plantarum* 59 (2): 373–381. https://doi.org/10.1007/s10535-015-0491-4.

Yang, S.L., Lan, S.S., and Gong, M. (2009). Hydrogenperoxide-induced proline and metabolic pathway of its accumulation in maize seedlings. *Journal of Plant Physiology* 166: 1694–1699.

Zengin, F. (2015). Effects of exogenous salicylic acid on growth characteristics and biochemical content of wheat seeds under arsenic stress. *Journal of Environmental Biology* 36: 249.

Zhao, F.J., McGrath, S.P., and Meharg, A.A. (2010). Arsenic as a food chain contaminant: mechanisms of plant uptake and metabolism and mitigation strategies. *Annual Review of Plant Biology* 61: 535–559. https://doi.org/10.1146/annurev-arplant-042809-112152.

Zhao, Y. (2010). Auxin biosynthesis and its role in plant development. *Annual Review of Plant Biology* 61: 49–64. https://doi.org/10.1146/annurevarplant-042809-112308.

Zheng, M.Z., Li, G., Sun, G.X. et al. (2013). Differential toxicity and accumulation of inorganic and methylated arsenic in rice. *Plant and Soil* 365: 227–238. https://doi.org/10.1007/s11104-012-1376-3.

Zvobgo, G., Hu, H., Shang, S. et al. (2015). The effects of phosphate on arsenic uptake and toxicity alleviation in tobacco genotypes with differing arsenic tolerances. *Environmental Toxicology and Chemistry* 34: 45–52. https://doi.org/10.1002/etc.2776.

Zvobgo, G., Lwalaba, J.L.W., Sagonda, T. et al. (2019). Alleviation of arsenic toxicity by phosphate is associated with its regulation of detoxification, defense, and transport gene expression in barley. *Journal of Integrative Agriculture* 18 (2): 381–394. https://doi.org/10.1016/S2095-3119(18)61955-0.

13

Strategies to Reduce the Arsenic Contamination in the Soil–Plant System

Mohammad Mehdizadeh[1], Waseem Mushtaq[2], Shahida Anusha Siddiqui[3,4], Samina Aslam[5], Duraid K.A. AL-Taey[6], Koko Tampubolon[7], Emad Jafarzadeh[8], and Anahita Omidi[9]

[1] *Department of Agronomy and Plant Breeding, University of Mohaghegh Ardabili, Ardabil, Iran*
[2] *Allelopathy Laboratory, Department of Botany, Aligarh Muslim University, Aligarh, Uttar Pradesh, India*
[3] *Department of Biotechnology and Sustainability, Technical University of Munich (TUM), Straubing, Bavaria, Germany*
[4] *DIL e.V.–German Institute of Food Technologies, D-Quakenbrück, Lower Saxony, Germany*
[5] *Department of Chemistry, Women University Multan, Multan, Punjab, Pakistan*
[6] *Department of Horticulture, Faculty of Agriculture, University of AL-Qasim Green, AL-Qasim, Babylon, Iraq*
[7] *Program Study of Agrotechnology, Faculty of Agriculture and Animal Husbandry, Universitas Tjut Nyak Dhien, Medan, Sumatera Utara, Indonesia*
[8] *Department of Toxicology and Pharmacology, Faculty of Pharmacy, Tehran University of Medical Sciences (TUMS), Tehran, Iran*
[9] *Department of GIS and Remote Sensing, Faculty of Geography, University of Tehran, Tehran, Iran*

13.1 Introduction

Arsenic is an abundant element in the earth's crust and soil environment (Bhadauria 2019). Therefore, large populations in the world are exposed to this element and its toxic compounds. According to the US Environmental Protection Agency (EPA) classification, arsenic is one of the most toxic carcinogenic compounds and the cause of many chronic diseases in humans (Sheikhi et al. 2021). Numerous factors, including agricultural activities and especially the application of fertilizers and chemical pesticides, mineral activities related to the extraction of metallogenic minerals as well as natural processes such as seasonal precipitation and rock erosion, are involved in the release of toxic arsenic compounds into the environment (Smedley and Kinniburgh 2002). There is about 2–5 $mg\,kg^{-1}$ soil of arsenic in different soils. The main part of it is blocked in the form of arsenopyrite. Arsenic has low solubility in water. Therefore, it usually has no toxic effects on humans or other organisms.

Groundwater contamination with arsenic and its compounds occurs through the leaching of rocks and minerals by rainfall or irrigation systems in agricultural ecosystems and then releasing arsenic-containing minerals into water sources (Henke 2009). The long half-life and stability of arsenic in the soil environment have increased the severity of environmental concerns related to this element. According to Yu et al. (2003), Bangladesh annually witnesses millions of protests against the use of arsenic and its destructive effects

Arsenic in Plants: Uptake, Consequences and Remediation Techniques, First Edition.
Edited by Prabhat Kumar Srivastava, Rachana Singh, Parul Parihar, and Sheo Mohan Prasad.

(Yu et al. 2003). Due to its high toxicity in the environment, arsenic has a high potential to endanger human and animal health and agroecosystems' safety (Chakraborty et al. 2014). Arsenic is available in organic and inorganic forms in the soil environment; however, the inorganic form is more common and could be found in water, air, and soil (Moreno-Jiménez et al. 2012). Exposure to arsenic and different arsenic species could have a range of destructive and irreversible effects on humans, animals, and other living microorganisms (Chung et al. 2014). Depending on various factors such as arsenic concentration, plant species, plant age, climatic and soil conditions, different ranges of adverse effects include reducing photosynthesis, enzymatic activity, growth inhibition, cell division, and other biological and physiological processes, which could be found in plants exposed to arsenic. The carcinogenic effect is the most critical adverse direct effect of arsenic associated with human health (Shah et al. 2020). Water contamination by arsenic is one of the most environmental concerning issues; however, many farmers, growers, and other people are not aware of this occurrence's potential adverse effects, such as the entrance of arsenic and its compounds into the food chain (Huq et al. 2006). As the most toxic forms of arsenic are arsine gas, breathing air in arsenic-contaminated areas can also directly affect humans' health and other living organisms. This phenomenon occurs more severely in areas with wind erosion. Groundwater sources and water used in agriculture may contain small levels of arsenic that could affect human life by entering the food chain (Sohrabi et al. 2020). However, several strategies, such as phytoremediation and degradation of arsenic by using microorganisms (bacteria, fungi, etc.), the addition of organic fertilizers to soils, bioleaching, and physical and chemical methods to reduce the destructive effects and modify arsenic-induced pollution in the environment. Table 13.1 represents some adverse effects of arsenic and arsenic species commonly detected in different environments. This chapter illustrates the fate, toxicity, and degradation of arsenic and introduces some effective strategies to reduce arsenic contaminations in soil–plant systems.

13.2 Arsenic

Arsenic is categorized as a metalloid (semimetal) element with a molecular weight of 74.92159 $g\,mol^{-1}$ found in solid form at 20 °C, and its oxidation states include −3, 0, +3, and +5 (WHO 2001). This element is considered a significant constituent for more than 245 different minerals (O'Neill 1995). Arsenic is known as a toxic substance, and some of its compounds have been used as insecticides and herbicides with limited use. Soil texture and organic matter content, pH, soil microorganism populations, and redoximorphic potential are among the key factors affecting the bioavailability and mobility of arsenic in different soils (Bhumbla and Keefer 1994). Arsenic is not normally soluble in water while depending on pH, the salt forms of arsenic show a wide range of solubility. Arsenic and arsenic-containing compounds are usually found in trace quantities in soil, rocks, air, water sources, and solid crystalline or powder forms. Generally, a large number of arsenic-containing compounds have been identified with different properties depending on their chemical composition and molecular structure. Some of the most crucial arsenic compounds are presented in Table 13.1. Arsenic and its compounds have been used for centuries in cases such as medicinal uses (Zhao et al. 2017), pest and weed control in some cropping systems

Table 13.1 Toxicity of arsenic and some commonly detected arsenic species.

Name	CAS number	Toxicity and adverse effects	Reference
Arsenic	7440-38-2	Human cord blood DNA methylation	Broberg et al. (2014)
		Health status and functional performance of children	Hsueh et al. (2016)
		Kidney cancer mortality	Yan et al. (2010)
		Mortality from pulmonary tuberculosis	Smith et al. (2011)
		Effect on rice (*Oryza sativa* L.) growth	Panaullah et al. (2009)
		Effect on rice (*Glycine max* L) growth	Neidhardt (2021)
		Effects on membranes and chlorophyll of mungbean (*Phaseolus aureus* Roxb.)	Malik et al. (2012)
		Effects on *Sinorhizobium-Medicago sativa* symbiotic	Pajuelo et al. (2008)
Arsines	7784-42-1	Humans' intravascular hemolysis	Danielson et al. (2006)
Arsenite	7784-46-5	Effects on earthworms (*Eisenia fetida*)	Wang et al. (2018b)
Arsenate	10048-95-0	Freshwater amphipod (*Gammarus pulex*)	Vellinger et al. (2012)
Monomethylarsonic acid	124-58-3	Carcinogenic effect in rats and mice	Arnold et al. (2003)
Arsenobetaine	64436-13-1	Effects on planktons	Caumette et al. (2012)
Gallium arsenide	1303-00-0	Effects on human lung and kidney	Tanaka (2004)
Indium arsenide	1303-11-3	Effects on human reproductive organs	Tanaka (2004)
Arsenochline	39895-81-3	Effect on mice respiration	Kaise et al. (1992)
Dimethylarsinic acid	75-60-5	Carcinogenic effect in rats	Kuroda et al. (2004)
Trimethylarsine	593-88-4	Effects on rats' embryos	Irvin and Irgolic (1995)
Arsenous acid	13464-58-9	Potentially fatal arrhythmias	Au et al. (2013)
Arsenic sulfide	56320-22-0	Gastric and colon cancer	Tan et al. (2018)
Arsenic selenide	12626-31-2	Effects on the lungs of hamsters	Tanaka et al. (1994)

(Tampubolon et al. 2020), wood preservation (Jones et al. 2019), and mining and glass production activities (Zhao et al. 2017). The other usages of arsenic and its compounds include some pigment production, leather preservation, chemical industries, painting, ceramic production, etc. Hajalilou et al. (2011) found the highest concentrations of arsenic (2000 $\mu g\,l^{-1}$) in mining areas of the North-West of Iran. If arsenic pollution in mining areas is not managed, the transfer of these pollutants through wind and water sources to non-polluted places could have occurred. Stachnik et al. (2020) reported the enhancement of arsenic concentrations due to gold mining activities in Poland. Accordingly, anthropogenic sources and human activities have the most significant impact on creating and exacerbating environmental pollution caused by arsenic and arsenic compounds.

13.3 Arsenic Use in Agricultural Soils

The health of agricultural soils plays an increasingly important role in producing crops and food and soil microorganisms and is directly related to human health. So, soil resources' contamination, especially soils of agricultural ecosystems, is one of the most critical effects of arsenic pollution (Mishra et al. 2014). From a nutritional point of view, arsenic is considered a nonessential element for plants, and even its presence in plants could have adverse effects and toxicity. Soil physicochemical properties, climate conditions, and soil microorganism communities' activity are the main factors influencing the bioavailability of arsenic to plants in the soil environment. Arsenic-based pesticides are primary sources of arsenic and arsenic compounds' contamination in agricultural soils and agroecosystems. Arsenic and arsenic-containing compounds in soils can be transferred to plant roots through water uptake, following which storage and accumulation of these compounds in fruit or other plants' parts.

On the other hand, the consumption of agricultural products contaminated with different concentrations of arsenic could directly affect human health. Inhibition of growth and photosynthesis, chlorosis, necrosis, and even plant death could be occurred due to exposure to high concentrations of arsenic in agricultural soils (Rosas-Castor et al. 2014). According to Robinson et al. (2006), aquatic plants such as *Callitriche stagnalis* and *Myriophyllum propinquum* can accumulate high concentrations of arsenic (more than $1\,g\,kg^{-1}$ dry weight of the plant). The presence of arsenic in the concentration of $15\,mg\,kg^{-1}$ soils affects the growth and yields of some vegetables such as spinach and, consequently, endangers human food safety (Aysha et al. 2017). Different plant species show different responses to arsenic exposure. Accumulation of arsenic compounds in plants such as rice has been observed abundantly. It has always been discussed (Li et al. 2014), while arsenic accumulation does not occur in plants such as tomatoes (Rodriguez-Iruretagoiena et al. 2015).

13.4 Arsenic Fate in Soil

The high importance of arsenic-induced environmental pollution is considered a significant problem. Therefore, it is essential to understand the factors affecting the fate, persistence, and bioavailability of arsenic and its compounds in different environments, especially soil. Arsenic could be present in the soil as oxides, chlorides, hydroxides, and sulfides. The main part of arsenic and arsenic compounds in the soil environment is sorbed to the soil matrix's surface. Generally, arsenic adsorption by soil is affected by several factors, the most important of which are:

- pH
- The effect of calcium
- The effect of anion competition
- The effect of ionic strength

The absorption of arsenic in different soils significantly affects the mobility, toxicity, and viability of arsenic in the soil environment. On the other hand, soil's physicochemical properties significantly affect the absorption of arsenic in the soil. Seiter (2009) found that the behavior and the sorption of arsenic in different soils were entirely dependent on texture, pH, and organic matter content. Zhang and Selim (2005) and Jiang et al. (2005) investigated the adsorption of arsenic. Its compounds in different soils observed that soil properties, especially soil texture, significantly affected arsenic adsorption. The process of arsenic adsorption-desorption has many complexities. Studies have shown that arsenic adsorption occurs in two distinct phases. The first phase occurs over a short time, while the second phase occurs over a very long time and may continue for years (Arai and Sparks 2002; O'Reilly et al. 2001). Leaching is another process that affects the fate of arsenic in the soil, affected by soil properties. Leaching is called the vertical transmission of chemicals to lower soil layers (Mehdizadeh and Mushtaq 2020). Benavente et al. (2020) reported the leaching of arsenic in a pH-dependent manner, where the concentration of arsenic in leaching residues increased 200 times by increasing the soil pH.

Microorganisms present in the soil have the potential to degrade chemicals in the soil environment. This process is known as biodegradation or microbial degradation. Yang et al. (2020) found that *Pseudomonas vancouverensis* strain m318 effectively improved the phytoremediation of arsenic in field trials.

13.5 Toxicity of Arsenic on Humans, Animals and Plants

International Agency for Research on Cancer (IARC) evaluated the carcinogenic effects of arsenic and its compounds in 1973 and found sufficient evidence for the carcinogenicity of these chemicals to humans (Group 1) and rare evidence for carcinogenicity in animals. Many factors are involved in arsenic toxicity in different environments, the most important of which are chemical structures and oxidative states. Arsenic is present in the soil environment as an inorganic form and could be transported via the food chain and could have toxic effects. The toxicity of arsenic and arsenical compounds is generally dependent on the form (inorganic or organic) and the oxidation state of these compounds. Arsenic has two important forms: inorganic with more toxic effects and organic with less toxic effects. Monomethylarsonic (MMA), dimethylarsinic acid (DMA), and trimethylarsine (TMA) are among the known toxic forms of arsenic with lethal dose (LD50) values of 1.8, 1.2, and $10\,g\,kg^{-1}$, respectively (Kaise and Fukui 1992). Cytotoxicity, genotoxicity, and epidemiological toxicity are among the toxic forms of arsenic in humans (Shankar et al. 2014). Different types of cancers are well-documented by other researchers due to human and animal exposure to arsenic (Yoshida et al. 2004; Tapio and Grosche 2006; Kumar et al. 2021). As the arsenic present in drinking water enters the human body, it enters the blood vessels by passing through the gastrointestinal tract and reaches the vital organs of the body through the bloodstream and could cause many disorders such as gastrointestinal, neurological, hormonal, and metabolic disorders as well as cardiovascular diseases (Ahamed et al. 2006; Argos et al. 2010). The inhibition of the genome replication process has been reported (Sinha et al. 2013). In Bangladesh, herbicides containing arsenical compounds killed cattle within a few hours (Eisler 1988).

Plants can also be affected by different concentrations of arsenic depending on the growth stage and plant species. As arsenic is not considered an essential element for plants, an arsenic concentration below 3.6 mg kg^{-1} does not affect plants' growth and development (Gebel 1997). However, higher concentrations of arsenic in the soil can lead to toxicity, growth inhibition, or plant death. Pandey et al. (2020) evaluated the effects of arsenic on Osa-miR156j expression in rice crops and found a significant impact on all physiological growth stages. Translocation and distribution of arsenic in plants are crucial for the toxicity of this compound in plants. Different plant parts showed various responses to the distribution of arsenic. Generally, stems, roots, and leaves have higher arsenic concentrations than seeds or fruits (Peryea 2001). Adriano and Weber (2001) reported a higher penetration to the cuticle and toxicity value of arsenite than arsenic. Accumulation of arsenic in aquatic microorganisms and algae has been reported (Ma et al. 2001). Reduction in fresh and dry biomass, total chlorophyll, and carotenoids content of *Vigna mungo* L. was significantly affected by arsenate (Srivastava and Sharma 2014). The low concentration of arsenic in the agroecosystems (0.2 mg l^{-1}) negatively impacted rice yield and yield components of rice. However, the radish showed less sensitivity to arsenic and was affected by 100 mg kg^{-1} soil (Abedin et al. 2002). Khan et al. (2009) found that the root and shoot biomass of *Brassica juncea* was significantly affected by 25 μM arsenic during 96 hours.

13.6 Strategies to Reduce the Arsenic Contamination in the Soil–Plant System

Due to the threat to human health, animals, and agricultural crops and the environmental hazards associated with arsenic and its compounds in the environment, finding effective strategies to eliminate or reduce the adverse effects of arsenic is of particular importance. In general, there are several strategies for detoxification and mitigation of arsenic in different environments. Agricultural management, chemical, and biotechnological methods, bioremediation of arsenic by microbial and fungal communities, accumulation and inactivation of arsenic in the cells of some plants, and the application of organic and mineral fertilizers are among the most important of these strategies.

13.6.1 Agricultural Management for Detoxification and Mitigation of Arsenic

Agricultural management referred to some agronomic practices such as management of the irrigation water, selecting arsenic-tolerant crops, and applying different fertilizers. Arsenic contamination of water resources is typically one of the most likely ways to introduce arsenic to agricultural ecosystems. Therefore, irrigation water management and implementing a proper irrigation system could effectively reduce pollution caused by arsenical compounds. Islam et al. (2017) evaluated the impacts of different irrigation systems and rice genotypes on reducing arsenic in rice plants and found the significant effect of irrigation and genotypes on arsenic rice accumulation. Their results revealed that using an alternate wetting and drying irrigation system, the arsenic concentration in rice was less than for a continuous flooding irrigation system. Li et al. (2019) reported the same results for the effectiveness of alternate wetting and drying irrigation system on arsenic uptake rice. On the other hand, the effect of moisture on arsenic uptake by plants and soil is undeniable and can directly affect the bioavailability of arsenic. Accordingly, irrigation management with the approach of saving water consumption and proper irrigation water timing and using genotypes with low water requirements could be considered an effective strategy to reduce the effects of arsenic in agroecosystems.

Phosphorus is one of the main elements in plant nutrition and one of the leading chemical fertilizers. This element is known as a chemical analog for arsenic (Strawn 2018) and can affect arsenic absorption by plants. Therefore, the management of chemical fertilizers, especially phosphorus-based fertilizers, is of particular importance in managing arsenic-induced pollution in agricultural soils. The effectiveness of phosphorus application to the soil in reducing arsenate uptake by rice plants was reported by Lee et al. (2016). Phosphorus fertilizer application increased the bioavailability of arsenic for *Brassica juncea* L. plant by increasing in desorption of arsenic (Bolan et al. 2013). Pigna et al. (2009) evaluated phosphate's effects on the arsenic uptake by *Triticum durum* L. They found that P fertilization led to a decrease in arsenic toxicity by minimizing the translocation of arsenic in different parts of wheat plants. Wang et al. (2018a) reported the significant effects of exogenous phosphate on the liability and phytoavailability of arsenic for *Lactuca sativa* plant.

13.6.2 Biotechnological Method

Naturally, in plant cells, there are several defense barriers with complex enzymatic structures to effectively fight against external hazards, biotic and abiotic stresses, and reactive oxygen species. Superoxide dismutase, ascorbate peroxidase, catalase, dehydroascorbate reductase, glutathione reductase, and peroxidase are among the main enzymes involved in plant defense systems (Hasanuzzaman and Fujita 2013). Biotechnological methods with enzymatic and genomic approaches could be considered an efficient strategy for reducing arsenic and its compounds in agroecosystems. Accordingly, different researches have been performed to understand the enzymatic mechanisms involved in the decomposition, toxification, transportation, and detoxification of arsenic and different soils and plants. Kertulis-Tartar et al. (2009) reported increased catalase activity in *Pteris vittata* as a defending response to arsenic exposure in the concentration of $10\,mg\,l^{-1}$. Shri et al. (2019) reported some enzymatic pathways for arsenic detoxification in rice plants, including arsenic methyltransferase, glutathione biosynthesis, and phytochelatin synthesis. The impact of glutathione S-transferase enzyme on the detoxification of arsenic in plants was reported (Kumar and Trivedi 2018). Ye et al. (2012) found transgenic rice could be used as a potential strategy for reducing arsenic accumulation in the plant by converting arsenic to a noneffective form. Dhankher et al. (2002) found that a gene named arsenate reductase (ArsC) γ Glutamyl cysteine synthetase (γ-ECS) was able to 2–3 fold arsenic accumulation in *Arabidopsis thaliana* tissues. Ye et al. (2011) reported *Pteris vittata* on the reduction of the total concentration of arsenic (3.5–11.4%) in the rice field.

13.6.3 Bioremediation

The term bioremediation refers to all biological methods for cleanup and removing polluted soil and water resources. These methods received tremendous attention as a promising technology for the remediation of contaminants in the environment due to their environmentally friendly properties.

13.6.3.1 Phytoremediation

Different plants can remove or change the nature of chemicals, toxins, and heavy metals entered into their cells, known as phytoremediation. They could be considered a cost-effective and eco-friendly strategy for reducing chemical pollution in agricultural ecosystems. Accumulation and inactivation of pollutants within different parts of plant cells, degradation, and transformation of molecules by plant cells are typical examples of phytoremediation. However, microbial communities associated with plant roots may also be involved in the chemical phytodegradation process. Some plants are inherently capable of inactivating and accumulating arsenic molecules. Some of these capable plants are including *Oriza sativa* (Syu et al. 2020; Abedi and Mojiri 2020), *Zea mays* L. (Baig et al. 2010), *Echinochloa cruss-galli* (Gani-Molla et al. 2010), *Brassica juncea* (Khan et al. 2009), *Talinum cuneifolium* (Sekhar et al. 2007), *Agrostis capillaris*, *Festuca rubra* and *Calamagrostis epigejos* (Dradrach et al. 2020), strawberry (Torres et al. 2020), *Atriplex atacamensis* (Vromman et al. 2011), and *Pteris vittata* (Huang et al. 2004). In soil contaminated by arsenic, 95 and 68% of arsenic were removed by *Pteris vittata* and *Nephrodium molle*, respectively (Hossain et al. 2006). Yadav et al. (2021) found a tissue and time-dependent behavior for accumulation of arsenic in root and shoot dry weight of rice. The arsenic accumulation was 5.4-, 4.7-, and 7.3-fold higher at 24 hours in roots, old leaves, and young leaves, respectively. Torres et al. (2020) reported a dose-dependent model for arsenic accumulation in roots and areal parts of the strawberry plant.

13.6.3.2 Microbial and Fungal Remediation

Microorganisms present in water and soil sources could be considered in the decomposition of arsenic. The degradation and alteration of arsenic by microbial communities are of particular importance. This process is entirely eco-friendly and cost-effective and leads to the cleansing of water and soil resources from the presence of chemical contaminants and heavy elements. Some of the gram-positive and gram-negative bacteria are resistant to arsenic and its compounds. In general, *ars* operon on plasmids is responsible for resistance to arsenic compounds (Ben Fekih et al. 2018). Microbial and fungal agents' bioremediation of arsenic could play a significant role in detoxifying and removing arsenic and its toxic compounds in agricultural soils. Some microorganisms could utilize soil contaminants as a source of nutrients by degrading and altering the chemical structure of their compounds in the soil. Oxidation and reduction are two important mechanisms of action for the detoxification of arsenic by bacterial communities (Macur et al. 2004). Turpeinen et al. (2002) reported microbes' impact on the transformation of arsenic to arsenite, MMA acid, and DMA acid in contaminated soils.

Numerous microorganisms such as bacterial and fungal communities have the potential for the biodegradation of arsenic in the agroecosystems. The effects of phosphorus application and arbuscular mycorrhizal fungi on arsenic were evaluated by Cozzolino et al. (2010). They found the fungal agent's significant effect along with phosphorus on reducing arsenic accumulation in root and shoot of plants. The long-term effects of *Pseudomonas vancouverensis* strain m318 for potential phytoremediation of arsenic evaluated by Yang et al. (2020). They found the effectiveness of *Pteris vittata* as a hyperaccumulator species along with this bacterium for the accumulation of arsenic in agricultural soil. Roy et al. (2013) used the baker yeast *Saccharomyces cerevisiae* as a biosorbate to remediate the Arsenic in the groundwater. Table 13.2 provides examples of arsenic bioremediation by soil microorganisms.

13.6.3.3 Addition of Fertilizers to Soils

Fertilization using organic or mineral amendments is a well-known practical strategy for mitigating and removing arsenical compounds in soil and water environments. One of the main mechanisms of arsenic remediation by using fertilizers is its adsorption-desorption process and reducing arsenic bioavailability in the environment. Zafeiriou et al. (2019) evaluated the effects of phosphate fertilizer on arsenic compounds in mining lands. They found that more than 50% of the potentially available arsenic was desorbed using this fertilizer in the soil. In contrast, Kim et al. (2019) reported the effectiveness of phosphate fertilizers on reducing the mobility of arsenic in soil under controlled conditions. Chen et al. (2008) studied the effects of nitrogen fertilizers on arsenic and found a significant reduction in arsenic uptake by rice plants due to nitrogen fertilizers.

13.6.3.4 Other Methods

The use of biosorbents such as iron oxide-based adsorbents is one of the current effective methods in reducing the environmental effects of arsenic, especially in aquatic environments. Xie et al. (2015) applied an aquifer iron-coating technology for managing the arsenic-contaminated groundwater and reported the high level of arsenic removal by using this method. Some plant-based low-molecular-weight peptides such as phytochelatins have the potential to bind some metals in plants. Schmoger et al. (2000) reported the use of phytochelatin for the detoxification of arsenic in different plants.

13.7 Conclusions

Environmental contaminations due to heavy metals, chemicals, and especially arsenical compounds, could directly affect human life, animal, plants, and microorganisms' safety. Therefore, it is essential to understand all the processes involved in the transfer and mobility, absorption, metabolism, leaching, evaporation, accumulation, and remediation of these compounds in different environments. This chapter summarizes the important properties of arsenic, the processes affecting its fate, and the side effects and toxicity of arsenic for humans and various plants. In the following, the most important strategies involving bioremediation, reducing accumulation, and removing arsenic from the environment are presented with emphasis on the latest studies. Accordingly, agricultural management such as irrigation and fertilization management practices, biotechnological perspective, and

Table 13.2 Arsenic remediation by using bacteria, fungi, and different amendments.

Pathway	Plant species/soil	Agents/amendments	Results	Reference
Bacteria	*Oryza sativa* L.	*Ochrobactrum tritici* As5	2.6 fold reduction in arsenic concentration	Moens et al. (2020)
		Ochrobactrum tritici SCII24	1.6 fold reduction in arsenic concentration	
		Serratia liquefaciens	28–52% reduction in available arsenic concentration	Cheng et al. (2020)
		Rhizobacteria	3.50–26.01% reduction in arsenic accumulation	Xiao et al. (2020)
		Arsenite-oxidizing bacteria	17% oxidation of arsenite after 60 days of incubation	Debiec-Andrzejewska et al. (2020)
	Hydrilla verticillata	Epiphytic bacteria	64.44% reduction in arsenic accumulation	Zhen et al. (2020)
Fungi	*Oryza sativa* L.	*Westerdykella aurantiaca*	28–37% reduction in arsenic accumulation	Verma et al. (2019)
		WaarsM gene of soil fungus *Westerdykella aurantiaca*	50–52% reduction in arsenic accumulation	Verma et al. (2018)
	Lens culinaris	Arbuscular Mycorrhizal Fungi	8–45 $mg\,kg^{-1}$ reduction in arsenic concentration in roots and shoots of lentil	Alam et al. (2019)
	Zea mays	*Glomus mosseae*	505 $\mu g\,pot^{-1}$ arsenic accumulation in roots	Yu et al. (2009)
	Glycine max L.	*Rhizophagus intraradices*	Reduction of arsenic accumulation from 7.8 to 6.0 $mg\,kg^{-1}$	Spagnoletti and Lavado (2015)
Different amendments	*Oryza sativa* L.	Addition of leonardite to the soil	$91.86 \pm 2.04\%$ efficiency in adsorbing initial arsenic concentration	Dolphen and Thiravetyan (2019)
	Soil	Addition of farmyard manure	Reduction of extractable soil arsenic from 16.5 to 0.87 $mg\,kg^{-1}$ by enhancing soil microbial activity	Das and Das (2020)
	Oryza sativa L. cv. Khitish	Applications of lathyrus + vermicompost + poultry manure	13.84–23.33% reduction of arsenic uptake by root, shoot, and leaf	Rahaman et al. (2011)
	Oryza sativa L.	Selenite amendment	27.5% reduction in arsenic accumulation in rice grain	Wan et al. (2018)
	Oryza sativa L.	Selenate amendment	144–179% reduction in arsenic accumulation in root and shoot	Kumar et al. (2016)

phytoremediation by focusing on microbial and fungal communities and applying different amendments are discussed as the most important strategies for reducing the effects of arsenic on agricultural ecosystems in this chapter. However, the presence of arsenic compounds in the environment and the behavior and subsequent processes related to these compounds are very complex; we hope that by reading this chapter, some of the ambiguities and questions of readers about the behavior of arsenic in the environment and effective strategies to reduce the toxicity and accumulation of these compounds in the environment will be answered.

References

Abedi, T. and Mojiri, A. (2020). Arsenic uptake and accumulation mechanisms in rice species. *Plants* 9 (2): 129.

Abedin, M.J., Cotter-Howells, J., and Meharg, A.A. (2002). Arsenic uptake and accumulation in rice (*Oryza sativa* L.) irrigated with contaminated water. *Plant Soil* 240: 311–319.

Adriano, D.C. and Weber, J.T. (2001). Influence of fly ash on soil physical properties and turfgrass establishment. *J. Environ. Qual.* 30: 596–601.

Ahamed, S., Sengupta, M.K., Mukherjee, A. et al. (2006). Arsenic groundwater contamination and its health effects in the state of Uttar Pradesh (UP) in upper and middle Ganga plain, India: a severe danger. *Sci. Total Environ.* 370 (2–3): 310–322.

Alam, M.Z., Hoque, M.A., Ahammed, G.J., and Carpenter-Boggs, L. (2019). Arbuscular mycorrhizal fungi reduce arsenic uptake and improve plant growth in *Lens culinaris*. *PLoS One* 14 (5): e0211441.

Arai, Y. and Sparks, D.L. (2002). Residence time effects on arsenate surface speciation at the aluminum oxide-water interface. *Soil Sci.* 167: 303–314.

Argos, M., Kalra, T., Rathouz, P.J. et al. (2010). Arsenic exposure from drinking water, and all-cause and chronic-disease mortalities in Bangladesh (HEALS): a prospective cohort study. *Lancet* 376 (9737): 252–258.

Arnold, L.L., Eldan, M., Gemert, M. et al. (2003). Chronic studies evaluating the carcinogenicity of monomethylarsonic acid in rats and mice. *Toxicol.* 190 (3): 197–219.

Au, W.Y., Fong, B.M., Tam, S., and Kwong, Y.L. (2013). Feasibility of oral arsenic trioxide treatment for acute promyelocytic leukemia during hemodialysis. *Ann. Hematol.* 92: 417–418.

Aysha, M.I.J., Zakir, H.M., Haque, R. et al. (2017). Health risk assessment for population via consumption of vegetables grown in soils artificially contaminated with arsenic. *Arch. Curr. Res. Int.* 10 (3): 1–12.

Baig, J.A., Kazi, T.G., Shah, A.Q. et al. (2010). Evaluating the accumulation of arsenic in maize (*Zea mays* L.) plants from its growing media by cloud point extraction. *Food Chem. Toxicol.* 48 (11): 3051–3057.

Ben Fekih, I., Zhang, C., Li, Y.P. et al. (2018). Distribution of arsenic resistance genes in prokaryotes. *Front. Microbiol.* 9: 2473.

Benavente, O., Hernández, M.C., Melo, E. et al. (2020). Stabilizing arsenic in copper heap leaching residues. *Metals* 10: 1242.

Bhadauria, R. (2019). Arsenic toxicity: an overview. *Horticult Int. J.* 3 (1): 20–22.

Bhumbla, D.K. and Keefer, R.F. (1994). Arsenic mobilization and bioavailability in soils. In: *Arsenic in the Environment. Part 1: Cycling and Characteristics. Advances in Environmental Science and Technology*, vol. 26 (ed. J.O. Nriagu), 5182. USA: John Wiley & Sons, Inc.

Bolan, N., Mahimairaja, S., Kunhikrishnan, A., and Choppala, G. (2013). Phosphorus – arsenic interactions in variable-charge soils in relation to arsenic mobility and bioavailability. *Sci. Total Environ.* 463–464: 1154–1162.

Broberg, K., Ahmed, S., Engström, K. et al. (2014). Arsenic exposure in early pregnancy alters genome-wide DNA methylation in cord blood, particularly in boys. *J. Dev. Orig. Health Dis.* 5 (4): 288–298.

Caumette, G., Kocha, I., and Reimer, K.J. (2012). Arsenobetaine formation in plankton: a review of studies at the base of the aquatic food chain. *J. Environ. Monit.* 14: 2841–2853.

Chakraborty, S., Alam, M.O., Bhattacharya, T., and Singh, Y.N. (2014). Arsenic accumulation in food crops: a potential threat in Bengal Delta plain. *Water Qual Expo Health* 6: 233–246.

Chen, X.P., Zhu, Y.G., Hong, M.N. et al. (2008). Effects of different forms of nitrogen fertilizers on arsenic uptake by rice plants. *Environ. Toxicol. Chem.* 27 (4): 881–887.

Cheng, C., Nie, Z.W., He, L.Y., and Sheng, X.F. (2020). Rice-derived facultative endophytic *Serratia liquefaciens* F2 decreases rice grain arsenic accumulation in arsenic-polluted soil. *Environ. Pollut.* 259: 113832. https://doi.org/10.1016/j.envpol.2019.113832.

Chung, J.Y., Yu, S.D., and Hong, Y.S. (2014). Environmental source of arsenic exposure. *J. Prev. Med. Public Health* 47 (5): 253–257.

Cozzolino, V., Pigna, M., Di Meo, V. et al. (2010). Effects of arbuscular mycorrhizal inoculation and phosphorus supply on the growth of *Lactuca sativa* L. and arsenic and phosphorus availability in an arsenic polluted soil under nonsterile conditions. *Appl. Soil Ecol.* 45: 262–268.

Danielson, C., Houseworth, J., Skipworth, E. et al. (2006). Arsine toxicity treated with red blood cell and plasma exchanges. *Transfusion* 46 (9): 1576–1579.

Das, S.K. and Das, S.K. (2020). Influence of phosphorus and organic matter on microbial transformation of arsenic. *Environ. Technol. Innov.* 19: 100930. https://doi.org/10.1016/j.eti.2020.100930.

Debiec-Andrzejewska, K., Krucon, T., Piatkowska, K., and Drewniak, L. (2020). Enhancing the plants growth and arsenic uptake from soil using arsenite-oxidizing bacteria. *Environ. Pollut.* 264: 114692. https://doi.org/10.1016/j.envpol.2020.114692.

Dhankher, O.P., Li, Y., Rosen, B.P. et al. (2002). Engineering tolerance and hyperaccumulation of arsenic in plants by combining arsenate reductase and γ-glutamylcysteine synthetase expression. *Nat. Biotechnol.* 20: 1140.

Dolphen, R. and Thiravetyan, P. (2019). Reducing arsenic in rice grains by leonardite and arsenic – resistant endophytic bacteria. *Chemosphere* 223: 448–454.

Dradrach, A., Karczewska, A., Szopka, K., and Lewińska, K. (2020). Accumulation of arsenic by plants growing in the sites strongly contaminated by historical mining in the sudetes region of poland. *Int. J. Environ. Res. Public Health* 17 (9): 3342.

Eisler, R. (1988). Arsenic hazards to fish, wildlife, and invertebrates: a synoptic review. *Biol. Rep.* 85: 1–12.

Gani-Molla, M.O., Islam, M.A., and Hasanuzzaman, M. (2010). Identification of arsenic hyperaccumulating plants for the development of phytomitigation technology. *J. Phytol.* 2: 41–48.

Gebel, T. (1997). Arsenic and antimony: comparative approach on mechanistic toxicology. *Chem. Biol. Interact.* 107 (3): 131–144.

Hajalilou, B., Mosaferi, M., Khaleghi, F. et al. (2011). Effects of abandoned arsenic mine on water resources pollution in north west of Iran. *Health Promot. Perspect.* 1 (1): 62–70.

Hasanuzzaman, M. and Fujita, M. (2013). Exogenous sodium nitroprusside alleviates arsenic-induced oxidative stress in wheat (*Triticum aestivum* L.) seedlings by enhancing antioxidant defense and glyoxalase system. *Ecotoxicol.* 22: 584–596.

Henke, K.R. (2009). *Arsenic. Environmental Chemistry, Health Threats and Waste Treatment.* West Sussex: John Wiley and Sons.

Hossain, A., Joardar, J.C., and Imamul Huq, S.M. (2006). Comparison of arsenic accumulation by two ferns: *Pteris vittata* and *Nephrodium molle*. *Dhaka Univ. J. Biol Sci.* 15: 95–103.

Hsueh, Y.M., Chen, W.J., Lee, C.Y. et al. (2016). Association of arsenic methylation capacity with developmental delays and health status in children: a prospective case – control trial. *Sci. Rep.* 6: 37287.

Huang, J.W., Poynton, C.Y., Kochian, L.V., and Elless, M.P. (2004). Phytofiltration of arsenic from drinking water using arsenic-hyperaccumulating ferns. *Environ. Sci. Technol.* 38: 3412–3417.

Huq, S.M., Joardar, J.C., Parvin, S. et al. (2006). Arsenic contamination in food-chain: transfer of arsenic into food materials through groundwater irrigation. *J. Health Popul. Nutr.* 24 (3): 305–316.

Irvin, T.R. and Irgolic, K.J. (1995). In-vitro prenatal toxicity of trimethylarsine, trimethylarsine oxide and trimethylarsine sulfide. *Appl. Organomet. Chem.* 9 (4): 315–321.

Islam, S., Rahman, M.M., Islam, M.R., and Naidu, R. (2017). Effect of irrigation and genotypes towards reduction in arsenic load in rice. *Sci. Total Environ.* 609: 311–318.

Jiang, W., Zhang, S.S., Shan, X.X. et al. (2005). Adsorption of arsenate on soils. Part 1: laboratory batch experiments using 16 chinese soils with different physiochemical properties. *Environ. Pollut.* 138: 278–284.

Jones, A.S., Marini, J., Solo-Gabriele, H.M. et al. (2019). Arsenic, copper, and chromium from treated wood products in the U.S. disposal sector. *Waste Manag.* 87: 731–740.

Kaise, T. and Fukui, S. (1992). The chemical form and acute toxicity of arsenic compounds in marine organisms. *Appl. Organomet. Chem.* 6: 155–160.

Kaise, T., Horigochi, Y., Fukui, S. et al. (1992). Acute toxicity and metabolism of arsenochline in mice. *Appl. Organomet. Chem.* 6: 369–373.

Kertulis-Tartar, G.M., Rathinasabapathi, B., and Maa, L.Q. (2009). Characterization of glutathione reductase and catalase in the fronds of two *Pteris ferns* upon arsenic exposure. *Plant Physiol. Biochem.* 47: 960–965.

Khan, I., Ahmad, A., and Iqbal, M. (2009). Modulation of antioxidant defence system for arsenic detoxification in Indian mustard. *Ecotoxicol. Environ. Saf.* 72: 626–634.

Kim, M.S., Min, H.G., Kim, J.G., and Lee, S.R. (2019). Estimating arsenic mobility and phytotoxicity using two different phosphorous fertilizer release rates in soil. *Agron.* 9: 111.

Kumar, A., Ali, M., Kumar, R. et al. (2021). Arsenic exposure in indo gangetic plains of Bihar causing increased cancer risk. *Sci. Rep.* 11: 2376.

Kumar, A., Dixit, G., Singh, A.P. et al. (2016). Selenate mitigates arsenite toxicity in rice (*Oryza sativa* L.) by reducing arsenic uptake and ameliorates amino acid content and thiol metabolism. *Ecotoxicol. Environ. Saf.* 133: 350–359.

Kumar, S. and Trivedi, P.K. (2018). Glutathione S-transferases: role in combating abiotic stresses including arsenic detoxification in plants. *Front. Plant Sci.* 9: 751.

Kuroda, K., Yoshida, K., Yoshimura, M. et al. (2004). Microbial metabolite of dimethylarsinic acid is highly toxic and genotoxic. *Toxicol. Appl. Pharmacol.* 198 (3): 345–353.

Lee, C.H., Wu, C.H., Syu, C.H. et al. (2016). Effects of phosphorous application on arsenic toxicity to and uptake by rice seedlings in As-contaminated paddy soils. *Geoderma* 270: 60–67.

Li, C., Carrijo, D.R., Nakayama, Y. et al. (2019). Impact of alternate wetting and drying irrigation on arsenic uptake and speciation in flooded rice systems. *Agric. Ecosyst. Environ.* 272: 188–198.

Li, Y.Y., Wang, H.B., Wang, H.J. et al. (2014). Heavy metal pollution in vegetables grown in the vicinity of a multimetal mining area in Gejiu, China: total concentrations, speciation analysis, and health risk. *Environ. Sci. Pollut. Res.* 21 (21): 12569–12582.

Ma, L.Q., Kumar, K.M., Tu, C. et al. (2001). A fern that hyperaccumulates arsenic. *Nature* 409: 579.

Macur, R., Jackson, C., Botero, L. et al. (2004). Bacterial populations associated with the oxidation and reduction of arsenic in an unsaturated soil. *Environ. Sci. Technol.* 38 (1): 104–111.

Malik, J.A., Goel, S., Kaur, N. et al. (2012). Selenium antagonises the toxic effects of arsenic on mungbean (*Phaseolus aureus* Roxb.) plants by restricting its uptake and enhancing the antioxidative and detoxification mechanisms. *Environ. Exp. Bot.* 77: 242–248.

Mehdizadeh, M. and Mushtaq, W. (2020). Biological control of weeds by allelopathic compounds from different plants: a bioherbicide approach. In: *Natural Remedies for Pest, Disease and Weed Control* (ed. C. Egbuna and B. Sawicka), 107–117. Academic Press.

Mishra, B.K., Dubey, C.S., Shukla, D.P. et al. (2014). Concentration of arsenic by selected vegetables cultivated in the Yamuna flood plains (YFP) of Delhi, India. *Environ. Earth Sci.* 72 (9): 3281–3291.

Moens, M., Branco, R., and Morais, P.V. (2020). Arsenic accumulation by a rhizosphere bacterial strain *Ochrobactrum tritici* reduces rice plant arsenic levels. *World J. Microbiol. Biotechnol.* 36: 23. https://doi.org/10.1007/s11274-020-2800-0.

Moreno-Jiménez, E., Esteban, E., and Peñalosa, J.M. (2012). The fate of arsenic in soil-plant systems. *Rev. Environ. Contam. Toxicol.* 215: 1–37.

Neidhardt, H. (2021). Arbuscular mycorrhizal fungi alleviate negative effects of arsenic-induced stress on crop plants: a meta-analysis. *Plants People Planet* 3 (5): 523–535.

O'Reilly, S.E., Strawn, D.G., and Sparks, D.L. (2001). Residence time effects on arsenate adsorption/desorption mechanisms on goethite. *Soil Sci. Soc. Am. J.* 65: 67–77.

O'Neill, P. (1995). *Arsenic. Heavy Metals in Soil.* Glasgow: Blackie Academic & Professional.

Pajuelo, E., Rodríguez-Llorente, I.D., Dary, M., and Palomares, A.J. (2008). Toxic effects of arsenic on *Sinorhizobium–Medicago sativa* symbiotic interaction. *Environ. Pollut.* 154 (2): 203–211.

Panaullah, G.M., Alam, T., Hossain, M.B. et al. (2009). Arsenic toxicity to rice (*Oryza sativa* L.) in Bangladesh. *Plant Soil* 317: 31.

Pandey, A.K., Gedda, M.R., and Verma, A.K. (2020). Effect of arsenic stress on expression pattern of a rice specific miR156j at various developmental stages and their allied co-expression target networks. *Front. Plant Sci.* 11: https://doi.org/10.3389/fpls.2020.00752.

Peryea, F.J. (2001). Gardening on lead and arsenic containing soils. Washington State coop. Extension Bulletin No. EB1884, Vol. 1.

Pigna, M., Cozzolino, V., Violante, A., and Meharg, A.A. (2009). Influence of phosphate on the arsenic uptake by wheat (*Triticum durum* L.) irrigated with arsenic solutions at three different concentrations. *Water Air Soil Pollut.* 197: 371–380.

Rahaman, S., Sinha, A.C., and Mukhopadhyay, D. (2011). Effect of water regimes and organic matters on transport of arsenic in summer rice (*Oryza sativa* L.). *J. Environ. Sci.* 23 (4): 633–639.

Robinson, B., Kim, N., Marchetti, M. et al. (2006). Arsenic hyperaccumulation by aquatic macrophytes in the Taupo volcanic zone, New Zealand. *Environ. Exp. Bot.* 58: 206–215.

Rodriguez-Iruretagoiena, A., Trebolazabala, J., Martinez-Arkarazo, I. et al. (2015). Metals and metalloids in fruits of tomatoes (*Solanum lycopersicum*) and their cultivation soils in the Basque Country: concentrations and accumulation trends. *Food Chem.* 173: 1083–1089.

Rosas-Castor, J.M., Guzman-Mar, J.L., Alfaro-Barbosa, J.M. et al. (2014). Evaluation of the transfer of soil arsenic to maize crops in suburban areas of San Luis Potosi, Mexico. *Sci. Total Environ.* 497: 153–162.

Roy, D., Gaur, P., Neeraj Verma, N. et al. (2013). Bioremediation of arsenic (III) from water using baker yeast *Sacchromyces cerevisiae. Int. J. Environ. Biorem. Biodeg.* 1 (1): 14–19.

Schmoger, M.E.V., Oven, M., and Grill, E. (2000). Detoxification of arsenic by phytochelatins in plants. *Plant Physiol.* 122 (3): 793–801.

Seiter, J.M. (2009). *The Fate and Speciation of Arsenic in Soils and Poultry Production Systems.* Newark, Delaware: University of Delaware.

Sekhar, K.C., Kamala, C.T., Chary, N.S., and Mukherjee, A.B. (2007). Arsenic accumulation by *Talinum cuneifolium* – application for phytoremediation of arsenic-contaminated soils of Patancheru, Hyderband, India. In: *Arsenic in Soil and Groundwater Environment: Biogeochemical Interactions, Health Effects and Remediation. Trace metals and other contaminants in environment*, vol. 9 (ed. P. Bhattacharya, A.B. Mukherjee, J. Bundschuh, et al.), 315–338. Amsterdam: Elsevier.

Shah, A.H., Shahid, M., Khalid, S. et al. (2020). Assessment of arsenic exposure by drinking well water and associated carcinogenic risk in peri-urban areas of Vehari, Pakistan. *Environ. Geochem. Health* 42: 121–133.

Shankar, S., Shanker, U., and Shikha (2014). Arsenic contamination of groundwater: a review of sources, prevalence, health risks, and strategies for mitigation. *Sci. World J.* 2014: https://doi.org/10.1155/2014/304524.

Sheikhi, S., Faraji, Z., and Aslani, H. (2021). Arsenic health risk assessment and the evaluation of groundwater quality using GWQI and multivariate statistical analysis in rural areas, Hashtroud, Iran. *Environ. Sci. Pollut. Res.* 28: 3617–3631.

Shri, M., Singh, P.K., Kidwai, M. et al. (2019). Recent advances in arsenic metabolism in plants: current status, challenges and highlighted biotechnological intervention to reduce grain arsenic in rice. *Metallomics* 11 (3): 519–532.

Sinha, D., Biswas, J., and Bishayee, A. (2013). Nrf2-mediated redox signaling in arsenic carcinogenesis: a review. *Arch. Toxicol.* 87 (2): 383–396.

Smedley, P.L. and Kinniburgh, D.G. (2002). A review of the source, behaviour and distribution of arsenic in natural waters. *Appl. Geochem.* 17 (5): 517–568.

Smith, A.H., Guillermo, M., Yan, Y. et al. (2011). Evidence from chile that arsenic in drinking water may increase mortality from pulmonary tuberculosis. *Am. J. Epidemiol.* 173 (4): 414–420.

Sohrabi, N., Kalantari, N., Amiri, V. et al. (2020). A probabilistic-deterministic analysis of human health risk related to the exposure to potentially toxic elements in groundwater of Urmia coastal aquifer (NW of Iran) with a special focus on arsenic speciation and temporal variation. *Stoch. Env. Res. Risk A.* https://doi.org/10.1007/s00477-020-01934-6.

Spagnoletti, F. and Lavado, R.S. (2015). The arbuscular mycorrhiza *Rhizophagus intraradices* reduces the negative effects of arsenic on soybean plants. *Agronomy* 5 (2): 188–199.

Srivastava, S. and Sharma, Y.K. (2014). Arsenic induced changes in growth and metabolism of black gram seedlings (*Vigna Mungo* L.) and the role of phosphate as an ameliorating agent. *Environ. Process* 1: 431–445.

Stachnik, Ł., Korabiewski, B., Raczyk, J. et al. (2020). Arsenic pollution in quaternary sediments and water near a former gold mine. *Sci. Rep.* 10: 18458.

Strawn, D.G. (2018). Review of interactions between phosphorus and arsenic in soils from four case studies. *Geochem. Trans.* 19 (1): 10.

Syu, C.H., Yu, S.H., and Lee, D.Y. (2020). Effect of applying calcium peroxide on the accumulation of arsenic in rice plants grown in arsenic-elevated paddy soils. *Environ. Pollut.* 266 (2): 115140.

Tampubolon, K., Purba, E., Basyuni, M., and Hanafiah, D.S. (2020). Application of monosodium methyl arsenate with diuron herbicide to control the characteristics of glyphosate-resistant Eleusine indica at oil palm plantations. *Bulg. J. Agric. Sci.* 26 (5): 1003–1012.

Tan, Z., Zhang, X., Kang, T. et al. (2018). Arsenic sulfide amplifies JQ1 toxicity via mitochondrial pathway in gastric and colon cancer cells. *Drug Des Devel Ther.* 12: 3913–3927.

Tanaka, A. (2004). Toxicity of indium arsenide, gallium arsenide, and aluminium gallium arsenide. *Toxicol. Appl. Pharmacol.* 198 (3): 405–411.

Tanaka, A., Hisanaga, A., Hirata, M. et al. (1994). Pulmonary toxicity of indium arsenide and arsenic selenide following repeated intratracheal instillations to the lungs of hamsters. *Appl. Organomet. Chem.* 8 (3): 265–271.

Tapio, S. and Grosche, B. (2006). Arsenic in the aetiology of cancer. *Mutat. Res. Genet. Toxicol. Environ. Mutagen.* 612: 215–246.

Torres, A.I.G., Giráldez, I., Martínez, F. et al. (2020). Arsenic accumulation and speciation in strawberry plants exposed to inorganic arsenic enriched irrigation. *Food Chem.* 315: 126215.

Turpeinen, R., Pantsar-Kallio, M., and Kairesalo, T. (2002). Role of microbes in controlling the speciation of arsenic and production of arsines in contaminated soils. *Sci. Total Environ.* 285: 133–145.

Vellinger, C., Parant, M., Rousselle, P., and Usseglio-Polatera, P. (2012). Antagonistic toxicity of arsenate and cadmium in a freshwater amphipod (*Gammarus pulex*). *Ecotoxicol.* 21: 1817–1827.

Verma, S., Verma, P.K., and Chakrabarty, D. (2019). Arsenic bio-volatilization by engineered yeast promotes rice growth and reduces arsenic accumulation in grains. *Int. J. Environ. Res.* 13: 475–485.

Verma, S., Verma, P.K., Meher, A.K. et al. (2018). A novel fungal arsenic methyltransferase, WaarsM reduces grain arsenic accumulation in transgenic rice (*Oryza sativa* L.). *J. Hazard. Mater.* 344: 626–634.

Vromman, D., Flores-Bavestrello, A., Šlejkovec, Z. et al. (2011). Arsenic accumulation and distribution in relation to young seedling growth in *Atriplex atacamensis* Phil. *Sci. Total Environ.* 412–413: 286–295.

Wan, Y., Camara, A.Y., Huang, Q. et al. (2018). Arsenic uptake and accumulation in rice (*Oryza sativa* L.) with selenite fertilization and water management. *Ecotoxicol. Environ. Saf.* 156: 67–74.

Wang, J., Zeng, X., Zhang, H. et al. (2018a). Effect of exogenous phosphate on the lability and phytoavailability of arsenic in soils. *Chemosphere* 196: 540–547.

Wang, Y., Wu, Y., Cavanagh, J. et al. (2018b). Toxicity of arsenite to earthworms and subsequent effects on soil properties. *Soil Biol. Biochem.* 117: 36–47.

WHO (2001). *Arsenic and Arsenic Compounds (Environmental Health Criteria 224)*, 2nde. Geneva: World Health Organization, International Programme on Chemical Safety.

Xiao, A.W., Li, W.C., and Ye, Z.H. (2020). The effect of plant growth-promoting rhizobacteria (PGPR) on arsenic accumulation and the growth of rice plants (*Oryza sativa* L.). *Chemosphere* 242: 125136. https://doi.org/10.1016/j.chemosphere.2019.125136.

Xie, X., Wang, Y., Pi, K. et al. (2015). In situ treatment of arsenic contaminated groundwater by aquifer iron coating: experimental study. *Sci. Total Environ.* 527–528: 38–46.

Yadav, P., Srivastava, S., Patil, T. et al. (2021). Tracking the time-dependent and tissue-specific processes of arsenic accumulation and stress responses in rice (*Oryza sativa* L.). *J. Hazard. Mater.* 406: 124307. https://doi.org/10.1016/j.jhazmat.2020.124307.

Yan, Y., Guillermo, M., Catterina, F. et al. (2010). Kidney cancer mortality: fifty-year latency patterns related to arsenic exposure. *Epidemiol.* 21 (1): 103–108.

Yang, C., Ho, Y.N., Inoue, C., and Chien, M.F. (2020). Long-term effectiveness of microbe-assisted arsenic phytoremediation by *Pteris vittata* in field trials. *Sci. Total Environ.* 740: 140137.

Ye, J., Rensing, C., Rosen, B.P., and Zhu, Y.G. (2012). Arsenic biomethylation by photosynthetic organisms. *Trends Plant Sci.* 17: 155–162.

Ye, W.L., Khan, M.A., McGrath, S.P., and Zhao, F.J. (2011). Phytoremediation of arsenic contaminated paddy soils with *Pteris vittata* markedly reduces arsenic uptake by rice. *Environ. Pollut.* 159: 3739–3743.

Yoshida, T., Yamauchi, H., and Fan Sun, G. (2004). Chronic health effects in people exposed to arsenic via the drinking water: dose-response relationships in review. *Toxicol. Appl. Pharmacol.* 198 (3): 243–252.

Yu, W.H., Harvey, C.M., and Harvey, C.F. (2003). Arsenic in groundwater in Bangladesh: a geostatistical and epidemiological framework for evaluating health effects and potential remedies. *Water Resour. Res.* 39: 1146–1162.

Yu, Y., Zhang, S., Huang, H. et al. (2009). Arsenic accumulation and speciation in maize as affected by inoculation with arbuscular mycorrhizal fungus *Glomus mosseae*. *J. Agric. Food Chem.* 57 (9): 3695–3701.

Zafeiriou, I., Gasparatos, D., Kalyvas, G. et al. (2019). Desorption of arsenic from calcareous mine affected soils by phosphate fertilizers application in relation to soil properties and as partitioning. *Soil Syst.* 3: 54.

Zhang, H. and Selim, H.M. (2005). Kinetics of arsenate adsorption-desorption in soils. *Environ. Sci. Technol.* 39: 6101–6108.

Zhao, Z.W., Chai, L.Y., Peng, B. et al. (2017). Arsenic vitrification by copper slag based glass: mechanism and stability studies. *J. Non-Cryst. Solids* 466-467: 21–28.

Zhen, Z., Yan, C., and Zhao, Y. (2020). Influence of epiphytic bacteria on arsenic metabolism in *Hydrilla verticillata*. *Environ. Pollut.* 261: 114232. https://doi.org/10.1016/j.envpol.2020.114232.

14

Arsenic Removal by Phytoremediation Techniques

Zahra Souri[1], *Hamidreza Sharifan*[2], *Letúzia Maria de Oliveira*[3], *and Lucy Ngatia*[3]

[1] *Department of Biology, Faculty of Science, Razi University, Kermanshah, Iran*
[2] *Department of Natural Science, Albany State University, Albany, GA, USA*
[3] *Center for Water Resources, College of Agriculture and Food Sciences, Florida A&M University, Tallahassee, FL, USA*

14.1 Arsenic Presence in the Environment

Arsenic (As) as a carcinogen is mostly found in air, soil, water, and biota (Chung et al. 2014). As contamination in potable water and food is a global problem. Asian countries, such as Bangladesh, India, Cambodia, and Vietnam, are reported the worst countries, with the rural and remote areas suffering from arsenic-polluted water (Palit et al. 2019). Exposure to inorganic As, essentially through potable water, can lead to serious health issues, such as cardiovascular, neurological, hematological, renal, and respiratory problems. Research has been conducted to provide economical and clean technologies for the removal of As in groundwater and soil.

Present in the environment through natural and anthropogenic sources, As concentration depends on the composition of the geological material. While As releases into the environment from natural sources, such rocks weathering, volcanic events, and bacterial volatilization (ATSDR 2021), use of pesticides/herbicides, preservatives of woods, fossil fuels consumption, and phosphate-based fertilizers are anthropogenic activities that increase the levels of As in soil and aquatic environments, which contribute to dispersion in air and water as suspended particles.

In potable water, the maximum contaminant level of As is $10\,\mu g\,l^{-1}$, established in 2001 by the Environmental Protection Agency (EPA). Most of the As in natural waters is present in inorganic forms as oxy anions or as neutral species. Inorganic As is found in different species as arsenate As(V) or arsenite As(III) species. Depending on redox potential (Eh) and pH of the water, As(III) holds higher toxicity and is more challenging to remove from the environment than As(V) (USEPA 2001). The predominant As(III) species are uncharged H_3AsO_3 with dissociation constants $pK_{a1} = 9$, while the primary As(V) species are monovalent H_2AsO^{-4} and divalent $HAsO_2{}^{-4}$ with dissociation constants of $pK_{a1} = 2.3$ and $pK_{a2} = 6.8$

Arsenic in Plants: Uptake, Consequences and Remediation Techniques, First Edition.
Edited by Prabhat Kumar Srivastava, Rachana Singh, Parul Parihar, and Sheo Mohan Prasad.

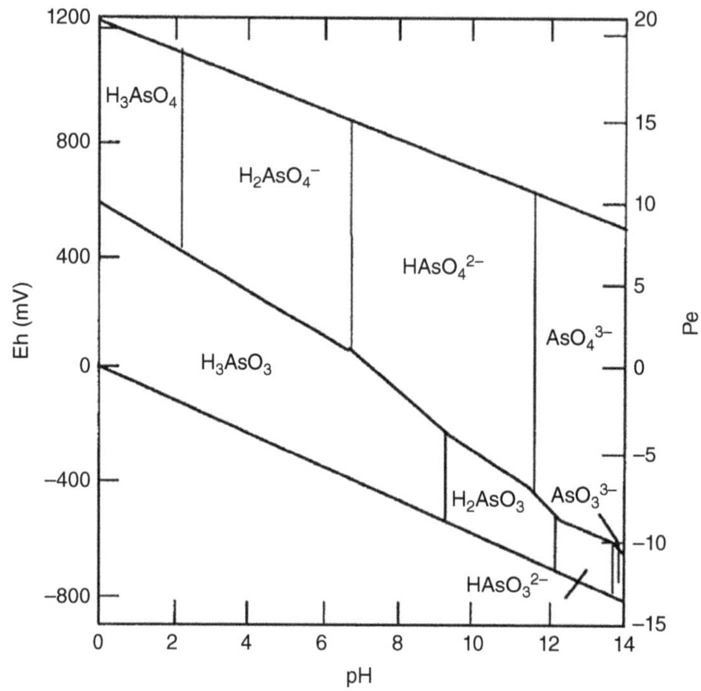

Figure 14.1 Eh-pH diagram of aqueous As species in the system As—O_2—H_2O at 25 °C and 1 bar total pressure.

(Mok and Wai 1990; Anderson and Bruland 1991; Shankar et al. 2014). Arsenate is favored in oxic waters and may be detected in covered pH range of most groundwater and surface water bodies as the oxyanions H_2AsO^{-4} (pH < 6.8) and $HAsO_2^{-4}$ (pH > 6.8), and As(III) is most found in anoxic waters, where the As neutral species (H_3AsO_3) predominates (pH < 9.2) (Figure 14.1) (Mok and Wai 1990; Anderson et al. 1991; Shankar and Shankar 2014).

The range of As levels found naturally in soil often varies between 0.1 and 40 mg kg^{-1}. However, As concentration in clean soils may reach up to 20 mg kg^{-1} while in contaminated soils it may be up to 600 mg kg^{-1} (Kabata Pendias 2011). In the soil As concentration depends on the parent material, background concentration, and soil type (Kabata-Pendias 2011). Soils originated from sedimentary rocks often contain 1.0–13 mg kg^{-1} of As and magmatic rocks may contain less than 0.5 to 2.6 mg kg^{-1} of As, (Kabata-Pendias 2011). As tends to accumulate on the earth surface due to atmospheric settlement and plant life cycle. In addition, oxides-anions and organic matter caused reduction in As mobility.

As from intact rocks often accumulate in the hydrated phases of oxy-hydroxides, mostly linked with sulfide (As sulfide) such as arsenopyrite, converting to As trioxide (Mandal and Suzuki 2002). In the soil As is also affected by pedogenesis, in which sandy soils indicate less As levels compared to organic soils (Mandal and Suzuki 2002; Kabata-Pendias 2011). In acidic soils, As is most bound to iron and aluminum oxides to form aluminum arsenate ($AlAsO_4$) and ferric arsenate ($FeAsO_4$), and in alkaline soils, As is bound with calcium, which predominantly forms calcium arsenate (Kabata-Pendias 2011).

14.2 Arsenic Contamination and its Effects on Human Health

Arsenic contamination causes adverse impacts on human health and nutritional values (Akter and Ali 2011). The guidelines of World Health Organization (WHO) set a value for As in potable water at 10 μg l^{-1} based on the typical 2 liter (l) daily water consumption (Chakraborti et al. 2010). In As-affected areas, it has been reported that on average adults may consume 3–4 l of water per day (Milton et al. 2006). Furthermore, farmers or other labor-intensive occupations may lead to 4–6 l daily water uptake (Chakraborti et al. 2010). In addition, significant amounts of As have been found in crops, specifically in rice and dietary vegetables, triggering their toxicity in consumers (Das et al. 2004). Serious health hazards could occur as a result of drinking As-contaminated water for long time (5–15 years). However, toxicity symptoms may appear in shorter time (i.e. 2.5 years) upon exposure to high levels of As contamination (Harvey et al. 2006).

After arsenic intake by humans, it is estimated that 50% of the arsenic is released in the urine (Das 1995), in addition small portions are excreted through the skin, feces, nail, and hair. Therefore, As in urine, skin, feces, nail, hair, and respiratory tissues indicates as an exhibit of the arsenic health threat to the population (Das et al. 1995; Goldsmith et al. 1972). For example, it was observed in Taiwan that for each 1 μg l^{-1} increment of As in groundwater, there was an increase of inorganic As metabolites in urine, hair, and toenail in the order of 1.0, 0.9, and 2.7 μg l^{-1}, respectively (Chen et al. 2005). Molecular, environmental, and epidemiological studies have explained biomarkers of As contact, susceptibility, and effects for As-related health hazards. As levels in nail, urine, and hair are measured as biomarkers for short exposure time (less than 365 days) internal dose, while skin hyperpigmentation and/or palmoplantar hyperkeratosis may be considered for long-term (more than a year) internal dose, and biologically effective dose exposure markers are shown by ratio of monomethylarsonic acid in overall metabolites of inorganic As detected in urine (Chen et al. 2005).

The As health hazard could range from cardiovascular problems to cancers of the kidney, bladder, lungs, and skin lesions (Hossain 2006; Shao et al. 2021; Tsai et al. 2021). Long-term contact with inorganic As from potable water has been reported to cause carcinogenic effects and vascular issues (Chen et al. 2005; Shao et al. 2021). It has been documented that 10% of humans who drink water that contains 500 μg l^{-1} As may succumb to cancer, including lung, bladder, and skin cancers (Smith et al. 2002). It has specifically been reported in Bangladesh that approximately 0.2 million population are highly prone to cancer development on an annual basis as a result of the consumption of As-contaminated water (Ahsan et al. 2006).

Different body organs in gastrointestinal, respiratory, renal, dermal, hepatic, nervous, hematopoietic, ophthalmic, and cardiovascular tissues are involved in subacute and acute toxicity of As (Chen et al. 1997a). However, long-term exposure to a nonlethal dose of As could result in chronic health threats with no specific organotropism (Chen et al. 1997b). In addition, the effects that target the uterus, the genitourinary tract, and other human tissues have been reported in the advanced stages of As toxicity (Jain and Ali 2000). Also, high levels of As in potable water have been reported to result in proliferation of spontaneous abortions and stillbirths as illustrated in Table 14.1 (Csanady and Straub 1995; Rahman et al. 2010; Rahman et al. 2007). A mother's As feeding is associated with the expression of

Table 14.1 Demographic analysis in response to arsenic concentration in urine and spontaneous abortion in 2002–2003 in Matlab, Bangladesh.

Age groups	Tested people (%)	Urine arsenic in gestational week 8 median ($\mu g\,l^{-1}$)	Spontaneous abortion or (95% CI)
<20	18	89	1.31 (0.88–1.96)
20–24	28	76	1
25–29	26	81	1.12 (0.77–1.62)
30–39	26	79	1.89 (1.34–2.65)
≥40	2	68	4.48 (2.25–8.90)

Source: Modified from Rahman et al. (2010).

fetal gene in newborn cord blood (Fry et al. 2007). Chakraborti et al. (2010) reported re-examination of more than 800 patients in Bangladesh who were diagnosed with arsenical skin lesions in the age group of 9–12 years (Pabna, Jessore, Kushtia, Chuadanga, Narayanaganj and Meherpur). The exhibition of arsenical skin lesions in people drinking As-contaminated water is considered a late manifestation of As toxicity (Jain and Ali 2000). Therefore, there are still difficulties in water management in developing countries including Bangladesh and West Bengal, India (Chakraborti et al. 2010). Such ongoing contaminations of natural waters, highlights the need for the water safety and accessibility. In addition, local contribution, co-operation, and education are highly important to control the effects of widespread As poisoning.

14.3 Arsenic Toxicity in Plants

In recent years, soil contamination by As containing molecules has been documented as a serious environmental threat, especially in the agricultural practices (Panda et al. 2010; Hettick et al. 2015; Cantamessa et al. 2020). Environmental events in soil might affect As speciation at liquid–solid interphases that affect plant' As uptake and As biogeochemical behavior in the soil crops microbiome (Rafiq et al. 2017; Abbas et al. 2018). Arsenic is extremely toxic to herbs and crops, leading to growth and development failure (Souri and Karimi 2017). The initial As exposure of plants occurs during its uptake by plants. Because As is a non-essential element, no specific mechanisms are developed in plants for its uptake. Therefore, its uptake takes place by sharing the similar protein transporters used for the uptake of essential elements by plants.

The As uptake rate depends on soil pH, organic substances, H_2O content, As level, speciation, uptake mechanisms, and plant species (Zhao et al. 2009; Finnegan and Chen 2012; Souri et al. 2017a). The As speciation in soil is linked to two common As forms As(III/V), and methylated As can affect As uptake via roots, which accumulate certain forms of As by designated pathways and transporters (Zhao et al. 2009; Gupta et al. 2011; Farooq et al. 2016). For example, in aerobic conditions, methylation of As by microbial activities to 1-ethylarsonic acid (MMA), 2-ethylarseinic acid (DMA), and 3-ethylarsine oxide (TMAsO) leads to its high bioavailability (Huang et al. 2011; Ye et al. 2012). The inorganic As is shown to have higher toxic effects compared to organic As. Also, compared to As(V), As(III) is more soluble and toxic. It has been demonstrated that As(III) is able to react with sulfhydryl (—SH) groups of proteins, resulting in inhibitory effects of cellular function and finally cellular death (Finnegan and Chen 2012; Nath et al. 2014; Souri et al. 2017a, 2017b; Abbas et al. 2018). In aerobic soils, As(V) is the highest available As species, whereas in flooded paddy soils, As(V) contains less ratio of total As (Huang et al. 2011; Jia et al. 2014). Phosphate transporters, including AtPht1 take up As(V);1/4 in *Arabidopsis* and OsPht1;8 (OsPT8) in rice (Shin et al. 2004; Wang et al. 2016b; Chen et al. 2017). An increase in phosphate level leads to reduced As(V) uptake in crops and vice versa (Meharg and Macnair 1992; Lihong and Guilan 2009; Karimi and Souri 2015). The reduction of As(V) to As(III) is achieved by the arsenate reductase (HAC) (Shi et al. 2016) (Figure 14.2). Under anaerobic conditions, As is mostly present as As(III) and its uptake is catalyzed by aquaporin proteins and nodulin-26-like-intrinsic proteins (NIPs) (Figure 14.2) (Chen et al. 2016; Zhao et al. 2009). Katsuhara et al. 2014 reported that in rice plant HvNIP1;2 and OsNIP3;3 are as arsenous acid-permeable NIPs. In addition, silicic acid-based transporters of Lsi1 (NIP2;1) and Lsi2 (Figure 14.2) enable the transport of arsenous acid in the roots of rice (Ma et al. 2008; Souri et al. 2017a). They both localize to the exodermal and endodermal cells of rice roots, however, on opposite sides, namely Lsi1 to the distal side and Lsi2 to the proximal sides (Ma et al. 2008). Both As(III) and Si pass through a similar mechanism for the root accumulation, and xylem loading occurs by OsLsi2 (Ma et al. 2008; Souri et al. 2017a). Mutants defective in either Lsi1 or Lsi2 are found to expressively lower the As accumulation in the shoot, including grains (Ma et al. 2008). Partial accumulation of As

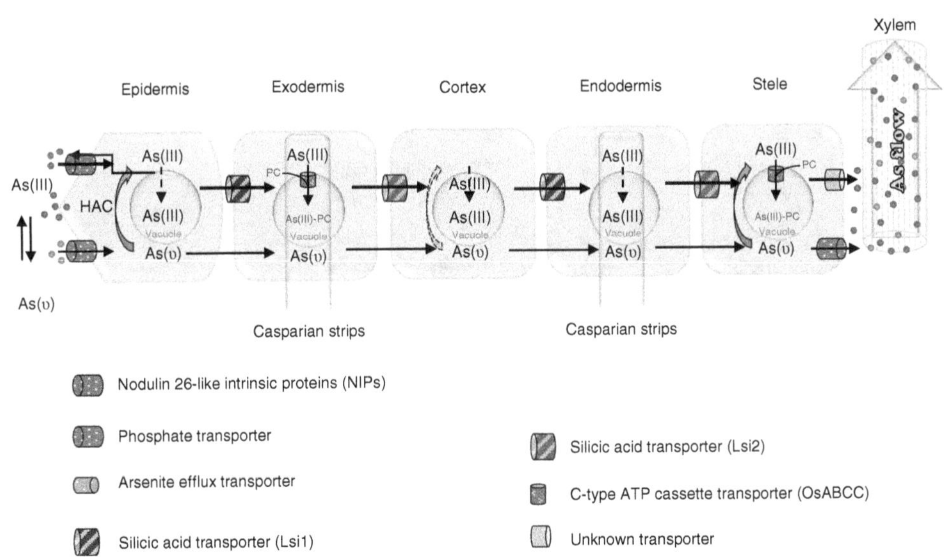

Figure 14.2 Overview of arsenic (As) uptake, transport, and translocation in rice. Nodulin 26-like intrinsic (NIP) aquaporins, phosphate transporter, and silicic acid transporters (Lsi1 and Lsi2) contribute to the uptake of arsenite (As(III)) and arsenate (As(V)). As(V) reduce to As(III) by arsenate reductase (HAC). Phytochelatin (PC) coordinates with As(III) to create molecules that sequester in vacuoles through C-type ATP cassette transporter (OsABCC). An unknown protein transporter is also responsible in As-(III) loading to xylem.

root tissues is released again, another part is sequestered in the root cell vacuole, while the remaining fraction is transported to the shoot. Recent finding stated that a C-type ATP cassette transporter in rice, OsABCC (Figure 14.2), may be responsible in As detoxification (Song et al. 2010) through sequestering As–PC inside the vacuoles, which hosts the cellular tonoplast and enhances the As resistance expressing PC synthase. Also, OsABCC1 drives a key role in preventing As transport to the rice grain (Song et al. 2010; Souri et al. 2017a).

Plant/crop growth is restricted through excessive As level in irrigation, soil, and also a gate to the trophic level; As in the foods causes more harm to human health (Abbas et al. 2018). It has also been demonstrated that As exposure decreases the leaf area, number of leaves, and fresh and dry biomass of plants (Figure 14.3) (Nath et al. 2014; Souri et al. 2018; 2020; 2021a; 2021b). The limitation of plant growth could be explained by As inhibitory effects of both cell

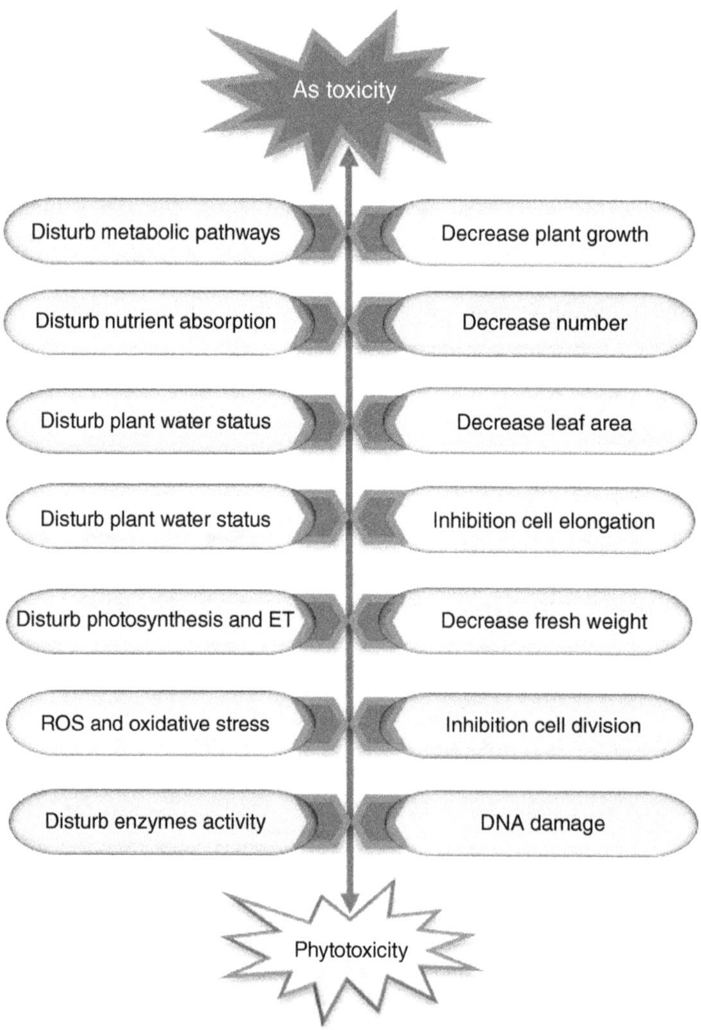

Figure 14.3 Potential detrimental effects of As exposure on plants that lead to phytotoxicity symptoms.

elongation and division. It has also reported that in crops growing in As-contaminated soils, As can disturb metabolic pathways like inhibit the nutrient uptake, interrupt water status (Figure 14.3), react with the enzymes' functional groups, and substitute the essential ions (Gunes et al. 2010; Karimi and Souri 2015; Khalid et al. 2017; Rafiq et al. 2017). Moreover, As stress is one of the driving factors that affects photosynthesis in terms of CO_2 absorption, electron transport, photophosphorylation, and the activity of enzymes (Duman et al. 2010; Abbas et al. 2018).

In general, As can cause the over-generation of reactive oxygen species (ROS) that causes progressive oxidative stress in plants (Figure 14.3). ROS induction in plants/crops is linked to the reduction process from As(V) to As(III) (Souri et al. 2017a; Abbas et al. 2018). On the other hand, when crops are exposed to toxic levels of As, they induce excessive amount of ROS, including superoxide anion (O_2^-), hydrogen peroxide (H_2O_2), and hydroxyl radical ($^{\cdot}OH$), which results in oxidation and dysfunction of important macro/micromolecules in plant cells (Mishra et al. 2019; Kaya et al. 2020; Souri et al. 2021a). In addition, ROS is dangerous for metabolic pathways that lead to permanent damage in key macromolecules of carbohydrate, DNA, lipids, and proteins (Mishra et al. 2019; Kaya et al. 2020; Souri et al. 2020, 2021a, 2021b). The oxidative stress in plants/crops is moderated through complex defensive mechanisms, which empower plant's immune system to counteract the ROS over-accumulation to protect the cells under oxidative stress (Karimi and Souri 2016; Kaya et al. 2020; Souri et al. 2021a). The ability of plants/crops to neutralize the toxic effects of ROS seems to be the main determinant of their tolerance to As stress (Souri et al. 2018; 2021a). Crops with high tolerance are able to lower As toxic levels by utilizing a number of mechanisms, which involve synthesizing the metal binding proteins such as glutathione (GSH) and PCs or excreting certain amounts of As from cells via specific transporters or compartmentalization (Souri et al. 2020, 2021a, 2021b).

14.4 Arsenic Attenuation by Phytoremediation Technology

Phytoremediation is a biological technique that uses plants/crops and associated microorganisms to eliminate or retain pollutants or lower their bioavailability in the environment (Berti and Cunningham 2000; Tangahu et al. 2011; Souri et al. 2017a). Phytoremediation is classified based on the pollutants, the types of plants/crops, site conditions, and level of clean-up required (Figure 14.4). There are four basic types of phytoremediation techniques: (i) phytoextraction – plants take up and accumulate contaminants into shoots, (ii) rhizofiltration – plants take up pollutants by plant roots from aqueous environment, (iii) phytostabilization – plants reduce the metal mobility in the soil, and (iv) phytovolatilization – harnessing plants/crops to extract contaminants from the environment and release them into the atmosphere in a volatile form (Macek et al. 2009). Advantages of using phytoremediation are (i) cost effectiveness of remediating groundwater and soils contaminated with metals, organics, and radionuclides and also it does not require huge equipment for installation, (ii) environment-friendly method, (iii) applicability – plants can be easily grown and it can be applied over a large land area, (iv) it prevents metal leaching and erosion through stabilizing contaminants, and (v) it improves soil fertility by recycling of nutrients in situ (Jacob et al. 2018; Wuana and Okieimen 2011).

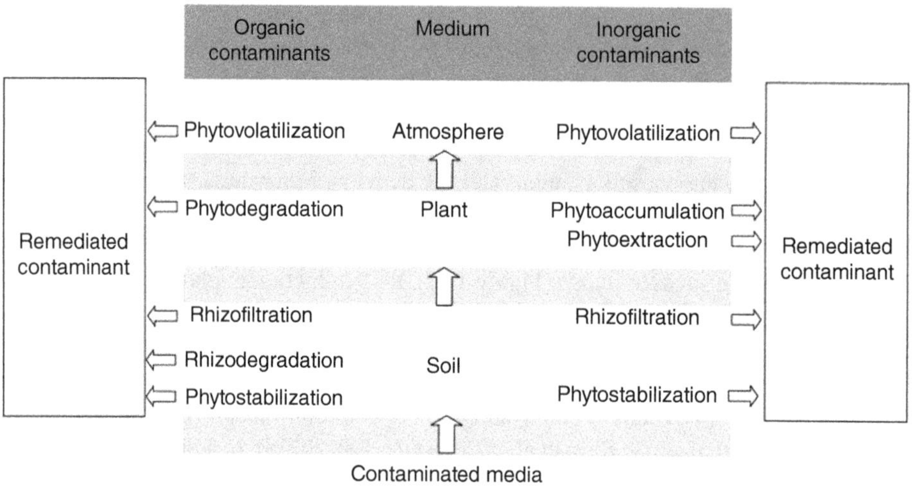

Figure 14.4 Uptake mechanisms on phytoremediation technology. *Source:* Tangahu et al. 2011/Public Domain CC BY 4.0.

14.5 Phytoextraction

Phytoextraction employs plants/crops for the extraction of contaminants from groundwater and soil. This process is the removal of metals, metalloids, radionuclides, and certain organic compounds by the ability of plants to accumulate these contaminants in the plant bodies (Souri et al. 2019; Wang et al. 2020). This process works by harvesting of shoots, which allows the metals to be removed from the soil; planting of one or more species that are As hyperaccumulators, and applying amendments to improve soil fertility.

Some plants/crops accumulate a large quantity of As with upgraded translocation from the plant root to aerial tissues. These plants do not display any noticeable stress symptoms when grown under high As contamination; however, hyperaccumulator plants have a low biomass production and low growth rate compared to non-tolerant plants. These plants are called hyperaccumulators that store large amounts of contaminants from groundwater and soil, even some of them that are not required for plant functioning, accumulating more As in aerial tissues (leaves and stems) compared to the roots (Souri et al. 2017a).

14.6 Arsenic Hyperaccumulation by Plants

Arsenic enters in the plant roots as AsO_4^{3-}, which occurs via high affinity phosphate transporter (DiTusa et al. 2016). When residing in plant cells, arsenate is reduced to arsenite by one of several different HAC (Mukhopadhyay and Rosen 2002). The remaining AsO_4^{3-} can be loaded into xylem by phosphate transporters (Ma et al. 2008; Rascio and Navari-Izzo 2011). Arsenite uptake occurs by nodulin intrinsic proteins (NIPs) (Souri et al. 2017a; Pommerrenig et al. 2020).

Pteris vittata is known for being an As hyperaccumulator (Ma et al. 2001); it can take up high amounts of As from water and soil and translocate it into the aerial parts (Pommerrenig et al. 2020). The As translocation in *P. vittata* takes place mostly as As(III) that accounts for more than 90% of As in the xylem (Su et al. 2008). Plants species such as *Adiantum capillus-veneris*, *Christella dentata*, and *Phragmites karka* has the ability to remove As from contaminated soils (Bhattacharya et al. 2021). The majority of the plants that accumulated As in their aboveground parts are not readily adaptable to different environmental conditions, and their slow growth rates restrict their remediation efficiency, especially in paddy soil (Bhattacharya et al. 2021).

14.7 Phytostabilization

Phytostabilization process uses tolerant plants to immobilize pollutants in the roots, limiting their mobility and bioavailability in the soil (Yan et al. 2020). This process occurs in precipitation or reduction of the pollutants in the rhizosphere that leads to assimilation and sequestration in root, or adsorption onto root cell walls, preventing the pollutants from translocating to aboveground plant tissues (Mendez and Maier 2008). There are many advantages for the removal of contaminants by phytostabilization process: prevent soil erosion and runoff, leaching, and soil dispersion (Raskin and Ensley 2000; Yan et al. 2020).

Arsenic phytostabilization plants should include a high As tolerance, a low As translocation factor, dense roots development, and large content of biomass (Yan et al. 2020). Plants that meet the mentioned requirements may be used for phytostabilization of As in contaminated soils, especially the plants found growing in mine tailing.

King et al. (2008) conducted a research with *Eucalyptus cladocalyx* in mining tails contaminated with As and treated with biosolid amendments in comparison with untreated tailings. The authors observed no effect on As availability or soil pH on this plant.

In a mine tailing in Arizona, a phytostabilization project for 3-years field basis was carried out; tailings were contaminated with As ($3\,g\,kg^{-1}$) from arsenopyrite (Hammond et al. 2020). The tails were modified in the top 20 cm with 100, 150, and $200\,g\,kg^{-1}$ of composted organic matter and cultivated with native plants. Authors found that less than 1% of total As was in water-extractable form. Association with secondary minerals, demonstrating that organic amendment associated with phytostabilization practices did not significantly enhance As mobilization over non-amended controls (Hammond et al. 2020), attenuated more than 99% of As(V) released through arsenopyrite weathering.

14.8 Phytovolatilization

In phytovolatilization process, plants take up contaminants, translocate to shoot, and turn them to volatile forms and subsequently release them to the atmosphere through transpiration (Raskin et al. 1997; He et al. 2001; Tangahu et al. 2011). The process takes place as growing crops absorb water and organic contaminants from the soil and transport them from the xylem. Some pollutants move through the vascular tissues to the leaves and evaporate or volatilize into the air. The advantage of the phytovolatilization process is that the pollutants are removed from the soil without further plant disposal (Yan et al. 2020). However, in urban areas, this technique is associated with health risk, due the exposure of contaminated air (Zhang et al. 2020).

14.9 Rhizofiltration

Rhizofiltration method uses plants rhizosphere to accumulate, concentrate, and precipitate pollutants from aqueous solutions. The soil's rhizosphere area is surrounded by the roots, which have a dense and diverse fauna. The contaminants in the rhizosphere zone are adsorbed onto the root surface and often are absorbed through the plant roots. In this process the contaminants do not have to be translocated above the ground.

Plants used for rhizofiltration are adapted to the pollutant in a hydroponically grown system before exposure to the pollutants. After the plants develop a large root system, the contaminant agents are added to the water source to adapt the plant. After the acclimation time, the plants are transferred into the polluted area. Plants are harvested after the roots of the plants take up the contaminant from polluted water (Dushenkov et al. 1995).

Crops that accumulate and tolerate high levels of chemical pollutants with easy operation and low associated cost would be the ideal plants for rhizofiltration process (Dushenkov et al. 1995). In experiments with *Eleocharis macrostachya* in the wetland system contaminated with As with different levels of As (0.1 and 0.5 $mg\,l^{-1}$), this plant is able to remove up to 99% of the As from the water and accumulate in the roots.

14.10 Novel Approaches of Phytoremediation Technology

14.10.1 Using Nanotechnology

Nanomaterials and nanocomposites have been a revolutionary advancement in arsenic removal process. Chemically, nanomaterials are provided from organic sources (i.e. quantum dots, nanotubes, and fullerene) or inorganic sources (i.e. zero-valent metals and metal oxides/salts). Their physicochemical properties include size range (<100 nm), shape, surface area, aggregation state, zeta potential, and density. Their unique properties classify them into three fundamental groups: (i) nanoparticulates, (ii) nanoplatelets, and (iii) nanofibers.

Nanoparticles (NPs) in phytoremediation must hold specific physicochemical and coatings characteristics; unique mechanical stability, electrical, and thermal conductivity (Rossi et al. 2019; Sharifan 2020) make them highly suitable for reduction of As into two primary forms of soil amendments and foliar application.

14.10.2 Nanoparticles in Soil

There have been many NPs (i.e. iron-zirconium, cerium-manganese, and iron-chromium) applied in As-contaminated soil to reduce the As uptake. For example, the copper oxide NPs significantly lowered the total As uptake in rice tissues (Wang et al. 2018, 2019).

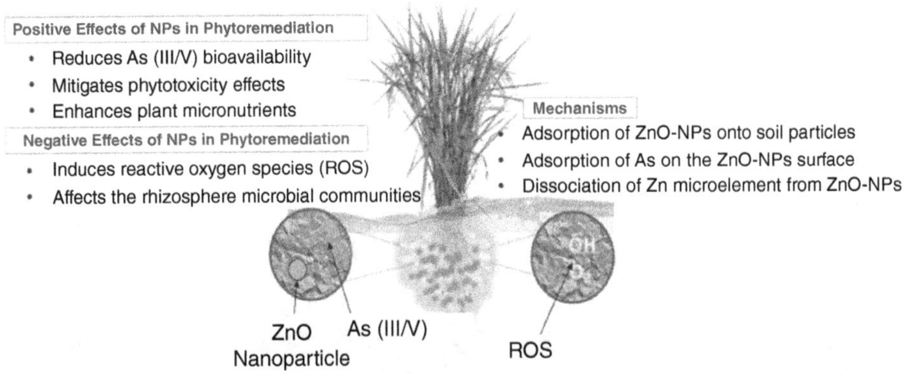

Figure 14.5 Positive and negative effects of presence of NPs in soil and the potential drying mechanism in nano-enabled phytoremediation process.

However, among many applied NPs, zinc oxide nanoparticles (ZnO-NPs) have been considered as efficient nanofertilizers and nanoadsorbents because of the prevalent Zn deficiency in farming soils worldwide (Milani et al. 2015; Dimkpa and Bindraban 2017). ZnO-NPs also may act as antimicrobial agents at greater concentrations. The multiple functions of ZnO-NPs make it an ideal candidate for a wide range of uses in agronomy (Sharifan et al. 2019). Ma et al. (2020b) revealed that ZnO-NPs are able to reduce the As accumulation in rice root and shoot by 39.5 and 60.2%, respectively. This significant reduction was attributed to the effects of ZnO-NPs in the reduction of inorganic As(III) and organic As species (Ma et al. 2020a).

In a similar study, Wu et al. 2020 demonstrated that 100 mg l^{-1} ZnO-NPs amendment in soil reduced arsenic amount in roots and shoots by 72 and 56% and improved the germination rate by 9%, total biomass by 66%, and elevated the chlorophyll level up to 40.1%, while the superoxide dismutase (SOD) and catalase (CAT) levels were increased by 23 and 61%, respectively (Wu et al. 2020). Figure 14.5 briefly describes the advantages and disadvantages of this technique, which needs further investigation. Despite phytostabilization of As in soil by ZnO-NPs, there is a new development in the application of NPs that mediates uptake of As in hyperaccumulators. For example, Souri et al. (2021b) found glutathione-modified superparamagnetic iron oxide NPs ($n\text{Fe}_3\text{O}_4$-GSH) lead to 56% greater As accumulation in *Isatis cappadocica* shoots compared to plants exposed to As alone (Souri et al. 2021b). However, phytoremediation of As using NPs has been demonstrated in vitro and in contaminated aqueous systems, but their application to remediate the As uptake in plants in large-scale practice needs further investigations (Figure 14.6).

14.10.3 Foliar Application of Nanoparticles

Foliar spraying of NPs is a proposed approach that affects the plant uptake of heavy metal(loids). For example, in contrast to soil amendment with NPs, the foliar spraying of TiO_2 NPs demonstrated to be more effective in reducing Cd toxicity in Maize (*Zea mays* L.) (Lian et al. 2020). Similarly, the silicon (Si) amendment of soil can promote the As

Figure 14.6 Phytoremediation site in Changsha, China, demonstrates the intercropping strategies in reduction of As and other heavy metals in a contaminated soil. The picture was taken in September 2019.

release from iron (Fe) (hydr)oxides via competing for the sorption sites with As. While the foliar application of Si NPs was proposed as an alternative technique to reduce As plant uptake (Liu et al. 2014), foliar of Si on rice at the tillering stage noticeably reduced the inorganic As levels in targeted rice tissue at as high as 61.4%. Additionally, foliar of the mixed solution of Si and surfactant-Tween 80 resulted in a reduction of inorganic As in rice grain by 49.2% (Zhang et al. 2020). Additionally, foliar application of SiO_2 NPs were found to maintain the cell integrity by increasing 77.4% of the thickness of the rice cell wall and reducing the ROS level in the cell, while reducing the As uptake (Cui et al. 2020).

14.10.4 Intercrops and Rotation Cultivation

Intercropping technology is an economical technique to increase the annual production from As-contaminated soil with less environmental risk. Hence, intercropping of As hyperaccumulators with economic crops has been in practice in slight or moderate As-contaminated soils. Certain species have a high potential for intercropping. For example, *P. vittata* was suggested as an ideal hyperaccumulator due to its ability to reduce the As mobility in the rhizosphere area of the intercropping system. There are a variety of species for intercropping purposes to reduce the As uptake by target crops. Therefore, As hyperaccumulators can take As from the soil, and safe products can then be produced by the intercropped crops. A variety of hyperaccumulators have been proposed for the intercropping techniques. Huang et al. (2021) found a significant reduction of As in rice cultivar by intercropping of water spinach species (*Ipomoea aquatica* Forsk), water celery species (*Oenanthe javanica*), and Guangdong white arrowhead species (*Sagittaria sagittifolia* L. var) (Huang et al. 2021). Crop rotation over the years with hyperaccumulators is a practiced technique that led to the reduction of As uptake in crops (Neu et al. 2020).

14.10.5 Irrigation Regime Management

Water management is an efficient and inexpensive technique to manipulate As uptake in plants, however slight is known about its effect on As transport and transformation in the soil–plant system. For instance, continuous flooding of the soil creates anaerobic conditions that change the physiochemistry of soil leading to alter the bioavailability of As. During flood irrigation regimes, the soil redox potential (Eh) naturally remains below −100 mV, which drives inorganic As(V) reduction to As(III), and simultaneously, Fe hydroxides shift to soluble ferrous Fe(II). As a result, the As bioavailability raises due to the higher As(III) mobility and higher As dissociation from Fe hydroxide and other soil minerals (i.e. metal oxide) (Li et al. 2015). Therefore, the alternate wetting and drying irrigation strategies have been practiced in some rice-growing regions as it leads to less As bioavailability during the aerobic condition when the drying period takes place; As(III) oxidizes to As(V) and soluble form of Fe(II) converts to Fe hydroxides that hold higher sites for As(V) adsorption (Xu et al. 2008; Sun et al. 2009a, 2009b). Moreover, this approach efficiently reduces the required irrigation water and mitigates emissions of greenhouse gases without compromising rice yield.

14.10.6 Soil Oxyanions Management

Agricultural practices are integrated into phytoremediation techniques by nitrogen and perchlorate management in soil that affect As accumulation in plants (Wang et al. 2016b). For example, nitrate can elevate the Eh in soil leading to an oxidizing condition, while ammonium can cause a lower soil Eh, possibly leading to reduction of Fe(III) and release of As on Fe hydroxide (Yamaguchi et al. 2011; Ding et al. 2014; Zhou et al. 2016). Studies reported the presence of nitrate and perchlorate in rice paddy soil has elevated the rice grain as high as 93% and reduced the total As in dry biomass of rice grains by 34–45% (Ma et al. 2020b). The potential effects of such oxyanions on arsenic plant uptake may differ for each plant species depending on applied concentration, chemical properties, and growth stages. For example, presence of phosphate significantly enhanced *P. vittata* growth when the As level was raised to 400 mg kg^{-1}, suggesting a minimum ratio of 1 : 2 (P/As) to promote the growth of *P. vittata* (Singh and Ma 2006). However, the oxyanions can modify the soil Eh and consequently the soil microbial community that calls for further studies.

References

Abbas, G., Murtaza, B., Bibi, I. et al. (2018). Arsenic uptake, toxicity, detoxification, and speciation in plants: physiological, biochemical, and molecular aspects. *International Journal of Environmental Research and Public Health* 15 (1): 59.

Agency for Toxic Substance and Disease Registry – ATSDR (2021). *Health and Human Services.* Atlanta, GA: ATSDR https://www.atsdr.cdc.gov/spl/index.html (accessed 2019).

Ahsan, H., Chen, Y., Parvez, F. et al. (2006). Arsenic exposure from drinking water and risk of premalignant skin lesions in Bangladesh: baseline results from the health effects of arsenic longitudinal study. *American Journal of Epidemiology* 163 (12): 1138–1148.

Akter, A. and Ali, M.H. (2011). Arsenic contamination in groundwater and its proposed remedial measures. *International Journal of Environmental Science and Technology* 8 (2): 433–443.

Anderson, L.C. and Bruland, K.W. (1991). Biogeochemistry of arsenic in natural waters: the importance of methylated species. *Environmental Science & Technology* 25: 420–427.

Berti, W.R. and Cunningham, S.D. (2000). Phytostabilization of metals. In: *Phytoremediation of Toxic Metals: Using Plants to Clean-Up the Environment* (ed. I. Raskin and B.D. Ensley), 71–88. New York: Wiley.

Bhattacharya, S., Sharma, P., Mitra, S. et al. (2021). Arsenic uptake and bioaccumulation in plants: a review on remediation and socio-economic perspective in Southeast Asia. *Environmental Nanotechnology, Monitoring & Management* 15: 100430.

Cantamessa, S., Massa, N., Gamalero, E., and Berta, G. (2020). Phytoremediation of a highly arsenic polluted site, using *Pteris vittata* L. and Arbuscular Mycorrhizal Fungi. *Plants* 9 (9): 1211.

Chakraborti, D., Rahman, M.T., Das, B. et al. (2010). Status of groundwater arsenic contamination in Bangladesh: a 14-year study report. *Water Research* 44: 5789–5802.

Chen, C.J., Chiou, H.Y., Huang, W.I. et al. (1997a). Systemic noncarcinogenic effects and developmental toxicity of inorganic arsenic. In: *Arsenic: Exposure and Health Effects* (ed. C.O. Abernathy, R.L. Calderon and W.R. Chappell), 124–134. London: Chapman & Hall.

Chen, C.J., Hsu, L.I., Wang, C.H. et al. (2005). Biomarkers of exposure, effect, and susceptibility of arsenic-induced health hazards in Taiwan. *Toxicology and Applied Pharmacology* 206: 198–206.

Chen, C.J., Hsueh, Y.M., Chiou, H.Y. et al. (1997b). Human carcinogenicity of inorganic arsenic. In: *Arsenic: Exposure and Health Effects* (ed. C.O. Abernathy, R.L. Calderon and W.R. Chappell), 232–242. London: Chapman & Hall.

Chen, Y., Fu, J.W., Han, Y.H. et al. (2016). High As exposure induced substantial arsenite efflux in As-hyperaccumulator *Pteris vittata*. *Chemosphere* 144: 2189–2194.

Chen, Y., Han, Y.H., Cao, Y. et al. (2017). Arsenic transport in rice and biological solutions to reduce arsenic risk from rice. *Frontiers in Plant Science* 8: 268.

Chung, J.Y., Yu, S.D., and Hong, Y.S. (2014). Environmental source of arsenic exposure. *Journal of Preventive Medicine and Public Health = Yebang Uihakhoe chi* 47 (5): 253–257. https://doi.org/10.3961/jpmph.14.036.

Csanady, M and Straub, I (1995). Health damage due to pollution in Hungary. In *Proceedings of the Rome Symposium*, Rome, Italy (September 1994). IAHS Publ. No. 233, pp. 1–11.

Cui, J., Li, Y., Jin, Q., and Li, F. (2020). Silica nanoparticles inhibit arsenic uptake into rice suspension cells via improving pectin synthesis and the mechanical force of the cell wall. *Environmental Science: Nano* 7 (1): 162–171.

Das, D. (1995). Arsenic species along with other metal/metalloid present and responsible for arsenic episode in ground water of West Bengal and a cheap technique to remove arsenic, thus making the ground water suitable for drinking and cooking. Ph.D. thesis. Jadavpur University. Calcutta, India.

Das, D., Chatterjee, A., Mandal, B.K. et al. (1995). Arsenic in ground water in six districts of West Bengal, India: the biggest arsenic calamity in the world. Part 2. Arsenic concentration in drinking water, hair, nails, urine, skin-scalesand liver tissues (biopsy) of the affected people. *Analyst* 120: 917–924.

Das, H.K., Mitra, A.K., Sengupta, P.K. et al. (2004). Arsenic concentrations in rice, vegetables, and fish in Bangladesh: a preliminary study. *Environment International* 30: 383–387.

Dimkpa, C.O. and Bindraban, P.S. (2017). Nanofertilizers: new products for the industry? *Journal of Agricultural and Food Chemistry* 66 (26): 6462–6473.

Ding, L.-J., Zhu, Y.-G., Ding, L.-J. et al. (2014). Nitrogen loss through anaerobic ammonium oxidation coupled to iron reduction from paddy soils in a chronosequence. *Environmental Science & Technology* 48 (18): 10641–10647.

DiTusa, S.F., Fontenot, E.B., Wallace, R.W. et al. (2016). A member of the phosphate transporter 1 (Pht1) family from the arsenic hyperaccumulating fern *Pteris vittata* is a high-affinity arsenate transporter. *New Phytologist* 209: 762–772.

Duman, F., Ozturk, F., and Aydin, Z. (2010). Biological responses of duckweed (*Lemna minor* L.) exposed to the inorganic arsenic species As(III) and As(V): effects of concentration and duration of exposure. *Ecotoxicology* 19: 983–993.

Dushenkov, V., Kumar, P.N., Motto, H., and Raskin, I. (1995). Rhizofiltration: the use of plants to remove heavy metals from aqueous streams. *Environmental Science & Technology* 29: 1239–1245.

Farooq, M.A., Islama, F., Ali, B. et al. (2016). Arsenic toxicity in plants: cellular and molecular mechanisms of its transport and metabolism. *Environmental and Experimental Botany* 132: 42–52. https://doi.org/10.1016/j.envexpbot.2016.08.004.

Finnegan, P.M. and Chen, W. (2012). Arsenic toxicity: the effects on plant metabolism. *Frontiers in Physiology* 3: 182.

Fry, R.C., Navasumrit, P., Valiathan, C. et al. (2007). Activation of inflammation/NF-kB signaling in infants born to arsenic-exposed mothers. *PLoS Genetics* 3: 2180–e2189.

Goldsmith, J.R., Deane, M., Thom, J., and Gentry, G. (1972). Evaluation of health implications of elevated arsenic in well waters. *Water Research* 6: 1133–1136.

Gunes, A., Inal, A., Bagci, E.G., and Kadioglu, Y.K. (2010). Combined effect of arsenic and phosphorus on mineral element concentrations of sunflower. *Communications in Soil Science and Plant Analysis* 41: 361–372.

Gupta, D.K., Srivastava, S., Huang, H.G. et al. (2011). Arsenic tolerance and detoxification mechanisms in plants. In: *Detoxification of Heavy Metals, Soil Biology*, vol. 30 (ed. I. Sherameti and A. Varma), 169–179. Berlin; Heidelberg: Springer-Verlag.

Hammond, C.M., Root, R.A., Maier, R.M., and Chorover, J. (2020). Arsenic and iron speciation and mobilization during phytostabilization of pyritic mine tailings. *Geochimica et Cosmochimica Acta* 286: 306–323.

Harvey, C.F., Ashfaque, K.N., Yu, W. et al. (2006). Groundwater dynamics and arsenic contamination in Bangladesh. *Chemical Geology* 228 (1-3): 112–136.

He, Y.K., Sun, J.G., Feng, X.Z. et al. (2001). Differential mercury volatilization by tobacco organs expressing a modified bacterial merA gene. *Cell Research* 11: 231–236.

Hettick, B.E., Cañas-Carrell, J.E., French, A.D., and Klein, D.M. (2015). Arsenic: a review of the element's toxicity, plant interactions, and potential methods of remediation. *Journal of Agricultural and Food Chemistry* 63 (32): 7097–7107.

Hossain, M.F. (2006). Arsenic contamination in Bangladesh – an overview. *Agriculture, Ecosystems & Environment* 113 (1–4): 1–16.

Huang, J.H., Hu, K.N., and Decker, B. (2011). Organic arsenic in the soil environment: speciation, occurrence, transformation, and adsorption behavior. *Water, Air, & Soil Pollution* 219: 401–415.

Huang, S., Zhuo, C., Du, X.Y., and Li, H.S. (2021). Remediation of arsenic-contaminated paddy soil by intercropping aquatic vegetables and rice. *International Journal of Phytoremediation* 1–9.

Jacob, J.M., Karthik, C., Saratale, R.G. et al. (2018). Biological approaches to tackle heavy metal pollution: a survey of literature. *Journal of Environmental Management* 217: 56–70.

Jain, C.K. and Ali, I. (2000). Arsenic: occurence, toxicity and speciation techniques. *Water Research* 34 (17): 4304–4312.

Jia, Y., Huang, H., Chen, Z., and Zhu, Y.G. (2014). Arsenic uptake by rice is influenced by microbe-mediated arsenic redox changes in the rhizosphere. *Environmental Science & Technology* 48: 1001–1007.

Kabata-Pendias, A. (2011). *Trace Elements in Soils and Plants*, 4e. Boca Raton, FL: CRC.

Karimi, N. and Souri, Z. (2015). Effect of phosphorus on arsenic accumulation and detoxification in arsenic hyperaccumulator, *Isatis cappadocica*. *Journal of Plant Growth Regulation* 34: 88–95.

Karimi, N. and Souri, Z. (2016). Antioxidant enzymes and compounds complement each other during arsenic detoxification in shoots of *Isatis cappadocica* Desv. *Chemistry and Ecology* 32: 937–951.

Katsuhara, M., Sasano, S., Horie, T. et al. (2014). Functional and molecular characteristics of rice and barley NIP aquaporins transporting water, hydrogen peroxide and arsenite. *Plant Biotechnology* 31: 213–219.

Kaya, C., Ashraf, M., Alyemeni, M.N. et al. (2020). Salicylic acid-induced nitric oxide enhances arsenic toxicity tolerance in maize plants by upregulating the ascorbate-glutathione cycle and glyoxalase system. *Journal of Hazardous Materials* 399: 123020.

Khalid, S., Shahid, M., Niazi, N.K. et al. (2017). Arsenic behaviour in soil-plant system: biogeochemical reactions and chemical speciation influences. In: *Enhancing Cleanup of Environmental Pollutants: Volume 2: Non-Biological Approaches* (ed. N.A. Anjum, S.S. Gill and N. Tuteja), 97–140. Berlin, Germany: Springer.

King, D.J., Doronila, A.I., Feenstra, C. et al. (2008). Phytostabilisation of arsenical gold mine tailings using four *Eucalyptus* species: growth, arsenic uptake and availability after five years. *Science of the Total Environment* 406 (1–2): 35–42. https://doi.org/10.1016/j.scitotenv.2008.07.054. Epub 2008 Sep 17. PMID: 18801558.

Li, N., Wang, J., and Song, W.-Y. (2015). Arsenic uptake and translocation in plants. *Plant and Cell Physiology* 57 (1): 4–13.

Lian, J., Zhao, L., Wu, J. et al. (2020). Foliar spray of TiO_2 nanoparticles prevails over root application in reducing Cd accumulation and mitigating Cd-induced phytotoxicity in maize (*Zea mays* L.). *Chemosphere* 239: 124794.

Lihong, W. and Guilan, D. (2009). Effect of external and internal phosphate status on arsenic toxicity and accumulation in rice seedlings. *Journal of Environmental Sciences* 21: 349–351.

Liu, C., Wei, L., Zhang, S. et al. (2014). Effects of nanoscale silica sol foliar application on arsenic uptake, distribution and oxidative damage defense in rice (*Oryza sativa* L.) under arsenic stress. *RSC Advances* 4 (100): 57227–57234.

Ma, J.F., Yamaji, N., Mitani, N. et al. (2008). Transporters of arsenite in rice and their role in arsenic accumulation in rice grain. *Proceedings of the National Academy of Sciences of the United States of America* 105: 9931–9935.

Ma, L.Q., Komar, K.M., Tu, C. et al. (2001). A fern that hyperaccumulates As: a hardy, versatile, fast-growing plant helps to remove as from contaminated soils. *Nature* 409: 579.

Ma, X., Sharifan, H., Dou, F., and Sun, W. (2020a). Simultaneous reduction of arsenic (As) and cadmium (Cd) accumulation in rice by zinc oxide nanoparticles. *Chemical Engineering Journal* 384: 123802.

Ma, X., Sun, K.Y., Dou, F. et al. (2020b). Impact of elevated nitrate and perchlorate in irrigation water on the uptake, speciation, and accumulation of arsenic in rice (*Oryza sativa* L.). *Water, Air, & Soil Pollution* 231 (6): 1–12.

Macek, T., Uhlik, O., Jecna, K. et al. (2009). Advances in phytoremediation and rhizoremediation. In: *Advances in Applied Bioremediation*, vol. 17 (ed. A. Singh, R. Kuhad and O. Ward), 257–277. Berlin, Heidelberg: Springer https://doi.org/10.1007/978-3-540-89621-0_14.

Mandal, B.K. and Suzuki, K.T. (2002). Arsenic round the world: a review. *Talanta* 58: 201–235.

Meharg, A.A. and Macnair, M.R. (1992). Suppression of the high-affinity phosphate uptake system: a mechanism of arsenate tolerance in *Holcus lanatus* L. *Journal of Experimental Botany* 43 (519): 524.

Mendez, M.O. and Maier, R.M. (2008). Phytostabilization of mine tailings in arid and semiarid environments – an emerging remediation technology. *Environmental Health Perspectives* 116: 278–283.

Milani, N., Hettiarachchi, G.M., Kirby, J.K. et al. (2015). Fate of zinc oxide nanoparticles coated onto macronutrient fertilizers in an alkaline calcareous soil. *PLoS One* 10 (5): e0126275.

Milton, A.H., Rahman, H., Smith, W. et al. (2006). Water consumption patterns in rural Bangladesh: are we underestimating total arsenic load? *Journal of Water and Health* 4: 431–436.

Mishra, S., Dwivedi, S., Mallick, S., and Tripathi, R.D. (2019). Redox homeostasis in plants under arsenic stress. In: *Redox Homeostasis in Plants. Signaling and Communication in Plants* (ed. S. Panda and Y. Yamamoto). Cham: Springer https://doi.org/10.1007/978-3-319-95315-1_9.

Mok, W.M. and Wai, C.M. (1990). Distribution and mobilization of arsenic and antimony species in the Coeur D'Alene River, Idaho. *Environmental Science & Technology* 24: 102–108.

Mukhopadhyay, R. and Rosen, B.P. (2002). Arsenate reductases in prokaryotes and eukaryotes. *Environmental Health Perspectives* 110: 745–748.

Nath, S., Panda, P., Mishra, S. et al. (2014). Arsenic stress in rice: redox consequences and regulation by iron. *Plant Physiology and Biochemistry* 80: 203–210.

Neu, S., Müller, I., and Dudel, E.G. (2020). Management of trace element-contaminated agricultural land by in situ stabilization combined with phytoexclusion over a three years crop rotation. *International Journal of Phytoremediation* 22 (10): 1059–1067.

Palit, S., Misra, K., and Mishra, J. (2019). Arsenic contamination in South Asian regions: the difficulties, challenges and vision for the future. In: *Separation Science and Technology*, vol. 11 (ed. S. Ahuja), 113–123. Academic Press https://doi.org/10.1016/B978-0-12-815730-5.00005-3.

Panda, S., Upadhyay, R., and Nath, S. (2010). Arsenic stress in plants. *Journal of Agronomy and Crop Science* 196: 161–174.

Pommerrenig, B., Diehn, T.A., Bernhardt, N. et al. (2020). Functional evolution of nodulin 26-like intrinsic proteins: from bacterial arsenic detoxification to plant nutrient transport. *New Phytologist* 225 (3): 1383–1396. https://doi.org/10.1111/nph.16217. Epub 2019 Nov 2. Erratum in: New Phytologist 2020 Dec;228(5):1690. PMID: 31550387.

Rafiq, M., Shahid, M., Abbas, G. et al. (2017). Comparative effect of calcium and EDTA on arsenic uptake and physiological attributes of *Pisum sativum*. *International Journal of Phytoremediation* 19: 662–669.

Rahman, A., Persson, L.-Å., Nermell, B. et al. (2010). Arsenic exposure and risk of spontaneous abortion, stillbirth, and infant mortality. *Epidemiology* 21: 797–804.

Rahman, A., Vahter, M., Ekstr€om, E.-C. et al. (2007). Association of arsenic exposure during pregnancy with fetal loss and infant death: a cohort study in Bangladesh. *American Journal of Epidemiology* 165: 1389–1396.

Rascio, N. and Navari-Izzo, F. (2011). Heavy metal hyperaccumulating plants: how and why do they do it? And what makes them so interesting? *Plant Science* 180: 169–181.

Raskin, I. and Ensley, B.D. (2000). *Phytoremediation of Toxic Metals*. New York: Wiley.

Raskin, I., Smith, R.D., and Salt, D.E. (1997). Phytoremediation of metals: using plants to remove pollutants from the environment. *Current Opinion in Biotechnology* 8: 221–226.

Rossi, L., Fedenia, L.N., Sharifan, H. et al. (2019). Effects of foliar application of zinc sulfate and zinc nanoparticles in coffee (*Coffea arabica* L.) plants. *Plant Physiology and Biochemistry* 135: 160–166.

Shankar, S., Shanker, U., and Shikha, U. (2014). Arsenic contamination of groundwater: a review of sources, prevalence, health risks, and strategies for mitigation. *The Scientific World Journal* 2014: 304524.

Shao, K., Zhou, Z., Xun, P., and Cohen, S.M. (2021). Bayesian benchmark dose analysis for inorganic arsenic in drinking water associated with bladder and lung cancer using epidemiological Data. *Toxicology* 152752. https://doi.org/10.1016/j.tox.2021.152752.

Sharifan, H. (2020). Alarming the impacts of the organic and inorganic UV blockers on endangered coral's species in the Persian Gulf: a scientific concern for coral protection. *Sustainable Futures* 2: 100017.

Sharifan, H. et al. (2019). Zinc oxide nanoparticles alleviated the bioavailability of cadmium and lead and changed the uptake of iron in hydroponically grown lettuce (*Lactuca sativa* L. var. Longifolia). *ACS Sustainable Chemistry & Engineering* 7 (19): 16401–16409.

Shi, S., Wang, T., Chen, Z. et al. (2016). OsHAC1;1 and OsHAC1;2 function as arsenate reductases and regulate arsenic accumulation. *Plant Physiology* 172: 1708–1719.

Shin, H., Shin, H.S., Dewbre, G.R., and Harrison, M.J. (2004). Phosphate transport in Arabidopsis: Pht1;1 and Pht1;4 play a major role in phosphate acquisition from both low- and high-phosphate environments. *The Plant Journal* 39: 629–642.

Singh, N. and Ma, L.Q. (2006). Arsenic speciation, and arsenic and phosphate distribution in arsenic hyperaccumulator *Pteris vittata* L. and non-hyperaccumulator *Pteris ensiformis* L. *Environmental Pollution* 141 (2): 238–246.

Smith, A.H., Lopipero, P.A., and Bates, M.N. (2002). Arsenic epidemiology and drinking water standards. *Science* 296: 2145–2146.

Song, W.-Y., Parka, J., Mendoza Cozatl, D. et al. (2010). Arsenic tolerance in Arabidopsis is mediated by two ABCC-type phytochelatin transporters. *Proceedings of the National Academy of Sciences of the United States of America* 107: 21187–21192. https://doi.org/10.1073/pnas.1013964107.

Souri, Z., Cardoso, A.A., da Silva, C.J. et al. (2019). Heavy metals and photosynthesis: recent developments. In: *Photosynthesis, Productivity and Environmental Stress*, 107–134. Wiley.

Souri, Z. and Karimi, N. (2017). Enhanced phytoextraction by As hyperaccumulator *Isatis cappadocica* spiked with sodium nitroprusside. *Soil and Sediment Contamination* 26 (4): 457–468.

Souri, Z., Karimi, N., and Ahmad, P. (2021a). The effect of NADPH oxidase inhibitor diphenyleneiodonium (DPI) and glutathione (GSH) on *Isatis cappadocica*, under Arsenic (As) toxicity. *International Journal of Phytoremediation* 21: 1–13.

Souri, Z., Karimi, N., and de Oliveira, L.M. (2018). Antioxidant enzymes responses in shoots of arsenic hyperaccumulator, *Isatis cappadocica* Desv., under interaction of arsenate and phosphate. *Environmental Technology* 39: 1316e1327.

Souri, Z., Karimi, N., Farooq, M.A., and Sandalio, L.M. (2020). Nitric oxide improves tolerance to arsenic stress in *Isatis cappadocica* desv. shoots by enhancing antioxidant defenses. *Chemosphere* 239: 124523.

Souri, Z., Karimi, N., and Luisa, M.S. (2017a). Arsenic hyperaccumulation strategies: an overview. *Frontiers in Cell and Development Biology* 5: 67.

Souri, Z., Karimi, N., Norouzi, L., and Ma, X. (2021b). Elucidating the physiological mechanisms underlying enhanced arsenic hyperaccumulation by glutathione modified superparamagnetic iron oxide nanoparticles in *Isatis cappadocica*. *Ecotoxicology and Environmental Safety* 206: 111336.

Souri, Z., Karimi, N., Sarmadi, M., and Rostami, E. (2017b). Salicylic acid nanoparticle (SANPs) improve growth and phytoremediation efficiency of *Isatis cappadocica* Desv. under arsenic stress. *IET Nanobiotechnology* 11 (6): 650–655.

Su, Y.H., McGrath, S.P., Zhu, Y.G., and Zhao, F.J. (2008). Highly efficient xylem transport of arsenite in the arsenic hyperaccumulator *Pteris vittata*. *New Phytologist* 180: 434–441.

Sun, W., Sierra, R., Fernandez, N. et al. (2009b). Molecular characterization and in situ quantification of anoxic arsenite-oxidizing denitrifying enrichment cultures. *FEMS Microbiology Ecology* 68 (1): 72–85.

Sun, W., Sierra-Alvarez, R., Milner, L. et al. (2009a). Arsenite and ferrous iron oxidation linked to chemolithotrophic denitrification for the immobilization of arsenic in anoxic environments. *Environmental Science & Technology* 43 (17): 6585–6591.

Tangahu, B.V., Sheikh Abdullah, S.R., Basri, H. et al. (2011). A review on heavy metals (As, Pb, and Hg) uptake by plants through phytoremediation. *International Journal of Chemical Engineering. 2011* 939161.

Tsai, T.L., Kuo, C.C., Hsu, L.I. et al. (2021). Association between arsenic exposure, DNA damage, and urological cancers incidence: a long term follow-up study of residents in an arseniasis endemic area of Northeastern Taiwan. *Chemosphere* 266: 129094.

USEPA (2001). National primary drinking water regulation; arsenic and clarifications to compliance and new source contaminants monitoring; final rule. Federal Register, 66, p. 6975.

Wan, X., Lei, M., and Chen, T. (2020). Review on remediation technologies for arsenic-contaminated soil. *Frontiers of Environmental Science & Engineering* 14: 1–1.

Wang, H.-Y., Wen, S.L., Chen, P. et al. (2016a). Mitigation of cadmium and arsenic in rice grain by applying different silicon fertilizers in contaminated fields. *Environmental Science and Pollution Research* 23 (4): 3781–3788.

Wang, P., Zhang, W., Mao, C. et al. (2016b). The role of OsPT8 in arsenate uptake and varietal difference in arsenate tolerance in rice. *Journal of Experimental Botany* 67: 6051–6059.

Wang, X., Sun, W., and Ma, X. (2019). Differential impacts of copper oxide nanoparticles and copper (II) ions on the uptake and accumulation of arsenic in rice (*Oryza sativa*). *Environmental Pollution* 252: 967–973.

Wang, X., Sun, W., Zhang, S. et al. (2018). Elucidating the effects of cerium oxide nanoparticles and zinc pxide nanoparticles on arsenic uptake and speciation in rice (*Oryza sativa*) in a hydroponic system. *Environmental Science & Technology* 52 (17): 10040–10047.

Wu, F., Fang, Q., Yan, S. et al. (2020). Effects of zinc oxide nanoparticles on arsenic stress in rice (*Oryza sativa* L.): germination, early growth, and arsenic uptake. *Environmental Science and Pollution Research* 27: 26974–26981.

Wuana, R.A. and Okieimen, F.E. (2011). Heavy metals in contaminated soils: a review of sources, chemistry, risks and best available strategies for remediation. International Scholarly Research Network. *Ecology* 1–21.

Xu, X., McGrath, S.P., Meharg, A.A., and Zhao, F.J. (2008). Growing rice aerobically markedly decreases arsenic accumulation. *Environmental Science & Technology* 42 (15): 5574–5579.

Yamaguchi, N., Nakamura, T., Dong, D. et al. (2011). Arsenic release from flooded paddy soils is influenced by speciation, Eh, pH, and iron dissolution. *Chemosphere* 83 (7): 925–932.

Yan, A., Wang, Y., Tan, S.N. et al. (2020). Phytoremediation: a promising approach for revegetation of heavy metal-polluted land. *Frontiers in Plant Science* 11: 359.

Ye, J., Rensing, C., Rosen, B.P., and Zhu, Y.G. (2012). Arsenic biomethylation by photosynthetic organisms. *Trends in Plant Science* 17: 155–162.

Zhang, Q., Kong, W., Wei, L. et al. (2020). Uptake, phytovolatilization, and interconversion of 2,4-dibromophenol and 2,4-dibromoanisole in rice plants. *Environment International* 142: 105888.

Zhang, S., Geng, L., Fan, L. et al. (2020). Spraying silicon to decrease inorganic arsenic accumulation in rice grain from arsenic-contaminated paddy soil. *Science of the Total Environment* 704: 135239.

Zhao, F.J., Ma, J.F., Meharg, A.A., and McGrath, S.P. (2009). Arsenic uptake and metabolism in plants. *New Phytologist* 181: 777–794.

Zhou, G.-W., Yang, X.-R., Li, H. et al. (2016). Electron shuttles enhance anaerobic ammonium oxidation coupled to iron (III) reduction. *Environmental Science & Technology* 50 (17): 9298–9307.

15

Arsenic Removal by Electrocoagulation

Aysegul Yagmur Goren[1] *and Mehmet Kobya*[2,3]

[1] *Department of Environmental Science and Engineering, Izmir Institute of Technology, Izmir, Turkey*
[2] *Department of Environmental Engineering, Gebze Technical University, Kocaeli, Turkey*
[3] *Department of Environmental Engineering, Kyrgyz-Turkish Manas University, Bishkek, Kyrgyzstan*

15.1 Introduction

One of the most important issues today and in the future is the provision of clean water. Today, above 2.50 billion publics worldwide meet their water needs from groundwater (Grönwall and Danert 2020). Recently, arsenic (As) pollution in groundwater resources has emerged as a severe problem in the world of concern due to its toxicity to human beings. It is predicted that almost 108 countries and greater than 230 million publics in these countries are facing As contamination through natural groundwater sources above the WHO permissible limit of $10\,\mu g\,l^{-1}$ (Podgorski and Berg 2020; Shaji et al. 2021). Among arsenic removal technologies, electrocoagulation (EC) has advantages over many removal technologies. In this study, the extent of As pollution in natural water sources, the advantages and disadvantages of As removal technologies, the As removal approach of the EC technology, and the effects of the factors affecting As reduction in the EC technology are discussed in this chapter.

15.2 Arsenic Contamination in Natural Waters

Podgorski and Berg (2020) estimated that the percentage of area and population significantly influenced by As contents in groundwater resources above $10\,\mu g\,As\,l^{-1}$ were 64 and 94% for Asia, 14 and 1.6% for South America, 10 and 0.60% for North America, 9 and 3.70% for Africa, 2 and 0.01% for Oceania, and 1 and 0.20% for Europe, respectively. Some countries/regions with a high risk of As contamination, populations, and As concentrations in groundwater in the world are given in Table 15.1. Argentina, the Ganges–Brahmaputra delta, Mexico, the Mekong River, the Pannonian Basin, Mongolia, the Red River deltas, the Indus Valley, and the United States are the most known areas facing groundwater As pollution. The southeast Asian countries such as Bangladesh, India (especially West Bengal),

Arsenic in Plants: Uptake, Consequences and Remediation Techniques, First Edition.
Edited by Prabhat Kumar Srivastava, Rachana Singh, Parul Parihar, and Sheo Mohan Prasad.

Table 15.1 Global As contamination, affected population, and As concentrations in groundwater.

Country/region	As concentration (mg l^{-1})	As-exposed population (million)	References
Bangladesh/Ganges, Brahmaputra, and Meghna	<1–4700	65.00	Ahmad et al. (2018)
India/West Bengal, Ganga Plains	<1–4200	41.00	Bhattacharya et al. (2013)
China/Huhhot-Hetao-Datong Basins	<1–1860	20.60	Guo et al. (2014)
Vietnam/Mekong and Red River Deltas (Hanoi)	<1–3050	9.50	Berg et al. (2001)
Myanmar/River basins (i.e. Ayeyarwady)	<10–630	8.80	Pincetti-Zúniga et al. (2020)
Pakistan/Punjab and Sindh provinces	0.1–2090	8.00	Shahid et al. (2018)
Argentina/Chaco-Pampean Plain	7–5280	5.60	Nicolli et al. (2012)
Argentina/La Pampa Province	<4–5300		
Canada/British Columbia, Quebec Canada/Saskatchewan, Nova Scotia, Alberta	<10–650 <10–5340	4.4	Mc-Guigan et al. (2010)
Nepal/Terai Region	1–1500	3.05	Brikowski et al. (2018)
Mexico/Rio Salado River, Comarca Lagunera	1–1880	2.50	Camacho et al. (2011)
USA/the Upper Midwest and Southwest States	<0.1–58	2.10	Ayotte et al. (2017)
Chile/Antofagasta, Rio Loa	10–860	1.80	Ferreccio and Sancha (2006)
Hungary/Great Hungarian Plain	0–560	1.80	Jones et al. (2008)
Peru/Western Amazonia Zone	0–700	0.88	De Meyer et al. (2017)
Afghanistan/Ghazni and Maidan Wardak	0–6240	0.70	Hayat and Baba (2017)
Cambodia/Mekong River Basin	1–1842	0.35	Phan et al. (2010)
Taiwan/South-West Coastal Zone	<1–3590	0.24	Guo et al. (2014)
Thailand/Ron Phibun District	1–9000	0.06	Jones et al. (2008)
Turkey/Kütahya-Emet and Simav, Bigadiç	0.5–10 700	0.03	Kobya et al. (2015)
South Africa/Johannesburg, Karoo aquifers	<10–6150	—	Abiye and Bhattacharya (2019)

China, Vietnam, Myanmar, Pakistan, Nepal, Afghanistan, and Thailand are most impacted by high As contamination (Polya and Middleton 2017). In addition, countries like Mexico, the southeast USA, Canada, Hungary, Turkey, and South Africa are also affected (Ravenscrof et al. 2009). In Bangladesh alone, over 57 million people are exposed to arsenic-contaminated water with levels above 3200 $\mu g\,l^{-1}$.

The principal sources of arsenic pollution (>90%) in the groundwater are generally geogenic (Shaji et al. 2021). Other major sources of concern for arsenic pollution are anthropogenic sources such as mine wastes, agricultural applications (pesticides and fertilizers), and fossil fuel combustion. Arsenic contamination in the world generally tends to be found in arid or semiarid regions, intensely reducing aquifers resulting from alluvium and deltaic, sedimentary basins including deltaic regions and mountain belts, tropical climate regions, tectonic regions, and areas with geothermal fluids and volcanic activities (Smedley and Kinniburgh 2002). The dissolution of As-containing rocks (water–rock interactions) and minerals may contribute As to the groundwater. As-rich groundwater resources are also located in geothermal fields (e.g. "hot springs") and mining areas. The most common reason for extensive natural As contamination is estimated to be the reduction of As-bearing Fe minerals in the occurrence of organic carbon and the discharge of adsorbed As from these As-bearing iron minerals.

The use of As containing groundwater sources for drinking water purposes may cause crucial health complications like skin lesions of melanosis, keratosis, and hyperkeratosis; cancer of the bladder, kidney, lung, and stomach; hypertension, Bowen's disease, peripheral vascular disease (black foot disease), and diabetes, among other diseases (Sinha and Prasad 2020). Due to its serious toxic effects, in 1993, the WHO reduced the permissible level of arsenic in drinking water from 50 to 10 $\mu g\,l^{-1}$ (WHO 2004). However, implementation of this value was started in the year 2001 by many authorities (European Union 1998; US EPA 2001). The United States Environmental Protection Agency (US EPA) even stated that this value should be decreased to 5 $\mu g\,l^{-1}$ in terms of human health (Hering et al. 1996). The most stringent arsenic standard in drinking water is currently implemented by Australia as 7 $\mu g\,l^{-1}$ and New Jersey (USA) as 5 $\mu g\,l^{-1}$ (Henke 2009). Based on the WHO and US EPA, it is scientifically possible to accomplish As concentrations of 5 $\mu g\,l^{-1}$ by conventional treatment, but the As removal process needs strict process control and optimization. Because of insufficient testing facilities for lower As concentrations, operating cost (OC) burden, and practical problems in terms of As removal to lower concentrations, some developing countries affected with As contamination still operate drinking water systems at present according to the former WHO standard (50 $\mu g\,As\,l^{-1}$).

The species of inorganic As varies depending on pH and redox potential (e.g. aerobic and anaerobic) in the natural environment. Generally, As is found in groundwater in the form of As (III) and in surface waters as As (V). The inorganic As (III) is 25–60 times further poisonous than As (V) and 70 times further poisonous than organic As species. As (V) in natural waters at pH 5.5–8.5 are negatively charged forms as $H_2AsO_4^-$ and $HAsO_4^{2-}$, but the neutral As (III) species as H_3AsO_3 is present in waters of pH < 9. Therefore, As (V) with most treatment methods is more readily and effectively removed than As (III). Therefore, oxidation of As (III) to As (V) with ozone, free chlorine, or hypochlorite is required before the removal process. As a result, the type of As (As (III), As (V) or both) in the groundwater to be treated, the pH of the water, and other ions (i.e. phosphate, silicate, bicarbonate, and chloride) in the water greatly affect the treatment process efficiency and process selection.

15.3 Advantages and Disadvantages of Main Arsenic Removal Technologies

Removal processes for As treatment can be categorized into five classes: conventional, innovative, enhanced, biological, and emerging processes (Camacho et al. 2011). Conventional processes involve precipitation, filtration after coagulation–flocculation with alum and iron chloride, softening with lime, adsorption, ion exchange, and membrane filtration (including reverse osmosis, nanofiltration, and electrodialysis). Enhanced processes are modifications of currently existing treatment methods (e.g. activated alumina, granular ferric hydroxides, and iron-exchanged natural zeolites). Emerging processes are particularly concentrated on EC and plant remediation. Innovative processes contain the practice of new flocculent and adsorbent materials (e.g. novel adsorbent materials, iron-doped activated carbons, and iron-based nanoparticles). Biological methods (bioremediation) remove the arsenic by biomethylation–demethylation, phytoremediation with sorption, coprecipitation, oxidation–reduction, and complexation.

The US EPA (2000), for As removal from aqueous solutions, are also reported as the best available technologies (BATs): activated alumina, coagulation–flocculation, ion exchange, softening with lime, microfiltration (MF), electrodialysis, and oxidation processes. In the adsorption process, As species are adsorbed on the surface of adsorbent agents through chemical and physical dynamism. Activated carbon, iron oxide-based filters, copper–zinc granules, granular ferric hydroxide, clay minerals (kaolinite and montmorill), and activated alumina are widely used adsorbents in the adsorption process for As removal from aqueous solutions. Furthermore, the chitosan with magnetic nanoparticles (MNPs), magnetic-clay (MtMag), $MnFe_2O_4/S(IV)$, TiO_2 nanoparticle, titanium xerogel, and katoite are used as modified or novel adsorbents to enhance As removal (Kloster et al. 2020; Nazari et al. 2020; Sha et al. 2020; Zhang et al. 2020; Barraque et al. 2021; Ding et al. 2021). The adsorbent surface area, surface charge, and aqueous solution pH significantly influence the As removal efficiency. Among these adsorbents, iron-based (hydrous ferric oxide and goethite) adsorbents have been studiessd widely for As removal. For having high As (V) adsorption ability of iron hydroxides, Hering et al. (1996, 1997) indicated that the coprecipitation using ferric salts and lime at pH values of 7–12 succeed by coagulation and sedimentation and/or filtration processes is an efficient and low-cost method for removing As in water. Besides, $FeCl_3$, $Fe_2(SO_4)_3$, and $Al_2(SO_4)_3$ are the most widely handled coagulant agents for As removal in the coagulation–flocculation process. Recently, the modified coagulant chemicals (cellulose with $FeCl_3$, titanium xerogel, biologically activated carbon, etc.) are used to enhance As removal performance of the coagulation process (Baskan and Pala 2010; Pramanik et al. 2016; Katsoyiannis et al. 2017; Inam et al. 2018; Kumar and Quaff 2019; Laskaridis et al. 2020; Wang et al. 2021). Overall, the type of the coagulant chemicals, coagulant dose, solution pH, and presence of the other ions are important operating parameters affecting As removal efficiency. Previous studies showed that iron-based coagulants have a high affinity for As and stability at a wide range of pH compared with aluminum-based coagulants. The optimum As removal performance of both aluminum and ferric sulfate is achieved in the range of 5.0–8.0. It has also been stated that the As(V) removal of 90.0% is observed using $20\,mg\,l^{-1}$ of iron chloride and $40\,mg\,l^{-1}$ alum (McNeill and Edwards 1995). The ion exchange process removes As from aqueous solutions electrostatically holding it on the surface of the resin. Commonly, commercial anion exchange resins are used for removal, but it is reported that the ion exchange process removes only negatively charged As (V) species. Therefore, it can be concluded that there is a necessity of As (III) oxidation to As (V) for sufficient treatment with the ion exchange process. As (III) is mainly neutral at $pH < 9.2$ and, thus, the treatment of As (III) is more challenging than As (V) without oxidizing agents including ozone, permanganate, hypochlorite, and hydrogen peroxide. Furthermore, the oxidation, membrane processes, biological processes, and hybrid processes have also been reported as promising methods for the removal of As, but they have some disadvantages such as high operational and maintenance costs, high amount of sludge formation, and production of toxic by-products (Rosales et al. 2020; Wan et al. 2020; Atabaki et al. 2021; Garcia-Costa et al. 2021; Şen et al. 2021; Yaqub and Lee 2021).

Table 15.2 summarizes the advantages and disadvantages of technologies reported in this section (Feenstra et al. 2007). Adsorption and chemical coagulation processes are

Table 15.2 Summary of treatment processes used to remove arsenic from aqueous solutions.

Treatment method	Merits	Demerits
Adsorption	• High As (V) removal efficiency • Easy operation • Successful for small-scale treatment operations • No need for additional chemicals	• Transfers the pollutant from the aqueous phase to another media • Need for the adsorbent regeneration • High cost for adsorbent regeneration agents • Suitable for waters with low arsenic concentration • Low removal efficiency with the presence of ions
Coagulation/ flocculation	• Applicable on household and community scale • Simple operation • Efficient As (V) removal • Low maintenance and operational costs	• Routine control is required • High sludge formation with a high disposal cost • Need for additional chemical for pH adjustment • Need for filtration • Need for As (III) pre-oxidation
Oxidation	• High As (III) oxidation performance • Simple process • Low operating cost	• Efficient for the only conversation of As (III) to As (V) • Toxic or carcinogenic pollutants may be formed • Need for an additional process for further treatment
Ion exchange	• Easy operation • High arsenic sequestration capacity • Low sludge production	• High operational and maintenance cost • Limited ion exchange resin life • Low removal efficiency in presence of other ions

(Continued)

Table 15.2 (Continued)

Treatment method	Merits	Demerits
Biological processes	• Environmentally friendly • Cost-effective	• Needs for high operating time • Pollutants may have effects on the growth of microbes • Toxic pollutants may be formed by microbes • Needs acclimatization of microbes or plants before use
Membrane processes	• Useful for all contaminants • Successful for household use	• High operational and maintenance costs • High energy consumption • High amount of concentrated water formation • Low As (III) removal efficiency • Membrane fouling and scaling

some of the cost-effective and easily operated As removal technologies. The adsorption process is successfully treating As-contaminated water sources to achieve drinking water standards in countries facing insufficient energy supply and lack of experienced persons for operation. Besides, the adsorption processes do not require chemical addition compared with other processes. On the other hand, the application of the adsorption process in large-scale treatment plants is lacking for As removal due to the difficulty of adsorbent usage at high amounts resulting in high material cost and hard operation. The coagulation method is a simple technology for the treatment of waters with a high amount of As. This process has significant application potential in developing and/or undeveloped countries due to the low installation, maintenance, and OCs. The most important shortcoming with coagulation is that control of As-containing sludge is needed and this increases the cost of the system. The effectiveness of the ion exchange process for As removal is sensitive to As species in aqueous solution and other water components. Moreover, the high operating and maintenance costs and limited resin life are fundamental drawbacks of ion exchange processes. The membrane process is reported as a promising method for As removal. However, the high operating, maintenance, and capital costs, and production of concentrated or brackish water are significant demerits of this process. Therefore, the membrane processes may be used less frequently in large-scale applications than other treatment processes. Recently, the biological processes as an environmentally friendly method have also been studied for As treatment. But it has some demerits such as the formation of toxic by-products by microbes, long operating time, the need for acclimatization of microbes, and sensitivity of microbes to pollutants.

15.4 As Removal Mechanism with EC

EC as a promising water/wastewater treatment process has gained considerable interest in recent years (Hakizimana et al. 2017; Moussa et al. 2017). The EC is a low-cost and environmentally promising process for the treatment of aqueous solutions containing a number of pollutants such as suspended solids, oils, organic matter (Biological oxygen demand [BOD] and Chemical oxygen demand [COD]), heavy metals, anions, dyes, and some organic compounds. Because it has some advantages (i.e. not adjusted with an acid or base, no need for an additional pre-oxidation step, a relatively easy system to operate, low maintenance costs, produces effluent with lower total dissolved solids (TDS) for the lack of chemical additives, produces very little sludge, compact treatment facility, relatively economical, and greatly reduced membrane fouling) over conventional treatment technologies (Moussa et al. 2017). It also showed that it is an alternative and effective arsenic removal method as it provides high As removal (>99%) from water resources (Song et al. 2017; Kobya et al. 2020).

In the EC method, a voltage is applied between two metal (usually iron or aluminum) electrode materials, controlling the dissolution of the anode to aqueous Me^{n+}. In the simplest EC reactor, which consists of an anode (generally Fe and Al) and a cathode (i.e. Al, Fe, titanium, and stainless steel) in an aqueous solution, oxidation on the anode electrode and reduction reactions on the cathode electrode take place when electric current is applied to the electrodes.

$$Me_{(s)} \rightarrow Me^{n+} + n\bar{e} \quad (\text{at anode}) \tag{15.1}$$

$$2H_2O + 2\bar{e} \rightarrow H_{2(g)} + 2OH^- \quad (\text{at cathode}) \tag{15.2}$$

The EC includes the formation of active coagulant species through metal anode dissolution (Me: Al or Fe) with simultaneous production of OH^- ions and H_2 gas by reduction of water on cathode surface (Kobya et al. 2015). Using Al anode electrode in EC reactor, main reactions in anode and cathode are:

$$Al_{(s)} \rightarrow Al^{3+} + 3\bar{e} \quad (\text{at anode}) \tag{15.3}$$

$$3H_2O + 3\bar{e} \rightarrow \frac{3}{2}H_{2(g)} + 3OH^- \quad (\text{at cathode}) \tag{15.4}$$

Furthermore, using the Fe anode electrode in the EC reactor, firstly, iron oxidizes to ferrous ion, and then based on the anode material potential, it may oxidize to ferric ion. The main reactions in anode are:

$$4Fe_{(s)} \rightarrow 4Fe^{2+} + 8\bar{e} \tag{15.5}$$

$$4Fe_{(s)} \rightarrow 4Fe^{3+} + 12\bar{e} \tag{15.6}$$

$$Fe^{2+} \rightarrow Fe^{3+} + \bar{e} \tag{15.7}$$

Main reaction for Fe cathode:

$$2H_2O + 2\bar{e} \rightarrow H_{2(g)} + 2OH^- \tag{15.8}$$

Fe (II) in bulk solution was rapidly oxidized to Fe (III) with supplied oxygen during the process:

$$4Fe^{2+} + 2H_2O + O_2 \rightarrow 4Fe^{3+} + 4OH^- \tag{15.9}$$

$$Fe^{2+} + 1.5H_2O + 0.25O_2 \rightarrow \gamma - FeOOH_{(s)} + 2H^+ \tag{15.10}$$

The metal ions (Al^{3+}, Fe^{2+}, and Fe^{3+}) react with OH^- ions produced by electrolysis of water on the cathode electrode to produce the corresponding form of monomeric and polymeric metallic (oxyhydr)oxide complex species ($Me_n(OH)_m^{3n-m}$, transform finally into solid $Me(OH)_3$), which enhance the production of the flocs by destabilization of pollutants. These metal cations also act as coagulants and then, after all precipitation in the formation of metal (oxyhydr)oxide, precipitates like HFO, $Al(OH)_{3(s)}$, $Fe(OH)_{3(s)}$, and $\gamma-FeOOH$ in the range of pH 4.5–9.0 (Kobya et al. 2020). The produced flocs may be removed from the solution surface by scraping when the H_2 bubbles formed on the cathode material allow flotation of generated flocs in the EC reactor and promote turbulence in the reactor (where these are concentrated, collected, and removed) or these flocs settle depending on their density:

$$Me^{n+} + OH^- \rightarrow Me(OH)_{n(s)} \tag{15.11}$$

$$Al^{3+} + 3OH^- \rightarrow Al(OH)_{3(s)} \tag{15.12}$$

$$Al^{3+} + 3H_2O \rightarrow Al_2O_{3(s)} + 6H^+ \tag{15.13}$$

$$Fe^{3+} + 3OH^- \rightarrow Fe(OH)_{3(s)} \tag{15.14}$$

The iron and aluminum oxides (especially Fe(III) (oxyhydr)oxides) are known to have a strong adsorption affinity for arsenic. Therefore, surface complex adsorption is mainly responsible for As removal using Fe and Al species. Coprecipitation through the surface of arsenic-metal flocs could also take place on the surface of metal (oxyhydr)oxide precipitates. Oxidation of As (III) to As (V) and subsequent surface complexation with metal (oxyhydr)oxides or adsorption on generated metal-oxy (hydroxides) surfaces were suggested as the mechanism of As (III) removal by EC using Fe or Al electrodes (Kobya et al. 2020). These reactions are presented in the equations below:

$$2FeOOH_{(s)} + H_2AsO_4^- \rightarrow \left[(FeO)_2 HAsO_4^-\right]_{(s)} + H_2O + OH^- \tag{15.15}$$

$$3FeOOH_{(s)} + HAsO_4^{2-} \rightarrow \left[(FeO)_3 AsO_4^-\right]_{(s)} + H_2O + 2OH^- \tag{15.16}$$

$$Fe(OH)_{3(s)} + AsO_4^{3-} \rightarrow \left[Fe(OH)_{3(s)} * AsO_4^{3-}\right]_{(s)} \tag{15.17}$$

$$Al(OH)_{3(s)} + HAsO_4^{2-} \rightarrow \left[Al(OH)_{3(s)} * HAsO_4^{2-}\right]_{(s)} \tag{15.18}$$

$$Al_2O_{3(s)} + HAsO_4^{2-} \rightarrow \left[Al_2O_3 * HAsO_4^{2-}\right]_{(s)} \tag{15.19}$$

$$mAl^{3+} + (3m-n)OH^- + nHAsO_4^{2-} \rightarrow \left[Al_m(OH)_{(3m-n)}(HAsO_4)_{n(s)}\right] \tag{15.20}$$

The separation step is the step in which aggregated arsenic-metal (oxyhydr)oxide precipitates or flocs are separated from water through filtration or decantation after flocs have had time to settle or some other method of separating the flocs from treated water. As expected, the produced sludge after the EC process consists of a significant amount of As and iron or aluminum related to the electrode material type. Goren and Kobya (2021) investigated the effects of anions on As removal from groundwater using 3D Al electrodes. Morphological analyses were performed to define the characteristics of the sludge at the end of the EC method and they reported that the aluminum hydroxide products emerged to be aggregated particles (Figure 15.1a). Moreover, Energy Dispersive X-Ray Analysis (EDX) results showed that the O, Al, P, and Al contents were found to be 33.6, 24.2, 9.8, and 12.9%, respectively (Figure 15.1b).

15.5 Operating Parameters Affecting Arsenic Removal Through EC

The EC operation mode, operating parameters, operational cost of the system, and optimum As treatment efficiencies of the articles reviewed in this chapter are presented in Table 15.3. The As removal potential of EC technology is significantly affected by the operating parameters including electrode shape and material, electrode connection mode and distance, initial pH, aeration rate, applied current, arsenic speciation and concentrations, and operating time. These operational parameters are commonly studied to optimize the As removal and operational cost of the EC method. In the following sections, the effects of operational parameters were summarized in detail to demonstrate the potential achieved by optimizing operational parameters in the EC process.

15.6 Electrode Shape and Material

The electrode shape and material are very notable parameters on EC process performance (Kobya et al. 2020). The As removal using EC with different electrode materials and shapes is compared and reviewed in Table 15.3. Fe and Al are the mostly performed electrode materials for the As removal in water by the EC method. Fe electrodes are commonly preferred due to their cost-effectiveness and less toxicity when compared with other electrode materials used in the EC process. Furthermore, iron electrodes can be utilized in wide pH ranges owing to the formation of $Fe(OH)_x$ species. The As removal and cost of EC method with Fe plates were in the range of 75–99.4% and 0.017–1.55 US\$ m^{-3}, whereas the As removal and cost were in the range of 85.0–99.0% and 0.098–0.584 US\$ m^{-3} for the EC process with Al plate electrodes. Mendoza-Chávez and coworkers studied As treatment from synthetic wastewater and groundwater using the EC method using the Fe plate shape anode (Mendoza-Chávez et al. 2020). The optimum operating parameters were found as 0.2 A of applied current and operating time of 7 minutes with the As removal of 90.0% in 100 $\mu g\,l^{-1}$ of As containing synthetic solution. Furthermore, the highest As removal from real groundwater was found to be 85.2%

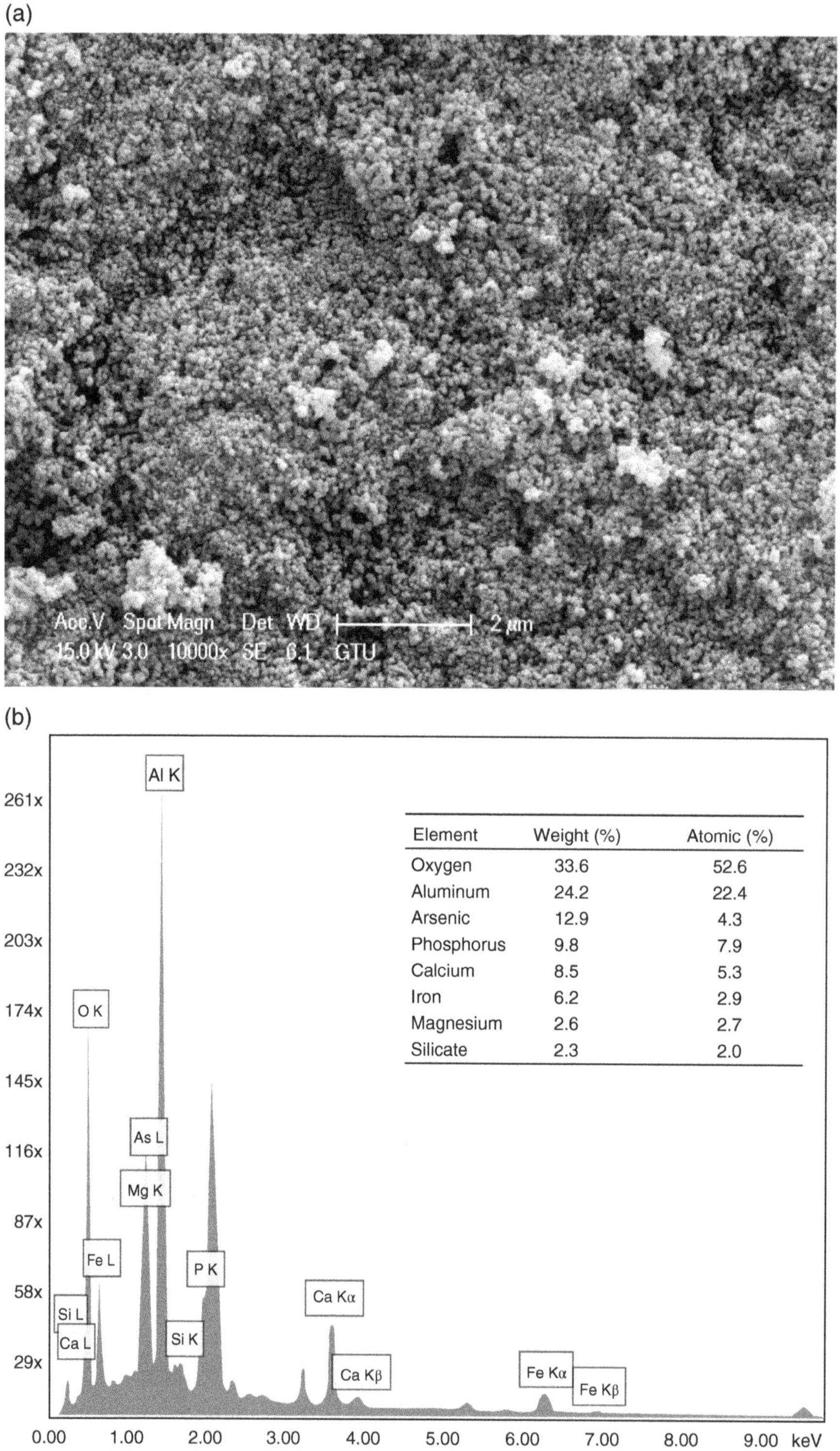

Element	Weight (%)	Atomic (%)
Oxygen	33.6	52.6
Aluminum	24.2	22.4
Arsenic	12.9	4.3
Phosphorus	9.8	7.9
Calcium	8.5	5.3
Iron	6.2	2.9
Magnesium	2.6	2.7
Silicate	2.3	2.0

Figure 15.1 Surface morphology with (a) Scanning electron microscope (SEM) and (b) EDS spectrum of sludge sample.

under optimum operating conditions. The As treatment in groundwater by the EC process with upflow reactor using Al plate was performed. The optimum As removal efficiency, OC, and energy consumption (ENC) were found to be 94.3%, 0.358 US\$ m^{-3}, and 1.64 kWh m^{-3} at a flow of 1.2 cm s^{-1}, pH of 7.56, initial As content of 23 μg l^{-1}, current of 7 A m^{-2}, and operating time of 60 minutes, respectively (Castaneda et al. 2019). In a separate study, the As (III) removal using Fe and Al electrodes from drinking water was studied (Kobya et al. 2011). At a current of 2.5 A m^{-2}, the highest As removal was found to be 94.1% for 150 μg l^{-1} of As concentration with the OC of 0.0054 US\$ m^{-3} at solution pH of 6.5 and operating time of 2.5 minutes using Fe plate, while the maximum As removal efficiency of 93.1% with the cost of 0.0073 US\$ m^{-3} was monitored when the EC system conducted with Al plate electrode at neutral pH (7.0) and conducting time of 4 minutes. The highest As removal and the minimum OC were observed with Fe electrodes. Naturally, 145 μg l^{-1} As containing groundwater treatment performance of packed bed EC using Al and Fe scrap electrode materials was studied (Omwene et al. 2019). The highest As removal of 93.0% was achieved with both anodes at a solution pH of 7.2. Furthermore, findings revealed that the EC process using Al scrap electrodes was more expensive than Fe plate electrodes by 0.181 \$ m^{-3} due to the increased operating time and applied current. Moreover, the Al—Fe-hybrid electrodes are widely used to improve the As removal efficiency of EC technology compared to the pure electrodes (Table 15.2). These hybrid electrodes are providing significant As removal efficiency owing to higher floc formation with larger metal hydroxide surface area for As adsorption. For instance, the As removal efficiency of 100% has been obtained using Fe—Al hybrid plate electrode at pH of 5.0, current of 4.5 A m^{-2}, and operating time of 15 minutes (Lopez-Guzman et al. 2019). Kobya and coworkers also studied the removal of As (V) using continuous flow EC with Fe—Al hybrid plate electrodes (Kobya et al. 2014). The optimum As (V) removal efficiency was found to be 96.1% at initial pH of 7.0, initial As (V) content of 150 μg l^{-1}, current of 2.5 A m^{-2}, and operating time of 3 minutes using hybrid electrodes.

Furthermore, the application of different electrode shapes such as scrap, ball, and rod have been investigated to enhance As removal performance of EC process from aqueous solutions compared to plate shape electrodes (Table 15.3). The surface area of the electrodes was increased with the application of scrap and ball-shaped electrodes (Figure 15.2). The treatment of groundwater with 200 μg l^{-1} of As (V) using a specifically designed aerated EC container with 3D Al ball electrodes was considered and the highest removal percentage of 99.2% was achieved at pH of 7.5, current of 0.3 A, and conducting time of 12 minutes with OC of 0.583 \$ m^{-3} (Goren et al. 2020a). The effect of anode level in the container was examined and they reported that the surface of electrodes for ball size of 7.5 mm was found to be 0.032, 0.076, and 0.182 m^2 for column heights of 2, 5, and 8 cm, respectively. The As (III) removal was enhanced with increasing Al ball surface area with the increased formation of aluminum hydroxide species. In addition, the Al and Fe waste scraps as an electrode material are favorable considering waste management and its cost-effectiveness as well as significant As removal efficiency with relatively low OC (Omwene et al. 2019).

Table 15.3 Summary of EC processes used to remove arsenic from aqueous solutions.

EC reactor mode	Electrode material and shape	As species	Initial pH	Initial As concentration ($\mu g\,l^{-1}$)	i, j, or U (A, $A\,m^{-2}$, or V)	Operating time (min, h)	Removal efficiency (%)	Operating cost ($ m^{-3})	References
CFM	Fe-plate	As (III)	7.1	40	5 A	30 $l\,min^{-1}$	99.0	0.002	Parga et al. (2005)
CFM	Fe-plate	As (III) + As (V)	7.0	100 000	120 $A\,m^{-2}$	120 min	98.0	—	Hansen et al. (2007)
BM	Fe—Al hybrid plate	As (III)	4.0	13 400	300 $A\,m^{-2}$	60 min	99.6	—	Gomes et al. (2007)
BM	Fe-plate	As (III)	4.0	15 000	75 $A\,m^{-2}$	30 min	99.3	—	Thella et al. (2008)
CFM	Fe-plate	As (III)	8.3	38.2 ± 2.01	1.9 $A\,m^{-2}$	30 min	96.0	0.0154	Kuan et al. (2009)
CFM	Fe-plate	As (V)	7.2	133	15–45 $A\,m^{-2}$	0.8–10.5 min	99.0	—	García-Lara et al. (2009)
BM	Fe-plate	As (V)	7.0	100 000	50–150 $A\,m^{-2}$	50 min	94.0	—	Balasubramanian et al. (2009)
BM	Fe-rod	As (III) + As (V)	6.5	50	282 A	0.5 $l\,min^{-1}$	80–95.0%	—	Lakshmanan et al. (2010)
BM	Fe—Al hybrid plate	As (V)	7.0	500–1500	20 $A\,m^{-2}$	30 min	98.4	—	Vasudevan et al. (2010)
CFM	Fe-plate	As (V)	7.2	1000	20 V	60 min	75	1.00	Kumar and Goel (2010)
BM	Fe-plate	As (III)	6.5	150	2.5 $A\,m^{-2}$	2.5 min	94.1	0.0054	Kobya et al. (2011)
BM	Al-plate	As (III)	7.0	150	2.5 $A\,m^{-2}$	4 min	93.5	0.0073	Kobya et al. (2011)
BM	Fe-rod	As (III) + As (V)	7.0	100 000	220 A	15–90 min	99.1	—	Wan et al. (2011)

BM	Fe—Al hybrid plate	As (V)	5.0–8.0	2000–5000	3 V	12 min	99.9	—	Ali et al. (2012)
CFM	Al-plate	As (III)	5.0	5.4	57.8 A m^{-2}	90 min	85.0	—	Mohora et al. (2012)
BM	Fe—Al hybrid plate	As (III) + As (V)	6.0	10 000	48 A m^{-2}	10 mi	99.9	—	Daniel and Rao (2012)
BM	Fe-plate	As (III)	7.4	112.3	5.64 A m^{-2}	5 min	93.9	0.076	Kobya et al. (2013)
BM	Fe-plate	As (V)	7.0	100	—	16 h	95.0	0.120	Molgora et al. (2013)
CFM	Fe—Al hybrid plate	As (V)	7.0	150	2.5 A m^{-2}	3 min	96.1	0.0104	Kobya et al. (2014)
BM	Fe-plate	As (III) and As (V)	7.1	266	2.1 V	400 C/L	98.2	0.84–1.04	Amrose et al. (2014)
BM	Fe-ball	As (III) + As (V)	7.6	285	0.3 A	20 min	99.3	1.55	Kobya et al. (2015)
BM	Fe-ball	As (V)	7.2	150	0.5 A	1.2 min	99.2	0.031	Sık et al. (2015)
BM	Fe—Al hybrid plate	As (III)	7.0	500	47 A m^{-2}	2 min	99.9	0.0782	Song et al. (2017)
BM	Fe-ball	As (V)	7.5	200	0.3 A	12 min	95.8	0.546	Sık et al. (2017)
BM	Fe-ball	As (III) + As (V)	7.5	200	0.15 A	3 min	95.1	0.041	Kobya et al. (2018)
BM	Al-ball	As (V)	7.03	613.4	0.29 A	10.5 min	99.0	0.442	Goren et al. (2018)
BM	Fe-ball	As (III)	8.5	50	0.05 A	4.94 min	99.0	0.01	Demirbas et al. (2019)
BM	Al-scrap	As (III) + As (V)	7.42	145	0.1 A	30 min	93.0	0.181	Omwene et al. (2019)

(Continued)

Table 15.3 (Continued)

EC reactor mode	Electrode material and shape	As species	Initial pH	Initial As concentration ($\mu g\,l^{-1}$)	*i*, *j*, or *U* (A, $A\,m^{-2}$, or V)	Operating time (min, h)	Removal efficiency (%)	Operating cost ($\$\,m^{-3}$)	References
BM	Fe-scrap	As (III) + As (V)	7.42	145	0.05 A	8 min	93.0	0.017	Omwene et al. (2019)
CFM	Al-plate	As (total)	7.56	23	$7\,A\,m^{-2}$	60 min	94.3	0.387	Castaneda et al. (2019)
BM	Fe-plate	As (V)	7.0	100	$0.431\,A\,m^{-2}$	30 min	99.4	0.0488	Das and Nandi (2019)
CFM	Al-plate	As (III)	7.0	550	$10\,A\,m^{-2}$	95 min	98.8	0.358	Thakur et al. (2019)
CFM	Al-plate	As (total)	8.5	50.4	$7\,A\,m^{-2}$	—	85.1	0.14	Sandoval et al. (2019)
BM	Fe—Al hybrid plate	As (total)	5.0	80	$4.5\,A\,m^{-2}$	15 min	100	—	Lopez-Guzman et al. (2019)
BM	Al-ball	As (III)	7.5	200	0.3 A	12 min	99.2	0.583	Goren et al. (2020a)
BM	Fe—Al hybrid plate	As (III)	7.0	10 000	0.46 A	2 min	94.5	0.0013	Khan et al. (2020)
BM	Al-plate	As (III)	6.0	1000	5 V	60 min	99.0	-	Babu et al. (2020)
CFM	Al-plate	As (total)	7.6	22	$10\,A\,m^{-2}$	$1.2\,cm\,s^{-1}$	93.1	1.093	Castaneda et al. (2020)
CFM	Fe—Al hybrid plate	As (III)	7.5	200	$9.9\,A\,m^{-2}$	50 min	95.2	0.99	Nyangi et al. (2020)
BM	Fe-plate	As (V)	—	100	0.2 A	7 min	90.0	—	Mendoza-Chávez et al. (2020)
BM	Al-ball	As (III) + As (V)	7.5	200	0.15 A	10.5 min	98.6	0.098	Goren and Kobya (2021)

BM, batch mode operation; CFM, continuous flow mode operation; C/L, charge loading; *i*, utilized current; *j*, current density; *U*, applied voltage.

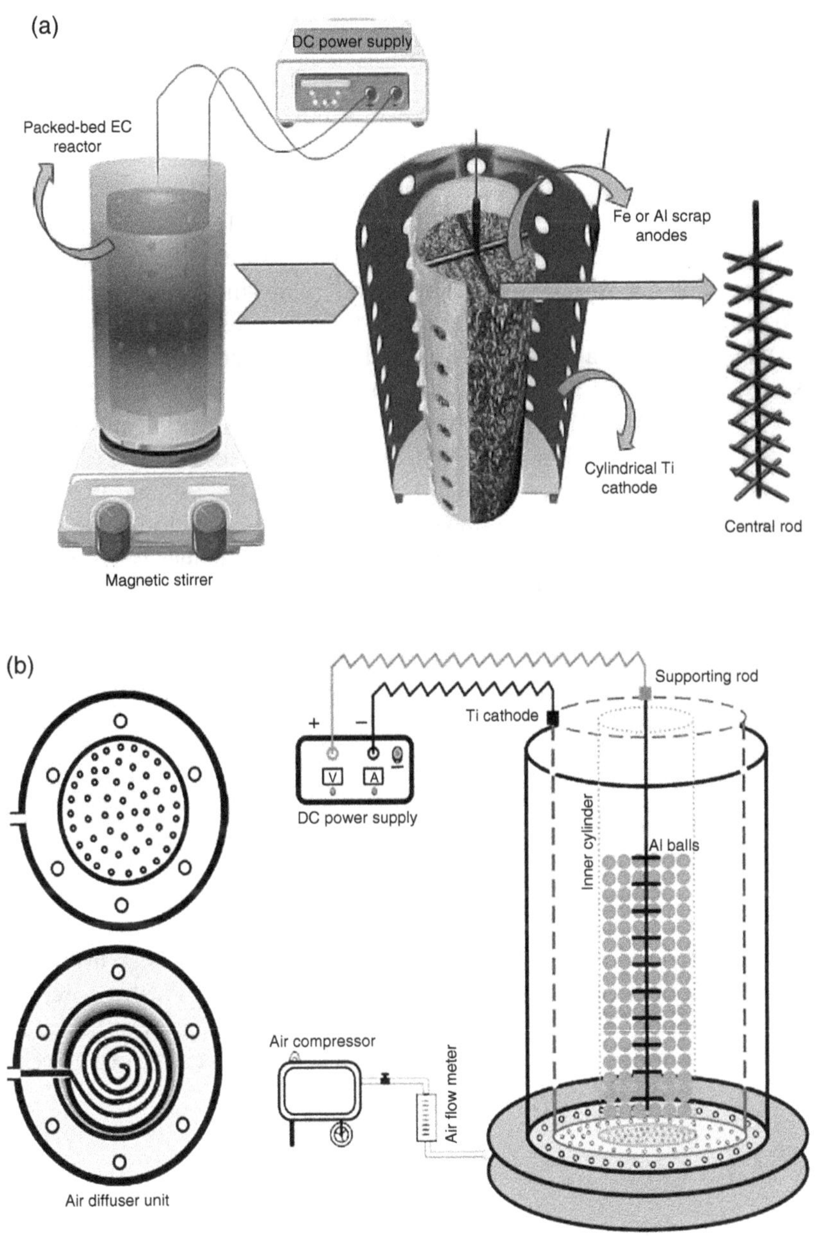

Figure 15.2 EC reactor configurations with scrap (a) and ball (b)-shaped electrodes for As removal.

15.7 Solution pH

The pH is key operational variable that influences the efficiency of EC method considering metal hydroxide formation, rate of dissolution of electrode material, and arsenic speciation. According to Eqs. (15.4) and (15.8), the pH of the aqueous media increases with increasing conducting time related to production of OH^- ions on the cathode surface during the EC experiments. Furthermore, the As speciation directly depends on pH of solutions. The As (III) is originated as H_3AsO_3, $H_2AsO_3^-$, $HAsO_3^{2-}$, and AsO_3^{3-} at low pH levels (Smedley and Kinniburgh 2002). The As (III) species of H_3AsO_3 is dominated at $pH < 9.2$, while the negative As (III) species of $H_2AsO_3^-$, $HAsO_3^{2-}$, and AsO_3^{3-} are dominated at $9.2 < pH < 12$ (Can et al. 2014). However, the As (V) species is formed as AsO_4^{3-} at $pH > 12.4$, $H_2AsO_4^-$ at 3.6 < pH 7.2, and $HAsO_4^{2-}$ at $7.2 < pH < 12.4$ (Vaclavikova et al. 2008). Consequently, at pH in the range of 4–10, As (V) species are dominant and negatively charged, while the As (III) species are neutral. Thus, the reduction of As (V) species is estimated to be higher than As (III) species due to the adsorption of negatively charged As (V) species by metal hydroxides. In addition, the speciation of the metal hydroxides is influenced by pH of the solution. The metal hydroxides are formed as hydrous ferric oxide HFO, $Al(OH)_{3(s)}$, amorphous $Fe(OH)_{3(s)}$, lepidocrocite, and goethite metal hydroxides in the range of pH 4.5–9.0 (Kobya et al. 2020). The initial pH values for As treatment using EC method were reported in the range of 4.0–8.5 (Table 15.3). In a study, Goren et al. (2020a) investigated the impact of pH (5.5, 6.5, 7.5, and 8.5) of the solution on As (III) + As (V) removal efficiency using aerated EC container containing Al balls. The highest As removal was achieved at pH value of 7.5. This was most probably due to the easy formation of aluminum hydroxide flocs at pH 7.5 and then increment of the As coagulation in the solution. Furthermore, the As removal percentage of 99.9% was achieved at pH of 6.0 using Fe—Al hybrid electrodes under optimum operating conditions (Daniel and Prabhakara-Rao 2012).

15.8 Effect of Applied Current

In EC technology, the formation rate of metal ions is fundamentally managed by the current and treatment time of the EC system. The current density ($j = i/S_{electrode}$, $A\,m^{-2}$; current per active electrode area) influences the reaction rate, size of the metal hydroxide flocs, and formation of bubbles in a bulk solution (Yilmaz et al. 2008). Furthermore, the ENC, electrode consumption (ELC), and OC in the EC process is significantly affected by current density as represented with the following equations (Amrose et al. 2014; Omwene et al. 2019):

$$\mathrm{ELC}\left(\mathrm{kg\ m^{-3}}\right) = \frac{i t_{\mathrm{EC}} M_{\mathrm{Me}}}{z_{\mathrm{Me}} F v} \tag{15.21}$$

$$\mathrm{ENC}\left(\mathrm{kWh\,m^{-3}}\right) = \frac{U i t_{\mathrm{EC}}}{v} \tag{15.22}$$

$$\mathrm{OC}\left(\$\ \mathrm{m^{-3}}\right) = \alpha \mathrm{ENC} + \beta \mathrm{ELC} \tag{15.23}$$

where, i is utilized current (A), t_{EC} is treatment time (sec or hour), M_{Me} is molecular weight of the metal ($g\,mol^{-1}$), z_{Me} is electron number, F is Faraday's constant (96 487 °C), U is voltage (V), v is water volume (m^3 or l), α (US\$ kWh^{-1}) and β (US\$ kg^{-1} electrode) is electrical energy and electrode prices.

Furthermore, charge loading (q) is another critical variable to optimize the EC process that is directly affected by applied current (Amrose et al. 2013). The effect of i on q level in the EC system is presented as follows:

$$q\left(C/L\right)=\frac{it_{\mathrm{EC}}}{v} \tag{15.24}$$

The high applied current values may result in elevated charge loading in the EC process, which may also cause an increment in the amount of residue metal hydroxide in the solution. The increasing metal hydroxide amount in the solution may cause excess sludge formation and reduce the uniform mixing in the EC reactor. Moreover, excess charge loading increases the ELC, ENC, and OC of the system due to these parameters and is affected linearly by charge loading as seen in Eqs. (15.21)–(15.23), respectively. Kobya et al. (2016) studied on assessment of operational parameters related to charge loading in the EC process for As removal from potable water. They reported that the effluent As concentration as a function of utilized current decreased with increasing time and charge loading (Figure 15.3a and b).

At different applied currents (0.075–0.30 A), the As removal efficiency of the EC process with Al anodes was investigated (Goren et al. 2020a). The As removal improved with rising current and achieved the recommended level of $10\,\mu g\,l^{-1}$ by WHO at supplied current of 0.30 A and operating time of 12 minutes as presented in Figure 15.4. This was most probably due to the increasing dissolved aluminum concentration with the increase in supplied current and the amount of As interacting with more aluminum hydroxide; $Al(OH)_{3(s)}$ and thus the As removal efficiency was increased. Furthermore, the amount of sludge in solution was found to be 0.081 and $0.255\,kg\,m^{-3}$ for supplied current of 0.075 and 0.30 A, respectively. As conclusion, the high applied current values could not be economical considering overall EC performance, although it is significantly promoting the dissolution of electrodes and removal efficiency As from solution.

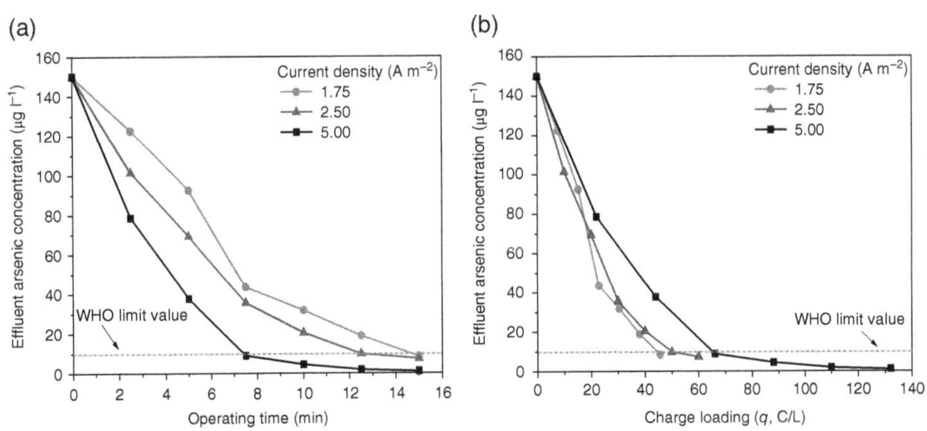

Figure 15.3 Effect of utilized current on the As treatment with (a) time and (b) charge loading (operating variables: initial pH of 6.5 and As content of $150\,\mu g\,l^{-1}$).

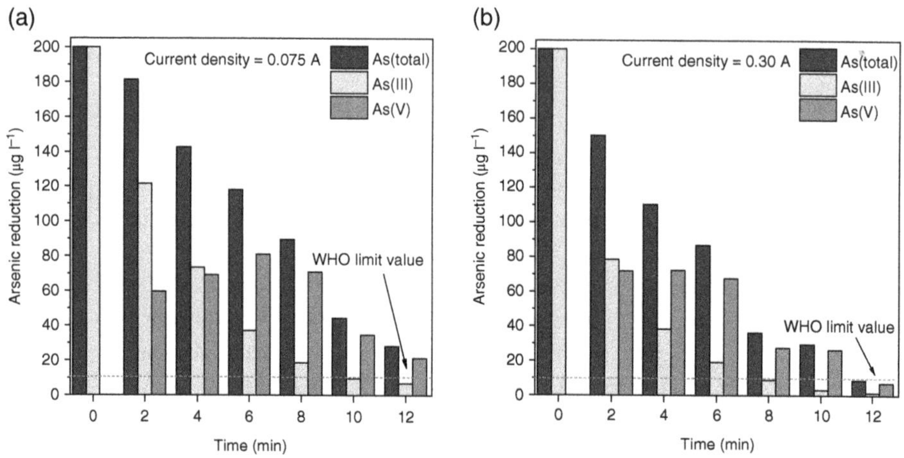

Figure 15.4 Arsenic reductions at different applied current values: (a) 0.075 A and (b) 0.30 A.

15.9 Optimization of EC Arsenic Removal Process

As mentioned in this chapter, the effectiveness of the EC process is directly related to operating parameters such as electrode shape and material, electrode connection mode and distance, initial pH, aeration rate, applied current, arsenic speciation and concentrations, and operating time. However, the proper selection of these operating conditions is one of the major challenges in the utilization of the EC process for As removal. Commonly, the experimental runs are performed analyzing one factor, while other parameters are sustained as constant. This procedure has several disadvantages such as it requires a significant number of experimental runs which resulted in high OCs, causes waste of time, and is insufficient for the investigation of the combined effect of operational variables. On the other hand, the statistical analysis of data with experimental design methodologies is a promising procedure to collect a significant amount of information from the extracted data with a small number of experimental runs. The material consumption and laboratory work time is decreasing owing to a reduction in the number of experimental runs in experimental design models. Furthermore, the assessment of statistical significance and accuracy of the operating parameter effects being studied as well as a determination of combined effects of parameters can be evaluated using mathematical models through experimental design. Therefore, the response surface methodology (RSM) has drawn considerable attention in the optimization, modeling, and experimental design of treatment processes. According to literature studies, RSM can be effective in the design and optimization of the EC processes for As removal (Goren et al. 2020a; Goren and Kobya 2021).

Goren and Kobya (2021) studied RSM to optimize the removal of As by the EC process with Al electrodes. For, Box–Behnken design (BBD) method, the authors identified phosphate, silica, bicarbonate, fluoride, boron, and operating time as six effective operational factors to optimize As removal efficiency, OC, ELC, and ENC as four responses. They specified two desired targets in the optimization process. The BBD was conducted within the specified operating parameter ranges for maximum As removal efficiency and minimum OC.

Using this target, with phosphate (C_{PO4}—$_P$) concentrations of 1.0 mg l^{-1}, silica (C_{SiO3}—$_{Si}$) concentration of 26.0 mg l^{-1}, bicarbonate (C_{HCO3}) concentration of 651.5 mg l^{-1}, fluoride (C_F) concentration of 2.0 mg l^{-1}, boron (C_B) concentration of 9.9 mg l^{-1}, and operating time of 10.5 minutes, they achieved As removal percentage of 98.6%, OC of 0.098 \$ m^{-3}, ELC of 0.0124 kg m^{-3}, and ENC of 0.411 kWh m^{-3}. In the second target, they set operating parameters in specified ranges to achieve the effluent As concentration of 9.9 μg l^{-1}. For this target, the maximum As removal efficiency of 95.1%, ENC of 0.790 kWh m^{-3}, ELC of 0.0225 kg m^{-3}, and OC of 0.23 \$ m^{-3} are achieved.

Demirbas et al. (2019) investigated the optimization of As (III) treatment in groundwater with RSM using the EC process. For the BBD method, they considered pH, utilized current, As (III) content, treatment time, Fe ball electrode diameter, column height in EC system, and aeration rate as the independent operating parameters for maximizing the As removal and minimizing the OC. For this scenario, they reported that the maximum As removal of 99% and minimum cost of 0.01 \$ m^{-3} were achieved at the pH of 8.5, utilized current of 0.05 A, treatment time of 5 minutes, Fe ball diameter of 7.5 mm, and airflow of 9.98 l min^{-1} for As (III) concentration of 50 μg l^{-1}. They concluded that the RSM was a promising tool to optimize the operational parameters of the EC process considering the highest removal and lowest cost. In a separate study, Song et al. (2017) employed RSM for evaluation of supplied current, initial pH, airflow, and conducting time on As removal efficiency and ENC using EC method with Fe—Al hybrid materials. The highest As removal efficiency of 99.9% and minimum ENC were observed at pH of 7.0, current density of 0.47 A dm^{-2}, airflow of 0.32 l min^{-1}, and conducting time of 20 minutes.

15.10 Cost of EC Arsenic Removal Method

As with other treatment processes, OC is another key parameter affecting the efficient usage of EC method for As removal. The OCs of the EC process under different operational variables in literature studies are presented in Table 15.3. The OC values can determine using various indirect and direct cost points such as ELC, ENC, sludge treatment, chemical consumption, maintenance, and labor costs (Kobya et al. 2014). Among these OC points, ENC and ELC costs are considered as principal cost points, as shown in Eq. (15.23). Based on the equation, the optimization of operating time and applied current is the major factor to minimize the OC of the EC method with the decrease in energy and ELC. RSM can be used in the optimization of the operating variables to achieve the minimum OC and maximum As removal efficiency in the EC process (Goren et al. 2020b). The effects of cations on As removal and OC of the EC method with Al balls were investigated using BBD. The highest As removal and lowest OC values were found to be 99.9% and 0.0332 US\$ m^{-3} at operational variables (utilized current of 0.15 A, pH of 7.5, Al ball electrode size of 7.5 mm, aeration of 6 l min^{-1}, and operating time of 17 minutes for As content of 200 μg l^{-1}), respectively. Removal of As (III) and As (V) using EC reactor using Fe electrodes was 96% (current of 0.30 A, 14 minutes of conducting time, and OC of 0.612 US\$ m^{-3}) and 95.8% (current of 0.30 A, 12 minutes of EC time, and OC of 0.546 US\$ m^{-3}) at operating conditions: pH of 7.5, initial As of 200 μg l^{-1}, iron ball diameter of 7.5 mm, ball's height of 7.5 cm, and feed airflow of 6 l min^{-1} (Şık et al. 2017). In a separate study, Parga et al. (2005) removed As with a percentage above 99% from natural groundwater (25–50 μg As l^{-1}, 600–4000 μS cm^{-1} of conductivity, and 5.5–7.1 of pH) using a novel EC process with air supply. This approved the decrease in the retention time, decreasing the remaining mass and, thus, decreasing the OC (0.002 US$ m^{-3} for this well-producing 2500 m^3 day^{-1}) of the process. Mólgora et al. (2013) performed treatment of As from tap water (100 μg As l^{-1}) using EC-Fe electrodes + MF hybrid process at continuous mode. The As removal efficiency of >97.0% and OC of the 0.12 US$ m^{-3} (0.076 \$ m^{-3} for EC and 0.0393 \$ m^{-3} for MF) were obtained under optimum operational variables (electrode space of 0.4 cm, pH of 7, Fe dosage of 4 mg l^{-1}, and operating time of 16 hours). Whereas, the OC of the coagulation + filtration process (operating conditions: initial pH of 7, Fe dosage of 4 mg l^{-1}, and operating time of 26 hours) is 0.066 US$ m^{-3}. A large portion of the OC of the EC-MF process results from the ENC cost (0.044 \$ m^{-3}). Naturally, arsenic-contaminated groundwater samples (~145 μg As l^{-1}) were removed from arsenic by a 0.50 l packed-bed EC reactor using Al and Fe scraps (Omwene et al. 2019). The OCs were calculated as 0.017 US$ m^{-3} for Fe scrap anodes and 0.181 US$ m^{-3} for Al scrap anodes.

Several operational parameters are reported to affect the OC (Amrose et al. 2013). It was noticed that the OCs were in the range of 0.002–1.0, 0.0073–1.09, 0.01–1.55, 0.098–0.583, and 0.001–0.99 US$ m^{-3} for Fe plate, Al plate, Fe ball, Al ball, and Fe-Al hybrid electrodes, respectively (Table 15.3). Addy et al. (2011) and Amrose et al. (2013) demonstrated that an EC reactor with Fe plate anodes (ECAR: Electrochemical Arsenic Remediation process; prototype reactor of 100 l) may decrease As below 10 μg l^{-1} in groundwater, Bangladesh (93–510 μg l^{-1}) and Cambodia (80–760 μg As l^{-1}), OC (including electricity and consumable iron) of this process was calculated as 0.22–0.31 US$ m^{-3}. Also, OC per person was estimated as 0.0038 US$ $person^{-1}$ day^{-1}, and waste produced of 120 g $person^{-1}$ $year^{-1}$. Field trial results of two identical 600 l of ECAR reactors in West Bengal (groundwater containing ~266 μg As l^{-1}) were obtained >98% of arsenic removal and total OC was assessed as 0.83–1.04 US$ m^{-3} (Amrose et al. 2014). Furthermore, the OC increases with the high initial As concentrations. On the other hand, the high solution conductivity in the solution significantly decreases the OC of the process. Consequently, the OC of the EC method for As removal is not only associated with applied current and operating time, but also related to initial As concentration, the conductivity of As-contaminated water, and electrode material. The OCs of BATs for As removal reported by WHO as 2.39–3.20 US$ m^{-3} for activated alumina, 0.054 US$ m^{-3} for oxidation-filtration, 0.73–1.70 US$ m^{-3} for coagulation-filtration, 1.20 US$ m^{-3} for granulated ferric hydroxide/oxide, 3.40 US$ m^{-3} for ion exchange, and 3.72 US$ m^{-3} for reverse osmosis (Ahmed 2004; Rahman et al. 2018). It can be concluded that the EC process is promising and cost-effective when the OC of other treatment technologies for As removal compared with the EC process. Considering the locations of arsenic contamination in the world to safe water access in rural areas, when the energy requirement of the EC process is met from alternative energy sources, it will be a very cost-effective process.

15.11 Merits and Demerits

The EC process is an essential and significant technology for As removal from aqueous solutions. Easy operation and simple equipment, low investment and OC, non-requirement of additional chemicals, low sludge production, easy floc formation, high TDS removal efficiency, and high As removal efficiency are significant advantages of this process.

Furthermore, the EC method is accomplished to treat As from water without the pre-oxidation step. This process has also the potential to remove other pollutants (e.g. iron, manganese, fluoride, natural organic matter (NOM), nitrate, and heavy metals) besides arsenic from groundwater and surface water sources. On the other hand, generation of toxic treatment sludge, high electricity price in some areas, and passivation on the electrodes are the main drawbacks of the EC process. The development of electrode type and materials, the EC process configuration, process integration with other wastewater treatment processes, optimization of operating parameters, energy, and ELC still need to be improved with detailed work. Besides, the current researches have been focused on the removal of specifically As from aqueous solutions, while very little research has been done on As removal from natural water sources in presence of other ions. Therefore, investigation on As removal using the EC process in presence of other ions may facilitate predicting the performance of the EC process before actually using them in real applications. Overall, these developments may make the EC process a more cost-effective, successful, environmentally friendly, and sustainable alternative technology for As removal from aqueous solutions.

15.12 Conclusions

Recent studies have shown EC to be an effective arsenic removal process to achieve the limit value of $10\,\mu g\,l^{-1}$ by WHO for safe water for As, which is one of the most toxic elements derived from the natural environment. The EC process for As removal from groundwater and surface water has attracted considerable attention due to its significant benefits such as high removal efficiency (>99%), cost-effectiveness, easy operation, low sludge production, and less environmental impact. Oxidation of As (III) to As (V) without using any external chemicals and arsenic removal is possible with the EC process. Furthermore, the EC process does not produce toxic by-products, concentrated water, and excess sludge formation as in the oxidation, chemical coagulation, membrane process, and adsorption process, respectively. This paper summarizes the occurrence of As in natural waters, As removal mechanism in the EC system, and effects of operational variables on EC in terms of OC and As removal performance, essentially including electrode type and shape, initial pH of the solution, and supplied current. The As removal in water using the EC process is in the range of 75.0–100%. The As removal efficiency of the EC method with Al—Fe hybrid electrodes has been more effectual than pure Al and Fe electrodes due to the effective adsorption reactions. Furthermore, the applied current can be considered as a significant operating parameter in influencing directly the As removal efficiency and OC of the EC system. The OC of EC is in the range of 0.001–1.55 \$ m^{-3}. At this point, ENC is the most important drawback of the EC process. To overcome this virtual drawback, environmentally friendly energy resources like geothermal, solar, wind, and tidal energy can be used. The application of these renewable energy sources could also reduce the cost and footprint of the EC method for As removal. Overall, the EC technology could be investigated further to enhance their performance as an environmentally friendly technology for the treatment of As containing water sources. Moreover, the efficiency of the EC method in presence of other contaminants may also be investigated to understand the process efficiency under real conditions in future studies. The development of hybrid processes with the EC process for the efficient As removal can be considered as an environmentally friendly option.

References

Abiye, T.A. and Bhattacharya, A. (2019). Arsenic concentration in groundwater: archetypal study from South Africa. *Groundwater for Sustainable Development* 9: 100246.

Addy, S. Gadgil, A.J. van Genuchten, C.M. Li, L. (2011). Locally affordable arsenic remediation for rural South Asia using electrocoagulation. 35th WEDC International Conference, Loughborough, UK (6–8 July 2011), p. 8.

Ahmad, S.A., Khan, M.H., and Haque, M. (2018). Arsenic contamination in groundwater in Bangladesh: implications and challenges for healthcare policy. *Risk Management and Health Policy* 11: 251–261.

Ahmed, F. (2004). Arsenic contamination of groundwater in South and East Asian countries. The World Bank, Volume I and II Policy and Technical Reports, World Bank, Washington, DC, Report no. 31303.

Ali, I., Gupta, V.K., Khan, T.A., and Asim, M. (2012). Removal of arsenate from aqueous solution by electro-coagulation method using Al—Fe electrodes. *International Journal of Electrochemical Science* 7: 1898–1907.

Amrose, S., Gadgil, A., Srinivasan, V. et al. (2013). Arsenic removal from groundwater using iron electrocoagulation: effect of charge dosage rate. *Journal of Environmental Science Health A* 48: 1019–1030.

Amrose, S.E., Bandaru, S.R., Delaire, C. et al. (2014). Electro-chemical arsenic remediation: field trials in West Bengal. *Science of the Total Environment* 488: 539–546.

Atabaki, N., Shaharuddin, N.A., Ahmad, S.A. et al. (2021). Assessment of water mimosa (*Neptunia oleracea* Lour.) morphological, physiological, and removal efficiency for phytoremediation of arsenic-polluted water. *Plants* 9: 1500.

Ayotte, J.D., Medalie, L., Qi, S.L. et al. (2017). Estimating the high-arsenic domestic-well population in the conterminous United States. *Environmental Science Technology* 51: 12443–12454.

Babu, D.S., Kadaverugu, R., Nidheesh, P.V., and Kumar, M.S. (2020). Importance of chloride addition on arsenite removal by aluminium electrocoagulation. *ChemistrySelect* 5: 10567–10573.

Balasubramanian, N., Kojima, T., and Srinivasakannan, C. (2009). Arsenic removal through electrocoagulation: kinetic and statistical modeling. *Chemical Engineering Journal* 155: 76–82.

Barraque, F., Montes, M.L., Fernandez, M.A. et al. (2021). Arsenate removal from aqueous solution by montmorillonite and organo-montmorillonite magnetic materials. *Environmental Research* 192: 110247.

Baskan, M.B. and Pala, A. (2010). A statistical experiment design approach for arsenic removal by coagulation process using aluminum sulfate. *Desalination* 254: 42–48.

Berg, M., Tran, H.C., Nguyen, T.C. et al. (2001). Arsenic contamination of groundwater and drinking water in Vietnam: a human health threat. *Environmental Science Technology* 35: 2621–2626.

Bhattacharya, P., Mukherjee, A., and Mukherjee, A.B. (2013). Groundwater arsenic in India: source, distribution, effects and alternate safe drinking water sources. In: *Reference Module in Earth Systems and Environmental Sciences*. USA: Elsevier ISBN: 9780124095489. https://doi.org/10.1016/B978-0-12-409548-9.09342-8.

Brikowski, T.H., Smith, L.S., and Neku, A. (2018). Groundwater arsenic in Nepal: occurrence and temporal variation. In: *Groundwater of South Asia* (ed. A. Mukherjee). Singapore: Springer Hydrogeology https://doi.org/10.1007/978-981-10-3889-1_23, ISBN: 978-981-10-3888-4.

Camacho, L.M., Gutierrez, M., Alarcon-Herrera, M.T. et al. (2011). Occurrence and treatment of arsenic in groundwater and soil in northern Mexico and southwestern USA. *Chemosphere* 83: 211–225.

Can, B.Z., Boncukcuoglu, R., Yilmaz, A.E., and Fil, B.A. (2014). Effect of some operational parameters on the arsenic removal by electrocoagulation using iron electrodes. *Journal of Environmental Health Science Engineering* 12–95.

Castaneda, L.F., Coreno, O., and Nava, J.L. (2019). Arsenic and hydrated silica removal from groundwater by electrocoagulation using an up-flow reactor in a serpentine array. *Journal of Environmental Chemical Engineering* 7: 103353.

Castaneda, L.F., Coreno, O., and Nava, J.L. (2020). Simultaneous elimination of hydrated silica, arsenic and phosphates from real groundwater by electrocoagulation using a cascade-shaped up-flow reactor. *Electrochimca Acta* 331: 135365.

Daniel, R. and Prabhakara-Rao, A. (2012). An efficient removal of arsenic from industrial effluents using electro-coagulation as clean technology option. *International Journal of Environmental Research* 6: 711–718.

Das, D. and Nandi, B.K. (2019). Arsenic removal from tap water by electrocoagulation: investigation of process parameters, kinetic analysis, and operating cost. *Journal of Dispersion Science and Technology* https://doi.org/10.1080/01932691.2019.1681280.

de Meyer, C.M.C., Rodriguez, J.M., Carpio, E.A. et al. (2017). Arsenic, manganese and aluminum contamination in groundwater resources of Western Amazonia (Peru). *Science of the Total Environment* 607–608: 1437–1450.

Demirbas, E., Kobya, M., Oncel, M.S. et al. (2019). Arsenite removal from groundwater in a batch electrocoagulation process: optimization through response surface methodology. *Separation Science and Technology* 54: 775–785.

Ding, W., Zheng, H., Sun, Y. et al. (2021). Activation of $MnFe_2O_4$ by sulfite for fast and efficient removal of arsenic (III) at circumneutral pH: involvement of Mn(III). *Journal of Hazardous Materials* 403: 123623.

European Union (1998). Council directive 98/83/EC on the quality of water intended for human consumption. *Official Journal of the European Communities, L* 330 (32): 32–54.

Feenstra L. van Erkel J. and Vasak L. (2007). Arsenic in groundwater: overview and evaluation of removal methods. IGRAC (International Groundwater Resources Assessment Centre) Report, The Netherlands-Utrecht, Report nr. SP 2007-2, December 2007, p. 23.

Ferreccio, C. and Sancha, A.M. (2006). Arsenic exposure and its impact on health in Chile. *Journal of Health, Population and Nutrition* 24: 164–175.

Garcia-Costa, A.L., Sarabia, A., Zazo, J.A., and Casas, J.A. (2021). UV-assisted catalytic wet peroxide oxidation and adsorption as efficient process for arsenic removal in groundwater. *Catalysis Today* 361: 176–182.

García-Lara, A., Montero-Ocampo, C., and Martínez-Villafañe, F. (2009). An empirical model for treatment of arsenic contaminated underground water by electrocoagulation process employing a bipolar cell configuration with continuous flow. *Water Science and Technology* 60: 2153–2160.

Gomes, J.A., Daida, P., Kesmez, M. et al. (2007). Arsenic removal by electrocoagulation using combined Al—Fe electrode system and characterization of products. *Journal of Hazardous Materials* 139: 220–231.

Goren, A., Oncel, M., Demirbas, E. et al. (2018). Removal of arsenate by electrocoagulation reactor using aluminum ball anode electrodes. *Water Practice and Technology* 13: 753–763.

Goren, A.Y. and Kobya, M. (2021). Arsenic removal from groundwater using an aerated electrocoagulation reactor with 3D Al electrodes in the presence of anions. *Chemosphere* 263: 128253.

Goren, A.Y., Kobya, M., and Oncel, M.S. (2020a). Arsenite removal from groundwater using aerated electrocoagulation reactor with Al ball electrodes: human health risk assessment. *Chemosphere* 251: 126363.

Goren, A.Y., Kobya, M., Şık, E. et al. (2020b). Combined influence of some cations on arsenic removal by an air-injection EC reactor using aluminum ball electrodes. *Desalination and Water Treatment* 178: 240–253.

Grönwall, J. and Danert, K. (2020). Regarding groundwater and drinking water access through a human rights lens: self-supply as a norm. *Water* 12: 419.

Guo, H., Wen, D., Liu, Z. et al. (2014). A review of high arsenic groundwater in Mainland and Taiwan, China: distribution, characteristics and geochemical processes. *Applied Geochemistry*. 41: 196–217.

Hakizimana, J.P., Gourich, B., Chafi, M. et al. (2017). Electrocoagulation process in water treatment: a review of electrocoagulation modeling approaches. *Desalination* 404: 1–21.

Hansen, H.K., Nunez, P., Raboy, D. et al. (2007). Electrocoagulation in wastewater containing arsenic: comparing different process designs. *Electrochimca Acta* 52: 3464–3470.

Hayat, E. and Baba, A. (2017). Quality of groundwater resources in Afghanistan. *Environmental Monitoring Assessment* 189: 318.

Henke, K. (2009). Appendix E: regulation of arsenic: a brief survey and bibliography. In: *Arsenic: Environmental Chemistry, Health Threats and Waste Treatment* (ed. K.R. Henke), 545–557. UK: Wiley ISBN: 978-0-470-02758-5.

Hering, J.G., Chen, P.Y., Wilkie, J.A., and Elimelech, M. (1997). Arsenic removal from drinking water during coagulation. *Journal of Environmental Engineering* 123: 800–807.

Hering, J.G., Chen, P.Y., Wilkie, J.A. et al. (1996). Arsenic removal by ferric chloride. *Journal of American Water Works Association* 88: 155–167.

Inam, M.A., Khan, R., Park, D.R. et al. (2018). Influence of pH and contaminant redox form on the competitive removal of arsenic and antimony from aqueous media by coagulation. *Minerals* 8: 574.

Jones, H., Visoottiviseth, P., Bux, M.K. et al. (2008). Case reports: arsenic pollution in Thailand, Bangladesh, and Hungary. *Reviews of Environmental Contamination and Toxicology* 197: 163–187.

Khan, S.U., Farooqi, I.H., Usman, M., and Basheer, F. (2020). Energy efficient rapid removal of Arsenic in an electrocoagulation reactor with hybrid Fe/Al electrodes: process optimization using CCD and kinetic modeling. *Water* 12: 2876.

Kloster, G.A., Valiente, M., Marcovich, N.E., and Mosiewicki, M.A. (2020). Adsorption of arsenic onto films based on chitosan and chitosan/nano-iron-oxide. *International Journal of Biological Macromolecules* 165: 1286–1295.

Kobya, M., Akyol, A., Demirbas, E., and Oncel, M. (2014). Removal of arsenic from drinking water by batch and continuous electrocoagulation processes using hybrid Al—Fe plate electrodes. *Environmental Progress and Sustainable Energy* 33: 131–140.

Kobya, M., Demirbas, E., Gebologlu, U. et al. (2013). Optimization of arsenic removal from drinking water by electrocoagulation batch process using response surface methodology. *Desalination and Water Treatment* 51: 6676–6687.

Kobya, M., Demirbas, E., and Ulu, F. (2016). Evaluation of operating parameters with respect to charge loading on the removal efficiency of arsenic from potable water by electrocoagulation. *Journal of Environmental Chemical Engineering* 4: 1484–1494.

Kobya, M., Gebologlu, U., Ulu, F. et al. (2011). Removal of arsenic from drinking water by the electrocoagulation using Fe and Al electrodes. *Electrochimica Acta* 56: 5060–5070.

Kobya, M., Ozyonar, F., Demirbas, E. et al. (2015). 'Arsenic removal from groundwater of Sivas-Şarkişla Plain, Turkey by electrocoagulation process: comparing with iron plate and ball electrodes. *Journal of Environmental Chemical Engineering* 3: 1096–1106.

Kobya, M., Sik, E., Demirbas, E. et al. (2018). Optimization of some cations for removal of arsenic from groundwater by electrocoagulation process. *Environmental Engineering Management Journal* 17: 1079–1093.

Kobya, M., Soltani, R.D.C., Omwene, P.I., and Khataee, A. (2020). A review on decontamination of arsenic-contained water by electrocoagulation: reactor configurations and operating cost along with removal mechanisms. *Environmental Technology and Innovation* 17: 100519.

Katsoyiannis, I.A., Tzollas, N.M., Tolkou, A.K. et al. (2017). Use of Novel Composite Coagulants for Arsenic Removal from Waters-Experimental Insight for the Application of Polyferric Sulfate (PFS). *Sustainability* 9: 590.

Kuan, W., Hu, C., and Chiang, M. (2009). Treatment of As (V) and As (III) by electrocoagulation using Al and Fe electrode. *Water Science and Technology* 60: 1341–1346.

Kumar, I. and Quaff, A.R. (2019). Comparative study on the efectiveness of natural coagulant aids and commercial coagulant: removal of arsenic from water. *International Journal of Environmental Science and Technology* 16: 5989–5994.

Kumar, N.S. and Goel, S. (2010). Factors influencing arsenic and nitrate removal from drinking water in a continuous flow electrocoagulation (EC) process. *Journal of Hazardous Materials* 173: 528–533.

Lakshmanan, D., Clifford, D.A., and Samanta, G. (2010). Comparative study of arsenic removal by iron using electrocoagulation and chemical coagulation. *Water Research* 44: 5641–5652.

Laskaridis, A., Sarakatsianos, I., Tzollas, N., and Katsoyiannis, I.A. (2020). Simultaneous removal of arsenate and chromate from ground- and surface- waters by iron-based redox assisted coagulation. *Sustainability* 12: 5394.

Lopez-Guzman, M., Alarcon-Herrera, M.T., Irigoyen-Campuzano, J.R. et al. (2019). Simultaneous removal of fluoride and arsenic from well water by electrocoagulation. *Science of the Total Environment* 678: 181–187.

Mc-Guigan, C.F., Hamula, C.L., Huang, S., and Le, X.C. (2010). A review on arsenic concentrations in Canadian drinking water. *Environmnetal Reviews* 18: 291–307.

McNeill, L.S. and Edwards, M. (1995). Soluble arsenic removal at water treatment plants. *Journal – American Water Works Association* 87: 105–113.

Mendoza-Chávez, C.E., Carabin, A., Dirany, A. et al. (2020). Statistical optimization of arsenic removal from synthetic water by electrocoagulation system and its application with real arsenic-polluted groundwater. *Environmental Technology* https://doi.org/10.1080/09593330.2020.1732472.

Mohora, E., Rončević, S., Dalmacija, B. et al. (2012). Removal of natural organic matter and arsenic from water by electrocoagulation/flotation continuous flow reactor. *Journal of Hazardous Materials* 235: 257–264.

Mólgora, C.C., Domínguez, A.M., Avila, E.M. et al. (2013). Removal of arsenic from drinking water: a comparative study between electrocoagulation-microfiltration and chemical coagulation-microfiltration processes. *Separation and Purification Technology* 118: 645–651.

Moussa, D.T., El-Naas, M.H., Nasser, M., and Al-Marri, M.J. (2017). A comprehensive review of electrocoagulation for water treatment: potentials and challenges. *Journal of Environmental Management* 1986: 24–41.

Nazari, A., Nakhaei, M., and Yari, A.R. (2020). Arsenic Adsorption by TiO_2 nanoparticles under conditions similar to groundwater: batch and column studies. *International Journal of Environmental Research* https://doi.org/10.1007/s41742-020-00298-7.

Nicolli, H.B., Bundschuh, J., Blanco, M.C. et al. (2012). Arsenic and associated trace-elements in groundwater from the Chaco-Pampean plain, Argentina: results from 100 years of research. *Science of the Total Environment* 429: 36–56.

Nyangi, M.J., Chebude, Y., Kilulya, K.F., and Andrew, M. (2020). Simultaneous removal of fluoride and arsenic from water by hybrid Al—Fe electrocoagulation: process optimization through surface response method. *Separation Science and Technology* https://doi.org/10.1080/01496395.2020.1837877.

Omwene, P.I., Çelen, M., Öncel, M.S., and Kobya, M. (2019). Arsenic removal from naturally arsenic contaminated ground water by packed-bed electrocoagulator using Al and Fe scrap anodes. *Process Safety and Environmental Protection* 121: 20–31.

Parga, J.R., Cocke, D.L., Valenzuela, J.L. et al. (2005). Arsenic removal via electrocoagulation from heavy metal contaminated groundwater in La Comarca Lagunera Mexico. *Journal of Hazardous Materials* 124: 247–254.

Phan, K., Sthiannopkao, S., Kim, K.W.W. et al. (2010). Health risk assessment of inorganic arsenic intake of Cambodia residents through groundwater drinking pathway. *Water Research* 44: 5777–5788.

Pincetti-Zúniga, G.P., Richards, L.A., Tun, Y.M. et al. (2020). Major and trace (including arsenic) groundwater chemistry in central and southern Myanmar. *Applied Geochemistry* 115: 104535.

Podgorski, J. and Berg, M. (2020). Global threat of arsenic in groundwater. *Science* 368: 845–850.

Polya, D.A. and Middleton, D.R.S. (2017). Chapter 1: Arsenic in drinking water: sources and human exposure. In: *Best Practice Guide on the Control of Arsenic in Drinking Water*', https://doi.org/10.2166/9781780404929_001 (ed. P. Bhattacharya, D.A. Polya and D. Jovanovic), July 2017, 1–23. UK: IWA Publishing.

Pramanik, B.K., Pramanik, S.K., and Suja, F. (2016). Removal of arsenic and iron removal from drinking water using coagulation and biological treatment. *Journal of Water and Health* 14: 90–96.

Rahman, M.A., Lee, S.G., Kabir, M.Y. et al. (2018). Arsenic calamity of groundwater in Southeast Asian countries: a public health hazard, socioeconomic effects, and mitigation technologies. *Australian Journal of Science and Technology* 18: 46–53.

Ravenscroft, P., Brammer, H., and Richards, K. (2009). *Arsenic Pollution: A Global Synthesis*. West Sussex, UK: John Wiley & Sons.

Rosales, M., Garcia, A., Fuenzalida, V.M. et al. (2020). Unprecedented arsenic photo-oxidation behavior of few- and multi-layer $Ti_3C_2T_x$ nano-sheets. *Applied Materials Today* 20: 100769.

Sandoval, M.A., Fuentes, R., Nava, J.L. et al. (2019). Simultaneous removal of fluoride and arsenic from groundwater by electrocoagulation using a filter-press flow reactor with a three-cell stack. *Separation and Purification Technology* 208: 208–216.

Şen, F., Altıok, E., Cyganowski, P. et al. (2021). Reclamation of RO permeate and concentrate of geothermal water by new chelating resins having *N*-methyl-d-glucamine ligands. *Separation and Purification Technology* 254: 117558.

Sha, L., Zou, Z., Qu, J. et al. (2020). As (III) removal from aqueous solution by katoite ($Ca_3Al_2(OH)_{12}$). *Chemosphere* 260: 127555.

Shahid, M., Niazi, N.K., Dumat, C. et al. (2018). A meta-analysis of the distribution, sources and health risks of arsenic-contaminated groundwater in Pakistan. *Environmental Pollution* 242: 307–319.

Shaji, E., Santosh, M., Sarath, K.V. et al. (2021). Arsenic contamination of groundwater: a global synopsis with focus on the Indian Peninsula. *Geoscience Frontiers* 12: 101079.

Şık, E., Demirbas, E., Goren, A.Y. et al. (2017). Arsenite and arsenate removals from groundwater by electrocoagulation using iron ball anodes: influence of operating parameters. *Journal of Water Process Engineering* 18: 83–91.

Sık, E., Kobya, M., Demirbas, E. et al. (2015). Removal of As(V) from groundwater by a new electrocoagulation reactor using Fe ball anodes: optimization of operating parameters. *Desalination and Water Treatment* 56: 1177–1190.

Sinha, D. and Prasad, P. (2020). Health effects inflicted by chronic low-level arsenic contamination in groundwater: a global public health challenge. *Journal of Applied Toxicology* 40: 87–131.

Smedley, P.L. and Kinniburgh, D.G. (2002). A review of the source, behaviour and distribution of arsenic in natural waters. *Applied Geochemistry* 17: 517–568.

Song, P., Yang, Z., Zeng, G. et al. (2017). Electrocoagulation treatment of arsenic in wastewaters: a comprehensive review. *Chemical Engineering Journal* 317: 707–725.

Thakur, L.S., Goyal, H., and Mondal, P. (2019). Simultaneous removal of arsenic and fluoride from synthetic solution through continuous electrocoagulation: operating cost and sludge utilization. *Journal of Environmental Chemical Engineering* 7: 102829.

Thella, K., Verma, B., Srivastava, V.C., and Srivastava, K.K. (2008). Electrocoagulation study for the removal of arsenic and chromium from aqueous solution. *Journal of Environmental Science Health A* 43: 554–562.

US-EPA (United States-Environmental Protection Agency) (2000). *Technologies and Costs for Removal of Arsenic from Drinking Water*. Washington, DC: United States-Environmental Protection Agency.

US-EPA (United States-Environmental Protection Agency) (2001). National primary drinking water regulations: arsenic and clarifications to compliance and new source contaminants monitoring. Fed. Register, 66:14:6975. 22 January.

Vaclavikova, M., Gallios, G.P., Hredzak, S., and Jakabsky, S. (2008). Removal of arsenic from water streams: an overview of available techniques. *Clean Technologies and Environmental Policy* 10: 89–95.

Vasudevan, S., Lakshmi, J., and Sozhan, G. (2010). Studies on the removal of arsenate by electrochemical coagulation using aluminum alloy anode. *Clean–Soil, Air, Water* 38: 506–515.

Wan, P., Yuan, M., Yu, X. et al. (2020). Arsenate removal by reactive mixed matrix PVDF hollow fiber membranes with UIO-66 metal organic frameworks. *Chemical Engineering Journal* 382: 122921.

Wan, W., Pepping, T.J., Banerji, T. et al. (2011). Effects of water chemistry on arsenic removal from drinking water by electrocoagulation. *Water Research* 45: 384–392.

Wang, X., Xu, J., Xu, M. et al. (2021). High-efficient removal of arsenite by coagulation with titanium xerogel coagulant. *Separation and Purification Technology* 258: 118047.

WHO (World Health Organization) (1993). *Guidelines for Drinking Water Quality: Remediations*, vol. 1, 3. Geneva, Switzerland: WHO.

WHO (World Health Organization) (2004). *Guidelines for Drinking Water Quality, 1, 3*. Geneva, Switzerland: WHO Press.

Yaqub, M. and Lee, S.H. (2021). Experimental and neural network modeling of micellar enhanced ultrafiltration for arsenic removal from aqueous solution. *Environmental Engineering Research* 26: 190261.

Yilmaz, A., Boncukoglu, R., Kocakerim, M.M. et al. (2008). Boron removal from geothermal waters by electrocoagulation. *Journal of Hazardous Materials* 153: 146–151.

Zhang, C., Wu, B., Pan, B. et al. (2020). Deep removal of arsenite from water with no need for pre-oxidation or in-line oxidation. *Chemical Engineering* 401: 126046.

16

Developments in Membrane Technologies and Ion-Exchange Methods for Arsenic Removal from Aquatic Ecosystems

Muhammad Bilal Shakoor[1], Israr Masood ul Hasan[2], Sajid Rashid Ahmad[1], Mujahid Farid[3], Muzaffar Majid[1], Irshad Bibi[4], Asim Jilani[5], Tanzeela Kokab[1], and Nabeel Khan Niazi[4]

[1] *College of Earth and Environmental Sciences, University of the Punjab, Lahore, Pakistan*
[2] *State Key Laboratory for Modification of Chemical Fibers and Polymer Materials, College of Environmental Science and Engineering, Donghua University, Shanghai, China*
[3] *Department of Environmental Sciences, University of Gujrat, Gujrat, Pakistan*
[4] *Institute of Soil and Environmental Sciences, University of Agriculture Faisalabad, Faisalabad, Pakistan*
[5] *Center of Nanotechnology, King Abdul Aziz University, Jeddah, Saudi Arabia*

16.1 Introduction

Arsenic is a toxic metalloid that becomes a part of the food chain through various pathways. Dietary sources of arsenic include fish, shellfish, poultry, and many other kinds of seafood. The concentration of arsenic in seafood is considered less toxic because it is found in organic form. But arsenic poisoning due to its contamination in drinking water is posing various health effects (Amen et al. 2020; Hussain et al. 2021). There are many countries where arsenic concentration is too high. These countries include Taiwan, Bangladesh, Nepal, and India. According to the World Health Organization (WHO), in 1984, the survey concentration of arsenic was 50 mg litter^{-1} in India and Bangladesh (Khan et al. 2000). The chemical nature of arsenic is very complex, which makes its treatment difficult.

Due to the complexity of arsenic, it exists in two different forms in groundwater. The first one is trivalent form and the second one is pentavalent form. In the oxidized environment, it is present as arsenate (As(V)) and in the reduced environment, it is present in the form of arsenite (As(III)). While arsenic in the organic state is known as MMA, which stands for monomethyl-arsonate, and DMA as dimethyl-arsenate (Kumar et al. 2020). The most toxic form of arsenic is arsenite, which is nearly sixty times more lethal than arsenate. Moreover, arsenite is also more mobile than arsenate.

There are natural mechanisms, which support the spread of arsenic throughout groundwater (Nickson et al. 1998). There are generally two mechanisms, which are responsible for the spread of arsenic in groundwater. These include aerobic and anaerobic. In the aerobic mechanism, the oxidative decomposition of the ores and minerals of arsenic occurs, which makes

Arsenic in Plants: Uptake, Consequences and Remediation Techniques, First Edition.
Edited by Prabhat Kumar Srivastava, Rachana Singh, Parul Parihar, and Sheo Mohan Prasad.

its spread in groundwater. While deposition and reductive dissolution of ferric hydroxides, manganese oxide and aluminum occur at the coatings of aquifers, where they release arsenic in groundwater. These mineral compounds are very reactive in nature. So, these oxidize easily in the water (Chowdhury et al. 2000; Nickson et al. 2000). Natural spread is not only the source of arsenic pollution in water, but also anthropogenic factors matter more (Ali et al. 2020). These include colors and dyes, pesticides, wood preservatives, oil and coal-burning plants, pharmaceutical works, disinfectants, cement works, electronic industries, glassware production, ore production and processing, galvanizing, metal treatment, and ammunition industries (Smedley and Kinniburg 2002). The poisoning of arsenic is very lethal.

As much as the matter of removal technologies of arsenic, there are many emerging techniques, which can aid in this process. These include oxidation, use of ferric salts and aluminum alum, filtration and coagulation, adsorption, membrane filtration methods, natural and modified zeolites and clays, use of the zerovalent iron, ion-exchange method by using ion-exchange resins, and by polymeric ligands.

Among all the treatment technologies, membrane filtration and ion-exchange methods are considered to be efficient and cost-effective for the removal of arsenic. The major benefit of the ion-exchange method is that it does not depend on pH and is a physical-chemical process. The other advantage of ion exchange is that as it uses resins, it is easy to regenerate resin by simple washing with sodium chloride solution only. Other technologies for arsenic removal are highly costly and require special skills to carry them with precautions, as they are also sensitive techniques.

16.2 Arsenic Chemistry, Sources, and Distribution in Water

Water is one of the most necessary substances for all living creatures on the earth for their survival and growth. Water pollution has evolved as vital fear for many unindustrialized country people that have been affected by both natural and anthropogenic sources of contaminants; thus, researchers believe that the availability of harmless hygienic drinking water is considered a potential substitute for sustainable development. It is well believed that till 2025, the world population will have grown-up ~8.45 billion people, which will lead to 60% of the total population (5.1 billion), and will have to live in those areas of the world that will be faced with adequate to dangerous food availability and water scarcity. With exponential industrial growth and climatic changes, there are various bacteriological, waterborne toxins, physiological, and chemical that should be addressed through the 3Rs (reduce, reuse, and recycle) method using low-cost and inventive techniques. Recently, heavy metals and their derivatives have become major concerning areas of scientists due to the fact that an increase in population and industrialization leads to the burning of fossil fuels, and mining and oil purifying release large amounts of heavy metals into the environment. In view of the above-mentioned concerned areas particularly, arsenic (As) is a lethal chemical element that permits special attention owing to its high relative profusion and carcinogenic properties along with cadmium (Cd), mercury (Hg), and palladium (Pd) (Aksu 2005; Basu et al. 2014; Ebrahimi et al. 2014).

Arsenic is a highly toxic, ubiquitous, and carcinogenic metalloid. As a metalloid, As has both metallic and nonmetallic properties. Its chemistry is a very extensive subject. It is a redox-sensitive element and its mobility, form of As, and distribution depend on several key factors, such as oxidation-reduction process, pH, aquatic chemistry, ionic species, and microbial activity. Although As contamination is reported worldwide in the environment but recent research revealed the poisoning effect of As in groundwater. It is an alarming and threatening situation for human health (Ghosh et al. 2019). It is a metalloid that is present as an oxyanion complex in surface water (Xu et al. 2002). It hardly ever occurs in a free state; it is present in big minerals, along with oxygen (O), sulfur (S), iron (Fe), and other metals. Generally, As is present in groundwater in two forms (aerobic and anaerobic) – arsenate (As (V)) as in aerobic condition while arsenite (As (III)) as in anaerobic condition.

Arsenic-prominent allotropes are black, metallic grey, and yellow. As is present in the Group VA of the periodic table of elements. It has four oxidation states (−3, 0, +3, and +5) with −3 being most contaminated and +5 being least venomous. As discussed above, As has only two oxidation forms, i.e. +3 and +5 in surface water (Baig et al. 2012). In fact, in groundwater, As (III) is more toxic than As (V). Arsenic mobilization occurs at the pH (6.5–8.5) specifically under both reducing and oxidizing environments present in groundwater (Islam-Ul-Haq et al. 2007). The As (III) mobility is also more than As (V). This high mobility fact shows the sorption of As (III) to the mineral surface is less than that of As (V) anions. Besides these inorganic forms, As is also present in organic forms such as MMA and dimethylarsinate (DMA) (Figure 16.1). The maximum permissible As level in groundwater is below $10\,\mu g\,l^{-1}$, in soil, it is $5\,mg\,kg^{-1}$ and in open seawater, it is $1–2\,\mu g\,l^{-1}$ (Chakrabarti et al. 2018).

Over the years, several natural (groundwater contact with mineral deposits and organic matter complexion) (Hao et al. 2018) and anthropogenic (mining, wood preservation, surface water contamination with fertilizers, pesticides, insecticides, fungicides, fossil fuels burning, and industrial discharge) (Shankar and Shankar 2014) sources are considered for As contamination in groundwater. Industrial discharge and disposal of As chemicals may also cause As contamination in drinking water. It is noteworthy that surface water As contamination is much higher from natural sources than anthropogenic sources.

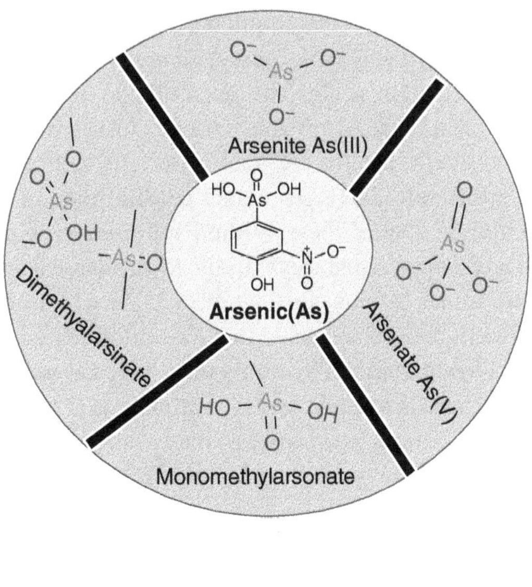

Figure 16.1 The composition of arsenic compounds in the natural environment.

Arsenic (As) concentration level is from 1–1000 $\mu g\,l^{-1}$ in surface water depending on the soil detoxification properties. Globally, As concentration in natural water ranges from 2.0×10^{-4} to >5 $mg\,l^{-1}$ (Wang and Mulligan 2006). It ranks as the 20th most abundant element on the earth's crust (0.0005% of the total earth's crust) with semimetallic properties. Arsenic concentration is much higher in finer-grained argillaceous phosphate and sediments than in rocks which is from 0.5 to 2.5 mg/kg (Mohan and Pittman 2007). Various conditions are considered for As accumulation in the water environment, such as weathering reactions, volcanic emissions, geochemical reactions, and biological activity. Other anthropogenic activities are fertilization, mining activities, herbicides, fossil fuel burning, As pesticides use, and some additional inputs to crops (Mohan and Pittman 2007).

Exact knowledge of As release source to water environment will help identify treatment method and safe drinking water recourses. In many As-contaminated areas, aerated strata, which were not exposed to oxidation before, increased by the high seasonal demand of water by which groundwater table drops. These aerated strata contain arsenian pyrite (FeS_2 containing As in trace to atom percent levels) and decomposition of the arsenian pyrite by which As is released into the aquifer. The colloidal/solid phase with natural pH mostly contains a majority of Fe (III) that bear As into water. These solid forms bearing As in water are a major cause of transport to an aquifer. Appelo et al. (2002) explained the mechanism of As releasing from sediment into surface water. Arsenic contamination is very severe as it is found worldwide such as in south-east Asia: Pakistan, Mongolia, China, Bangladesh, Vietnam, Thailand, Cambodia, West Bengal –India, Nepal, and Taiwan; in central and south America: Argentina, Mexico, and Chile and in North America (the USA and Canada), in Australia, and in Europe: Romania, Greece, Spain, Finland, Hungary, Italy, and Croatia (Biterna et al. 2007).

16.3 Health Implications of Arsenic

It is found that >200 minerals such as orpiment (As_2S_3), realgar (AsS), and arsenopyrite (FeAsS) contain arsenic and it causes so many severe health impacts on human and animal health. In many countries, As concentration in drinking water is more than the recommended permissible limit. According to the World Health Organization (WHO), 10 $\mu g\,l^{-1}$ is the maximum permissible limit for surface water. Two hundred and fifty million peoples of the world health are under serious concern to As contamination (Brammer and Ravenscroft 2009). Deliberate consumption of As may cause severe health problems in children. Out of these, around 110 million people live in these ten countries – Vietnam, Laos, Nepal, Bangladesh, India, Myanmar, Pakistan, Cambodia, Taiwan, and China, which are under serious consideration of As contamination (Brammer and Ravenscroft 2009). Recent research revealed that As-contaminated groundwater irrigation causes equally serious health hazards by eating those crops as well as for sustainable development of agricultural soils in affected areas (Mahimairaja et al. 2005; Ma et al. 2013). Prolonged exposure to As-containing food causes so many serious health issues such as pregnancy complications, lung and skin cancer, neurological disorders, kidney failure, and cardiovascular.

A precise mechanism of As-induced toxicity is yet needed to be fully understood as it is a complex mechanism and can alter by its solubility and oxidation state. Additionally, several internal and external factors such as genetics, length and frequency, exposure amount, biological species, nutritional factors, and sex also need to keep under serious consideration (Siddique et al. 2020). High consumption of As may also interfere in cell metabolism and in normal enzymatic activity.

16.4 Membrane Technologies

Membranes can be defined as synthetic materials which contain billions of holes or pores that act as specific barriers; these membranes are selective for some particles to pass through while rejecting other constituents. Developed methods for As removal are ion exchange, coagulation, adsorption, oxidation, chemical filtration, electrocoagulation, and membrane filtration. Membrane filtration technology gains much attention from researchers in the field of separation technology due to its easy operation, environmentally friendly, cost-effective, no sludge formation, simplicity, and versatility (Bolisetty et al. 2017; Siddique et al. 2020). This process required a driving force like potential difference/pressure difference. This membrane removes not only As but also dissolved minerals and bacteria that cause toxic diseases. Currently, this technology is considered best in all due to its easy operation and no sludge formation. However, flux reduction, dissolved solids in liquid wastes, and membrane fouling are excluded by the membrane system (Hao et al. 2018). Herein, we discussed two types of membrane methods that can control the As removal – high- and low-pressure membranes. In high-pressure membranes, contaminants are removed by chemical diffusion and in low-pressure membranes, mainly by physical sieving. In high-pressure membranes, we discussed reverse osmosis (RO) and nanofiltration, whereas, in low-pressure membranes, micro- and ultrafiltration (UF) were discussed.

16.4.1 High-Pressure Membranes

16.4.1.1 Reverse Osmosis

RO is defined as the rate at which water is diffusing, which is very low and has high osmotic pressure. The RO is proficient in the removal of salts, heavy metalloids (e.g. As), bacteria, turbidity, all dissolved substances, and organic matter from groundwater (Sarkar and Paul 2016). This membrane has tiny pores which are specific for the removal of the above-mentioned constituents. In the RO system, both low pH and temperature are key factors for As removal. The RO has further two types: thin-film composite (TFC or TFM) membrane and cellulose triacetate (CTA) membrane. TFC membrane nature is semipermeable and used in the desalination process. TFM membrane is a molecular sieve constructed on the base of two or more layers in the form of film. CTA membrane is most famous, used in hydration technologies (Ike et al. 2017). Removal efficiency for As (III) is lower, i.e. 70% than As (V) 90% when CTA membrane is used (Ghosh (Nath) 2019). Akin (2011) investigated the operational parameter for As removal via RO and found that As removal was mainly affected by pH and operation pressure.

16.4.1.2 Nanofiltration

Nanofiltration (NF) is a membrane liquid separation process that uses a bicycle pump for the removal of As and organic matter at low pressure. Polymer thin film is used for NF. This membrane is permeable and has a capital cost with a low energy requirement (Bolisetty et al. 2019). In this technique, selective constituents are rejected on basis of their charge and size. NF not only removes microbes but also removes color from water, humic and fulvic acid, natural minerals, and natural organic matter that cause hardness in water. NF provides two types of rejection – high and low rejection. High rejection has multivalent ions (calcium) while low rejection has monovalent ions (chloride). Typical properties of NF are molecular weight cut-off >100 Da (generally 300–500 Da), water permeability (5–50 l m^{-2} h^{-1} bar^{-1}), and pore size about 2 nm (Mulder 2012). NF has larger pores than RO but smaller than MF and UF. NF technique is used in industries and for food processing, e.g. dairy, lactose in fermentation, dye production, and also suitable for domestic supply (Table 16.1). Figoli (2010) stated that the As removal efficiency was increased with an increase in pH with a decrease of As feed concentration and operating temperature. They also stated that transmembrane had no or very little effect.

16.4.2 Low-Pressure Membrane

16.4.2.1 Microfiltration

Microfiltration (MF) is a physical process, which has a porous membrane that is used for the separation of bacteria, reduction of turbidity, and suspended constituents from liquid (Shih 2005). MF is used in purifying separating macromolecules, cold sterilization of beverages, pharmaceuticals, suspended particles from solution, clarifying wine (beer), fruit juices, petroleum refining, concentrating, liquid pharmaceutical, colloids, and medicines. The driving force in MF is the pressure difference. The size of the pore is 0.1 micron. MF removes all bacteria from water, protozoa, algae, sediment, but viruses remain in the water. MF membranes are made of many materials like ceramic, cellulose acetate, sliver, polyether sulfone (PES), an organic polymer (polypropylene), metal alloys, and silver. In this process, particularly As particulate matter is removed, which is very rare and as a result, the MF technique is clubbed with the coagulation-flocculation process which leads to an increase in molecular weight of particles (Ghurye and Clifford 2004).

Table 16.1 Nanofiltration uses.

Industry	Uses
Bulk chemistry	Product polishing and recovery of catalyst
Medicine	Extract lipids and amino acids from cell culture
Oil and petroleum chemistry	Remove tar component from feed purification
Natural essential oil and similar products	Fractionation of crude extract
Fine chemistry and pharmaceuticals	Solvent recovery and solvent exchange

16.4.2.2 Ultrafiltration

UF is a membrane filtration that uses suction or pressure for the removal of turbidity caused by protein (syrup and honey), bacteria, and suspended solid and colloidal matter (Table 16.2). The semipermeable membrane is used in UF through which fluid passes. This membrane retains only larger molecules that have low osmotic pressure. Due to extraordinary force, this filtrate is known as glomerular filtrate. UF is located in the kidney as a barrier between Bowman's capsule (glomerular capsule) and blood. UF is a volume of fluid that is removed from the patient. This membrane uses a hollow fiber membrane for the purification of water. UF provides a huge external area for particles to adhere to. This method depends on the size and charge of particles. UF is used for the separation of milk protein through the fine membrane, manufacture of dairy products and cheese, treatment of effluents in industries, and laboratory-grade manufacture. This membrane can also be used for effluents from the paper pulp mill, concentration and separation of enzymes, radiocarbon dating of bone collagen, enzymes recovery, solvent exchange of protein (via diafiltration), and the concentration of tomato paste and sucrose. Two types of materials used in UF are organic and inorganic polymers. In organic polymers, polycarbonate, polytetrafluoroethylene, and polyvinyl chloride are used, while in inorganic polymers, glass, alumina, titania, ceria, and zirconia are used. Brandhuber and Amy (2001) reported the mechanism of As removal and the effect of charge on the UF membrane. They stated that As removal was mainly due to an interaction between As ions and negatively charged membranes. They found 53% of As (III) elimination and 65% of As (V) removal.

Table 16.2 Membrane filtration types and their properties.

Ultrafiltration	Reverse osmosis	Microfiltration
Low pressure (2–6 bar)	High pressure (10–30 bar)	Low pressure (2–4 bar)
Chemical tolerance pH 1–13	pH 2–11	pH 1–13
Stream sterilization membrane available	Stream sterilization not possible	Stream sterilization possible
High temperature up to 80 °C	45 °C max	High temperature possible
Low energy consumption	High energy consumption	Low energy consumption
Hygienic module designs available	Module not as hygienic	Hygienic designs available
High resistance to oxidizing agents	Limited resistance to oxidizing agents	High resistance to oxidizing agents
High recovery (up to 95%)	Low recovery (50–80%)	100% recovery
Operates on difficult colloidal water	Required extensive pretreatment of colloids	Rapidly fouled by colloids giving high replacement costs

16.5 Ion Exchange

Ion exchange is a very ancient technique, which was observed and documented 100 years ago in history. From ancient times, this method is in use for softening water for its application at a wider scale. Now, this process has become essential in the industries and also important for many technical processes. There is a wide range of materials, which have the ability to exchange ions and can be either synthetic or natural, organic and inorganic, amphoteric, cationic, and anionic. It is a stoichiometric method of chemical reaction, which is reversible. In this technique, an ion from salt solution is exchanged with an ion of the solid phase in order to balance the electroneutrality. The most commonly used materials for this purpose are organic resins, inorganic 3-D matrix, and often hybrid materials of new generations.

This method involves the exchange of ions between two electrolytes in an electrolytic solution. This term is widely and commonly used to define the singularity of electrolytic solutions, which involves the refinement of the solid polymeric or ion exchangers in mineral form in the aqueous solutions. Furthermore, it is also used for "separation" and "purification" methods. There are various types of ion exchangers which include soil humus, zeolite, montmorillonite, and clay. Various resins are also used, which has the ability to exchange ions and these are called "functionalized porous gel" or "polymer gel." There are three major categories of ion exchangers – (i): cationic ion exchangers; has the ability for positively charged ion species, (ii): anionic ion exchangers; has the ability for negatively charged ion species, (iii): amphoteric; are those who can exchange both positive and negative ions simultaneously and more efficiently in the solutions having ion-exchange resins and often the treated solution is passed through other ion exchanger-containing solutions. Due to differences in the chemical structures of various kinds of ion exchangers, the exchange of ions would be indiscriminate and have a variable preference of binding with the ion exchangers. The rate and latent for the exchange of ions will be liable to the charge of the ion, its structure, and size.

The exchange of ions occurs in synthetically prepared resin beds. The water-containing arsenic is passed from the beds of sold resins where the arsenic ions in the aqueous phase are exchanged in equivalent amounts with the ions present in the solid phase. From polluted water, the arsenic ions are adsorbed over the resin beds. The ions which exchanged smoothly with the ions of As are passed over the resin beds in order to make the resin bed reusable.

There are two major types of resin ion exchangers. (i): anion exchanger resins; TiO_2 on the amberlite XAD-7 is anionic exchanger. (ii): cation exchanger resins; iron, lanthanum, and cerium are the cationic exchangers (Mohan and Pittman 2007; Sarkar and Paul 2016).

There are various factors, which govern the ion-exchange process, which further includes: the quality of resin, type of resin, and concentration of arsenic. Due to the high affinities of some ions with the resins, such as sulfates, carbonates, and bicarbonates, the exclusion of arsenic is greatly hindered. In order to overcome the effect of sulfates, carbonates, and bicarbonates, Cu resin was synthesized as a chelating agent and appropriate at pH 6.0–8.0 (An et al. 2010). Polymers with metals have the ability to extract both types of arsenic (III and V) by hindering the effect of other ions (Dambies 2004). Another type of alternative for As (V) removal was suggested by LeMire et al. (2010), which included the use of iron-containing exchange beads. This method considers the other factors of As removal, such as pH, level of arsenic in the feed, adsorption ability, competition, temperature, and content of iron present.

16.5.1 Ion-Exchange Resins

Generally, ion-exchange resins are used to remove the unnecessary anions and cations from bodies of water. The mechanism resembles the mechanism of activated alumina which is based on "adsorption." Synthetic resins have a high ion-exchange capacity as compared to natural resins. Most importantly, synthetic resins can be reused and regenerated oftentimes. Artificially prepared ion-exchange resins not only have the ability to remove anions such as sulfates and nitrates, but also have the efficiency to exclude arsenic.

Acidic resins negatively charged in nature have cations of Na ions for exclusion of cationic species by ion exchange. On the other hand, positively charged resins with chloride presence remove the anionic species.

Some anionic exchange resins with strong bases are available, which are used commercially to remove arsenic from water at a rate of one microgram per liter of water, while naturally occurring arsenite particles are not excluded by this process (Edwards et al. 1998; Ficklin 1983). Therefore the higher arsenic removal efficiency can be obtained by coupling the ion-exchange process with the oxidation process as a pretreatment to alter arsenite into arsenate.

Mostly, ion-exchange resin surfaces are treated with HCl and aid with the chloride ions as the chloride ion has the ability to detach arsenate ions due to the stronger basic properties of arsenate (Ghurye et al. 1999). Some ion-exchange resins also contain bromide and acetate ions (Edwards et al. 1998). Purolite A-505 and Relite A-490 are anion-exchange resins with a strong base that removes almost 99% arsenate in the situation of laboratory (Korngold et al. 2001).

The major drawback of the ion-exchange method is that practically the uptake capacity of resins for arsenate ions is altered because other anions present in the solution try to contend with arsenate ions to attach at the adsorption site and become dominant here.

16.5.2 Polymeric Ligand Exchangers

The US EPA has conducted various researches to identify the innovative and cost-effective treatment methods of arsenic removal from water and also to lower its standards. Helfferich introduced the idea of polymeric ligand exchangers (PLEs), which is a cost-effective and novel technique (Helfferich 1961, 1962). PLEs have two major components: (i) copper and iron transition metal ions, which are restrained to functional groups of host resin and (ii) cross-linked resin of host. Due to the presence of transition metals at the terminal functional group, ion exchange occurs according to Lewis acid–base interaction and forms a metal–ligand complex. Electrostatic interaction also occurs between the anion and fixed metal ion. Electrostatic interaction was common for uptake of anions in the conventional ion-exchange resins. According to the conventional ion-exchange method, the selection of ions is determined by the basicity of anions. While in PLE, selection occurs according to the strength of the ligand. Arsenate and phosphate are strong ligands, which have the ability to be accepted by PLE even in the presence of competitive sulfates and chloride ions.

Properties of resins:

- Polymeric resins must not utilize all the available sites which are present for the metal attachment. Metal ion attachment occurs at the ligand site according to Lewis acid–base interaction, by forming a bond in the aqueous phase.
- The functional groups of the polymer must not be charged always as positively charged particles would not react readily with the ligand to form electrostatic interaction with the ligand of the target in the solution.
- PLE should have the ability to hold the ions of metal tightly with the ligand during the process of ion exchange in order to prevent the seepage of metal in the solution.

Some of the neutral molecules of arsenite cannot be separated by using this method. That is why, to enhance the removal efficiency of arsenic, this method is joined with the pretreatment oxidation step. In this way, arsenite is converted into arsenate, which can be removed easily afterward.

Some resins are nontoxic, insoluble, and can resist chemicals and the polymer is not dissolved or hydrolyzed with the acid and base solutions. Such resins include polyglycidyl methacrylate-based chelating resins, iminodiacetic acid (IDA) resins, sulfonic acid, and polyhydroxyalkanoates (PHA) resins which are used preferably as they can do best in cation exchange. Biopolymer gels or chelating resins (polystyrene) and macroporous substances can also be used.

Alginate and chitosan biopolymers have less mechanical and chemical resistance. They do act after the immobilization of metal ions and work as cross-linking agents (Draget et al. 1992). Metal-containing resins PLE are used for the elimination of arsenic and categorized based on the metal used during the impregnation process.

16.5.3 Fe-Loaded Resins

Literature shows that the separation of arsenic is mostly done by Fe-loaded resins. The first work on it was documented by Yoshida et al. (1978). *O*-hydroxy-benzyl-nitrilo-diacetic iron-loaded resins were used commercially for arsenic separation from groundwater. According to their research, arsenate was adsorbed at a pH range of 3.6–5.5 while arsenite was adsorbed at a pH of 8.5.

The Fe-LDA and Fe (III)-(lysine-*N*,*N*-diacetic acid) were used to remove arsenic from polluted water (Matsunaga et al. 1996). Maximum arsenate adsorption occurred between 2 and 4 pH while adsorption of arsenite was maximum at 9 pH. The effective separation of arsenite and arsenate ions occurred at the addition of NaOH solution of 0.1 M. At pH 7 or neutral, the removal efficiency was too low for arsenate. There are some functionalized polymers, which are used to exclude both types of As (III & V). These include Fe (III)-loaded poly(hydroxamic) acid resin (Fe (III)-PHA) and Chelex 100 (Bio-Rad) (Fe (III)-IDA) which is iminodiacetic in nature (Atzei et al. 2001). By using these, arsenite sorption is very low as compared with arsenate.

Some resins containing bis(2-picolyamine) functional group (Dow XFS-4195) having iron III are a weak basic chelating material, can eliminate arsenate at pH 5, while arsenite elimination occurs at pH 10 (Suzuki et al. 2001).

Alginate (ALG) is a naturally occurring biopolymer with the loading of Fe (III), PLE (Fe (III)-ALG). This substance helps in the maximum removal of arsenite at a pH range of 3–4 (Min and Hering 1999). Efficient arsenate removal is possible in the presence of cation-exchange resin of strong acid, in addition to sulfate and phosphate intrusive ions. A microporous bio-Rad AGMP-50, the resin of sulfonic acid was formed (Guenegou et al. 1998). Precipitation of iron occurs into the molecules of a polymer by treating the resin with ferric chloride solution having a very acidic pH of 1. Then this is eluted with a 0.1 M solution of sodium hydroxide, which is highly basic. Whereas, adsorption concentration was variable for both types of arsenic – for arsenate, it was $11.2\,g\,l^{-1}$ at pH of 5.4, while arsenite was $10.9\,g\,l^{-1}$ at pH of 10.7. Iron (III) with chlorine can efficiently remove arsenate from industrial wastewater having a concentration of $1\,mol\,l^{-1}$. Thus, for industrial wastewater, iron (III)-loaded resins are more effective than basic ion-exchange resins.

16.5.4 Cu(II)-Loaded Resins

The Cu-containing resins were formed by adding (DOW-2N) containing picolylamine group, which is a less strong chelating agent with the copper (II) (Ramana and Sengupta 1992). According to researchers, copper resins have greater specifications and a strong affinity with arsenate than iron-loaded resins (Ramana and Sengupta 1992). Copper (II) has opposite properties to iron resins. Cu(II) is strong Lewis acid with more metal adsorption capacity (Irving and Williams 1953).

The chelating agent present in the copper resin works very effectively at pH 8.5 to detach the arsenate even in the presence of sulfate ions which are very competitive in nature. It is noted that iron-loaded resins greatly affect the Cu(II) resins for the discretion of arsenic in an effective way. Iron IDA can separate arsenate while copper IDA cannot do so due to variations in the electronic configuration of both metals (Atzei et al. 2001). DOW 3N is another copper PLE, which comprises more than one picolylamine group as a functional group than DOW 2N (Zhao and Sengupta 1998). When sulfate, nitrate, bicarbonate, and chloride are present, PLE gives uncommon selectivity for arsenate ions than strong base anion (SBA) resin (An et al, 2005). Copper and iron polychelates of Schiff base resins were made also for efficient removal of arsenic and it was observed that copper resins are much better than iron resins (Mohanty et al. 2013). Cu(II) containing functional groups with mesoporous sorbents of silica was also prepared by Fryxell, which has even more affinities as compared to SBA for arsenate and chromate than nitrate and sulfate (Fryxell et al. 1999).

Many other resins such as Zr(IV) in addition to iron- and copper-loaded resins are used (Atzei et al. 2001). Others include La(III) (Trung et al. 2001), Mo(VI) (Himeno et al. 1999), and Ce(IV) (Haron et al. 1997). All of them were introduced to eliminate arsenic from drinking water.

16.6 Conclusion

A number of remediation technologies have been applied for the removal of arsenic from contaminated water. Among them, membrane filtration and ion exchange-based technologies are proved to be very efficient. The application of membrane-based removal methods is highly investigated in the research field. Currently, this technology is considered the best owing to its easy operation and no sludge formation. However, flux reduction, dissolved solids in liquid wastes, and membrane fouling are some potential problems in the membrane system. Ion-exchange resins are widely used for the removal of arsenic from water. This technique involves the exchange of ions between two electrolytes in an electrolytic solution. Some anionic exchange resins having a strong base are used commercially to remove arsenic from water. Although, higher arsenic removal efficiency is obtained by coupling the ion-exchange process with the oxidation process as a pretreatment to transform arsenite into arsenate, thus showing the importance of the ion-exchange process in combination with the other method. Both the methods described above have some pros and cons but are refined to make them suitable in the available conditions. The basic aims of any method should be to reduce operational/maintenance costs, resolve sludge issues, and enhance arsenic removal performance in a user-friendly manner.

Acknowledgments

The authors are thankful to the University of the Punjab, Lahore, Pakistan. Dr. Nabeel Khan Niazi is thankful to the University of Agriculture Faisalabad, Pakistan.

References

Akin, I., Arslan, G., Tor, A. et al. (2011). Removal of arsenate [As (V)] andarsenite [As(III)] from water by SWHR and BW-30 reverse osmosis. *Desalination* 281: 88–92.

Aksu, Z.M. (2005). Application of biosorption for the removal of organic pollutants: a review. *Process Biochemistry* 40: 997–1026.

Ali, S., Rizwan, M., Shakoor, M.B. et al. (2020). High sorption efficiency for As (III) and As (V) from aqueous solutions using novel almond shell biochar. *Chemosphere* 243: 125330.

Amen, R., Bashir, H., Bibi, I. et al. (2020). A critical review on arsenic removal from water using biochar-based sorbents: the significance of modification and redox reactions. *Chemical Engineering Journal* 396: 125195.

An, B., Fu, Z., Xiong, Z. et al. (2010). Synthesis and characterizationof a new class of polymeric ligand exchangers for selective removal of arsenatefrom drinking water. *Reactive and Functional Polymers* 70 (8): 497–507.

An, B., Steinwinder, T.R., and Zhao, D. (2005). Selective removal of arsenate from drinking water using a polymeric ligand exchanger. *Water Research* 39: 4993–5004.

Appelo, C.A.J., Van der Weiden, M.J.J., Tournassat, C., and Charlet, L. (2002). Surface complexation of ferrous iron and carbonate on ferrihydrite and the mobilization of arsenic. *Environmental Science & Technology 36*: 3096–3103.

Atzei, D., Ferri, T., Sadun, C. et al. (2001). Structural characterization of complexes between iminodiacetate blocked on styrene-divinylbenzene matrix (Chelex 100 resin) and Fe(III), Cr(III), and Zn(II) in solid phase by energy-dispersive X-ray diffraction. *Journal of the American Chemical Society* 123: 2552–2558.

Baig, J.A., Kazi, T.G., and Elci, L. (2012). Biosorption characteristics of indigenous plant material for trivalent arsenic removal from groundwater: equilibrium and kinetic studies. *Separation Science and Technology* 47: 1044–1054.

Basu, A., Saha, D., Saha, R. et al. (2014). A review on sources, toxicity and remediation technologies for removing arsenic from drinking water. *Research on Chemical Intermediates* 40: 447–485.

Biterna, M., Arditsoglou, A., and Tsikouras, D. (2007). Arsenate removal by zero valent iron. Batch and column tests. *Journal of Hazardous Materials* 149 (3): 548–552.

Bolisetty, S., Peydayesh, M., and Mezzenga, R. (2019). Sustainable technologies for water purification from heavy metals: aeview and analysis. *Chemical Society Reviews* 48: 463–487.

Bolisetty, S., Reinhold, N., Zeder, C. et al. (2017). Effcient purification of arsenic-contaminated water using amyloidcarbon hybrid membranes. *Chemical Communications* 53: 5714–5717.

Brammer, H. and Ravenscroft, P. (2009). Arsenic in groundwater: a threat to sustainable agriculture in South and South-East Asia. *Environment International* 35: 647–654.

Brandhuber, P. and Amy, G. (2001). Arsenic removal by charged ultra filtration mem-brane – influences of membrane operating condition sand water qualityonar– senic rejection. *Desalination* 40: 1–14.

Chakrabarti, D., Singh, S.K., Rashid, M. H., Rahman, M. M. (2018). Arsenic: occurrence in groundwater. In: *Encyclopedia of Environmental Health*. Elsevier, pp. 165–180 http://dx.doi.org/10.1016/ January 2018, B978-970-912-409548-409549.410634-409547.

Chowdhury, U.K., Biswas, B.K., Chowdhury, T.R. et al. (2000). Groundwater arsenic contamination in Bangladesh and West Bengal, India. *Environmental Health Perspectives* 108: 393–397.

Dambies, L. (2004). Existing and prospective sorption technologies for the removal ofarsenic in water. *Separation Science and Technology* 39 (3): 603–627.

Draget, K.I., Varum, K.J., Moen, E. et al. (1992). Chitosan cross-linked with Mo(VI) polyoxyanions: a new gelling system. *Biomaterials* 13 (9): 635.

Ebrahimi, R., Maleki, A., Shahmoradi, B. et al. (2014). Elimination of arsenic contamination from water using chemically modified wheat straw. *Desalination and Water Treatment* 51: 2306–2316.

Edwards, M., Patel, S., McNeill, L. et al. (1998). Considerations in arsenic analysis and speciation. *Journal American Water Works Association* 90 (3): 103–113.

Ficklin, W.H. (1983). Separation of As(III) and As(V) in ground waters by ion exchange. *Talanta* 30 (5): 371.

Figoli, A., Cassano, A., Criscuoli, A. et al. (2010). Influence of operating parameters on the arsenic removal by nano-filtration. *Water Research* 44: 97–104.

Fryxell, G.E., Liu, J., Hauser, T.A. et al. (1999). Design and synthesis of selective mesoporous anion traps. *Chemical Materials* 11: 2148–2154.

Ghosh (Nath), S., Debsarkar, A., and Dutta, A. (2019). Technology alternatives for decontamination of arsenic-rich groundwater–A critical review. *Environmental Technology and Innovation* 13: 277–303.

Ghurye, G., Clifford, D., and Tripp, A. (1999). Combined arsenic and nitrate removal by ion exchange. *Journal American Water Works Association* 91 (10): 85–96.

Ghurye, G., Clifford, D., and Tripp, A. (2004). Iron coagulation and direct microfiltration to remove arsenic from groundwater. *Journal of American Water Works Association* 96 (4): 143–152.

Guenegou, T., Tambute, A., Jardy, A., and Caude, M. (1998). Elimination of arsenic traces contained in liquid effluents by chromatographic treatment. *Analusis* 26: 352–357.

Hao, L., Liu, M., Wang, N., and Li, G. (2018). A critical review on arsenic removal from water using iron-based adsorbents. *RSC Advances* 8: 39545–39560.

Haron, M.J., Yunus, M.Z.W., Sukari, M.A. et al. (1997). Removal of arsenic(V) by cerium(III) complexed chelating ion exchanger. *The Malaysian Journal of Analytical Sciences* 3 (1): 193.

Helfferich, F.G. (1961). Ligand exchange: a novel separation technique. *Nature* 189: 1001–1002.

Helfferich, F.G. (1962). *Ion Exchange*. London and New York: McGraw-Hill Book Company.

Himeno, S., Hashimoto, M., and Ueda, T. (1999). Formation and conversion of molybdophosphate and arsenate complexes in aqueous solution. *Inorganica Chimica Acta* 284: 237–245.

Hussain, M.M., Wang, J., Bibi, I. et al. (2021). Arsenic speciation and biotransformation pathways in the aquatic ecosystem: the significance of algae. *Journal of Hazardous Materials* 403: 124027.

Ike, I.O., John, O., and Duke, M. (2017). Chapter 7. Nanoparticle incorporation into desalination and water treatment membranes–potential advantages and challenges. In: *Desalination Sustainability–A Technical, Socioeconomic, and Environmental Approach*, 261–303. The Netherlands: Elsevier https://doi.org/10.1016/B978-0-12-809791-5.00007-9.

Irving, H.M.N.H. and Williams, R.J.P. (1953). The stability of transition-metal complexes. *Journal of the Chemical Society* 3192–3210: https://doi.org/10.1039/JR9530003192.

Islam-Ul-Haq, M.A., Baig, D.N., Hayat, H. (2007). Groundwater arsenic contamination – a multi directional emerging threat to water scarce areas of Pakistan. Sixth International IAHS Groundwater Quality Conference, Fremantle, Western Australia (2–7 December 2007), p. 1.

Khan, A.H., Rasul, S.B., Munir, A. et al. (2000). Appraisal of a simple arsenic removal method for groundwater of Bangladesh. *Journal of Environmental Science and Health* 35 (7): 1021–1041.

Kumar, M., Goswami, R., Patel, A.K. et al. (2020). Scenario, perspectives and mechanism of arsenic and fluoride co-occurrence in the groundwater: a review. *Chemosphere* 249: 126126.

Korngold, E., Belayev, N., and Aronov, L. (2001). Removal of arsenic from drinking water by anion exchangers. *Desalination* 141: 81–84.

LeMire, L.E., Teixeira, M.A., and Reed, B.E. (2010). Removal of As(V) using an ironimpregnatedion exchange bead. *Separation Science and Technology* 45 (14): 2051–2063.

Ma, J., Zhu, Z., Chen, B. et al. (2013). One-pot, large-scale synthesis of magnetic activated carbon nanotubes and their applications for arsenic removal. *Journal of Materials Chemistry A* 1: 4662–4666.

Mahimairaja, S., Bolan, N.S., Adriano, D., and Robinson, B. (2005). Arsenic contamination and its risk management in complex environmental settings. *Advances in Agronomy* 86: 1–82.

Matsunaga, H., Yokoyama, T., Eldridge, R.J., and Bolto, B.A. (1996). Adsorption characteristics of As(III) and As(V) on iron(III) -loaded chelating resin having lysine-Nα, Nα -diacetic acid moiety. *Reactive and Functional Polymers* 29: 167.

Min, J.H. and Hering, J.G. (1999). Arsenic sorption by Fe(III)- doped alginate gels. *Water Research* 32 (5): 1544–1552.

Mohan, D. and Pittman, C.U. Jr. (2007). Arsenic removal from water/wastewater using adsorbents – a critical review. *Journal of Hazardous Materials* 142: 1–53.

Mohanty, D., Acharya, S., and Samal, S. (2013). Selective removal of toxic and heavy metal ions like arsenic and copper from drinking water by using novel chelating resins immobilized on silica gel. *Research Journal of Pharmaceutical, Biological and Chemical Sciences* 4 (1): 43–58.

Mulder, M. (2012). *Basic Principles of Membrane Technology*. Berlin, Germany: Springer Science & Business Media.

Nickson, R.T., McArthur, J.M., Burgess, W.G. et al. (1998). Arsenic poisoning in Bangladesh groundwater. *Nature* 395: 338.

Nickson, R.T., McArthur, J.M., Ravenscroft, P. et al. (2000). Mechanism of arsenic release to groundwater, Bangladesh and West Bengal. *Applied Geochemistry* 15 (4): 403–413.

Ramana, A. and Sengupta, A. (1992). Removing selenium(IV) and arsenic(V) oxyanions with tailored chelating polymers. *Journal of Environmental Engineering* 118 (5): 755–775.

Sarkar, A. and Paul, B. (2016). The global menace of arsenic and its conventional remediation – A critical review. *Chemosphere* 158: 37–49.

Shankar, S. and Shankar, U. (2014). Arsenic contamination of groundwater: a review of sources, prevalence, health risks, and strategies for mitigation. *Scientific World Journal* 2014: 304524.

Shih, M.-C. (2005). An overview of arsenic removal by pressure-drivenmembrane processes. *Desalination* 172: 85–97.

Siddique, T., Dutta, N.K., and Roy Choudhury, N. (2020). Nanofiltration for arsenic removal: challenges, recent developments, and perspectives. *Nanomaterials* 10: 1323.

Smedley, P.L. and Kinniburg, D.G. (2002). A review of the source, behavior and distribution of arsenic in natural waters. *Applied Geochemistry* 17 (5): 517–568.

Suzuki, T.M., Tanco, M.L., Tanaka, D.A.P. et al. (2001). Adsorption characteristics and removal of oxy-anions or arsenic and selenium on the porous polymers loaded with monoclinic hydrous zirconium oxide. *Separation Science and Technology* 36 (1): 103–111.

Trung, D.Q., Anh, C.H., Trung, N.X. et al. (2001). Preconcentration of arsenic species in environmental waters by solid phase extraction using metalloaded chelating resins. *Analytical Sciences* 17: 1219–1222.

Wang, S. and Mulligan, C.N. (2006). Occurrence of arsenic contamination in Canada: sources, behavior and distribution. *The Science of the Total Environment* 366: 701–721.

Xu, Y.-H., Nakajima, T., and Ohki, A. (2002). Adsorption and removal of arsenic (V) from drinking water by aluminum-loaded Shirasu-zeolite. *Journal of Hazardous Materials* 92: 275–287.

Yoshida, I., Ueno, K., and Kobayashi, H. (1978). Selective separation of As(III) and As(V) ions with iron(III) complex of chelating ion exchange resin. *Separation Science and Technology* 13 (2): 173.

Zhao, D. and SenGupta, A.K. (1998). Ultimate removal and recovery of phosphate from wastewater using a new class of polymeric exchangers. *Water Research* 32 (5): 1613–1625.

17

Arsenic Removal by Membrane Technologies and Ion Exchange Methods from Wastewater

Simranjeet Singh[1], Harry Kaur[2], Daljeet Singh Dhanjal[3], Praveen C. Ramamurthy[1], and Joginder Singh[3]

[1] *Interdisciplinary Centre for Water Research (ICWaR), Indian Institute of Science, Bengaluru, Karnataka, India*
[2] *Department of Biosciences and Bioengineering, Indian Institute of Technology, Roorkee, Uttarakhand, India*
[3] *Department of Biotechnology, School of Bioengineering and Biosciences, Lovely Professional University, Phagwara, Punjab, India*

17.1 Introduction

Recently, the reservoirs of natural water are continuously disappearing owing to depletion of groundwater, climatic changes, as well as environmental pollution. Furthermore, the growing population, i.e. 81 million per year, leads to more demand for freshwater, i.e. 65 billion cubic meters per year (Al-Jubouri and Holmes 2020; Bashir et al. 2019; Gautam et al. 2009). The main contaminants present in water are categorized as organic and inorganic. The organic contaminants consist of dyes, pesticides, and pharmaceutical products while inorganic contaminants consist of heavy metals and radioactive metals (Kumar et al. 2019; Singh et al. 2020; Vyas et al. 2021). Amongst all, heavy metals are significant pollutants, and their toxicity is a major part of ecological, environmental, and evolutionary imbalance (Choong et al. 2007). Various natural and anthropogenic activities are the sources of heavy metals in excess, which include mining, weathering of the earth's crust, urban runoff, and excessive use of pesticides and fertilizers (Galadima et al. 2021; Kumar et al. 2015; Singh et al. 2019, 2021). The toxicity of As has severely increased in water bodies due to natural phenomena as well as anthropogenic activities by humans (Corsini et al. 2014; Fazi et al. 2016). Freshwater contaminated with As is a chief health hazard to humankind. As is the 20 most abundant metal in the earth's crust, 14 in the seawater, and the 12 most abundant metal found in the human body, having semimetallic properties and playing a vital role in our body (Kaur et al. 2018; Singh et al. 2022). There is no denying the fact that As is necessary as a nutrient in small amounts, but it is highly poisonous if ingested in large quantities (Garelick et al. 2005; Greenleaf et al. 2006). The contamination of freshwater with As has many severe health effects involving hematological and cardiovascular effects, diabetes mellitus, as well as various neurological disorders. Moreover, numerous epidemiological studies have affirmed that chronic As poisoning leads to internal and skin

Arsenic in Plants: Uptake, Consequences and Remediation Techniques, First Edition.
Edited by Prabhat Kumar Srivastava, Rachana Singh, Parul Parihar, and Sheo Mohan Prasad.

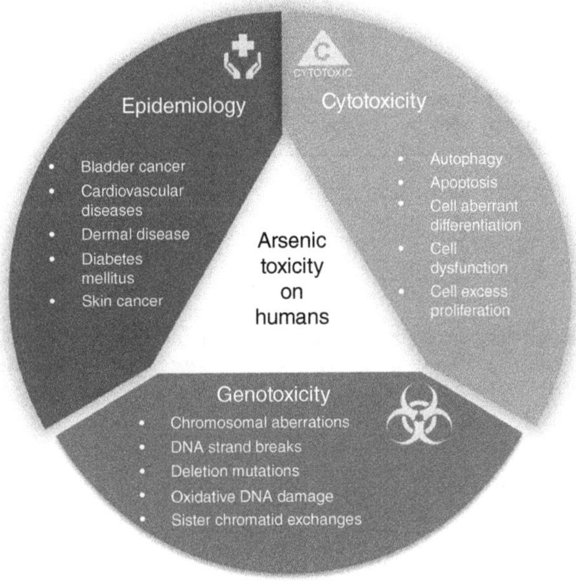

Figure 17.1 Arsenic toxicity of humans.

cancers (Hashim et al. 2011; Hristovski et al. 2008; Hu et al. 2012). Acute and subacute poisoning, such as acute gastrointestinal syndrome and acute paralytic syndrome, result from ingestion in large quantities (Figure 17.1).

The prevailing water treatment technologies that allow for the removal of As involve conventional removal technologies like coagulation, precipitation, filtration, and advanced technologies such as ion exchange, adsorption, and the use of physical separation processes such as membrane filtration (Irshad et al. 2020). Nowadays, membrane filtration technology has become popular and gained attention globally due to its versatility, clean operation, simplicity, no sludge generation, and the increasing availability of a wide range of membranes (Issa et al. 2011; Ito et al. 2012). This process does not require any chemicals and minimum energy requirements. The four most well-known pressure-driven membrane filtration processes are: (i) microfiltration (MF), (ii) ultrafiltration (UF), (iii) nanofiltration (NF), and (iv) reverse osmosis (RO) in descending order of particle size can be separated (Jadhav et al. 2015; Janin et al. 2009).

This book chapter briefly summarizes the role of different types of membrane filtration technologies such as MF, NF, RO, and UF for As removal. In addition to this, we tried to compile the flaws of both the technologies and challenges, which need to be investigated for developing sustainable approaches in the future.

17.2 Arsenic Removal Using Membrane Separation

Membrane filtration technology has emerged as an effective method for the removal of As from water because of its high removal efficiency, easy operation, and no sludge production. Table 17.1 provides information about different membrane filtration approaches

Table 17.1 List of different types of membrane filtration technology along with its distinct features.

Membrane filtration technology types	Applied pressure (bar)	Membrane pore size (nm)	Separation mechanism	Constituents removed
Microfiltration	0.1–2	100–1000	Steric exclusion	Bacteria and suspended solids
Nanofiltration	3–20	1–10	Steric exclusion and Donnan exclusion	Bacteria, multivalentions (partially), suspended solids, and viruses
Reverse osmosis	5–120	<1	Solution-diffusion	Bacteria, monovalentions, multivalent ions, suspended solids, and viruses
Ultrafiltration	0.1–5	11–100	Steric exclusion	Bacteria, suspended solids, and viruses

along with distinct characteristics. Different types of membrane filtration technology like MF, NF, RO, and UF used for As removal have been discussed below (Karakurt 2019; Kavakli et al. 2014; Koltuniewicz et al. 2004).

17.2.1 Microfiltration

MF is a pressure-dependent membrane technology in which particles of size ranging from 0.02 to 10 μm get separated from an aqueous solution (Charcosset 2012). The retention of solutes and particles takes place either due to adsorption in the membrane matrix or sieving. The large pore size of MF is effective in removing the particulate form of As but remains ineffective in removing a colloidal or dissolved form of As (Uddin et al. 2007). Therefore, in order to improve, the removal efficiency of As via MF membrane can only be achieved by increasing the size of particles bearing As (Shih 2005).

Coagulation and flocculation are methods that could be employed to augment the As particles, which will substantially improve the removal efficiency of As via MF membrane from water (Ng et al. 2004). Ferric chloride, as well as ferric sulphate, are predominantly used in flocculation, followed by MF with esters of cellulose nitrate and cellulose acetate with pore size ranging from 0.22 to 1.2 μm. The result revealed that the flocculation method, in combination with the MF method, is more effective in comparison to the flocculation method in combination with the sedimentation method for As removal (Uddin et al. 2007). Moreover, the removal efficiency of flocculation and MF method for As removal depends on the ability of As adsorption on the ferric complex and rejection of As flocs by MF membrane (Han et al. 2002). Usman et al. (2020) developed iron oxyhydroxide-based MF membrane made up of micro tetravalent manganese feroxyhyte (μTMF) and micro granular ferric hydroxide (μGFH) for removal of As (V). The result obtained showed the adsorption of As (V) on μTMF in contrast to μGFH. In case of high residence time, adsorption capacities for

As (V) were recorded 1.04 and 0.95 μg mg^{-1} for μTMF and μGFH, respectively. The overall result of the study indicates that the MF membrane of iron oxyhydroxides can be employed for the removal of As as well as water treatment (Usman et al. 2020).

17.2.2 Nanofiltration

NF membranes are extensively used for separating multivalent ions from monovalent ions. Additionally, it is possible to separate specific ions of the same valency by following ideal operating conditions and selecting the proper membrane (Hilal et al. 2004). In a few cases, NF membranes are also stated as "loose RO" owing to high water fluxes at low transmembrane pressure. These membranes are generally negatively charged and asymmetric at alkaline and neutral media (Nagy 2012). Hence, the anion separation depends not only on the diffusion rate via membrane and convection but also on repulsion between the surface groups and anions in the solution. The major advantage of introducing the extra mechanism of ion exclusion is that high ion rejections, as observed in RO, could also be obtained (Pérez-González et al. 2015).

Numerous studies have been conducted for removing As from drinking water via the NF membrane, and the majority of them have shown promising results. Brandhuber and Amy conducted the study by using three different NF membranes, i.e. 4040-UHA-ESNA (Hydranautics), HL-4040F1550 (DESAL), and NF70 4040-B (Film Tec) to assess the rejection of As. The result showed the 95% rejection of As (V) and 20–53% rejection of As (III) for all the membranes. Uras et al. also experimented with assessing the rejection potential of aromatic polyamide NF membrane (low pressure), i.e. ES-10 (Nitto-Denko Co. Ltd.). For this, groundwater was spiked with As (III) as well as As (V) and the rejection rates of 50–89% for As (III) and 87–93% for As (V) were recorded (Uddin et al. 2007). He et al. (2016) developed a thin-film nanocomposite (TFN) NF membrane composed of polyhedral oligomeric silsesquioxane (POSS). The NF membrane was evaluated for the removal of As and selenium from an aqueous solution. The result obtained from the study showed a high rejection of 97.4% for $HAsO_4^{2-}$, 93.9% for SeO_3^{2-}, and 96.5% for SeO_4^{2-} (He et al. 2016). Harfoush et al. also conducted the study to assess As removal efficiency from natural water by NF. The result obtained showed the maximum removal efficiency of 94% for As (V) at 28 °C and 7 bar pressure. Whereas, the maximum removal efficiency of 90% was recorded for As (III) at 28 °C and 5 bar pressure (Harfoush et al. 2018).

17.2.3 Reverse Osmosis

RO is one of the reliable methods for the desalination of water. This process involves the high rejection rate of ions and low molecular mass compounds. Additionally, this method is easy to control and can also be automated (Fritzmann et al. 2007). Various studies have been conducted for As removal by using RO technology. George et al. (2006) conducted a study to assess the potential of RO technology in As removal. On assessment, the post- and pre-RO concentrations of As were 87 and 443 μg l^{-1}, respectively. And after filtration, the reduction percentage of 79% for As was recorded (George et al. 2006). Abejón et al. (2015) also conducted the study to check the viability of an optimized RO system for removing As (V) with minimum cost as well as energy. The result obtained proposed that the 0.5 μg l^{-1}

As (V) concentration can be reduced from permeate water at a very reasonable rate (Abejón et al. 2015). Elick and his colleagues also conducted a study using four thin-film polyamide NFs (i.e. NF270, NF90) as well as RO, i.e. XLE, BW30 membranes that are arranged as flat-sheet. The initial concentration of As (III) was fixed at $100\,\mu g\,l^{-1}$ for all experiments. The result obtained showed the As (III) removal in the range of 97–99% for RO membranes in contrast to NF membranes (Elcik et al. 2016).

17.2.4 Ultrafiltration

UF is a low pressure-dependent approach used to eliminate colloids, macromolecules, and solutes of high molecular mass (Pal 2015). However, UF is not considered an effective method for As removal because of the large pore size of the membrane (Shih 2005). But, the electric repulsion-aided UF method has been found to show improved As removal efficacy in contrast to UF alone. Different studies have been conducted using the UF method for As removal (Siddique et al. 2020). In 2007, Iqbal and his colleague conducted a study to assess the removal efficiency of As (III) from groundwater with the help of cationic surfactant and micellar-enhanced ultrafiltration (MEUF). The As removal efficiency of 96, 94, 80, and 57% were recorded for hexadecyl pyridinium chloride (CPC), hexadecyltrimethylammonium bromide (CTAB), octadecyl amine acetate (ODA), and benzalkonium chloride, respectively. MEUF in conjugation with CPC or CTAB is claimed to feasible approach for effective removal of As (III) for groundwater (Iqbal et al. 2007). Another study was conducted to assess the effect of co-occurring inorganic solutes like HCO_3^-, H_4SiO_4, HPO_4^{2-}, and SO_4^{2-} in the water on the As (V) removal via cationic surfactant cetyl pyridinium chloride (CPC) attached to hydrophilic polyethersulfone (PES) flat UF membrane. The result obtained showed 78–100% As removal in the presence of cation surfactant and hydrophilic PES UF membrane (Dutta et al. 2012). Chatterjee and De also developed high flux polyacrylonitrile (PAN)–laterite-mixed matrix UF membrane for effective removal of As from groundwater in which the maximum adsorption capacity of about $1.4\,mg\,g^{-1}$ at 24.8 °C has been recorded for As removal by this selected UF membrane (Chatterjee and De 2015).

17.3 Arsenic Removal Using Ion Exchange Methods

Ion exchange is a reversible stoichiometric chemical reaction in which an ion from solution is swapped with an alike charged ion, associated with a fixed solid substance to balance the inclusive neutral charge (Koseoglu et al. 2011; Lata and Samadder 2016). Ion exchange is a recognized eco-friendly and advantageous technique due to its high selectivity, absence of sludge generation and the remediation of inorganic pollutants such as heavy metals based on their chemical characteristics (Lee et al. 2017).

17.3.1 Ion Exchange Resin

Ion exchange resins are recognized as appropriate intermediates for the ion exchange method to react with the ions of the same charge from a chemical solution, due to their specific selectivity for the exclusion of different ions. However, ion exchange resins are disadvantageous for their use in water remediation techniques for the reason that their kinetics of ion exchange was well-thought-out to be slow (Litter et al. 2010). Ion exchange resin is mainly practised to decrease H_2O hardness as well as to remove pollutants like arsenate, selenite, nitrate, etc., from wastewater, which utilizes habitually restored resins to reprocess and emphasize the exchanged ions (Liu et al. 2005; Mamindy-Pajany et al. 2011). Ion exchange removes negatively charged arsenate ions and do not remove arsenite ions from water because arsenite occurs in the form of an uncharged molecule, i.e. H_3AsO_3. In addition to this, the US Environmental Protection Agency has suggested precise ion exchange resins, particularly Cl^- ion form for the elimination of As (Ngomsik et al. 2005; Nicomel et al. 2015). During the exclusion of As metal ions, ion exchange resins are filled with Cl^- ions in exchange for the place where As-polluted H_2O is passing through the ion exchange vessel (Nigam et al. 2013). Elimination of As via the ion exchange resin method depends upon different factors, for instance, type of resin used, the initial amount of As present, total dissolved solids (TDS), and competing ions.

Ion exchange resins are supplemented with HCl to add Cl^- ions at the bottom, which are readily substituted by As, whose exclusion does not depend upon pH and substrate amount. However, ion exchange efficacy is altered by competitive ions such as SO_4^{2-}, TDS, selenium, fluoride, and nitrate (Litter et al. 2010). Ion exchange resins, i.e. Lewatit Sybron Ionac SR7, Lewatit FO 36, and Lewatit Mono plus M600 resins removed As efficiently via following Langmuir adsorption isotherm and pseudo-first-order kinetic approach (Karakurt 2019). Anionic exchange resins, i.e. IRA-900, IRA-400, and IRA-96 excluded As significantly among which IRA-900 exhibited maximum ability for the removal of As (Pallier et al. 2010). Hybrid ion exchange (HIX) media was synthesized by combining iron hydroxide nanomaterials with ion exchange resins. This HIX showed higher adsorption potential for arsenate (Hristovski et al. 2008). Ion exchange resins, i.e. Lewatit FO 36 and ArsenXnp, showed significant removal efficiency for As in which Lewatit FO 36 resin was more efficient as compared to ArsenXnp (Koseoglu et al. 2011). Two-hybrid (HY) resins; HY-Fe and HY-AgCl and Strong base anion exchange (SBAE), excluded significant amounts of inorganic As and organic As from wastewater (Issa et al. 2011).

17.3.2 Ion Exchange Fiber

Ion exchange fiber is a novel kind of ion exchange material, which has numerous benefits over ion exchange resins due to their properties like greater adsorption efficiency, higher accumulation ability, and also has several functional groups on the surface of the fiber. For example, polymer or inorganic hybrid fibrous ion exchangers formed by the fusion of hydrated Fe_2O_3 with ion exchanger fiber were practiced for As exclusion (Greenleaf et al. 2006; Vatutsina et al. 2007), and polyethylene-covered polypropylene and cellulose fibers were also utilized for As exclusion (Kavakli et al. 2014; Yu et al. 2013). New amine-doped acrylic ion exchange fiber exhibited a maximum adsorption ability of $205.32 \pm 3.57\,mg\,g^{-1}$ for As (pH 3.04) by following the pseudo-first-order kinetics, and the Redlich–Peterson adsorption isotherm model. Adsorption was quick in the first 20 minutes which attained equilibrium in 60 minutes and even after 9 cycles of use, the exclusion ability of this ion exchange fiber was more than 83% of the starting value (Lee et al. 2017). HIX fibers, which dispersed hydrated Fe_2O_3 nanomaterials exhibited better efficiency for the removal of As. These prepared hybrid fibers were effectively regenerated with 2% sodium hydroxide and 2% sodium chloride and used for immediate exclusion of both As^{5+} and As^{3+} (Greenleaf et al. 2006; Pandey et al. 2009).

17.4 Methods to Increase the Efficiency of Arsenic Removal

Elevating concentrations and carcinogenicity of As attained the focus of the scientists for the removal of As from various contaminated water sources. Therefore, several techniques like oxidation, adsorption, ion exchange, coagulation, membrane technologies, etc., are used for the removal of As from polluted groundwater or surface water efficiently (Figure 17.2) (Perez-Moreno et al. 2006; Salem Attia et al. 2014).

17.4.1 Oxidation

Arsenite is transformed to arsenate by the oxidation technique; however, it requires methods like adsorption, coagulation, or ion exchange to eliminate As from the solution. Oxidation is a prevalent method in anoxic water sources because arsenite is the predominant moiety of As at neutral pH (Singh et al. 2015). Oxidants should be carefully chosen for the easy exclusion of As from the aqueous solution, but this results in intricate remediation by releasing As residues whose disposal process is problematic (Nicomel et al. 2015). Fe-based photo-oxidation technique was efficaciously practiced to oxidize and exclude As from mine H_2O disposing of rigid rock Au, Ag, and Pb mines (Zaw and Emett 2002). Biological oxidation is another type of oxidation technique, which uses different microbes to form a complex by reacting with As. Pretreatment with immobilized bacteria removed 92% of As^{3+} efficiently from groundwater by oxidizing in a continuous flow bioreactor (Ito

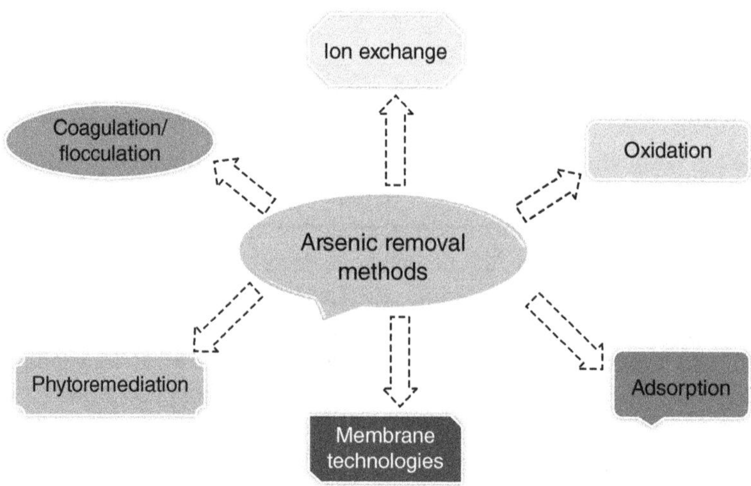

Figure 17.2 Different methods of increasing the efficiency of arsenic removal.

et al. 2012). Removal of As^{3+} by biological oxidation in drinking H_2O has been estimated which do not need reagent incorporation and also do not generate noxious substances. Arsenite oxidizing bacteria have been considered as an efficient, beneficial asset to mitigate disadvantages caused by chemical oxidation in biofilter experiments for As exclusion, which depend upon activated alumina, granular iron hydroxides and metallic iron adsorbents (Wan et al. 2010). Arsenite adsorption by *Aliihoeflea* sp. strain 2WW complexed with iron-based, goethite is a better technique for As exclusion from polluted groundwater in which bacterial biomass in combination with goethite eliminated 95% of As (Corsini et al. 2014).

17.4.2 Adsorption

As removal by adsorption is developed as an essential technique because this technique does not release unwanted substances and utilized materials are recyclable. Different adsorbents like surfactants, nanoparticles, agricultural waste, microbial biomass, iron-based soil amendment, etc., are practiced for metal removal (Hashim et al. 2011). Researchers formed nanoparticles to produce a promising adsorbent, which is active and robust for As exclusion. Nanoparticles show lower diffusion ability, which improves the influential role of nanoparticles as adsorbent, hence offering better metal removal efficiency in comparison to other macromolecules that have intramolecular diffusion, which, ultimately, reduces the adsorption frequency and accessible ability of the adsorbent (Sarkar and Paul 2016; Suman et al. 2018). Nanoparticles also exhibit certain specific characteristics like tiny dimensions, huge surface extent, greater reactivity, simple segregation, and several active sites, which enable improved As elimination efficacy in comparison to other accessible adsorbents. For metal removal, the ability of nanomaterials and micro-sized materials was analyzed where nanomaterials showed more promising results for metal exclusion as compared to micro-sized particles (Ungureanu et al. 2015; Vatutsina et al. 2007). Iron-based nanoparticles, i.e. goethite, hematite, and magnetite removed As significantly and concluded that As removal is associated with the Fe amount present in these nanoparticles (Mamindy-Pajany et al. 2011). Biosorption is also considered an efficient method for the remediation of heavy metals from wastewater, primarily due to the cell membrane organization, accommodating binding sites of functional groups, such as amino, hydroxyl, carboxyl, and sulfate, to form complex with metals through electrostatic interactions and ion exchange. Algae ability for the removal of As is likely to be improved as compared to other heavy metal cations (Ungureanu et al. 2015). In addition to this, plant biomass is also used for As adsorption, for example, biomass generated from *Momordica charantia* exhibited enhanced removal efficiency of 88% for As^{3+} (pH 9), an initial amount of $0.5\,mg\,l^{-1}$, and $5\,g\,l^{-1}$ adsorbent amount (Pandey et al. 2009).

17.4.3 Coagulation and Flocculation

Coagulation/flocculation is a primary practiced technique to make water clean for consumption. For example, Al and Fe salts are considered as promising coagulants for As exclusion from $mg\,l^{-1}$ levels (Pallier et al. 2010; Wickramasinghe et al. 2004; Yan et al. 2010). Effectual parting organization consequent to coagulation is essential to attain a significant inclusive As an exclusion, for example, amendments of conventional coagulation/flocculation methods can be achieved, i.e. pipe flocculation in the first phase and then straight sand filtration in the second phase (Yuan et al. 2006; Zouboulis and Katsoyiannis 2002). Coagulants' ability for As exclusion alters with the change in the pH value, for instance, under pH of 7.6, $Al_2(SO_4)_3$, and $FeCl_3$ coagulants showed similar efficiency for As exclusion from H_2O (Garelick et al. 2005). Still, at pH above 7.6, $FeCl_3$ has greater removal efficiency as compared to $Al_2(SO_4)_3$ and arsenate is eliminated more easily than arsenite (Nicomel et al. 2015; Yang et al. 2020). But still, Al-created coagulants were also capable of decreasing As amounts to lower than the highest quantity of $10\,g\,l^{-1}$ (Hu et al. 2012; Wan et al. 2010). Combining of $FeCl_3$ with permanganate as the coagulant elevated the co-precipitation of arsenite in which exclusion efficiency enhanced to 41–75.4% (Liu et al. 2005), whereas As exclusion capacity by using ferrate as coagulant reached around 98% at a pH range of 5–7 (Yuan et al. 2006).

17.4.4 Phytoremediation

Phytoremediation is an economical, environment-friendly, and inexpensive technique for As exclusion from polluted zones, which can be practiced both in situ and ex situ. This method is based on the utilization of plant species to uptake As in higher quantities in their various parts or to decline As toxicity (Suman et al. 2018). *Pteris vittata* has been extensively considered as a promising As hyperaccumulator; for example, *P. vittata* had high As uptake potential by absorbing around $5070\,mg\,kg^{-1}$ of As, where leaves of this fern accumulate the highest amount of As in comparison to leafstalks and roots, which is entirely dissimilar from that of normal plants (Fazi et al. 2016). Hyperaccumulators uptake heavy metals from the aqueous solution via accretion, translocation, dispersal, elimination, and osmoregulation with the transport of metal in the direction of the root side (Suman et al. 2018). Phytoremediation by using *Vallisneria denseserrulata* combined with *Batillus* sp. XZM accumulated and removed As significantly in higher amounts, where As was detoxified by amide, different functional groups present in the leaves of *V. denseerrulata* and with an As metabolization progression (Irshad et al. 2020). Colonization of *P. vittata* with *Cupriavidus basilensis* strain exhibited higher As accumulation of 171%, and strain *Cupriavidus basilensis* could enhance the richness of aioA-like genes in the rhizosphere (Yang et al. 2020). Good biosorption efficacy of submerged aquatic plant *Hydrilla verticillata* showed maximum absorption efficiency of 96.4% for As from drinking water, at pH 6.0, contact period of 5 hours, and starting amount of 100 ppb with 0.5 g of biomass $(100\,ml)^{-1}$ (Nigam et al. 2013).

17.5 Conclusion

The prevailing water treatment technologies that allow the removal of As involve conventional removal technologies like coagulation, precipitation, filtration, and advanced technologies such as ion exchange, adsorption, and the use of physical separation processes such as membrane filtration. These different membrane and ion exchange technologies are efficient and effective in the removal of As from water bodies. On the basis of the detailed analysis, it can be concluded that membrane technologies, including MF, NF, RO, and UF processes, can be efficiently used for As removal. Ion exchange is a recognized eco-friendly and advantageous technique due to its high selectivity, absence of sludge generation, and the remediation of inorganic pollutants such as heavy metals based on their chemical characteristics. Besides this, phytoremediation is also proved to be an environmentally friendly and inexpensive technique for As exclusion from in situ and ex situ polluted sites. Although there are certain shortcomings in the membrane and ion exchange technologies, researchers are working for the development of membrane and ion exchange technologies to make them sustainable options.

Acknowledgments

Dr. Simranjeet Singh is thankful to the ICWaR, Indian Institute of Science, Bengaluru for the financial assistance in the form of IOE-IISc Fellowship (Sr. No: IE/REAC-20-0134) and for providing laboratory and library facilities.

Conflict of Interest: The authors declare that they have no conflict of interest.

Compliance with Ethical Requirements: This article does not contain any studies with animals performed by any of the authors.

References

Abejón, A., Garea, A., and Irabien, A. (2015). Arsenic removal from drinking water by reverse osmosis: minimization of costs and energy consumption. *Sep. Purif. Technol.* 144: 46–53. https://doi.org/10.1016/j.seppur.2015.02.017.

Al-Jubouri, S.M. and Holmes, S.M. (2020). Immobilization of cobalt ions using hierarchically porous 4A zeolite-based carbon composites: ion-exchange and solidification. *J. Water Process Eng.* 33: 101059. https://doi.org/10.1016/j.jwpe.2019.101059.

Bashir, A., Malik, L.A., Ahad, S. et al. (2019). Removal of heavy metal ions from aqueous system by ion-exchange and biosorption methods. *Environ. Chem. Lett.* https://doi.org/10.1007/s10311-018-00828-y.

Charcosset, C. (2012). Microfiltration. In: *Membrane Processes in Biotechnology and Pharmaceutics* (ed. C. Charcosset), 101–141. Amsterdam: Elsevier https://doi.org/10.1016/B978-0-444-56334-7.00003-4.

Chatterjee, S. and De, S. (2015). Adsorptive removal of arsenic from groundwater using a novel high flux polyacrylonitrile (PAN)-laterite mixed matrix ultrafiltration membrane. *Environ. Sci. Water Res. Technol.* 1: 227–243. https://doi.org/10.1039/c4ew00075g.

Choong, T.S.Y., Chuah, T.G., Robiah, Y. et al. (2007). Arsenic toxicity, health hazards and removal techniques from water: an overview. *Desalination* 217: 139–166. https://doi.org/10.1016/j.desal.2007.01.015.

Corsini, A., Zaccheo, P., Muyzer, G. et al. (2014). Arsenic transforming abilities of groundwater bacteria and the combined use of *Aliihoeflea* sp. Strain 2WW and goethite in metalloid removal. *J. Hazard. Mater.* 269: 89–97. https://doi.org/10.1016/j.jhazmat.2013.12.037.

Dutta, T., Bhattacherjee, C., and Bhattacherjee, S. (2012). Removal of arsenic using membrane technology – a review. *Int. J. Eng. Res. Technol.* 1: 1–23.

Elcik, H., Celik, S.O., Cakmakci, M., and Özkaya, B. (2016). Performance of nanofiltration and reverse osmosis membranes for arsenic removal from drinking water. *Desalin. Water Treat.* 57: 20422–20429. https://doi.org/10.1080/19443994.2015.1111812.

Fazi, S., Amalfitano, S., Casentini, B. et al. (2016). Arsenic removal from naturally contaminated waters: a review of methods combining chemical and biological treatments. *Rend. Lincei* 27: 51–58. https://doi.org/10.1007/s12210-015-0461-y.

Fritzmann, C., Löwenberg, J., Wintgens, T., and Melin, T. (2007). State-of-the-art of reverse osmosis desalination. *Desalination* 216: 1–76. https://doi.org/10.1016/j.desal.2006.12.009.

Galadima, M., Singh, S., Pawar, A. et al. (2021). Toxicity, microbial degradation and analytical detection of pyrethroids: a review. *Environ. Adv.* 5: 100105. https://doi.org/10.1016/J.ENVADV.2021.100105.

Garelick, H., Dybowska, A., Valsami-Jones, E., and Priest, N.D. (2005). Remediation technologies for arsenic contaminated drinking waters. *J. Soils Sediments.* https://doi.org/10.1065/jss2005.06.140.

Gautam, D., Gupta, B., Anjum, S., and Ikram, S. (2009). Arsenic removal from water: an overview of recent technologies PVA gels view project development of antimicrobial polyester sutures view project 7 arsenic removal from water: an overview of recent technologies author's copy arsenic removal from water: an IUP. *J. Chem.* 2: 1–52.

George, C.M., Smith, A.H., Kalman, D.A., and Steinmaus, C.M. (2006). Reverse osmosis filter use and high arsenic levels in private well water. *Arch. Environ. Occup. Heal.* 61: 171–175. https://doi.org/10.3200/AEOH.61.4.171-175.

Greenleaf, J.E., Lin, J.-C., and Sengupta, A.K. (2006). Two novel applications of ion exchange fibers: arsenic removal and chemical-free softening of hard water. *Environ. Prog.* 25: 300–311. https://doi.org/10.1002/ep.10163.

Han, B., Runnells, T., Zimbron, J., and Wickramasinghe, R. (2002). Arsenic removal from drinking water by flocculation and microfiltration. *Desalination* 145: 293–298. https://doi.org/10.1016/S0011-9164(02)00425-3.

Harfoush, M., Mirbagheri, S.A., Ehteshami, M., and Nejati, S. (2018). Arsenic removal from drinking water using low-pressure nanofiltration under various operating conditions. *Water Pract. Technol.* 13: 295–302. https://doi.org/10.2166/wpt.2018.042.

Hashim, M.A., Mukhopadhyay, S., Sahu, J.N., and Sengupta, B. (2011). Remediation technologies for heavy metal contaminated groundwater. *J. Environ. Manage.* https://doi.org/10.1016/j.jenvman.2011.06.009.

He, Y., Tang, Y.P., and Chung, T.S. (2016). Concurrent removal of selenium and arsenic from water using polyhedral oligomeric silsesquioxane (POSS)-polyamide thin-film nanocomposite nanofiltration membranes. *Ind. Eng. Chem. Res.* 55: 12929–12938. https://doi.org/10.1021/acs.iecr.6b04272.

Hilal, N., Al-Zoubi, H., Darwish, N.A. et al. (2004). A comprehensive review of nanofiltration membranes: treatment, pretreatment, modelling, and atomic force microscopy. *Desalination* 170: 281–308. https://doi.org/10.1016/j.desal.2004.01.007.

Hristovski, K., Westerhoff, P., Möller, T. et al. (2008). Simultaneous removal of perchlorate and arsenate by ion-exchange media modified with nanostructured iron (hydr)oxide. *J. Hazard. Mater.* 152: 397–406. https://doi.org/10.1016/j.jhazmat.2007.07.016.

Hu, C., Liu, H., Chen, G. et al. (2012). As(III) oxidation by active chlorine and subsequent removal of As(V) by Al 13 polymer coagulation using a novel dual function reagent. *Environ. Sci. Technol.* 46: 6776–6782. https://doi.org/10.1021/es203917g.

Iqbal, J., Kim, H.J., Yang, J.S. et al. (2007). Removal of arsenic from groundwater by micellar-enhanced ultrafiltration (MEUF). *Chemosphere* 66: 970–976. https://doi.org/10.1016/j.chemosphere.2006.06.005.

Irshad, S., Xie, Z., Wang, J. et al. (2020). Indigenous strain Bacillus XZM assisted phytoremediation and detoxification of arsenic in *Vallisneria denseserrulata*. *J. Hazard. Mater.* 381: 120903. https://doi.org/10.1016/j.jhazmat.2019.120903.

Issa, N.B., Rajaković-Ognjanović, V.N., Marinković, A.D., and Rajaković, L.V. (2011). Separation and determination of arsenic species in water by selective exchange and hybrid resins. *Anal. Chim. Acta* 706: 191–198. https://doi.org/10.1016/j.aca.2011.08.015.

Ito, A., Miura, J.I., Ishikawa, N., and Umita, T. (2012). Biological oxidation of arsenite in synthetic groundwater using immobilised bacteria. *Water Res.* 46: 4825–4831. https://doi.org/10.1016/j.watres.2012.06.013.

Jadhav, S.V., Bringas, E., Yadav, G.D. et al. (2015). Arsenic and fluoride contaminated groundwaters: a review of current technologies for contaminants removal. *J. Environ. Manage.* https://doi.org/10.1016/j.jenvman.2015.07.020.

Janin, A., Blais, J.F., Mercier, G., and Drogui, P. (2009). Selective recovery of Cr and Cu in leachate from chromated copper arsenate treated wood using chelating and acidic ion exchange resins. *J. Hazard. Mater.* 169: 1099–1105. https://doi.org/10.1016/j.jhazmat.2009.04.066.

Karakurt, S. (2019). Removal of carcinogenic arsenic from drinking water by the application of ion exchange resins. *Oncogen J.* 2: 1–8.

Kaur, P., Singh, S., Kumar, V. et al. (2018). Effect of rhizobacteria on arsenic uptake by macrophyte *Eichhornia crassipes* (Mart.) Solms. *Int. J. Phytoremediation* 20: 114–120. https://doi.org/10.1080/15226514.2017.1337071.

Kavakli, C., Akkaş Kavakli, P., Turan, B.D. et al. (2014). Quaternized dimethylaminoethyl methacrylate strong base anion exchange fibers for as(V) adsorption. *Radiat. Phys. Chem.* 102: 84–95. https://doi.org/10.1016/j.radphyschem.2014.04.011.

Koltuniewicz, A.B., Witek, A., and Bezak, K. (2004). Efficiency of membrane-sorption integrated processes. *J. Memb. Sci.* 239: 129–141. https://doi.org/10.1016/j.memsci.2004.02.037.

Koseoglu, P., Yoshizuka, K., Nishihama, S. et al. (2011). Removal of boron and arsenic from Geothermal water in Kyushu Island, Japan, by using selective ion exchange resins. *Solvent Extr. Ion Exch.* 29: 440–457. https://doi.org/10.1080/07366299.2011.573448.

Kumar, V., Singh, S., Bhadrecha, P. et al. (2015). Bioremediation of heavy metals by employing resistant microbial isolates from agricultural soil irrigated with industrial waste water. *Orient. J. Chem.* 31: 357–361. https://doi.org/10.13005/ojc/310142.

Kumar, V., Singh, S., Singh, A. et al. (2019). Assessment of heavy metal ions, essential metal ions, and antioxidant properties of the most common herbal drugs in Indian Ayurvedic hospital: for ensuring quality assurance of certain Ayurvedic drugs. *Biocatal. Agric. Biotechnol.* 18: 101018. https://doi.org/10.1016/J.BCAB.2019.01.056.

Lata, S. and Samadder, S.R. (2016). Removal of arsenic from water using nano adsorbents and challenges: a review. *J. Environ. Manage.* https://doi.org/10.1016/j.jenvman.2015.10.039.

Lee, C.G., Alvarez, P.J.J., Nam, A. et al. (2017). Arsenic (V) removal using an amine-doped acrylic ion exchange fiber: kinetic, equilibrium, and regeneration studies. *J. Hazard. Mater.* 325: 223–229. https://doi.org/10.1016/j.jhazmat.2016.12.003.

Litter, M.I., Morgada, M.E., and Bundschuh, J. (2010). Possible treatments for arsenic removal in Latin American waters for human consumption. *Environ. Pollut.* https://doi.org/10.1016/j.envpol.2010.01.028.

Liu, R.P., Li, X., Xia, S.J. et al. (2005). Effectiveness and mechanism of permanganate enhancing arsenite co-precipitation with ferric chloride. *Huanjing Kexue/Environ. Sci.* 26: 72–75.

Mamindy-Pajany, Y., Hurel, C., Marmier, N., and Roméo, M. (2011). Arsenic (V) adsorption from aqueous solution onto goethite, hematite, magnetite and zero-valent iron: effects of pH, concentration and reversibility. *Desalination* 281: 93–99. https://doi.org/10.1016/j.desal.2011.07.046.

Nagy, E. (2012). Nanofiltration. In: *Basic Equations of the Mass Transport Through a Membrane Layer* (ed. E. Nagy), 249–266. Amsterdam: Elsevier https://doi.org/10.1016/b978-0-12-416025-5.00010-7.

Ng, K.S., Ujang, Z., and Le-Clech, P. (2004). Arsenic removal technologies for drinking water treatment. *Rev. Environ. Sci. Biotechnol.* https://doi.org/10.1023/B:RESB.0000040054.28151.84.

Ngomsik, A.F., Bee, A., Draye, M. et al. (2005). Magnetic nano- and microparticles for metal removal and environmental applications: a review. *Comptes Rendus Chim.* 8: 963–970. https://doi.org/10.1016/j.crci.2005.01.001.

Nicomel, N.R., Leus, K., Folens, K. et al. (2015). Technologies for arsenic removal from water: current status and future perspectives. *Int. J. Environ. Res. Public Health* 13: 62. https://doi.org/10.3390/ijerph13010062.

Nigam, S., Gopal, K., and Vankar, P.S. (2013). Biosorption of arsenic in drinking water by submerged plant: *Hydrilla verticilata*. *Environ. Sci. Pollut. Res.* 20: 4000–4008. https://doi.org/10.1007/s11356-012-1342-x.

Pal, P. (2015). Introduction to the arsenic contamination problem. In: *Groundwater Arsenic Remediation* (ed. P. Pal), 1–23. Amsterdam: Elsevier https://doi.org/10.1016/b978-0-12-801281-9.00001-1.

Pallier, V., Feuillade-Cathalifaud, G., Serpaud, B., and Bollinger, J.C. (2010). Effect of organic matter on arsenic removal during coagulation/flocculation treatment. *J. Colloid Interface Sci.* 342: 26–32. https://doi.org/10.1016/j.jcis.2009.09.068.

Pandey, P.K., Choubey, S., Verma, Y. et al. (2009). Biosorptive removal of arsenic from drinking water. *Bioresour. Technol.* 100: 634–637. https://doi.org/10.1016/j.biortech.2008.07.063.

Pérez-González, A., Ibáñez, R., Gómez, P. et al. (2015). Nanofiltration separation of polyvalent and monovalent anions in desalination brines. *J. Memb. Sci.* 473: 16–27. https://doi.org/10.1016/j.memsci.2014.08.045.

Perez-Moreno, F., Prieto-Garcia, F., Rojas-Hernandez, A. et al. (2006). Study of arsenic removal with ionic exchange resins in drinking water from Zimapan, Hidalgo State, Mexico. *Rev. Metal.* 42: 391–395.

Salem Attia, T.M., Hu, X.L., and Yin, D.Q. (2014). Synthesised magnetic nanoparticles coated zeolite (MNCZ) for the removal of arsenic (As) from aqueous solution. *J. Exp. Nanosci.* 9: 551–560. https://doi.org/10.1080/17458080.2012.677549.

Sarkar, A. and Paul, B. (2016). The global menace of arsenic and its conventional remediation – a critical review. *Chemosphere* 158: 37–49. https://doi.org/10.1016/j.chemosphere.2016.05.043.

Shih, M.C. (2005). An overview of arsenic removal by pressure-driven membrane processes. *Desalination* 172: 85–97. https://doi.org/10.1016/j.desal.2004.07.031.

Siddique, T.A., Dutta, N.K., and Roy Choudhury, N. (2020). Nanofiltration for arsenic removal: challenges, recent developments, and perspectives. *Nanomaterials* 10: 1323.

Singh, R., Singh, S., Parihar, P. et al. (2015). Arsenic contamination, consequences and remediation techniques: a review. *Ecotoxicol. Environ. Saf.* 112: 247–270. https://doi.org/10.1016/j.ecoenv.2014.10.009.

Singh, S., Kapoor, D., Khasnabis, S. et al. (2021). Mechanism and kinetics of adsorption and removal of heavy metals from wastewater using nanomaterials. *Environ. Chem. Lett* 193 (19): 2351–2381. 10.1007/S10311-021-01196-W.

Singh, S., Khasnabis, S., Anil, A.G. et al. (2022). Multifunctional nanohybrid for simultaneous detection and removal of Arsenic(III) from aqueous solutions. *Chemosphere* 289: 133101. https://doi.org/10.1016/J.CHEMOSPHERE.2021.133101.

Singh, S., Kumar, V., Sidhu, G.K. et al. (2019). Plant growth promoting rhizobacteria from heavy metal contaminated soil promote growth attributes of *Pisum sativum* L. *Biocatal. Agric. Biotechnol.* 17: 665–671. https://doi.org/10.1016/J.BCAB.2019.01.035.

Singh, S., Kumar, V., Datta, S. et al. (2020). Physiological responses, tolerance, and remediation strategies in plants exposed to metalloids. *Environ. Sci. Pollut. Res.* 28: 40233–40248. https://doi.org/10.1007/S11356-020-10293-2.

Suman, J., Uhlik, O., Viktorova, J., and Macek, T. (2018). Phytoextraction of heavy metals: a promising tool for clean-up of polluted environment? *Front. Plant. Sci.* 9: 1476. https://doi.org/10.3389/fpls.2018.01476.

Uddin, M.T., Mozumder, M.S.I., Figoli, A. et al. (2007). Arsenic removal by conventional and membrane technology: an overview. *Indian J. Chem. Technol.* 14: 441–450.

Ungureanu, G., Santos, S., Boaventura, R., and Botelho, C. (2015). Arsenic and antimony in water and wastewater: overview of removal techniques with special reference to latest advances in adsorption. *J. Environ. Manage.* 151: 326–342. https://doi.org/10.1016/j.jenvman.2014.12.051.

Usman, M., Katsoyiannis, I., Rodrigues, J.H., and Ernst, M. (2020). Arsenate removal from drinking water using by-products from conventional iron oxyhydroxides production as adsorbents coupled with submerged microfiltration unit. *Environ. Sci. Pollut. Res.* 1–13: https://doi.org/10.1007/s11356-020-08327-w.

Vatutsina, O.M., Soldatov, V.S., Sokolova, V.I. et al. (2007). A new hybrid (polymer/inorganic) fibrous sorbent for arsenic removal from drinking water. *React. Funct. Polym.* 67: 184–201. https://doi.org/10.1016/j.reactfunctpolym.2006.10.009.

Vyas, P., Singh, S., Dhaka, V. et al. (2021). Rhizoremediation for efficient management of inorganic and organic contaminants in soil. In: *Microbe Mediated Remediation of Environmental Contaminants* (ed. A. Kumar, V.K. Singh, P. Singh and V.K. Mishra), 95–106. Sawston, United Kingdom: Woodhead Publishing https://doi.org/10.1016/B978-0-12-821199-1.00010-9.

Wan, J., Klein, J., Simon, S. et al. (2010). AsIII oxidation by *Thiomonas arsenivorans* in up-flow fixed-bed reactors coupled to As sequestration onto zero-valent iron-coated sand. *Water Res.* 44: 5098–5108. https://doi.org/10.1016/j.watres.2010.08.044.

Wickramasinghe, S.R., Han, B., Zimbron, J. et al. (2004). Arsenic removal by coagulation and filtration: comparison of groundwaters from the United States and Bangladesh. *Desalination* 169: 231–244. https://doi.org/10.1016/j.desal.2004.03.013.

Yan, L., Yin, H., Zhang, S. et al. (2010). Biosorption of inorganic and organic arsenic from aqueous solution by *Acidithiobacillus* ferrooxidans BY-3. *J. Hazard. Mater.* 178: 209–217. https://doi.org/10.1016/j.jhazmat.2010.01.065.

Yang, C., Ho, Y.N., Makita, R. et al. (2020). *Cupriavidus basilensis* strain r507, a toxic arsenic phytoextraction facilitator, potentiates the arsenic accumulation by Pteris vittata. *Ecotoxicol. Environ. Saf.* 190: 110075. https://doi.org/10.1016/j.ecoenv.2019.110075.

Yu, X., Tong, S., Ge, M. et al. (2013). Synthesis and characterization of multi-amino-functionalized cellulose for arsenic adsorption. *Carbohydr. Polym.* 92: 380–387. https://doi.org/10.1016/j.carbpol.2012.09.050.

Yuan, B.L., Li, K.L., Deng, L.L., and Zhang, Z.D. (2006). Removal of arsenic (III) by ferrate oxidation-coagulation from drinking water. *Huanjing Kexue/Environ. Sci.* 27: 281–284.

Zaw, M. and Emett, M.T. (2002). Arsenic removal from water using advanced oxidation processes. *Toxicol. Lett.* 133: 113–118. https://doi.org/10.1016/S0378-4274(02)00081-4.

Zouboulis, A. and Katsoyiannis, I. (2002). Removal of arsenates from contaminated water by coagulation-direct filtration. *Sep. Sci. Technol.* 37: 2859–2873. https://doi.org/10.1081/SS-120005470.

18

Methods to Detect Arsenic Compounds

Shraddha Mishra and Sanjay Kumar Verma

Department of Biological Sciences, Birla Institute of Technology and Science, Pilani, Rajasthan, India

18.1 Introduction

Arsenic (As) is a well-known metalloid causing deleterious effects on all life forms. It exerts potential health risks on human beings. In the soil, it is present in the inorganic forms majorly including the mineral forms or the organic forms accumulated in the living beings uptaking arsenic through any source from the environment. Arsenic is present in around 300 minerals on the earth, including sulfosalts, arsenites, arsenates, sulfides, arsenides, metal alloys, and native elements. Among all the mineral forms, sulfides including pyrite, arsenopyrite, realgar, etc., and arsenate minerals including scorodite, yukonite, and beudantite are the most prevalent forms. The other mineral types are formed during weathering. Organoarsenic compounds such as monomethylarsine (MMA), dimethylarsine (DMA), and trimethylarsine (TMA) are formed due to the methylation of inorganic arsenic, which is less toxic and can enter the food chain. Some other organic forms of organic arsenic such as lewisite, adamsite, and cacodylic acid, were developed during World War II or found in seafood (such as arsenobetaine or arsenocholine) (Shrivastava et al. 2015). Four predominant oxidation states of arsenic +5 [arsenate As (V)], +3 [arsenite (As (III)], 0 (elemental arsenic), and −3 (arsine gas or arsenides) exist in the natural environment among which As (V) and As (III) are found more abundantly.

However, both As (V) and As (III) are highly toxic forms of arsenic, but each has a different mechanism to impose the toxicity. As (V) acts as a chemical analog of phosphate and enters the cell through phosphate transporters. Inside the cell, it replaces the phosphate in several cellular reactions leading to the inhibition of metabolic processes and also hinders adenosine triphosphate (ATP) synthesis. As (III) reacts with the thiol and sulfhydryl groups of proteins leading to dysregulation or inhibition of that protein or enzyme. The toxicity order of different arsenic species includes As (III) on the topmost place followed by As (V), MMA, and DMA, where the toxicity of MMA and DMA is very less in comparison to As (V) and As (III) (Sankararamakrishnan and Mishra 2018).

Arsenic in Plants: Uptake, Consequences and Remediation Techniques, First Edition.
Edited by Prabhat Kumar Srivastava, Rachana Singh, Parul Parihar, and Sheo Mohan Prasad.

Arsenic is naturally present in the surrounding environment in some concentrations. However, various anthropogenic actions which involve mining, usage of wood preservatives and fertilizers, smelting, utilization of fossil fuels, chemotherapeutic drugs used in cancer treatment, and production of glass and semiconductors have resulted in increasing arsenic contamination in the environment.

The risk of arsenic exposure on humans has been increased in recent years, which may have severe consequences on human health. It causes many clinical manifestations, including cutaneous lesions, melanosis (hyperpigmentation), leukomelanosis (hypopigmentation), and keratosis. The carcinogenic effect of arsenic on humans includes pulmonary, skin, hepatic, bladder, and renal cancers. In addition to these, arsenic can also cause ischemic heart disease, atherosclerosis, impaired motor and cognitive functions, and hormonal dysregulation. However, a very low concentration of arsenic ($<3.6\,mg\,kg^{-1}$) exists in plants that are growing on the natural soil, and its high level imposes toxic effects on them. Arsenic induces phytotoxicity in plants by interfering with the metabolic processes leading to plant growth inhibition. Various symptoms of As-induced toxicity on plants include decreased plant height, inhibition of seed germination, necrosis, depressed tillering, reduction of root and shoot growth, the lower yield of fruit and grain, underdeveloped roots, shrunken yellow leaves, reduction in the amount of chlorophyll, and protein and reduced photosynthetic capacity. Considering the adverse effect of arsenic on human health, the permissible limit of arsenic concentration in drinking water has been established to $10\,\mu g\,l^{-1}$ by the WHO as well as the EPA above that is toxic to human beings (Gudlavalleti et al. 2017).

One of the major concerns of the world in the present scenario includes the contamination of soil and groundwater sources by arsenic. Drinking water contamination of arsenic has been reported in various countries (>70) mostly belonging to South Asia and Southeast Asia. However, some other countries, for instance, Bangladesh, India, China, Pakistan, Argentina, Chile, Hungary, Mexico, Vietnam as well as many places of the USA are suffering from elevated levels of arsenic ($As>50\,\mu g\,l^{-1}$) in groundwater. The study of As contamination, its detection, and remediation drew considerable attention in the last decades as contamination level of As revealed alarming situation in some places such as West Bengal in India and the neighboring country, Bangladesh (Ali et al. 2019). The global arsenic contamination problem has rendered it necessary to study the existing methods of arsenic detection in environmental samples, which will contribute to the quick implementation of various remediation techniques in those places. The study of various technologies, their advantages, and limitations can allow the development of more reliable and economical methods of arsenic detection, providing better sensitivity and selectivity. Several analytical methods, for example, atomic absorption spectroscopy (AAS) colorimetric, electrochemical, inductively coupled plasma (ICP)/mass spectrometry (MS), liquid chromatography (LC)/(MS/MS), atomic fluorescence spectrometry (AFS), fluorescence resonance energy transfer (FRET), etc., have been applied to detect low levels of arsenic in different environmental samples. In addition to these analytical techniques, various types of biosensors can also be used to detect arsenic in the environmental sample. In this chapter, we have discussed various techniques applied to arsenic detection.

18.2 Colorimetric Method

Colorimetric methods of arsenic detection are based on the principle of determining arsenic concentration in a solution by the addition of a coloring agent. One of the most employed colorimetric methods of arsenic detection is the molybdenum blue method. The basic principle of this method comprises the reaction of molybdenum with arsenate under acidic conditions, which leads to the formation of a yellow-colored heteropoly acid further reducing to molybdenum blue complex. The color intensity of this complex is measured using a spectrophotometer. Although this method detects the presence of arsenate, phosphate, and silicate in solution, it cannot differentiate arsenate and phosphate when both are present in a sample and thus is not specific to arsenate (Johnson and Pilson 1972). Several other chromogenic reagents such as leucocrystal violet, rhodamine B (RhB), azure B, 2-(5-bromo-2-pyridylazo)-5-di-ethylaminophenol (Br-PADAP), methylene blue (MB), variamine blue, and 2,4-dihydroxy benzophenone-2-amino thiophenol (BPBT) have also been used for the detection of As (V) and As (III). The same principle of colored compound formation upon reaction with arsenic and measurement of color intensity using spectrophotometer has been employed for its detection. In recent years, colorimetric methods based on nanoparticles (NPs), nanorods, and microfluidic systems have been developed for arsenic detection to improve the sensitivity up to $\sim 1.0\,\mu g\,l^{-1}$ (Lace et al. 2019; Nath et al. 2017; Priyadarshni et al. 2018).

18.3 Electrochemical Method

The heavy metal (such as arsenic) detection using electrochemical methods is based on the electrochemical properties of the metal-containing solution where the solution acts as a part of the chemical battery. The change in several electrical parameters, such as conductivity, resistance, current (I), potential (V), and I-V curve analysis, is applied for the quantification of heavy metal in the solution (Li et al. 2018). The heavy metals having cationic nature are electro-reduced to their metallic form in an electrode. Electrochemical methods can be divided into polarographic and voltamperometric methods. The subclasses of polarographic methods include direct current (DC), alternating current, square wave (SW) current, normal phase, and differential pulse (DP) polarography. The voltamperometric methods can be subdivided into linearly changing potential cyclic and inverse (Niedzielski and Siepak 2003). In both techniques, either microelectrodes or screen-printed electrodes (SPEs) can be applied. Among different voltammetric techniques developed for arsenic detection such as cyclic, linear sweep (LS), DP, and SW voltammetry, the latter two voltammetry techniques have been used as a potential tool to determine traces of arsenic due to improved detection limit, greater signal-to-noise along with an effective pre-concentration step. All these voltammetric techniques make use of stripping analysis which can be further divided into either anodic or cathodic stripping voltammetry, for example, differential pulse anodic stripping voltammetry (DPASV), SW cathodic stripping voltammetry (SWCSV), linear sweep anodic stripping voltammetry (LSASV), etc. Electrochemical methods have applied different types of electrodes including graphite, diamond, screen-printed

(SP), glassy carbon (GC), gold (Au), carbon paste (CP), silver (Ag), platinum (Pt), and mercury (Hg) electrode (E), which have been upgraded in recent years to improve the sensitivity, selectivity, stability, and reproducibility of the detection. Several modifications involved in the fabrication of these electrodes involve the application of biological molecules (e.g. arsenic-specific ligands, biomacromolecules, enzymes, and conductive polymers), nanomaterial-based modified electrodes (e.g. carbon, gold, platinum, silver, and iridium), and metal-oxide nanostructures. A limit of detection (LOD) of up to 0.00097 $\mu g\,l^{-1}$ has been achieved by incorporating these modifications in electrodes constructed to detect As (III) by electrochemical methods (Liu and Huang 2014). Arsenic detection via electrochemical method does not involve toxic arsine (AsH_3) gas hydride production, which distinguishes it from other arsenic detection methods. The inherent advantages of electrochemical methods include their independence on sample size to be measured, small size, low sample amount requirement [sample amount $\propto$ (surface\volume)$^{-1}$], and low interference due to the presence of other ions in solution (Yogarajah and Tsai 2015).

18.4 Method Based on FRET

Fluorescence resonance energy transfer (FRET), which is also known as Förster resonance energy transfer, is a fundamental phenomenon of nonradiative energy transfer between a fluorophore (donor) and a chromophore (acceptor) molecule. This transfer of energy from donor to acceptor molecule leads to the reduction of fluorescence intensity as well as the lifespan of donor molecule at the excited state while enhancement in the acceptor molecule's emission intensity (T.H. Förster 1965). The FRET efficiency (energy transfer from donor to acceptor) can be affected by various factors such as the closeness of donor and acceptor molecule (should be 1–10 nm), the refractive index of the solution, emission, and absorption spectrum of donor and acceptor molecule (should be overlapped), the orientation of electric dipole of the donor with respect to the acceptor (Hazarika and Rajkonwar 2019). FRET-based techniques can be used to measure the presence of an external agent as it may cause a change in any of the above-mentioned factors triggering a change in energy transfer efficiency. There have been several reports on the detection of arsenic using FRET-based techniques. The FRET between CdTe quantum dots (QDs) having a capping of mercaptopropionic acid (MPA) and fluorescent dye rhodamine 6G (R6G) has been used to detect the As (III) where CdTe QDs act as a donor and transfer energy to R6G dye acceptor molecules. The As (III) ions bind with glutathione (GSH) molecules (could be attached to the surface of the QDs and decrease the FRET efficiency) and detach it from the surface resulting in the recovery of FRET efficiency. The LOD of 0.45 $\mu g\,l^{-1}$ for As (III) is provided by this system (Tang et al. 2015). Acriflavine (Acf) and RhB dye have also been used to develop a FRET-based radiometric sensor where Acf behaves as an energy donor and RhB as an acceptor. The Acf and RhB make a proper donor-acceptor pair because their emission and absorption spectra overlap with each other in the absence of As (V). However, under the As (V) presence, there is an increase in the fluorescence of RhB and a decrease in the fluorescence of Acf due to the allocation of energy from Acf to RhB. In this process, the As (V) helps in the formation of H-bonding and electrostatic interaction between two cationic dyes (Acf and RhB) causing them to come under close proximity resulting in an increase in the FRET efficiency. The concentration of As (V) in the liquid phase is directly related to the increase in the FRET. This simple sensor provides a LOD of 10 μg l for the As (V) (Saha et al. 2017). The technique involving laser-induced fluorescence resonance energy transfer (LIFRET) has also been applied for As (III) detection utilizing 1,8-naphthalimide (NA) and RhB dyes as donor and acceptor, respectively. In the presence of As (III), the FRET spectrum gets shifted toward the lower wavelength that is directly proportional to the As (III) concentration in the sample. This technique offers a convenient, portable, and cost-effective way of As (III) detection providing a LOD of 10 $\mu g\, l^{-1}$ (Hazarika and Rajkonwar 2019).

18.5 Method Based on SPR

The presence of arsenic and other heavy metals in a sample can be detected using the surface plasmon resonance (SPR)-based method. This method involves the formation of plasmonic waves between the metal surface and the layers of the dielectric medium leading to the detection of even small disturbances in the dielectric medium (Al-Rekabi et al. 2019). The groundwater arsenic can be screened on-site using a cost-effective and simple device consisting of a sensor probe made up of thiol-containing organic compounds that allow the detection of arsenic concentration below 10 $\mu g\, l^{-1}$ (Forzani et al. 2007). Signal-enhanced SPR-based detection has allowed a LOD of 0.075 $\mu g\, l^{-1}$ for As (III). It is an ion-selective method, which involves a hybrid system consisting of radially porous silica particles and a thin film of gold. The mesopores which are radially oriented inside the silica particle can increase their size giving rise to a larger surface area exposing receptors to bind with the target molecule, which leads to improved sensitivity. The As (III) present in the sample forms the electrostatic complex together with the amine functional group present on the surface of porous silica particles causing changes in the SPR profile of the thin gold film that can be detected using this method (Choi et al. 2014). Some metal NPs, for example, gold (Au) and silver (Ag) NPs, display the property of localized surface plasmon resonance or LSPR which arise from the combined oscillations of the electrons of the NPs present in the conduction band where the interaction of NPs with ultraviolet (UV)-visible radiation leads to the combined oscillations of the electrons. The LSPR property of NPs can be utilized to construct a chemical sensor using AuNPs modified with lauryl sulfate. The arsenic present in the sample interacts with NPs causing LSPR band shift which occurs due to the coupling effect between the particles. Due to the shift in the LSPR band, the pink color of AuNPs changes to blue, which can be detected visually. The LOD offered by these selective and simple sensors is up to 2 $\mu g\, l^{-1}$ (Shrivas et al. 2015). Several prism-based SPR sensors have also been developed containing a coating of ferric oxide-magnetite-reduced graphene oxide ($Fe_2H_2O_4$–Fe_3O_4–rGO) nanocomposite film. These sensors confer improved sensitivity and selectivity for arsenic detection with the LOD of 0.1 $\mu g\, l^{-1}$ (Al-Rekabi et al. 2019).

18.6 Method Based on Spectrometry

The arsenic detection based on spectrometry technique includes methods such as AAS, ICP-MS, AFS, electrothermal (ET) vaporization-ICP-MS, ET-AAS, hydride generation (HG)-AFS, HG-AAS, etc., which follows the principles of spectrometry for arsenic detection. The advantages of these methods include high throughput sample analysis, reproducibility, and low detection limit (<10 μg l^{-1}).

Several oxidation states of an element can be present in the environmental samples to be analyzed. The prerequisite for different analytical methods is to allow the detection of all those elemental forms with high precision and accuracy. The identification and separation of different ionic forms of an element are termed speciation. It requires each form of the element to be determined independently with no interference without causing the change in the property of the original sample. If there is the unavailability of such a method, speciation can be performed by combining different analytical techniques such as spectroscopy with different chromatographic techniques. A lot of spectrometric techniques have been combined with different chromatographic separation and pre-concentration techniques to allow the detection of the analyte to be more selective and sensitive. Due to the varying physicochemical properties of the arsenic compounds, different separation/pre-concentration methods have been combined with sensitive detection methods for arsenic speciation among which combining chromatographic separation methods with a detection system is the most common one. While selecting the separation technique, the physicochemical properties of the different forms of the analyte such as oxidation state, volatility, and polarity are considered which sometimes require a combination of two or more methods. The detection method can be selected based on the concentration level of the different elemental forms existing in the sample, considering the matrix type and its composition. However, it is very difficult to quantify any one species/form of the metal present in the sample due to its low concentration which may not fall under the detection limit of the analytical technique available (Terlecka 2005). To overcome this problem, several hyphenated techniques have been exploited for the arsenic speciation such as a combination of high-performance liquid chromatography (HPLC) or ion chromatography (IC) with ICP-MS, combining IC with HG-ICP atomic emission spectrometry (HG-ICP-AES) or electrospray MS (ES-MS). Among these combined techniques, gas chromatography (GC), LC, HG, and capillary electrophoresis have been utilized for the separation of different species of arsenic. The HG separation technique has the advantage of being easily connected to a wide variety of detection systems such as AAS, ETAAS, AFS, and ICP-MS, etc. We can select any of the combinations of these methods depending on the requirement of the sample to be analyzed. However, preserving the stability of different arsenic forms during sample preparation is still a challenge while performing speciation analysis (Anawar 2012; Niedzielski et al. 2002). Various spectrometry-based arsenic detection methods have been discussed next in detail.

18.6.1 Atomic Absorption Spectrometry

AAS is an analytical technique that functions on the principle of the absorption of electromagnetic radiation in the UV and visible regions. It is employed for the detection of trace metals. Detection of metals/heavy metals using this technique involves dissociation of the elemental form of metal into atoms by applying thermal energy not exceeding 3000 °C. High-temperature treatment of the metal-containing sample leads to the excitation of monoatomic forms of the element into the higher energy level and when they come back to the ground level, the emission of radiation of a particular wavelength occurs, which depends on the nature of the element. However, still, gaseous atoms of metal are present at the ground state in abundance that can absorb the radiant energy of a specific resonance wavelength. This principle of absorption is exploited by AAS to detect the metal concentration in a sample by relating it with a standard solution that contains the known concentration of the analyte to be detected. The measurement of the concentration of analyte in the sample is based on the principle of Beer–Lambert Law stating that "absorbance is directly proportional to the concentration of analyte sample in a cell" (Oti Wilberforce).

AAS has been a method of choice for arsenic detection since the 1970s due to its user-friendly instrumentation, cheapness, high sensitivity, good precision, and low interference caused due to the presence of other metals during analysis. Various pre-concentration techniques such as liquid–liquid, solid–phase, cloud point, and homogeneous liquid–liquid extractions (LLE, SPE, CPE, and HLLE) have been applied before the analysis of an analyte using AAS. The microextraction (ME) techniques utilized in pre-concentration include liquid-phase, hollow fiber liquid-phase, dispersive liquid–liquid ME technique based on the CPE (LLME, HFLPME, and DLLME). The SPE coupled with DLLME utilizing diethyl dithiophosphate for chelation has also been applied for the pre-concentration based on the floating organic drop solidification (Anawar 2012). In addition to these, nanostructures such as carbon nanotube, Mg—Al—Fe(NO_3^-) ternary layered double hydroxide, and aluminum oxide NPs have also been exploited in various SPE-based methods of arsenic speciation and pre-concentration. Depending on the derivatization methods used in As detection, the method of AAS has been divided into HG and ET/graphite furnace (GF) AAS.

18.6.1.1 Hydride Generation Atomic Absorption Spectrometry

HG technique is used to detect the trace level of As in a sample involving the reaction of sodium tetrahydroborate with arsenic compounds in acidic conditions, which results in the production of various arsines (AsH_3) depending on the type of arsenic compound. The inert carrier (argon, helium) transports this released arsine gas to either atomizer (in AAS or AFS) or an excitation source (ICP). The formation of hydrides of arsine can be described by the following reaction:

$$NaBH_4 + 3H_2O + HCl \rightarrow H_3BO_3 + NaCl + 8H^- \xrightarrow{As^{m+}} AsH_n + H_2 \uparrow$$

HG can be linked to several detection techniques such as ICP-MS AAS, AFS, and ICP-MS in order to improve its sensitivity along with the elimination of the spectral and chemical interferences. Despite these advantages, the HG technique has limitations of its applicability for detecting only volatile gases and single elements, the requirement of strict control over the reaction conditions, interference by certain elements reducing the efficiency, and requirement of labor (Anawar 2012).

18.6.1.2 Electrothermal/Graphite Furnace Atomic Absorption Spectrometry

Electrothermal Furnace Atomic Absorption Spectrometry (ETAAS), which is also known as Graphite Furnace Atomic Absorption Spectrometry (GFAAS), is an effective spectroscopy technique applied to detect As (III) and As (V) in the environmental sample. In the GF technique, the absorption of the freely available atoms is measured, which are produced from the heat-treated small graphite tube having sample deposition. This method is capable of eliminating the organic matrix before the atomization of the sample by gradually increasing the temperature. However, most of the GFAAS-based arsenic detection methods needed pre-concentration to improve the LOD. Several chemical modifiers have been employed for the stabilization of volatile elements so that matrix components present in a sample can be volatilized using higher permitted pyrolysis temperature leading to reduced matrix interference and increased accuracy (Hung et al. 2004). The most appropriate modifiers include the nitrate of Ni, Cu, Pd, and Mg. Mg-Pd nitrate is a widespread modifier for As-containing samples giving rise to a pyrolysis temperature of 1400 °C in the presence of NaCl and K_2SO_4 while a pyrolysis temperature of 1300 °C is offered by the Ni-nitrate. However, arsenic and zirconium modifiers are used to obtain a better detection limit and stabilization providing pyrolysis temperature as high as 1500 °C. All these modifiers form a combination of oxides stable at a higher temperature. Some disadvantages of using these chemical modifiers include poor performance while detecting As in organic matrices directly using Pd-Mg, Pd, Rh, W, and Ag modifiers and interference caused due to reaction of Pd^{2+} with matrix constituents containing Cl^- when $Pd(NO_3)_2$ is used as a modifier. To overcome these problems, several other modifiers such as La, colloidal Pd, a combination of Zr and Ir, a combination of W with noble metals (e.g. W-Rh, W-Ru, and W-Ir), Ag and Au NPs have been used to eliminate the problem of interference caused due to former modifiers (Anawar 2012).

18.6.2 Atomic Fluorescence Spectrometry

AFS is used as a quantitative elemental analysis technique since the 1960s and has become a routine analytical technique nowadays. The basic principle of this method is radiative excitation of gaseous atoms to a higher energy level using a source of light followed by emission of photons by a deactivation process, which can be measured as fluorescence signal. Due to the involvement of both excitation and de-excitation, this method can be distinguished from the AAS (Butcher 2005). AFS has been exploited for arsenic detection because it is cost-effective and provides better sensitivity. However, to get the more sensitive and specific analysis using this technique, it is linked with the HG technique. A combination of more than two techniques such as coupling of HG double-channel AFS (HG-DC-AFS) along with the online solid-phase extraction (HG-DC-AFS) has been used to detect inorganic As and Sb in the water sample simultaneously. In this method, speciation is performed using single-walled carbon nanotubes that are filled in a micro-column to generate and retain ammonium pyrrolidine dithiocarbamate complexes of As (III) and Sb (III) online (Wu et al. 2011). Multichannel HG-AFS can simultaneously detect As, Bi, Se, and Te, imparting higher accuracy and precision (Zhang et al. 2011). To detect a very low concentration of arsenic in a sample, HG-cryotrapping (CT)-AFS coupled with the flame-in-gas-shield (FIGS) atomizer technique has been used, which is a cost-effective technique (Musil et al. 2014).

18.6.3 Inductively Coupled Plasma Techniques

ICP techniques are in use since the 1960s for the detection of arsenic. In this technique, plasma (ionized gas) is used for the ionization and atomization of the entire forms of the analyte in the sample treated with acid. The atomized sample is then detected by means of either an optical emission spectrometer (OES) or mass spectrometer (MS) detector, where MS provides better ionization efficiency and sensitivity (Colon et al. 2011).

18.6.3.1 Inductively Coupled Plasma Mass Spectrometry

ICP-MS is an analytical technique allowing quantitative determination of multiple elements simultaneously providing a wide detection range. The principle of ICP-MS involves the dissociation of analytes present in a sample into its parent atom or ion using the argon plasma source, which further encounters the plasma, and isolated based on their mass to charge (m/z) ratio with the help of a quadrupole or magnetic sector analyzer present in the mass spectrometer. The analysis of the liquid sample is performed by pumping it through a peristaltic pump into the nebulizer followed by the introduction of the gaseous or aerosol form of the sample to the plasma torch producing an abundance of ions with low backgrounds. This technique provides a good detection limit in parts-per-trillion ranges (Helaluddin et al. 2016). Different isotopes of an atom can also be determined using ICP-MS with better precision, making it more applicable than ICP-AES. However, some of the problems that occur during the detection of an analyte using ICP-MS include interference due to the $^{40}Ar^{35}Cl$ dimer formed during the direct nebulization process from the Ar. To overcome this problem, diffusion cell or collision technology has been used, which allows the collision of the dimers with low molecular weight gases like H_2, He, etc., after the ion selection leading to the splitting of polyatomic species (Colon et al. 2011).

ICP-MS technique can be interfaced with a variety of pre-concentration, separation, speciation, or derivatization methods which include different chromatographic procedures like ion exchange chromatography (IXC) or ion exclusion chromatography (IEC), HPLC, capillary electrophoresis, and HG technique which provides higher sensitivity and selectivity in As detection.

18.6.3.2 Inductively Coupled Plasma/Optical Emission Spectrometry

The spectroscopic technique ICP-OES involves the excitation of atoms and ions by a radiofrequency (RF) discharge leading to the spontaneous emission of photons. Analysis of the sample using ICP-OES is performed by introducing the aerosolized liquid sample into the center of the plasma, where an elevated temperature of around 10 000 K is maintained for atomization. However, solid samples require acid digestion before their injection into the instrument. The generated gaseous plasma-free atoms are converted into the ions using the energy followed by their excitation into the higher energy state. These exited ions emit the photons of a specific wavelength while coming back to the ground state, which helps in the identification and quantification of the element present in the sample as the number of photons emitted by the metal ion is directly related to its concentration in the tested sample. This technique utilizes several methods to introduce the sample such as nebulization, HG, laser ablation, and electrothermal vaporization (Helaluddin et al. 2016).

18.7 Biosensor for Arsenic Detection

A biosensor contains biological molecules such as an enzyme, whole-cell, antibody, DNA, etc., as a recognition element; a physical transducer such as optical, electrochemical, or thermal; and an amplification unit that leads to the generation of data in the quantitative or semiquantitative form, which corresponds to the concentration of the target analyte in the sample (Kaur et al. 2015; Park et al. 2013; Singh et al. 2008). Many biosensors have been developed with improved detection limits and specificity to detect heavy metals such as arsenic taking advantage of very strong and specific binding between the target analyte and recognition element. These sensing devices generate either an enhanced or reduced biochemical signal with the help of the biorecognition element, which is later translated into a coherent signal by the detector. Biosensors can be classified into two distinct categories of whole cell-based biosensors and cell-free/biomolecules-based biosensors based on the biorecognition element.

18.7.1 Whole Cell-Based Biosensor

The fabrication of the whole cell-based biosensor employs the use of the cellular metabolic process such as metal resistance for its design. For example, whole cell-based arsenic biosensors have been constructed by utilizing the ability of repressor protein arsR to bind As (III) with high specificity. After binding with As (III), it detaches from the promoter site leading to the consequent synthesis of the reporter gene (Diesel et al. 2009; Merulla et al. 2013). The basic functional mechanism of the whole cell-based biosensors has been depicted in Figure 18.1. The whole cell-based biosensors make use of the live microorganism as the recognition element for biosensing. The response which can be of different types such as alterations in cellular metabolism, expression of genes, and pH is quantified using different techniques such as electrochemical and optical sensing. Based on the reporter proteins involved in arsenic detection, whole cell-based biosensors can be broadly classified into green fluorescent protein (GFP), bioluminescence/luciferase, and β-galactosidase/lacZ-based biosensors.

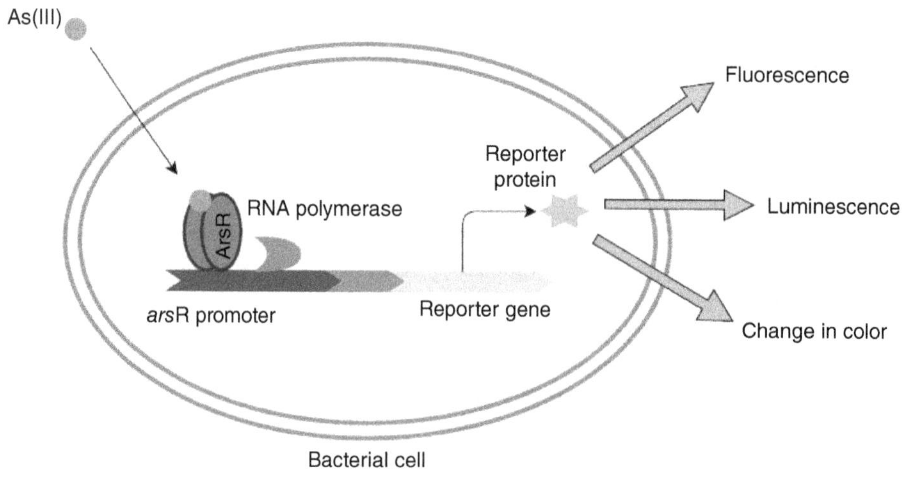

Figure 18.1 Schematic representation of a whole-cell biosensor involving reporter protein-based detection using color change, luminescence, or fluorescence signal.

18.7.1.1 Green Fluorescent Protein-Based Biosensor

The GFP, having a molecular weight of 26.9 kDa is frequently used in molecular biology as a reporter protein exhibiting green fluorescence under blue or UV light. It has been used for arsenic whole-cell biosensor development for a long time because of its ability to show response without requiring any substrate that makes the sensing more economical in comparison to luciferase- and β-galactosidase-based biosensors (Kaur et al. 2015; Upadhyay et al. 2018). The first study of the GFP-based biosensor has been conducted by Roberto et al. (2002), which involved the fusion of GFP from *Aequorea victoria* to a gene harboring operator/promoter of ars operon, arsR gene, and a few elements of arsD from p1RC120. This combined gene construct was further transformed into *Escherichia coli* (*E. coli*) (p1RC1140). The induction period to get stable fluorescence was 12 hours providing a LOD of 1 $\mu g\, l^{-1}$.

The functional mechanism of the GFP-based biosensor involves the detachment of arsR from the operator/promoter site in the presence of As (III) which triggers the expression of GFP gene leading to the emission of green fluorescence corresponding to the analyte concentration in the sample. Several studies have been done to make improvements in GFP-based biosensors in the past years. Truffer et al. (2014) have constructed a biosensor comprising a microfluidics polydimethylsiloxane (PDMS) platform. On the microfluidics PDMS platform, the biosensor (*E. coli* strain 1598 carrying pPROBE-arsR-ABS plasmid) encapsulated inside the agarose bead has been placed and As (III) detection takes place in two parallel channels provided. Each parallel channel of the automated electronic system performs measurement at an interval of 20 minutes providing a LOD of 10 μg/l at 120 minutes. Some variants of GFP, such as enhanced GFP (EGFP), have also been employed as a reporter protein for arsenic detection because of better signal production than wild-type GFP (Scaria et al. 2009). A study by Gudlavalleti et al. (2017) involves the whole-cell biosensor construction by transforming plasmid pJSKV51 into the *E. coli* DH5α cells. Plasmid pJSKV51 carries a promoter/operator region (in-frame), complete arsR gene, a fragment of ARSD gene, and EGFP gene. The constructed biosensor strain detected 5–100 $\mu g\, l^{-1}$ of As (III) by producing the fluorescence response in the presence of As (III) ions after the activation of upstream gene expression by arsR protein. Elcin and Öktem (2019) have developed a GFP-based *E. coli* biosensor to detect both inorganic forms of arsenic [As (III) and As (V)] present in water. The construction of this biosensor involved the transformation of *E. coli* MG1655 strain with a sensor plasmid. The operator or promoter region of *ars* operon, *ars*R gene of the *E. coli* plasmid R773, and promoter-less GFP*uv* gene (reporter) which is a mutant form of GFP are transcriptionally fused to construct the sensor plasmid. This biosensor offers a LOD of 10 $\mu g\, l^{-1}$ for As (III) after 4 hours of induction. However, As (V) showed comparatively longer induction time and less sensitivity as compared to As (III) due to the competitive interference by phosphate ions.

The advantages of green fluorescent-based biosensors include real-time detection of the target analyte to be measured, no need for any substrate and cell disruption, not influenced by the growth phase of the microbe and internal reducing agents, and simplicity of detection. These advantages have augmented the application of the GFP gene in biosensor development. However, they have the disadvantage of less sensitivity in comparison to the β-galactosidase-based system.

18.7.1.2 Bioluminescence/Luciferase-Based Biosensor

Taking advantage of the luminescent property of the luciferase gene, the first luminescent whole cell-based biosensor for arsenic detection was developed by Tauriainen et al. (1999). Most of the luciferase-based biosensors employ the use of arsenic-sensing property of the arsR protein (Kaur et al. 2015). The mechanism involves the joining of a reporter gene (here *lux* gene) with arsR gene either by removing all other genes present in the operon or by replacing a portion of it followed by the expression of the reporter gene. In the presence of arsenic, the promoter is turned on leading to the production of reporter proteins which can be detected form of light. The *lux* system comprises five genes *luxCDABE*. The *luxA* and *luxB* genes are translated into the luciferase enzyme responsible for the bioluminescent response. The response is generated by the conversion of long-chain aldehydes into fatty acids due to the action of the luciferase enzyme along with the production of photons in the blue-green range (490 nm) of the visible light. The other three multienzyme genes *luxC*, *luxD*, and *luxE* recycle fatty acid into the substrate aldehyde employing reductase, transferase, and synthetase activity. The *lux* fusion for the detection of environmental arsenic can be constructed having either the *luxAB* complex or the entire *luxCDABE* cassette. An aldehyde substrate decanal must be supplied to the cells if only *luxAB* genes are used. However, if the entire *luxCDABE* cassette is utilized a self-contained bioluminescent response is generated without adding any substrate exogenously. Generation of the signal or response requires the sensing of As (III) by arsR (Soangra et al. 2015). The advantages of luciferase-based biosensors include better sensitivity, broad dynamic range, and simplicity with a LOD up to 0.0007 $\mu g\,l^{-1}$ (Bakhrat et al. 2011; Park et al. 2013).

18.7.1.3 β-galactosidase/*lacZ*-based biosensor

The gene coding for β-galactosidase protein is employed for arsenic biosensing giving quantitative and reproducible results. These biosensors have also been constructed taking advantage of the arsenic-binding property of arsR protein along with the enzymatic property of β-galactosidase coded by the *lacZ* gene. The first lacZ-based whole-cell biosensor was constructed by Scott et al. (1997) for which *E. coli* was genetically manipulated to produce β-galactosidase upon exposure to As (III) and Sb (III). The plasmid pBGD23 including promoter or operator region (P), *ars*R gene, and *lac*Z gene was used to transform *E. coli* strain JM109. The *ars*R and *lac*Z genes lie repressed in the absence of As (III) due to the binding of arsR with the promoter region which blocks the expression of these genes. When As (III) is present, arsR releases from the promoter site due to the conformational change allowing the expression of the gene coding for β-galactosidase which is detected by electrochemical methods. To quantify the expression of the β-galactosidase gene, *p*-aminophenyl β-D-galactopyranoside (PAPG) substrate was used and the product *p*-aminophenol (PAP) formed due to the activity of β-galactosidase was quantified electrochemically. Due to the progressive advancement in biosensor technology, various other lacZ-based whole-cell biosensors have been developed offering several advantages such as simple analysis methods, less time requirement, user-friendly nature, higher sensitivity, applicability in harsh environmental conditions, and cost-effectiveness. Novel approaches introduced in the construction of lacZ-based biosensors include a pH-based *E. coli* system, which takes advantage of the capability of wild-type *E. coli* to ferment lactose and other sugars by "mixed acid fermentation pathway" and producing a large number of acidic products. Engineered bacteria lacking a part of the enzyme β-galactosidase have been developed to detect the lacZ-bearing plasmids. The presence of As (III) and lactose leads to the expression of lacZ that allow the fermentation to take place generating acidic products that cause a decrease in pH, which can be measured. However, in the absence of As (III), lacZ is not expressed and thus no change in pH occurs (Joshi et al. 2009). A color-based bacterial biosensor has also been developed for the detection of As consisting of *E. coli* DH5α bearing pAs-lacZ. This construct exhibits the expression of the β-galactosidase gene in the presence of As (III) causing blue color formation in the presence of X-gal substrate, which is measured by taking optical density at 595 nm using a microplate reader. This biosensor is cost-effective and user-friendly (Huang et al. 2015). With the incorporation of various modifications in the development of these biosensors, a LOD of up to $0.2\,\mu g\,l^{-1}$ has been achieved (Wackwitz et al. 2008).

18.7.1.4 Whole-Cell Biosensor Based on Other Approaches

In addition to the genes mentioned above that are most frequently used for the construction of whole cell-based biosensors, several other genes such as the carotenoid biosynthesis gene (*crt*A), *ypf* gene, and *xyl*E gene encoding catechol-2,3-dioxygenase have also been applied for biosensor development. The construction and mechanism of these biosensors have been discussed in detail next.

***18.7.1.4.1 Carotenoid Biosynthesis Gene,* crtA** In addition to the GFP, lacZ, and luciferase-based whole-cell biosensors, some biosensors have been developed based on alternative approaches such as new whole-cell arsenite biosensors having carotenoid biosynthesis gene and *Rhodopseudomonas palustris* no. 7 containing *crt*A gene (Yoshida et al. 2008). The latter was constructed using the operator/promoter region of *ars* operon, *E. coli ars*R gene, and the gene for phytoene dehydrogenase enzyme (*crt*I). All these genes are transformed into the blue-green mutant *R. palustris* no. 711 containing deleted *crt*I. The detection process employs a colorimetric assay involving catalytic conversion of carotenoid phytoene, which is colorless into the red carotenoid lycopene by the activity of *crt*I in the presence of As (III). This conversion does not require molecular oxygen and the product can be visualized either directly or with the help of a spectrophotometer. This is the first carotenoid-based biosensor for detecting arsenic that is practically applied in the analysis of the natural sample.

18.7.1.4.2* ypf *Gene-Based Biosensor The phiYFP reporter protein-based biosensor was built for the As (III) and As (V) detection. It has stability in the changing pH environment and is appropriate for the production of cell lines that can retain transfected genes for a longer time (Wachter and Remington 1999). The phiYFP gene was obtained from the jellyfish *Phialidium* sp. and is considered as an improved version of the fluorescent protein occurring naturally. This was the first *ypf* gene-based whole-cell biosensor containing recombinant plasmid which is constructed using *ypf* gene, a promotor (As-resistant), and *ars*R gene obtained from the *E. coli* DH5α genome. The plasmid was introduced into *E. coli* DH5α for the construction of arsenic-resistant whole-cell biosensor (WCB-11). This biosensor produced time and dose-dependent yellow fluorescence upon the exposure of As (III) and As (V) (Hu et al. 2010).

18.7.1.4.3 Spore-Based Biosensor A spore-based biosensor was developed by Joshi et al. (2009) to overcome the requirement of freeze-drying and long-time storage of the cells. The property of *B. subtils* to form endospores during starvation is used for the development of this biosensor. An endospore is a resting state of the cell that can be stored even at elevated temperatures in dried form for many years. For this purpose, the *Pseudomonas putida xyl*E gene, which translates into catechol-2,3-dioxygenase, *ars* promoter, and *ars*R gene from *B. subtilis* were fused and inserted into the pVK168 vector, which can be maintained in both *B. subtilis* and *E. coli* cells. The stored endospore of biosensor strain was used as inoculum during the experiment. The biosensing assay involved the sensing of As (III) by arsR protein leading to the expression of the *xyl*E gene which is translated into the catechol-2,3-dioxygenase enzyme. This enzyme reacts with catechol to produce 2-hydroxy-*cis*,*cis*-muconic semialdehyde (HCMS), which is a bright yellow-colored compound (Wicke et al. 2018).

18.7.2 Cell-Free/Biomolecules-Based Biosensor

Although whole-cell biosensors provide an excellent detection system, some of their disadvantages, such as non-feasibility and non-reproducibility, ensued the development of cell-free biosensors for the detection of a target molecule of interest. Several cell-free biosensors have been developed for arsenic detection making use of either the specific interaction of arsenic with a variety of biomolecules such as DNA, RNA, and proteins, or competitive inhibition of enzymatic reaction by the arsenic (Merulla et al. 2013). The biosensors based on cell-free technology have several advantages: (i) These do not have any biosafety problem because there is no release of GMO; (ii) They can be stored for a longer time using freeze-drying technology providing higher stability; (iii) Generation of quicker response upon reaction with the testing material due to its direct interaction with the target analyte; (iv) Conferring better sensitivity with the LOD in nanomolar range; (v) This biosensor system has chemical tolerance allowing it to operate under the toxic conditions; (vi) Selective in nature as interference due to the presence of other analytes can be reduced by the system; (vii) Robust and does not change its structure and function as a result of disturbance due to environmental factors (Zhang et al. 2020). This type of biosensors can be broadly classified into DNA, aptamer, and protein-based biosensors.

18.7.2.1 DNA-Based Biosensor

A DNA-base biosensor uses DNA as a biorecognition component to determine the specific binding of the target molecule (e.g. heavy metals) with DNA using electrochemical (e.g. carbon electrode) transduction, thus also called electrochemical DNA biosensor (Tencaliec et al. 2006). Nucleic acid biosensors such as DNA biosensors provide a better alternative to the whole cell-based biosensor because of their high affinity toward the target molecules such as heavy metals leading to the strong interaction between them. This interaction could be electrostatic, occurring between the phosphate group of DNA having a negative charge and positively charged metal ion, the interaction of metal ion with the major or minor grooves of DNA which is termed as binding interaction and interaction caused due to the intercalation of metal ion between the DNA's stacked base pairs (Arora et al. 2007; Kaur et al. 2015). To study the interaction between heavy metals and DNA, either native state DNA, such as *calf thymus* DNA, double-stranded DNA/ds-DNA, or denatured DNA, such as single-stranded DNA/ss-DNA have been used involving electrochemical techniques, such as voltammetry and chronopotentiometry for detection. These electrochemical techniques provide high sensitivity, improved selectivity and signal-to-noise ratio, cost-effective and fast detection (Babkina et al. 2004; Oliveira et al. 2008; Tencaliec et al. 2006; Turdean 2011; Wong et al. 2007).

The first DNA-based biosensors for arsenic detection have been reported by Ozsoz et al. (2003). It was constructed using dsDNA and ssDNA from the *calf thymus* along with a small oligonucleotide containing 17 nt (Probe A) as a biorecognition element for As (III) detection. The damage caused due to oxidative stress after the interaction of As (III) with guanine in DNA was detected electrochemically utilizing differential pulse voltammetry (DPV) having CP electrode. The binding of As (III) and dsDNA in the solution phase was studied by potentiometric stripping analysis utilizing a pencil graphite electrode which can be reused further. This is the first report of determining the interaction of As (III) with dsDNA, ssDNA, and short oligonucleotide. After this, many studies have been done on the development of DNA-based arsenic biosensors. The most recent study includes the adsorption of fluorescein amidites-labeled (FAM-labeled) DNA oligonucleotide on iron oxide NPs. The adsorption of DNA oligonucleotide occurs via phosphate backbone that causes fluorescence quenching. However, the As (V), which shares similar solution chemistry with phosphate leads to the displacement of adsorbed DNA from the iron oxide NP causing an increase in the fluorescence. Even though this is a novel way of utilizing the interaction between DNA and analyte, it requires excess salt to avoid strong charge repulsion between DNA and iron oxide NPs and is also affected by the interference caused due to the higher amount of phosphate ion (Liu and Liu 2014).

18.7.2.2 Aptamer-Based Biosensors

Aptamers are artificial ssRNA or ssDNA nucleotides that have been reorganized *in vitro* to selectively bind with the target molecules (Kaur et al. 2015). The biosensors employing aptamers for the selective molecular recognition of target analytes are referred to as aptamer-based biosensors or aptasensors. Based on the detection system, aptamer-based biosensors can be classified into the wide categories of fluorescence-based, colorimetric, and electrochemical aptasensor, though some other types are also present. The unique recognition capacity of aptamers has been exploited by many scientists to construct biosensors for arsenic detection. It also offers the advantages of quick response, easy fabrication, and high sensitivity (Zhang et al. 2018). Three consecutive aptamer-based arsenic biosensors were developed by Wu et al. (2012b) based on the affinity of aptamer Ars-3 [constructed employing in vitro selection of arsenic-binding aptamers using systematic evolution of ligands by exponential enrichment (SELEX) techniques] toward the arsenic leading to the aggregation of gold NPs (AuNPs), which is induced by a cationic polymer Table 18.1.

The aggregation of AuNPs is associated with the alteration in the absorption or resonance scattering of these NPs. The first biosensor development includes a cationic polymer poly-diallyl dimethylammonium (PDDA) which is soluble in water, salt-induced aggregation of AuNPs, and hybridization of Ars-3 by means of electrostatic interaction. In the presence of As (III), the formation of aptamer-As (III) complex takes place causing PDDA induced aggregation of AuNPs which leads to the wine red to blue color change. This color

Table 18.1 Different analytical and biosensor-based methods used for arsenic detection, their detection limit, and disadvantages.

Analytical methods	Detection limit ($\mu g\ l^{-1}$)	Disadvantages
Colorimetric	1–20	The error-prone result, sample preparation is tedious, interference due to the ions of Pb, Ni, S, and P
Electrochemical	0.0008–0.0375	Interference of Cu, expensive and require trained persons to operate
FRET-based	0.45–10	Construction of suitable donor–acceptor pair is difficult, detection is sensitive to the fluctuations in the local environment, low signal-to-noise ratio
SPR-based	0.075–10	
Spectrometry-based		
AAS-based	0.001–30	String interference due to other ions, require matrix modifiers, which are costly
AFS-based		
ICP-based	0.0004–8	The analysis is expensive, required trained persons, interference from ArCl
Biosensors		
Whole cell-based biosensors		
GFP-based biosensors	1–615	Autofluorescence, low detection limit, lag-time for stable fluorescence
Bioluminescence/luciferase-based biosensor	0.0007–4000	Low permeability, the requirement of ATP and O_2
β-galactosidase/*lacZ*-based biosensors	0.2–100	Dependent on the substrate and less penetrability
Cell-free/Biomolecules-based biosensor		
DNA-based biosensors	0.05–77 000	Poor detection limit, non-specificity
Aptamer-based biosensors	0.2–6	
Protein-based biosensors	0.015–692	Interference due to other metals, low storage stability, non-specificity

change can be detected calorimetrically. While in the absence of As (III), the hybridization of PDDA with Ars-3 takes place making it unavailable to induce AuNPs aggregation and thus no change in the solution's color takes place (Wu et al. 2012b). The construction of the second biosensor involved the use of cetyltrimethylammonium bromide (CTAB) which is a cationic surfactant to obtain better sensitivity utilizing the same strategy applied in the first biosensor development (Wu et al. 2012a). The third biosensor was constructed using NP assembly of Ars-3-crystal violet (CV) as the biorecognition element. In the presence of As (III), a change in the size of NPs is induced causing a shift in the intensity of Rayleigh scattering spectra (RSS). The interaction between As (III) and Ars3-CV complex involves its three hydroxyl (—OH) groups forming inter- or intramolecular hydrogen bonding that leads to a change in average particle size. Other aptamer-based arsenic biosensors are based on the SP carbon electrode, Au-Ag core-shell NPs, flower-like nanospheres of MoS2 coated with carboxylic polypyrene, and molecular switch (triple helix)-based cascade target recycling amplification strategy assisted by exonuclease III. A label-free impedimetric aptasensor employs an arsenic-specific ArsS aptamer immobilized on a gold electrode. Another label-free approach was applied to check the electrochemical capability of the aptamer-based biosensors where the resistance across the electrode caused due to charge-transfer was studied with the help of a redox probe. Under the increasing concentration of As (III), a decrease in the resistance was observed due to the interaction of As (III) with the ArsS aptamer. Similar behavior of aptamer was observed by Cui et al. (2016) when the DPV technique was applied for the detection.

18.7.2.3 Protein-Based Biosensors

Protein-based arsenic biosensors include all the biosensors containing enzymes, peptides, or metallothionines as biosensing elements. The fabrication of these types of biosensors is based on the strong interaction of As (III) with sulfur thiolates in the proteins because of its thiophilic nature. The presence of arsenic makes the sulfur of cysteine amino acids in protein susceptible to oxidation. The interaction of As (III) with the sulfhydryl group of proteins follows four mechanisms including inhibition, coordination, oxidative coupling, and oxidation (Kaur et al. 2015). Most of the protein-based biosensors utilizing enzymes as the sensory element are based on the inhibition mechanism. For example, the acetylcholinesterase-based biosensor is an amperometric biosensor that in the presence of As (III) exhibits the inhibition of acetylcholinesterase activity immobilized on graphite electrode (Stoytcheva et al. 1998). As the concentration of As (III) increases, there is a decrease in the level of thiocholine oxidation current corresponding to the As (III) concentration. The drawbacks of this sensor include its non-specificity and interference due to heavy metals such as Cd^{2+}, Cu^{2+}, and Fe^{3+}. Several other biosensors have also been constructed using enzymes and small peptides as sensing elements such as acid phosphatases (Cosnier et al. 2006; Sanllorente-Mendez et al. 2012), polyphenol oxidases (Cosnier et al. 2006), arsenite oxidase (Male et al. 2007), cytochrome C (Fuku et al. 2012), and L-cysteine (Sarkar et al. 2010)-based biosensors. All of these biosensors have limitations such as non-specificity, interference from other components, and requirement of pretreatment. Apart from these, fluorescently labeled phytochelatin-like peptide sequences have been used as sensing elements for selective binding and differentiation of As (III) and As (V) species in the wastewater sample. The binding of As (III) or As (V) with the sensing element carrying the peptide causes fluorescence quenching of the attached naphthyl group (Parker et al. 2005).

Metallothioneins (MTs) have been widely studied for arsenic biosensor development. MT is a low molecular weight (approximately 6500 Da) protein comprising almost 30% thiol-containing cysteine residue. It was characterized by Kagi and Valle in 1960 as a protein with higher cysteine and heavy metal content and thus named metallothionein. MT stores and transports essential metals such as Cu and Zn and protects the cell against reactive oxygen species. The strong metal-binding property, simplicity, and low cost of MTs have implicated their role in biosensor development (Smith and Nordberg 2015). To construct a biosensor using MTs, it has been adsorbed on paper discs, which are placed over the SP carbon electrodes where the binding of MTs with As (III) has been detected electrochemically. However, still more studies are required to develop more sensitive and efficient MT-based biosensors (Irvine et al. 2017). QDs having CdSe/ZnS core-shell and capping of cysteine-rich ligand GSH have been employed in the construction of As-biosensors to measure As (III) in surface water. The binding of As (III) to MTs was detected by taking fluorescence emission spectra of QDs, which can also be recorded by mobile phone cameras (De Villiers et al. 2015). The inherent As (III) binding property of engineered flavin mononucleotide (FMN) chromophore-based fluorescent protein sensing light-oxygen-voltage domain (iLOV Protein) has also been used to construct biosensor for As (III) detection. The mechanisms of sensing involve the quenching of fluorescence signal generated by iLOV that exhibits linear relation with the As (III) concentration. Substitution of the larger amino acid residue of protein (Asn) to the smaller (Ser) around the chromophore created additional space for the effective binding of As (II) to the iLOV protein increasing its sensitivity (Ravikumar et al. 2017). Recently, the phage display technique has been used to screen the As (III)-binding peptides. Based on the screening, a peptide sequence having T–Q–S–Y–K–H–G residues was selected that exhibited very high specificity for As (III). It also had the highest frequency of occurrence and best selectivity. The binding of As (III) with the designed peptide disallowed the aggregation of AuNps due to the unavailability of peptide-inducing aggregation. A LOD of $4\,\mu g\,l^{-1}$ was achieved using this sensing method. In comparison to the complicated interaction occurring in aptamer-based AuNPs aggregation, this approach provides an easy operation. All the reports on protein-based arsenic biosensors show the potential application of arsenic-binding proteins in the development of As-biosensors, which can be employed in different arsenic-contaminated sites (Yang et al. 2018)

18.8 Conclusion

Various analytical and biosensor-based techniques have been developed to detect arsenic in diverse environmental samples. Several extraction/pre-concentration methods have been applied for the treatment of different environmental samples in order to extract As from it. This chapter provides an insight into various arsenic detection methods, their limitations, recent advancement in the previously existing technique, and recently developed arsenic detection methods. It also describes different types of biosensors for arsenic detection and their functional mechanism. Based on the information provided in this chapter, future possibilities to enhance the sensitivity and selectivity of different As detection methods, including biosensors-based methods, can be explored.

References

Ali, W., Rasool, A., Junaid, M., and Zhang, H. (2019). A comprehensive review on current status, mechanism, and possible sources of arsenic contamination in groundwater: a global perspective with prominence of Pakistan scenario. *Environ. Geochem. Health* 41: 737–760.

Al-Rekabi, S.H., Mustapha Kamil, Y., Abu Bakar, M.H. et al. (2019). Hydrous ferric oxide-magnetite-reduced graphene oxide nanocomposite for optical detection of arsenic using surface plasmon resonance. *Opt. Laser Technol.* 111: 417–423.

Anawar, H.M. (2012). Arsenic speciation in environmental samples by hydride generation and electrothermal atomic absorption spectrometry. *Talanta* 88: 30–42.

Arora, K., Prabhakar, N., Chand, S., and Malhotra, B. (2007). Immobilization of single stranded DNA probe onto polypyrrole-polyvinyl sulfonate for application to DNA hybridization biosensor. *Sens. Actuators B Chem.* 126: 655–663.

Babkina, S., Ulakhovich, N., and Zyavkina, Y.I. (2004). Amperometric DNA biosensor for the determination of auto-antibodies using DNA interaction with Pt (II) complex. *Anal. Chim. Acta* 502: 23–30.

Bakhrat, A., Eltzov, E., Finkelstein, Y. et al. (2011). UV and arsenate toxicity: a specific and sensitive yeast bioluminescence assay. *Cell Biol. Toxicol.* 27: 227–236.

Butcher, D.J. (2005). Atomic fluorescence spectrometry. In: *Encyclopedia of Analytical Science*, 2e (ed. P. Worsfold, A. Townshend and C. Poole), 221–228. Oxford: Elsevier https://doi.org/10.1016/B0-12-369397-7/00034-0.

Choi, Y., Kwak, H., and Hong, S. (2014). Quantification of arsenic(iii) in aqueous media using a novel hybrid platform comprised of radially porous silica particles and a gold thin film. *Anal. Methods* 6: 7054–7061.

Colon, M., Hidalgo, M., and Iglesias, M. (2011). Arsenic determination by ICP-QMS with octopole collision/reaction cell. Overcome of matrix effects under vented and pressurized cell conditions. *Talanta* 85: 1941–1947.

Cosnier, S., Mousty, C., Cui, X. et al. (2006). Specific determination of As (V) by an acid phosphatase – polyphenol oxidase biosensor. *Anal. Chem.* 78: 4985–4989.

Cui, L., Wu, J., and Ju, H. (2016). Label-free signal-on aptasensor for sensitive electrochemical detection of arsenite. *Biosens. Bioelectron.* 79: 861–865.

De Villiers, C.A., Lapsley, M.C., and Hall, E.A.H. (2015). A step towards mobile arsenic measurement for surface waters. *Analyst* 140: 2644–2655.

Diesel, E., Schreiber, M., and van der Meer, J.R. (2009). Development of bacteria-based bioassays for arsenic detection in natural waters. *Anal. Bioanal. Chem.* 394: 687–693.

Elcin, E. and Öktem, H.A. (2019). Whole-cell fluorescent bacterial bioreporter for arsenic detection in water. *Int. J. Environ. Sci. Technol.* 16: 5489–5500.

Förster, T.H. (1965). *Modern Quantum Chemistry*, Part III. NewYork: Academic Press.

Forzani, E.S., Foley, K., Westerhoff, P., and Tao, N. (2007). Detection of arsenic in groundwater using a surface plasmon resonance sensor. *Sens. Actuators B Chem.* 123: 82–88.

Fuku, X., Iftikar, F., Hess, E. et al. (2012). Cytochrome c biosensor for determination of trace levels of cyanide and arsenic compounds. *Anal. Chim. Acta* 730: 49–59.

Gudlavalleti, R.H., Bose, S.C., Verma, S.K. et al. (2017). A novel fluorometric bio-sensing-based arsenic detection system for groundwater. *IEEE Sens. J.* 17: 5391–5398.

Hazarika, S. and Rajkonwar, C. (2019). Laser induced fluorescence resonance energy transfer for detection of arsenic in water. *Laser Phys.* 29: 015601.

Helaluddin, A.B.M., Khalid, R.S., Alaama, M., and Abbas, S.A. (2016). Main analytical techniques used for elemental analysis in various matrices. *Trop. J. Pharm. Res.* 15: 427–434.

Hu, Q., Li, L., Wang, Y. et al. (2010). Construction of WCB-11: a novel phiYFP arsenic-resistant whole-cell biosensor. *J. Environ. Sci.* 22: 1469–1474.

Huang, C.W., Wei, C.C., and Liao, V.H. (2015). A low cost color-based bacterial biosensor for measuring arsenic in groundwater. *Chemosphere* 141: 44–49.

Hung, D.Q., Nekrassova, O., and Compton, R.G. (2004). Analytical methods for inorganic arsenic in water: a review. *Talanta* 64: 269–277.

Irvine, G.W., Tan, S.N., and Stillman, M.J. (2017). A simple metallothionein-based biosensor for enhanced detection of arsenic and mercury. *Biosensors* 7: 14.

Johnson, D.L. and Pilson, M.E. (1972). Spectrophotometric determination of arsenite, arsenate, and phosphate in natural waters. *Anal. Chim. Acta* 58: 289–299.

Joshi, N., Wang, X., Montgomery, L. et al. (2009). Novel approaches to biosensors for detection of arsenic in drinking water. *Desalination* 248: 517–523.

Kaur, H., Kumar, R., Babu, J.N., and Mittal, S. (2015). Advances in arsenic biosensor development – a comprehensive review. *Biosens. Bioelectron.* 63: 533–545.

Lace, A., Ryan, D., Bowkett, M., and Cleary, J. (2019). Arsenic monitoring in water by colorimetry using an optimized leucomalachite green method. *Molecules* 24: 339.

Li, S., Zhang, C., Wang, S. et al. (2018). Electrochemical microfluidics techniques for heavy metal ion detection. *Analyst* 143: 4230–4246.

Liu, B. and Liu, J. (2014). DNA adsorption by magnetic iron oxide nanoparticles and its application for arsenate detection. *Chem. Commun.* 50: 8568–8570.

Liu, Z.-G. and Huang, X.-J. (2014). Voltammetric determination of inorganic arsenic. *TrAC* 60: 25–35.

Male, K.B., Hrapovic, S., Santini, J.M., and Luong, J.H. (2007). Biosensor for arsenite using arsenite oxidase and multiwalled carbon nanotube modified electrodes. *Anal. Chem.* 79: 7831–7837.

Merulla, D., Buffi, N., Beggah, S. et al. (2013). Bioreporters and biosensors for arsenic detection. Biotechnological solutions for a world-wide pollution problem. *Curr. Opin. Biotechnol.* 24: 534–541.

Musil, S., Matousek, T., Currier, J.M. et al. (2014). Speciation analysis of arsenic by selective hydride generation-cryotrapping-atomic fluorescence spectrometry with flame-in-gas-shield atomizer: achieving extremely low detection limits with inexpensive instrumentation. *Anal. Chem.* 86: 10422–10428.

Nath, P., Priyadarshni, N., and Chanda, N. (2017). Europium-coordinated gold nanoparticles on paper for the colorimetric detection of arsenic(III, V) in aqueous solution. *ACS Appl. Nano Mater.* 1: 73–81.

Niedzielski, P. and Siepak, M. (2003). Analytical methods for determining arsenic, antimony and selenium in environmental samples. *Pol. J. Environ. Stud.* 12: 653–667.

Niedzielski, P., Siepak, M., Siepak, J., and Przybylek, J. (2002). Determination of different forms of arsenic antimony and selenium in water samples using hydride generation. *Pol. J. Environ. Stud.* 11: 219–224.

Oliveira, S., Corduneanu, O., and Oliveira-Brett, A. (2008). In situ evaluation of heavy metal – DNA interactions using an electrochemical DNA biosensor. *Bioelectrochemistry* 72: 53–58.

Oti Wilberforce, J.O. (2016). Review of principles and application of AAS, PIXE and XRF and their usefulness in environmental analysis of heavy metals. *IOSR J. Appl. Chem.* 9: 15–17. e-ISSN.2278-5736.

Ozsoz, M., Erdem, A., Kara, P. et al. (2003). Electrochemical biosensor for the detection of interaction between arsenic trioxide and DNA based on guanine signal. *Electroanalysis* 15: 613–619.

Park, M., Tsai, S.L., and Chen, W. (2013). Microbial biosensors: engineered microorganisms as the sensing machinery. *Sensors* 13: 5777–5795.

Parker, K.J., Kumar, S., Pearce, D.A., and Sutherland, A.J. (2005). Design, synthesis and evaluation of a fluorescent peptidyl sensor for the selective recognition of arsenite. *Tetrahedron Lett.* 46: 7043–7045.

Priyadarshni, N., Nath, P., and Nagahanumaiah, C.N. (2018). DMSA-functionalized gold nanorod on paper for colorimetric detection and estimation of arsenic (III and V) contamination in groundwater. *ACS Sustain. Chem. Eng.* 6: 6264–6272.

Ravikumar, Y., Nadarajan, S.P., Lee, C.S., and Yun, H. (2017). Engineering an FMN-based iLOV protein for the detection of arsenic ions. *Anal. Biochem.* 525: 38–43.

Roberto, F.F., Barnes, J.M., and Bruhn, D.F. (2002). Evaluation of a GFP reporter gene construct for environmental arsenic detection. *Talanta* 58: 181–188.

Saha, J., Roy, A.D., Dey, D. et al. (2017). Development of arsenic(v) sensor based on fluorescence resonance energy transfer. *Sens. Actuators B Chem.* 241: 1014–1023.

Sankararamakrishnan, N. and Mishra, S. (2018). A comprehensive review on various analytical methods for the determination of inorganic and organic arsenic in environmental samples. *Environ. Contam.* 21–41.

Sanllorente-Mendez, S., Dominguez-Renedo, O., and Arcos-Martinez, M.J. (2012). Development of acid phosphatase based amperometric biosensors for the inhibitive determination of As(V). *Talanta* 93: 301–306.

Sarkar, P., Banerjee, S., Bhattacharyay, D., and Turner, A.P. (2010). Electrochemical sensing systems for arsenate estimation by oxidation of l-cysteine. *Ecotoxicol. Environ. Saf.* 73: 1495–1501.

Scaria, J., Ramachandran, S., Jain, P., and Verma, S. (2009). Construction and testing of EGFP based bacterial biosensor for the detection of residual tetracyclines in milk and water. *Res. J. Microbiol.* 4: 104–111.

Scott, D.L., Ramanathan, S., Shi, W. et al. (1997). Genetically engineered bacteria: electrochemical sensing systems for antimonite and arsenite. *Anal. Chem.* 69: 16–20.

Shrivas, K., Shankar, R., and Dewangan, K. (2015). Gold nanoparticles as a localized surface plasmon resonance based chemical sensor for on-site colorimetric detection of arsenic in water samples. *Sens. Actuators B Chem.* 220: 1376–1383.

Shrivastava, A., Ghosh, D., Dash, A., and Bose, S. (2015). Arsenic contamination in soil and sediment in India: sources, effects, and remediation. *Curr. Pollut. Rep.* 1: 35–46.

Singh, M., Verma, N., Garg, A.K., and Redhu, N. (2008). Urea biosensors. *Sens. Actuators B Chem.* 134: 345–351.

Smith, D.R. and Nordberg, M. (2015). Chapter 2 – general chemistry, sampling, analytical methods, and speciation*. In: *Handbook on the Toxicology of Metals*, 4e (ed. G.F. Nordberg, B.A. Fowler and M. Nordberg), 15–44. San Diego, CA: Academic Press https://doi.org/10.1016/B978-0-444-59453-2.00002-0.

Soangra, R., Majumder, C., and Roy, P. (2015). Whole cell arsenic biosensor – a cheap technology for bioavailable arsenic (As) determination. *EJAET* 2: 52–61.

Stoytcheva, M., Sharkova, V., and Magnin, J.P. (1998). Electrochemical approach in studying the inactivation of immobilized acetylcholinesterase by arsenate (III). *Electroynalysis* 10: 994–998.

Tang, G., Wang, J., Li, Y., and Su, X. (2015). Determination of arsenic(iii) based on the fluorescence resonance energy transfer between CdTe QDs and Rhodamine 6G. *RSC Adv.* 5: 17519–17525.

Tauriainen, S., Virta, M., Chang, W., and Karp, M. (1999). Measurement of firefly luciferase reporter gene activity from cells and lysates using *Escherichia coli* arsenite and mercury sensors. *Anal. Biochem.* 272: 191–198.

Tencaliec, A.M., Laschi, S., Magearu, V., and Mascini, M. (2006). A comparison study between a disposable electrochemical DNA biosensor and a Vibrio fischeri-based luminescent sensor for the detection of toxicants in water samples. *Talanta* 69: 365–369.

Terlecka, E. (2005). Arsenic speciation analysis in water samples: a review of the hyphenated techniques. *Environ. Monit. Assess.* 107: 259–284.

Truffer, F. et al. (2014). Compact portable biosensor for arsenic detection in aqueous samples with *Escherichia coli* bioreporter cells. *Rev. Sci. Instrum.* 85: 015120.

Turdean, G.L. (2011). Design and development of biosensors for the detection of heavy metal toxicity. *Int. J. Electrochem.* 2011: 1–15.

Upadhyay, L.S.B., Kumar, N., and Chauhan, S. (2018). Minireview: whole-cell, nucleotide, and enzyme inhibition-based biosensors for the determination of arsenic. *Anal. Lett.* 51: 1265–1279.

Wachter, R.M. and Remington, S.J. (1999). Sensitivity of the yellow variant of green fluorescent protein to halides and nitrate. *Curr. Biol.* 9: R628–R629.

Wackwitz, A., Harms, H., Chatzinotas, A. et al. (2008). Internal arsenite bioassay calibration using multiple bioreporter cell lines. *Microb. Biotechnol.* 1: 149–157.

Wicke, N., Radford, D.S., and French, C.E. (2018). A simple chromogenic whole-cell arsenic biosensor based on *Bacillus subtilis*. *bioRxiv* 395178.

Wong, E.L., Chow, E., and Gooding, J.J. (2007). The electrochemical detection of cadmium using surface-immobilized DNA. *Electrochem. Commun.* 9: 845–849.

Wu, H. et al. (2011). Simultaneous speciation of inorganic arsenic and antimony in water samples by hydride generation-double channel atomic fluorescence spectrometry with on-line solid-phase extraction using single-walled carbon nanotubes micro-column. *Spectrochim. Acta Part B At. Spectrosc.* 66: 74–80.

Wu, Y., Liu, L., Zhan, S. et al. (2012a). Ultrasensitive aptamer biosensor for arsenic (III) detection in aqueous solution based on surfactant-induced aggregation of gold nanoparticles. *Analyst* 137: 4171–4178.

Wu, Y., Zhan, S., Wang, F. et al. (2012b). Cationic polymers and aptamers mediated aggregation of gold nanoparticles for the colorimetric detection of arsenic (III) in aqueous solution. *Chem. Comm.* 48: 4459–4461.

Yang, T., Zhang, X.X., Yang, J.Y. et al. (2018). Screening arsenic(III)-binding peptide for colorimetric detection of arsenic(III) based on the peptide induced aggregation of gold nanoparticles. *Talanta* 177: 212–216.

Yogarajah, N. and Tsai, S.S.H. (2015). Detection of trace arsenic in drinking water: challenges and opportunities for microfluidics. *Environ. Sci. Water Res. Technol.* 1: 426–447.

Yoshida, K., Inoue, K., Takahashi, Y. et al. (2008). Novel carotenoid-based biosensor for simple visual detection of arsenite: characterization and preliminary evaluation for environmental application. *Appl. Environ.* 74: 6730–6738.

Zhang, L., Guo, W., and Lu, Y. (2020). Advances in cell-free biosensors: principle, mechanism, and applications. *J. Biotechnol.* 15: 2000187.

Zhang, N., Fu, N., Fang, Z. et al. (2011). Simultaneous multi-channel hydride generation atomic fluorescence spectrometry determination of arsenic, bismuth, tellurium and selenium in tea leaves. *Food Chem.* 124: 1185–1188.

Zhang, W., Liu, Q.X., Guo, Z.H., and Lin, J.S. (2018). Practical application of aptamer-based biosensors in detection of low molecular weight pollutants in water sources. *Molecules* 23: 344.

19

An Overview on Emerging and Innovative Technologies for Regulating Arsenic Toxicity in Plants

Arun Kumar, Pradeep Kumar Yadav, and Anita Singh

Center of Advanced Study in Botany, Institute of Science, Banaras Hindu University, Varanasi, Uttar Pradesh, India

19.1 Introduction

Arsenic (As), ranked as the twentieth most abundant element on the earth, is a metalloid that occurs in various redox-sensitive chemical forms and is categorized as a class-1 carcinogen by the US Environmental Protection Agency (Li et al. 2017). The compounds of As can be grouped into three forms: organic As compounds, inorganic As compounds, and arsine gas. The most common oxidation states of As are arsenate (III), and arsenite (V). The trivalent inorganic As compounds include arsenic trioxide, sodium arsenite, and As trichloride while pentavalent forms include arsenic acid, arsenic pentoxide, lead arsenate, calcium arsenate, etc. Organic forms of As are methylarsonic acid, arsanilic acid, dimethylarsinic acid and arsenobetaine. As is distributed ubiquitously in air, soil, and water. The principal natural sources of As are volcanic activities, weathering of rocks, and discharge from hot springs. Anthropogenic activities such as mining, smelting, coal combustion, use of wood preservatives, and use of As-composed pesticides and fertilizers also substantially contribute to the elevation of As level in the environment (Adriano 1986; Marin et al. 1992).

Arsenic is not an essential element for most of the living organisms. The prolonged usage of wastewater contaminated with As for irrigation purposes causes increased levels of As in the soil and further leads to accretion in the plants (Roychowdhury et al. 2005). The excess contamination of soil with As imposes stress on the plant which results in the inhibition of growth, altered physiological, and metabolic activities, and even death (Stoeva and Bineva 2003). Food crops and water contaminated with As are the chief sources of As ingestion in human beings. It causes many health-associated risks to humans including skin lesions, cancer, cardiovascular, and kidney-related problems (Ratnaike 2003). Apart from posing a serious health challenge, As toxicity has also raised a threat to food security.

Therefore, it is of utmost need to build efficient strategies to minimize the dreadful influences of As on plants and there of to consumers. The first and foremost step is the physicochemical treatment of As-contaminated waters to prevent furthermore contamination of the soil with the heavy metal. However, this technique is only effective against high levels

Arsenic in Plants: Uptake, Consequences and Remediation Techniques, First Edition.
Edited by Prabhat Kumar Srivastava, Rachana Singh, Parul Parihar, and Sheo Mohan Prasad.

of As in wastewater besides being highly costly (Sarkar and Paul 2016). The nanomaterials and metalorganic frameworks (MOFs) are emerging adsorbent materials to reduce the level of As (Zhao 2020). The alternative cost-effective technique involves the use of living organisms, a strategy named bioremediation wherein the tolerant plant or microbes accumulate the As present in the soil, and water (Hayat et al. 2017). However, this method is effectively very slow and requires proper disposal of the As-accumulated plant. Moreover, high concentrations of As inhibit the growth of the accumulator plant. Nitric oxide (NO), the gaseous signaling molecule, is known to serve different purposes in plants along with alleviating abiotic stress effects (Arasimowicz et al. 2007; Neill et al. 2008). NO-mediated stress mitigation is postulated to be because of its free radical nature that scavenges reactive oxygen species (ROS) (Lamattina et al. 2003) and also its property to serve as an antioxidant inducer (Groß et al. 2013) by posttranslational modifications (Grennan, 2007; Tanou et al. 2009). Superficial administration of NO in the form of sodium nitroprusside is reported to provide tolerance against several heavy metals (Zhang et al. 2011a; Singh et al. 2016). The other growth regulators are also having an essential role in several physiological processes to mitigate metal toxicity (Metwally et al. 2003; Zhou et al. 2009).

With the above context, this chapter is also an overview of emerging and innovative technologies that are specifically related to the regulation of As (As) toxicity in plants. It will provide in-depth knowledge about the mechanism behind each technology and its pros and cons so that it will be easier for the decision-maker to select the most sustainable technology for regulating As toxicity.

19.2 Uptake of Arsenic

Arsenic does not have any essential role in the growth of plants and, hence, plants do not have any particular route assigned for its uptake and translocation. The entry of As into the plant system occurs via the transport system present for the entry of essential nutrients (Figure 19.1). Too much of As interferes in the uptake and distribution of the essential nutrient elements causing their deficiency in the plant system (Mahimairaja et al. 1994). The uptake of As in the plant system is primarily performed by the root absorption system. As occurs in three major forms in the soil for the plant system to absorb, namely, arsenite [As (III)], arsenate [As (V)], and methylated As (monomethyl As acid [MMA], and dimethyl As acid [DMA]). The transformation of arsenate to arsenite and further to MMA and DMA is performed by the soil microorganisms. As (V) is the most dominant phytoavailable form of As occurring in aerobic soils (Takahashi et al. 2004). Its uptake is known to occur by the phosphate uptake system owing to its analogous nature to that of phosphate (Zhao et al. 2010) and the phosphate transporter (PHT) proteins load them in the xylem vessels for their transportation (Macnair Mehargand 1990; Zhao et al. 2010). The uptake of As (V) is independent of the level of Pi in the soil. This was confirmed by Shin et al. 2004 in *Arabidopsis thaliana* where two high-affinity isoforms of Pht1 (AtPht1;1 and AtPht1;4) mediated uptake of both As (V) and Pi from soils having both low and high content of phosphorus (Shin et al. 2004). The role of PHT in As uptake is much supported when the loss of function of AtPht1;5 reduced As (V) toxicity in the plant. Similarly, the overexpression of AtPht1;7 resulted in increased accumulation of As (V) in the reproductive tissues of

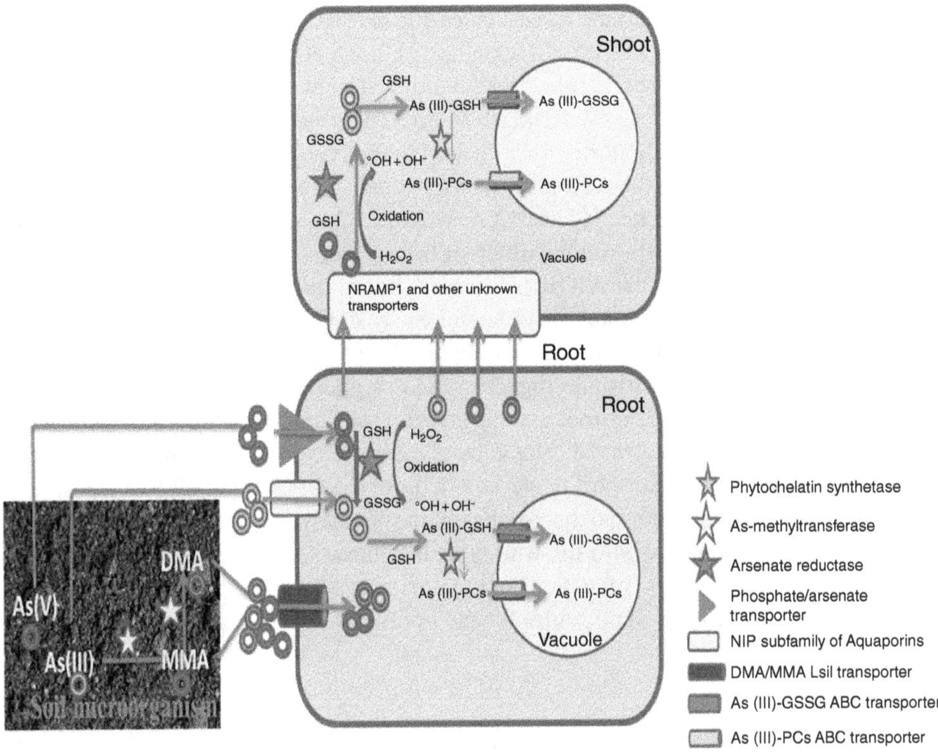

Figure 19.1 Schematic representation of uptake of different forms of As in a plant's parts.
Source: Modified from Abbas et al. 2018, MDPI.

Arabidopsis (LeBlanc et al. 2013). The high-affinity Pht1 transporters were also seen to aid in the uptake of As (V) in *Oryza sativa* L. OsPht1;8 has a high affinity for both Pi and As (V). It is expressed in the root system as well as the shoot system. The As (V) uptake and translocation get increased in rice when OsPht1;8 is overexpressed (Jia et al. 2011; Wu et al. 2011). Some authors (Castrillo et al. 2013; Wang et al. 2014) have reported the involvement of several WRKY-associated transcription factors such as WRKY6 and WRKY45 in the modulation of As (V) uptake by regulating the expression of AtPht1;1.

However, in flooded soil, where anerobic condition is attained, the Pi transport pathway plays a limited role for the As uptake because majorly As (III) is present in such soils (Xu et al. 2008). In such cases, transport of As (III) is facilitated by the plant aquaporins (Mukhopadhyay et al. 2014; Li et al. 2014). The As (III) is a neutral molecule having a diameter of 411 pm, whose uptake is brought up by a subclass of aquaporins - nodulin 26-like intrinsic proteins (NIPs). As (III) being similar to silicic acid is then introduced in the xylem by the silicon – (Si^-) uptake pathway for further translocation. Bienert et al. (2008) demonstrated that the transport of As (III) across cell membranes is carried by NIP5;1 and NIP6;1 in *A. thaliana*, NIP2;1and NIP3;2 in *O. sativa*, and by NIP5;1 and NIP6;1 in *Lotus japonicus*. Several others like AtNIP7;1, AtNIP1;1 AtNIP1;2, and AtNIP3;1 are also involved in the uptake of As (III) (Ma et al. (2008); Kamiya et al. 2009; Xu et al. 2015). found that the expression of OsNIP1;1, OsNIP2;2, and OsNIP3;1 is too weak for the uptake of As (III) by the rice roots as compared to OsNIP2;1.

Soil microorganisms convert the As pesticides into several methylated forms of As mainly into DMA and MMA (Ye et al. 2012; Huang 2014). Raab et al. (2007) and Finnegan and Chen (2012) determined the uptake rate for the different species of As in rice in the following order: As (III)>As (V)>DMA>MMA. The organic forms of As (MMA and DMA) are uptaken relatively slower as compared to the inorganic forms of As species due to their lower affinity with the As transporters (Jia et al. 2012). However, it was found that the accumulation of these organic forms of As was much higher (10–90% of total As) in the rice grains (Zhao et al. 2010). Among different transporters, NIP2;1 is responsible for the uptake of several molecules like As (III), silicic acid, DMA, and MMA (Ma and Yamaji 2006). Compared to MMA, DMA is more mobile and easily uptaken as seen in rice and castor bean plants (Carey et al. 2011). It has been reported that as compared to the uptake of As (III), the uptake of DMA is 10 times higher in the rice grains (Carey et al. 2010). Rahman et al. (2011) observed in the rice root that the glycerol transport pathway is also responsible for the uptake of DMA and MMA. The aqua-glyceroporins present in the plasma membrane responsible for the transport of glycerol helps in the DMAA and MMAA transportation. Li et al. (2009) have reported that for the uptake of MMA and DMA, OsLsi1 is involved in the rice roots. Compared with the wild type, in mutant forms of rice, where OsLsi1 is mutated, there were 80–50%, reductions in the uptake of MMA and DMA, respectively (Li et al. 2009). So, there are several routes for the uptake of As depending upon their forms and soil conditions.

19.3 Arsenic Toxicity on Plants

The uptake of Arsenic (As) shows several deleterious impacts on the morphological and physiological processes in the plants (Figure. 19.2). Among the two inorganic forms of As, arsenite is reported to be more noxious for the plant than arsenate (Hughes et al. 2011). Based upon the transporter, arsenate (AsV) hampers phosphate metabolisms like ATP synthesis and phosphorylation and processes with the replacement of phosphate group, whereas arsenite affects the structure and function of enzymes or proteins by binding to their thiol group. The interference in the normal physiological processes like photosynthesis and respiration leads to the generation of ROS and reactive nitrogen species (RNS). These are responsible for the disruption in the signaling pathways that consequently affect the growth and other functional activities of the plants (Mawia et al. 2020).

Malik et al. 2011 observed a reduction in height of *Cicer arietinum* L. when irrigated with As-contaminated water. Similarly, seedlings of *O. sativa* L. (Vromman et al. 2013) showed a remarkable decline in its root and shoot growth in the presence of As. Plant exposure to As also results in the suppression of the number of leaves, leaf area, and biomass of the plants (Nath et al. 2014). As is reported to incur several negative consequences such as impeded nutrient absorption, disturbed photosynthetic machinery, disruption of plant water status, altered enzyme function, and ion replacement from ATP, decreased level of amino acids (Malik et al. 2011; Khalid et al. 2017; Rafiq et al. 2017a,2017b).

The light-harvesting system gets affected due to the reduction in chlorophyll concentration, suppression of the major events of photosynthesis, reduction in PS II activity that overall disturbs the heat dissipation capacity of a plant (Anjum et al. 2011; Chandrakar 2016;

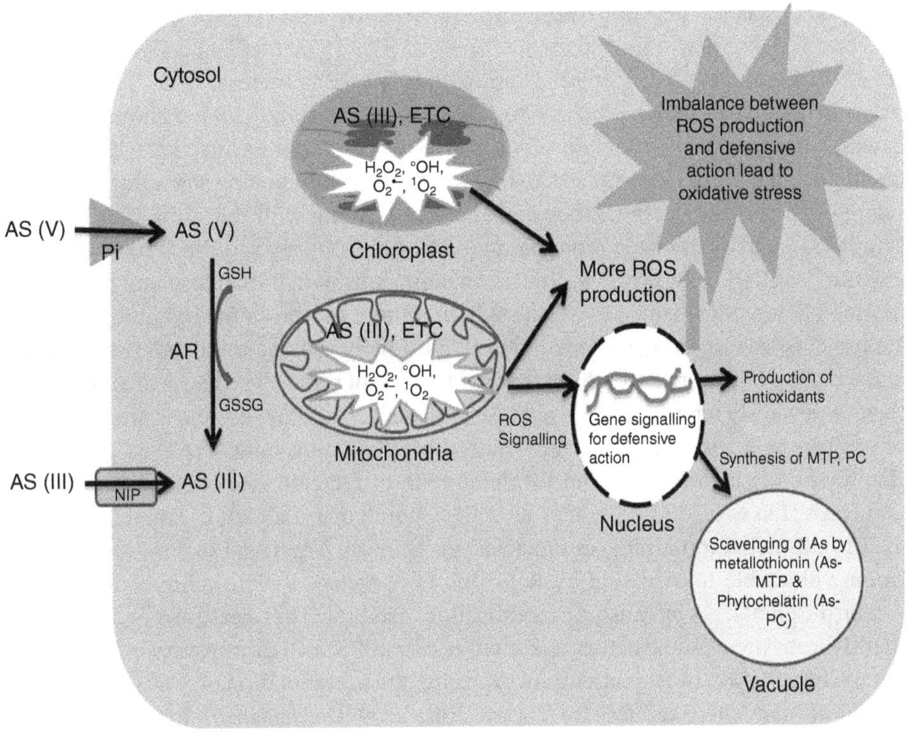

Figure 19.2 Schematic representation of As (As) toxicity in plants. *Source:* Modified from Hasanuzzaman et al. 2020.

Debona et al. 2017). Among different photosynthetic parameters, the ratio of Fv/Fm provides an overall photochemical efficiency of both the reaction centers, i.e. PSII and PSI. Under As stress, the ratio of Fv/Fm gets inhibited due to a decrease in the activity of the PSII reaction center that affects the quantum fluorescence of chlorophylls located in the antenna molecule (Farooq et al. 2021). The two cultivars of *B. napus* (ZS758 and ZD622) under As stress showed an increase in non-photochemical quenching (NPQ), due to damage in the photosynthetic system (Farooq et al. 2021).

Arsenic (V), being similar to the Pi molecule, disrupts the phosphate-dependent metabolic processes such as ATP generation, nucleic acid metabolism, phospholipid metabolism, and phosphorylation processes (Chen and Finnegan 2012; Mawia et al. 2020). Replacement of Pi with As (V) is reported to be the main route for As-induced toxicity in plants (Chen and Finnegan 2012). Arsenate is also able to disrupt the biochemical processes by reacting with Pi or Pi-ester. The damage to the cellular membranes due to disturbed nutrient and water uptake under As stress leads to reduced stomatal conductance and affects plant transpiration (DeOliveira et al. 2013). The membrane damage is reported to occur due to the increased formation of malondialdehyde (MDA), along with the leakage of key cellular electrolytes (Singh et al. 2007). The conversion of As (V) to As (III) and the inhibition of key enzymes results in the formation of ROS (Sharma 2012). The reduction of As (V) to As (III) is subsequently followed by a redox-driven methylation process. The methylated As species are known to deliver redox-active iron (Fe) species from ferritin, which later release harmful superoxide species ($O_2^{\cdot-}$) via the Haber–Weiss reaction (Flora 2011). Kaur et al. (2012) reported approximately 15–28% increase of $O_2^{\cdot-}$, and 23–82% increase of H_2O_2 level under exposure to 50 μM As concentration after 12 hours of exposure, which reached 104–216% after 48 hours of exposure. As stress affects the metabolism of starch and sugar also (Gramss 2012). It converts the non-reducing sugar (sucrose) into reducing sugar (hexose) form (Dubey and Jha 2004). Along with this, inhibition in the α and β amylase activity has been reported under As stress (Dubey and Jha 2004).

Arsenite [As (III)] is known to be 100 times more detrimental than arsenate [As (V)] in disrupting plant metabolism. As (III) has a high affinity for the sulfhydryl groups present in the proteins leading to membrane damage and finally causing cell death (Meharg and Whitaker 2002). The decrease in the content of total plant protein as a cause of As exposure has been extensively reported (Singh et al. 2006; Ismail 2012). The decline in the activity of proteases can be attributed to the ROS-caused protein damage, which leads to reduced availability of raw materials needed for the growth of radicle in As exposed seedlings of several crops (Liu et al. 2005; Singh et al. 2006; Choudhury et al. 2011; Ismail 2012; Kaur et al. 2012). Proteins containing amino acids, particularly Arg, His, Lys, Pro, Thr, and Trp are more vulnerable to oxidation by ROS (Job et al. 2005). ROS also has a detrimental impact on the plant DNA by causing a modification of its base. The release of 8-oxoguanosine (8-OHdG) from the DNA structure is known to cause G:C to T:A transversion mutations (Okamoto et al. 2008). In response to As exposure, the accumulation of 8-OHdG has been evaluated in many plant tissues (De Vizcaya-Ruiz 2009; Crohns 2010). ROS also leads to single-strand breaks in DNA (Kligerman et al. 2010). Stress-responsive genes and defensive mechanisms of plants varied with the different forms of As (As (V); As (III)). Compared to As (III), As (V) leads to more upregulation or downregulation of genes controlling the oxidative stress responses. In presence of As (III), the expressions of the glutaredoxin (*Os01g26912*) gene get increased to manage the defensive mechanism (Rai et al. 2011). Under As (V) stress, most of the sulfate transporters (OsSultr1;1) in the sensitive rice variety (IET-4786) are found to be downregulated (Tripathi et al. 2012). By using oligonucleotide microarrays Abercrombie et al. (2008) have analyzed the transcriptional response of *A. thaliana* exposed to As (V) stress. It is found that under As (V) stress, the activity of Cu/Zn superoxide dismutase (SOD) gets induced and the function of Fe-SOD gets repressed.

In total, the Omics study (transcriptome, proteome, and metabolome) of the whole plant gives an idea about the functional activities in response to any environmental changes. Tripathi et al. (2012) used "Arsenomics" approach to study the variation in the transcriptome, proteome, and metabolome of plants under As stress. In response to As (V) stress, metallothioneins, heat-shock proteins, sulfate transporters, multidrug and toxic compound extrusion (MATE) transporters are found to be upregulated in rice plants, whereas Pi transporter, Zn transporter, amino acid transporters, aquaporin gene, and peroxidases (PODs) get downregulated (Yu et al. 2012). The transcriptomes of plants under As (III) stress have been studied in detail by sequencing technology (Yu et al. 2012). They have reported that the expression of genes responsible for lipid metabolism and phytohormone signaling get increased under As (III) stress. The tolerant and sensitive varieties of *Brassica juncea* were analyzed by a comparative biochemical and transcriptional profiling. Compared to the sensitive variety, in tolerant variety, sulfate transporters and auxin and jasmonate biosynthesis pathway genes get upregulated and ethylene biosynthesis and cytokinin-responsive genes get downregulated (Srivastava et al. 2009).

Peña-Garcia et al. (2021) have performed a genome-wide association study to characterize the *Arabidopsis* F-box protein gene *AT2G16220* (As Stress-Related F-box (ASRF). The seedlings in which this gene has been mutated (*asrf* mutant) showed less tolerance toward As (As V) toxicity. The *asrf* mutant seedlings showed a reduction in their growth due to more production of ROS compared to the wild type. Further, with the transcriptomics study, it has been observed that in the *asrf* mutant seedlings, the phosphate homeostasis gets altered suggesting the role of *ASRF* in the maintenance of phosphate and cellular homeostasis under As stress (Peña-Garcia et al. 2021). These genetic regulations lead to several physiological and ultrastructural changes in the important organelles. The chloroplast structure of maize cultivars was studied by microscopy under As stress ($40\,mg\,l^{-1}$). The sensitive cultivar showed distorted photosynthetic lamellae along with bleached, randomly, and loosely arranged thylakoids (Du et al. 2017). The yield of plants also gets affected by As stress. There was an 80% reduction in the yield of the plant when grown under As-contaminated soil ($60\,mg\,kg^{-1}$). The reduction in the yield is due to a decrease in the number of the tiller and mature grains (Das et al. 2013). Zhao et al. (2009) have reported that with the reduction in the chlorophyll content of plant seedlings, the growth and biomass get decreased under As stress. Overall, As stress affects each metabolism including carbohydrate, lipid, and protein metabolism to regulate the yield and growth of plants (Chen and Finnegan 2012). So, to maintain the yield of plants under As stress, there should be the implementation of several strategies that are discussed in detail in the next section.

19.4 Remediation Strategies of Arsenic Toxicity in Plants

19.4.1 With the Application of Signaling Molecules and Phytohormones

Signaling molecules participate in the development of plants and providing protection against stress conditions. To mitigate the As stress also, the signaling molecules are applied in several plants as discussed further. Singh et al. (2017a) reported that the deleterious impact of AS (III) on the seed germination of rice got reduced upon treatment with signaling molecules, such as nitric oxide supplementation under As stress partially restored the overall growth of the plant to the control level, particularly shoot length while complete restoration of the root was reported. In wheat raised under As exposure, Fujita and Hasanuzzaman (2013) reported that with the application of nitric oxide donor the plant was able to combat with the stress in a better way. Likewise in *Vicia faba,* exogenous treatment of nitric oxide increased the seed yield, phytohormone levels, photosynthetic pigments, and mineral uptake that altogether reduced the As toxicity. The positive effect of nitric oxide in chlorophyll production has been reported by many authors (Liu et al. 2011; Zhang et al. 2011). Nitric oxide is known to modulate the iron metabolism to increase the chlorophyll level (Zhang et al. 2012). The reduction of catalase (CAT) and ascorbate peroxidase (APX) activity due to As-caused reduction of Fe levels was also reversed by the application of exogenous nitric oxide by increasing the availability of iron to plants

(Kermanian and Namdjoyan 2013). The treatment of nitric oxide significantly reduced the production of H_2O_2 and thiobarbituric acid reactive substances (TBARS) in As (III)-treated plants. Approximately, 90% of As (III) accumulated in roots of As exposed plants got reduced to 35% with the application of nitric oxide. Further, the application of nitric oxide reduced the translocation of As (III) up to 50% from the roots to the shoots (Singh et al. 2017a). This effect is brought by the activation of ABC transporters by the nitric oxide signaling molecules that sequester more of the heavy metal into the root vacuoles. Moreover, nitric acid treatment also downregulated the *OsLsi2* gene, which reduced the uptake of As in the plant.

To know more about the mechanism behind the ameliorative role of nitric oxide for As toxicity, the expression patterns of stress-responsive genes are identified (Singh et al. 2017a). The application of nitric oxide in presence of As stress modulated the expression of genes including glutathione S-transferases (GSTs), cytochrome, PODs, glutaredoxins (GRXs) P450s, and heat shock proteins (HSPs). To regulate the GST enzymes, the *Os03g0283200* (*OsGSTL1*) and *Os03g0283100* (*OsGSTL2*) genes are found to be upregulated with nitric oxide application unlike in control plants (Singh et al. 2017a). The upregulation of OsGRX9 (*Os02g0618100*) and *OsGRX4* (*Os01g368900*) genes showed the protective role of glutathione in the regulation of As stress. They help by modulating the expression of aquaporins and by maintaining the level of GSH via the S-glutathionylation process. It ultimately affects the uptake of As (III) in the rice plants (Singh et al. 2017a). The cytochrome P450 (*Os08g0105700*, *Os01g0627400*, and *Os02g0569900*) genes and HSP genes (*HSPs*, *OsHSP17*, *OsHSP18.0-CII*, and *OsHSP26* (*Os01g0136200*, *Os01g0184100*, and *Os03g0245800*)) are found to be significantly upregulated with the application of NO donor, i.e. Sodium nitroprusside (SNP) (Singh et al. 2017a). Under As toxicity, the application of nitric oxide controls the metabolism of nitrogen, amino acid, and other hormones. These processes regulate the growth and development of the plants under stress conditions (Khan et al. 2016). Singh et al. (2017a) have also reported the upregulation of nitrate reductase gene (*Os08g0468100*) under SNP treatment (NO donor). Compared to the control treatment, combined application of As (III)+SNP brought upregulation in the gene *Os04g0516600* (L-allo-threonine aldolase), and *Os03g0196600* (serine acetyltransferase) that encode for glycine and cysteine amino acids, respectively. They help in maintaining the cysteine and glycine pool to counterbalance the As toxicity. Along with this, the genes responsible for the metabolism of phytohormones like auxin (*Os01g0802700*; *OsPIN9*), cytokinin (*Os05g0311801*; *isopentenyl transferase*: *IPT7*) and gibberellins (*Os03g0618300*; *gibberellin 20 oxidase 2*) were upregulated in NO-treated plants exposed to As stress (Singh et al. 2017a). So, overall plants regulate As toxicity by activating their signaling cascade to increase the synthesis of both signaling molecules and phytohormones. Along with this, calcium molecule also acts as a signaling molecule and plays an important role in regulating As stress by decreasing the level of oxidative biomarkers as well as by declining the uptake of As in the plants (Rahman et al. 2015). The application of calcium also reduced oxidative damage by enhancing antioxidant components of the plant, mainly the superoxide dismutase (SOD) and CAT, and lowered the production of ROS. Calcium also enhanced the amount of AsA, and AsA/Dehydroascorbic acid (DHA) ratio, which is the key component required for the reduction of ROS in the AsA-GSH cycle wherein AsA acts as a substrate for the APX to detoxify H_2O_2. Calcium also increased the GSH/Oxidized glutathione (GSSG) ratio by reducing the oxidation of GSH to GSSG. Calcium supplementation also increased the monodehydroascorbate reductase (MDHAR) and dehydroascorbate reductase (DHAR) activity in As-stressed plants. The GSH, MDHAR, and DHAR altogether are important in regenerating AsA (Rahman et al. 2015). Several other types of research involving calcium application deciphered that calcium is effective in reducing metal-induced oxidative damage and restoring metal-degraded photosynthetic pigments (Tian and Zhang 2011; Ahmad et al. 2015). The overproduction of methylglyoxalase (MG) in As-exposed plants was also put under control by an increase in the activity of Gly I and Gly II brought by calcium supplementation (Rahman et al. 2015). With these, the exogenous application of H_2S, a gaseous signaling molecule, improves photosynthesis and growth and reduces ROS production, and oxidative damage induced by As (V). H_2S enhances the AsA-GSH activity to protect from the As (V) induced ROS production. H_2S application reduces the ROS generation and protects the stability of the cellular membranes, which further prevents the entry of As in the plant system. H_2S application also increases the biomass of the plants in comparison with the control plants under As (V) stress (Singh et al. 2015a). H_2S application has also been found in reducing the uptake of heavy metals and increasing the antioxidant defence activity (Li et al. 2013; Sun et al. 2013).

Along with the signaling molecules, several plant hormones also have a role in the regulation of As toxicity (Raghu et al. 2014; Soares et al. 2016). The 24-epibrassinolide (24-EBL), a brassinosteroid donor covers essential functions of the plant growth and development such as stem elongation, membrane stabilization, nutrient uptake, protein, and nucleic acid synthesis, activation of enzymes, photosynthesis, and xylem differentiation (Rady 2011). It also regulates the uptake of metal ions by regulating the radicular system of the roots. Further, EBL also promotes the synthesis of phytochelatin involved in the chelation of free metals and hence reduces the oxidative injury caused by them (Soares et al. 2016). In presence of As, replacement of phosphate group from ATP leads to inhibition of energy that inhibits the germination of radish plants. This reduction was reversed by the exogenous supplementation of EBL. Further, the damage of nucleic acids due to As toxicity in radish seedlings was also put under control (Raghu et al. 2014). As toxicity causes an increase in the levels of free proline. The addition of brassinosteroid, however, caused a further increase in proline levels (Raghu et al. 2014). The synthesized proline during stress conditions is suggested to serve as a reserve of organic nitrogen that is utilized for recovery (Trotel et al. 1989). Raghu et al. (2014) reported that the increased content of MDA due to As toxicity gets under control by the application of brassinosteroids in radish seedlings. Sharma et al. (2011) and Ramakrishna and Rao (2013a) also reported a decrease in the MDA levels by the application of brassinosteroid. Brassinosteroid application also increases the activity of antioxidative enzymes needed for scavenging ROS and restoring well growth of the radish seedlings under As stress (Raghu et al. 2014). 28-homobrassinolide, another donor of brassinosteroid, was also reported to regulate the ion uptake, reduce the accumulation of heavy metals and reduce the toxicity caused by them (Fariduddin et al. 2014). Application of 28-Homobrassinolide (HBL) reduced the accumulation of Zn in radish seedlings (Ramakrishna and Rao 2013a) and Ni in *B. juncea* (Sharma et al. 2008). The alleviation of the metal stress by 28-HBL is reported to involve the increase in the activity of glutathione and other antioxidative enzymes (Ramakrishna and Rao 2013b).

Methyl jasmonate (MJ) also has a vital role in reducing heavy metal stress in several plant species (Keramat et al. 2010; Yan et al. 2013). MJ treatment in rice plants is found to increase their dry weight, chlorophyll levels, carotenoid content, chlorophyll fluorescence at all the applied concentrations of As (Mousavi et al. 2020). The protection of the photosynthetic apparatus is because of the greater accumulation of secondary metabolites after the application of MJ (Memelink et al. 2001). Application of MJ in rice plants reduced the buildup of As in the roots and leaves and also controlled the decline in uptake of Fe in the leaves. Farooq et al. (2018) reported similar results in *B. napus* subjected to As toxicity. Farmer and Ryan (1990) showed MJ amendment increases the production of jasmonic acid in rice plants subjected to increased levels of As. The rise in jasmonic acid content is due to the conversion of MJ by an esterase enzyme into jasmonic acid. Jasmonic acid holds a reputable position in the plant defense system (Farmer and Ryan 1990). Dar et al. (2015) reported the importance of jasmonic acid in reducing oxidative stress against several heavy metals. Application of MJ reduced the formation of H_2O_2 and thereby decreased the degree of lipid peroxidation in As-treated rice plants (Mousavi et al. 2020). Yan et al. (2013) also reported similar results in *Capsicum frutescens* under Cd toxicity. The application of exogenous MJ increased the efficiency of antioxidants like SOD, POD, and APX in a concentration-dependent manner to combat the oxidative stress in As-treated rice plants (Mousavi et al. 2020). In response to As stress, the application of MJ caused an increase in the non-enzymatic antioxidant AsA. The DHA level increased in the rice plants under As exposure also got reduced from MJ application. Exogenous MJ application also increased the GSH content in As-treated rice plants. As exposure decreased the AsA/DHA and GSH/GSSG ratio while treatment with the application of MJ improved the ratios (Mousavi et al. 2020). Santino et al. (2013) explained the increase in antioxidative enzymes due to MJ supplementation occurs because of the translation or post-transcriptional modifications. MJ treatment brings an increase in GSH content by upregulating the genes entailed in its synthesis (Dar et al. 2015).

Arsenic exposure increases, the expression levels of *LS1* and *LS6* genes that belong to Nodulin 26-like intrinsic protein (NIP) III group aquaporin are involved in the uptake and transport of As. These genes get downregulated by the application of MJ in the rice plant (Mousavi et al. 2020). NRAMP1 and NRAMP5 facilitate the uptake of many metals like Fe, Mn, As, and Cd in plants (Curie et al. 2000; Ishimaru et al. 2012). As treatment increases the NRAMP1 expression which can be downregulated by MJ application (Mousavi et al. 2020). The transporter FRDL1 is involved in the xylem loading and transport of Fe to the shoot system. As treatment decreases FRDL1 expression in rice, involved in Fe translocation to the shoot system, in a dose-dependent manner. Mousavi et al. (2020) reported the importance of MJ application in reversing the decline in its expression. Application of MJ also increases the mRNA level of YSL2, involved in long-distance transport of Fe to combat its deficiency in presence of As stress (Mousavi et al. 2020). The endogenous level of abscisic acid is known to get increased under heavy metal exposure suggesting its involvement in the defense mechanism against stress (Hollenbach et al. 1997). Transcriptomic studies revealed that under As stress, the expression of abscisic acid (ABA) biosynthetic genes *OsNCED2* and *OsNCED3* in rice plants gets strongly upregulated along with the upregulation of four ABA signaling genes (Huang et al. 2012). Likewise, exogenous application of indole acetic acid (IAA) is found to be effective in improving the growth response in

B. juncea subjected to As toxicity (Srivastava et al. 2013). Co-application of selenium and IAA reduced As-driven stress in rice to a greater extent (Gupta and Pandey 2015). Yu et al. (2012) reported an increment in the level of phytohormones via the upregulation of their respective biosynthetic genes depicting their importance in alleviating heavy metal stress in plants.

19.4.2 With the Application of Nano Particles

Nanoparticles are materials of nanoscale dimensions that have a large surface area and high adsorption rate compared to their original substrate (Regis 1996; Roduner 2006). Nanotechnology deals with the manipulation, processing, and utilization of NPs synthesized in different ways. The sustainable use of nanotechnology in increasing agricultural production and protection from environmental stress has been studied by several researchers (Prasad et al. 2017; Zulfiqar et al. 2019; Usman et al. 2020). In context to this, certain nanoparticles of metal oxides are used for regulating As and other metal toxicities such as iron oxide, alumina, zinc oxide, copper oxide, and titanium oxide (Samadder and Lata 2016). The application of engineered nanomaterials in soil restricts the mobility of As. The synthesized nanomaterials can also be used for heavy metal remediation by modifying them as nano-adsorbents, nano-pesticides, and nano-fertilizers (Maity et al. 2020). The application of NPs entraps the fertilizers leading to a controlled and gradual liberation of nutrients into the soil (Khan and Rizvi 2017). Several studies have reported the effectiveness of NPs in controlling the toxicity of heavy metals in crops such as rice, wheat, soybean, and maize (Tripathi et al. 2016; Konate et al. 2017; Coman et al. 2019; Wang et al. 2019; Yan et al. 2021). The NPs are known to form a precipitate in the existence of heavy metals by virtue of their large adsorption capacity, hence lower bioavailability of the heavy metals for the plant uptake. Wang et al. (2018a) reported decreased accumulation of As in roots of rice seedlings when ZnO was applied along with As (III) (72%) or As (V) (68%). The author stated several reasons for the effect (i) ZnO NPS and As caused a negative effect on the root development, which led to a reduction in the As transporters. (ii) Production of ROS induced by the NP oxidizes the root exudates causing reduction of electrons required for As reduction. (iii) Adsorption of As on the ZnO NPs causing reduced mobility and bioavailability to the plant. However, ZnO was slightly ineffective in controlling the As (III) level in the shoot region as compared to As (V) in the rice shoot. It may be because of the difference in the redox status of the two As species and also both the species use different transporters for their uptake. Ma et al. (2020) performed an experiment on rice under As and Cd exposure and found that ZnO NP was effective in reducing the heavy metals from the roots by 39.5% for As, and 83.3% for Cd while 60.2% of As and 80.0% of Cd were reduced from the shoots compared to those without ZnO NP amendment and explained the mechanism same as Wang et al. (2018a). Moreover, the ZnO NPs affect germination, biochemical response, early growth, and As uptake (Wu et al. 2020a, 2020b), suggesting nanoparticles as effective agrochemicals to restrict As accumulation and enhance the growth of plants under metal stress conditions. Similar effects of As stress alleviation have been found with the application of iron NPs in rice. The application of nano-Fe_3O_4 and nano-zerovalent iron (nFe) effectively reduced the uptake and accumulation of As in rice seedlings. The reduced uptake of As resulted in decreased activity of antioxidants in rice seedlings (Huang et al. 2018).

The effect of CuO NPs in reducing the accumulation of As in rice has been described by Wang et al. (2019). The author reported a decline of As level by 23% in roots and by 45% in the shoots with the co-application of CuO NP with the reason being the same as with ZnO NP. The author further stated a synergistic phytotoxic effect involved in the reduced uptake of As between CuO NP and As. The report is supported by the fact that combined exposure of CuO NP and As had a considerably lower rate of water transpiration as compared with CuO treatment alone. The CuO NP application at $50\,mg\,l^{-1}$ reduced the accumulation of As by 35% compared to dehusked rice grains (Liu 2018). The adsorption of As (V) on the CuO NP surface causes reduced mobility and bioavailability of As which renders a decrease in the accretion of As in dehusked rice grains (Martinson and Reddy 2009) along with the decreased uptake of As due to the release of the root exudates (Conway et al. 2015; Anderson et al. 2017). However, the increased concentration of CuO NP applications led to high Cu content in the dehusked rice grains due to the presence of polyfunctional acids in the root exudates that increase the Cu bioavailability (Warren et al. 1995; Anderson et al. 2017). Compared to As (V), CuO NP is found to reduce the accumulation of As (III) to a greater extent. This is due to the occurrence of a redox reaction between As (III) and Cu^{2+} ions, which converts them to As (V) and Cu^{+} ions, respectively. The formation of ROS induced by CuO NP also leads to the oxidation of As (III) to As (V) (Wang et al. 2013, 2015). Zhou et al. (2015) studied the effect of MnO_2 NPs in ameliorating As toxicity in rice and reported a 65.4% decrease in As level in the rice grains. The reduction was due to the increased formation of As (V) by the enhanced oxidation of As (III), which later adsorbs to iron hydroxide present in the rhizosphere affecting its mobility and bioavailability. The formation of iron hydroxide escalates with the application of MnO_2 (Ehlert et al. 2014).

Silicon and arsenite share a common pathway for their uptake and translocation (Ma and Yamaji 2006; Limmer et al. 2018). Silicon, besides being a beneficial element to the plant, has also been found to diminish the accumulation of As (Li et al. 2009). The application of the NP form of silicon has been shown to regulate the accumulation and toxicity of As in rice. Liu et al. (2014) reported the benefits of foliar application of SiO_2 NP in controlling the toxic effects of As and its accumulation in rice grains. The foliar spray of SiO_2 reduced the As accumulation by 28% in pot experiments and 40% in field experiments. The reduction in As uptake is reported to occur due to the raised competition between Si and As (III) for the same Si transporter. The SiO_2 NP increased the activity of antioxidants and reduced the oxidative damage caused by As. In addition, the application of SiO_2 restricted the movement of As from the shoot to the rice grains by sequestering them into the cell walls in the shoot (Ma and Yamaji 2006). In recent research, Cui et al. (2020) discovered SiO_2 application rendering resistance against As toxicity by increasing cell wall thickness, pectin content, and maintaining cell integrity. Liang et al. (2003) reported that Si reduces lipid peroxidation by scavenging excess of the ROS generated under As stress. The application of SiO_2 in As-treated rice plants impeded the expression level of *Lsi1* and *Lsi2* genes, whereas the expression of *OsNIP1;1* and *OsNIP3;3* was upregulated in comparison to untreated ones.

The role of cerium oxide NPs on the uptake as well as the toxicity of As has been studied by Wang et al. (2018a). CeO_2 NPs reduced the amount of As (V) in the roots more than it did for As (III) in an unknown way. It is postulated that the root exudates that are excreted in the co-presence of As (V) and CeO_2 donate electrons, which reduce As (V) to As (III). Rossi et al. (2018) obtained similar results under co-treatment of CeO_2 NPs and Cd in

Glycine max. Titanium-based NPs have been reported by several authors in removing As from contaminated water (Deng et al. 2010; Ashraf et al. 2019; Maity et al. 2020). Three different types of TiO_2 NPs have been studied for their As stress alleviation in rice. (Wu et al. 2020a, 2020b) reported that the rutile form of TiO_2 (NRT) NP reduced As in the roots by 88.4%, whereas the anatase form of TiO_2 NP (NAT) reduced As in rice roots by 39.7% compared to the control. NRT application is reported to reduce 90% of As accumulation in the shoots, whereas NAT brought a 14% reduction compared to the control. Adsorption of As by TiO_2 is known to bring a reduction in As accumulation. Application of TiO_2 in the soil as an amendment can be a good strategy as You et al. (2018) reported that even a high dose of TiO_2 NP (2000 $mg\,kg^{-1}$) had insignificant effects on soil properties.

19.4.3 With the Application of Genetic Manipulations

Arsenic tolerance can be increased through transgenic techniques by the modification of genes regulating the uptake, sequestration, distribution, or by increasing the expression level of the existing gene, which regulates As tolerance. An increased tolerance toward As has been reported in the transgenic variety of *B. juncea* and *A. thaliana* showing overexpression of the PCS gene (Li et al. 2004; Gasic and Korban 2007). In the tobacco plant, Zanella et al. (2016) reported increased tolerance for As with the increased expression of the *Arabidopsis* phytochelatin synthase1 (*AtPCS1*) gene. The formation of Phytochelatin (PC) is related to the availability of its precursor molecule GSH (Guo et al. 2008). Hence, it is suggested to combine the enhanced expression of PCS1 along with GSH1 (glutamate-cysteine ligase 1) or GSH2 (glutathione synthetase 2) expression to maximize As tolerance. Hayashi et al. (2017) reported overexpression of *OsPCS1* exhibited lesser accretion of As in the rice grains, whereas knockdown of *OsPCS2* raised As (III) sensitivity in the plant. The As content in rice grains was restricted by the overexpression of *CdPCS1* that enhanced the activity and synthesis of PC in transgenic rice plants. Mandal (2015) reported overexpression of *ACR2* gene in transgenic tobacco plants reduced the level of As accumulation and thus increased As tolerance in the plant. Meadows (2014) suggested that manipulating HAC1 through genetic techniques could reduce the As level in the plant. HAC1 is an As (V) reductase present in root hairs and root epidermis, which ceases the translocation of As from roots to shoots. The overexpression of As (V) reductase in the roots resulted in a lesser accumulation of As in the rice grains compared to the wild plants (Shi et al. 2016; Xu et al. 2017).

Disruption of *Ls1* and *Ls2* gene activity reduces the root uptake of As (III) and its loading into the xylem (Ma et al. 2008). Similarly, Chen et al. (2017) reported that knockout of *OsNIP 3;2* reduced As accumulation in rice roots. Reduction of As level in rice has been reported by several authors by modulating the Pi transporter by which the uptake of As occurs. In a study, it was reported that the T-DNA mutants of OsPT1 resulted in a decline in transcription expression by 50% in the roots and by 70% in the shoots and led to a reduced level of As in the shoot, whereas the increased expression of *OsPT1* resulted in excessive buildup of As in rice (Kamiya et al. 2013). The tolerance capacity of the plant can also be manipulated by regulating the activity of the transcription factors (Clemens 2001). Mutation of the transcription factor OsPHF1 resulted in the retention of OsPT2 and OsPT8. OsPT8 mutation induced by ethyl methanesulfonate

(EMS) in rice lowered As (V) uptake in roots by 33–57% (Wang et al. 2016). Ye et al. (2017) reported a reduction in As (V) content (14–30%) by the mutation of OsPT4 through CRISPR/Cas-9 gene-editing technology. In rice, the transcription factor OsARM1 modulates *Lsi1*, *Lsi2*, and *Lsi6* expressions. Knockout of the gene rendered protection against the deleterious effects of As (III), whereas, in the overexpressed lines, a higher accumulation of As (III) was observed. Further, OsARM1 was also involved in the root-to-shoot translocation of As (Wang et al. 2017). Similarly, another transcription factor OsWRKY28 is known to regulate the accumulation of As (V) and Pi in rice. Knockout of this gene caused a decrease in the accretion of As (V) and Pi in the shoots (Wang et al. 2018b). The mutants showed a 16–31% reduction in As accumulation and a 28–34% reduction of Pi to the wild-type plants. *OsHAC1;1* and *OsHAC1;2* genes when overexpressed were seen to result in a reduction of As content in the plant and hence improved arsenate tolerance (Shi et al. 2016). Overexpression of OsHAC4 also declined the As level in roots (39.2–42.1%) and shoots (38.2–40%) as compared with wild varieties. Through gene knockout of OsHAC4, Xu et al. (2017) reported increased accumulation of As in shoots, reduction in the efflux of As (III), and a decline in the reduction of As (V). Verma et al. (2020a) showed that in two GRX family proteins, OsGrx_C7 and OsGrx_C2.1, overexpression enhances tolerance against As (III) and limits its accumulation in grains by the modulating NIP transporters engaged in As (III) uptake and expulsion. The overexpressed lines resulted in a 67% reduction in total As in unpolished rice, whereas in the case of polished rice, an 85% reduction in the accumulation of As was observed.

The transport of GSH and γ-EC from plastid to cytosol is regulated by the *OsCLT1* gene. GSH and PC involve in forming a complex with As (III) that gets sequestered in vacuoles. Therefore, modifying GSH and PC to increase As tolerance can be a useful strategy to minimize the harmful effects of heavy metals. Loss of function of *OsCLT1* gene subsided GSH and PC levels in the plants. OsCLT1 mutants, when treated with As (V), showed a 50% reduction in the level of As in the root while the shoot accumulated 60% higher As (V), whereas, under As (III) treatment, the shoot showed no significant difference in As accumulation as compared to wild plants while the reduction of As in the root was the same as for As (V) (Yang et al. 2016). Wang et al. (2020) identified that a cytosolic cysteine synthase gene in rice *OsOASTL-A1* can also influence As accumulation. The osoastl-a1 mutants showed 34–38% reduction in the accumulation of As in roots along with causing a decline in levels of cysteine, GSH, and PC. In *OSCLT1* mutants, the shoot showed a higher accumulation of As (36–68%) compared to the wild plants suggesting *OsOASTL-A1* mutation exhibits higher mobility of As from roots to shoots. However, no change in the content of As in grains between the osoastl-a1 mutants and wild-type plants was found. Overexpression of OsPRX38 in *A. thaliana* showed increased tolerance to As stress (Kidwai et al. 2019). Lignin deposition in the apoplastic region prevents the entry of As in the OsPRX38 mutants. The reduction of As content in the rice grains is brought by the transfer of As (III)-PC complex into the vacuoles, which is mediated by OsABCC1 for its sequestration. Song et al. (2010) reported that in OsABCC1 knockout mutants, the As level in rice grains increased and compared to the wild-type the mutants were more sensitive to As. Tang et al. (2019) reported that the As (III) translocation from roots to shoots in rice involves the role of the *OsABCC7* gene. In the knockout variants of the *OsABCC7* gene, As accumulation in the sap of xylem and in the shoot was observed to get reduced. However, the gene knockout lines showed no variation in the deposit of As in roots and in grains compared to the wild types. A gene from *Pteris vittate PvACR3;1*, which is a hyperaccumulator of As, when expressed in rice, did not allow the translocation of As from the roots to the shoots and into the grains as well, and, hence, resulted in 28–39% reduction in the accumulation of As compared to wild-type plants (Chen et al. 2019).

Extrusion of As in yeast takes place by *ACR3* (Wysocki et al. 1997). The expression of this gene in *A. thaliana* led to increased tolerance against As (Ali et al. 2012). Similarly, in plants, transporters of the NIP aquaporin family involve in extruding As (III) from the plant (Pommerrenig et al. 2015). Therefore, to increase As (III) extrusion, overexpression of gene *HAC1* along with *NIP1;1* is recommended. It has been reported that the sulfur limitation 1 (SLIM1) transcription factor imparts resistance against As. Jobe et al. (2021) exposed *slim1–2* mutants to As exposure and found severe oxidative stress in both the variants and loss of function in thiol and sulfate accumulation.

So, overall, with exposure to As, there are a lot of adjustments in the gene expression of plants (Castrillo et al. 2013; Fu et al. 2014) and transcriptional regulators such as WRKY6, WRKY45, and OsARM1 (arsenite-responsive myb1) (Castrillo et al. 2013; Wang et al. 2017) to manage As toxicity in the plants. All the strategies are summarised in Table 19.1.

19.5 Conclusion

The presence of As in the environmental components leads to several problems as it is one of the most toxic and carcinogenic metalloids. It has a negative impact on plants, animals, and human health. So, it is one of the most important issues to apply different approaches to manage such a dangerous encounter. Earlier, conventional methods have been applied but they are too expensive as well as they are not able to reduce the As toxicity significantly. Further, the application of new technologies like the use of NPs and genetic engineering work as a beacon of hope in managing this problem. The new tools like gene editing through CRISPR/Cas systems and engineered NPs are proving to be much more effective in reducing the accumulation of As in edible plants that consequently improve the food chain contamination. Along with this, further researches are needed to understand the whole mechanism and future perspectives of all techniques used for As remediation.

Acknowledgments

We are very grateful to the Head and Coordinator, Center of Advanced Study in Botany, Institute of Science, Banaras Hindu University, Varanasi for providing the necessary facilities. Banaras Hindu University is also thankfully acknowledged for providing financial assistance to Dr. A. Singh as PI under the IoE seed grant scheme (Dev. IoE Scheme No. 6031). The author, A. Kumar is also thankful to the CSIR–UGC, New Delhi (UGC-Rf. No. 974) for providing financial assistance as junior research fellowship.

Table 19.1 Effect of application of phytohormone, gene modification, signaling molecule, and nanoparticle on the accumulation and toxicity of As in different plants.

Type of amelioration	Name of the experimental plant	Concentration	Phytohormone/nanoparticle/gene modification/signaling molecule used	Effect on plant	Reference
Signaling molecule	*Oryza sativa* L.	100 μM	Nitric oxide	Increased the level of PCs and decreased As accretion in the whole plant.	Singh et al. (2016)
	Oryza sativa L.	50–100 $\mu g\ ml^{-1}$	GABA	Downregulated the expression of Lsi1 and Lsi2 genes and reduced As accretion in the seedlings.	Kumar et al. (2017)
	Oryza sativa L.	10–100 μM	Nitric oxide	Minimized ROS levels and As buildup in a plant.	Singh et al. (2017a)
	Oryza sativa L.	30 μM	Nitric oxide	Accumulation of As was reduced.	Singh et al. (2017b)
	Oryza sativa L.		Nitric oxide	In the MAPK signaling cascade, MAP kinase (MKK) acts upstream to the MAP kinases. Transcript analysis revealed the regulatory pattern of MKKs in arsenite-treated rice plants.	Rao et al. (2011)
Phytohormone	*Oryza sativa* L.		Jasmonic acid	Ja biosynthesis and signaling genes were upregulated.	Yu et al. (2012)
	Arabidopsis thaliana		Auxin	Improved tolerance against As (III) in aux1 mutants most probably through ROS signaling.	Krishnamurthy et al. (2013)
	Oryza sativa L.		Salicylic acid	Reduced oxidative stress damage, increased antioxidant activity, enhanced GSH, PC, and ASC activity.	Singh et al. (2015b)
	Oryza sativa L.	20 μM of Ne_2SeO_4, 3 μM Auxin	Selenium and auxin	Slowed chlorophyll degradation and reduced MDA formation.	Gupta and Pandey (2015)
	Oryza sativa L.	40 μM of SA	Salicylic acid	Minimized As accumulation in shoots by 27%.	Singh et al. (2017a)

	Wheat		Salicylic acid/ Brassinosteroid	An improved antioxidant system, gas exchange parameters, chlorophyll level, relative water content, membrane stability, and osmotic molecules.	Kobra et al. (2019)
	Oryza sativa L.	0.25 μM	Methyl jasmonate	Lowered As accumulation and enhanced plant tolerance.	Verma et al. (2020b)
Nano particle	*Oryza sativa* L.	$100\,mg\,l^{-1}$	CeO_2	As (V) levels reduced in the root.	Wang et al. (2018a)
	Oryza sativa L.	$0.4\,mg\,l^{-1}$	Fe_3O_4 and zero valent iron	Reduced As translocation.	Huang et al. (2018)
	Oryza sativa L.	$0.1–100\,mg\,l^{-1}$	CuO	Reduced As accumulation in rice grains.	Liu et al. (2018)
	Oryza sativa L.	$100\,mg\,l^{-1}$	CuO	As level declined in roots and shoots.	Wang et al. (2019)
	Oryza sativa L.	$100\,mg\,l^{-1}$	ZnO	As level declined in roots and shoots.	Ma et al. (2020)
	Oryza sativa L.	$10–200\,mg\,l^{-1}$	ZnO	Up to 72% reduction of total As in roots and 56% in the shoots.	Wu et al. (2020a)
Gene modification	*Oryza sativa* L.		Aquaporins (As (III) transport) Lsi1 (OsNIP2;1)	As level in roots and in straw got reduced.	Ma et al. (2008)
	Oryza sativa L.		Phosphate transporter (As (V) transport) OsPT1	Shoots showed reduced accumulation of As.	Kamiya et al. (2013)
	Oryza sativa L.		Arsenate reductase *OsHAC1;1* and *OsHAC1;2*	As (III) efflux increased; As level in the grains decreased; Increased As (V) tolerance.	Shi et al. (2016)
	Arabidopsis thaliana		Class III peroxidases (As metabolism *OsPRX38*	Enhanced plant tolerance.	Kidwai et al. (2019)
	Oryza sativa L.		Arsenite (AsIII) antiporter PvACR3;1	As level in roots increased while accumulation in de-husked grains reduced.	Chen et al. (2019)
	Oryza sativa L.		*O*-acetylserine (thiol) lyase (As metabolism) *OASTL-A1*	Increased mobility of As from the roots to the shoots.	Wang et al. (2020)

References

Abbas, G., Behzad, M., Irshad, B. et al. (2018). As uptake, toxicity, detoxification, and speciation in plants: physiological, biochemical, and molecular aspects. *Int. J. Environ. Res. Public Health* 15: 59.

Abercrombie, J.M., Halfhill, M.D., Ranjan, P. et al. (2008). Transcriptional responses of *Arabidopsis thaliana* plants to As (V) stress. *BMC Plant Biol.* 8: 87.

Adriano, D.C. (1986). Arsenic. In: *Trace Elements in the Terrestrial Environment.*, 308–321. New York: Springer.

Ahmad, P., Sarwat, M., Bhat, N.A. et al. (2015). Alleviation of cadmium toxicity in *Brassica juncea* L. (Czern. & Coss.) by calcium application involves various physiological and biochemical strategies. *PLoS One* 10 (1): e0114571.

Ali, W., Isner, J.C., Isayenkov, S.V. et al. (2012). Heterologous expression of the yeast arsenite efflux system ACR3 improves *Arabidopsis thaliana* tolerance to As stress. *New Phytol.* 194: 716–723.

Anderson, A., McLean, J., McManus, P. et al. (2017). Soil chemistry influences the phytotoxicity of metal oxide nanoparticles. *Int. J. Nanotechnol.* 14 (1–6): 15–21.

Anjum, S.A., Xie, X.Y., Wang, L.C. et al. (2011). Morphological, physiological and biochemical responses of plants to drought stress. *Afr. J. Agric. Res.* 6: 2026–2032.

Arasimowicz, M. and Floryszak-Wieczorek, J. (2007). Nitric oxide as a bioactive signalling molecule in plant stress responses. *Plant Sci.* 172: 876–887.

Ashraf, S., Siddiqa, A., Shahida, S. et al. (2019). Titanium-based nanocomposite materials for As removal from water: a review. *Heliyon* 5 (5): e01577.

Bienert, G.P., Thorsen, M., Schussler, M.D. et al. (2008). A subgroup of plant aquaporins facilitate the bidirectional diffusion of $As(OH)_3$ and $Sb(OH)_3$ across membranes. *BMC Biol.* 6: 26.

Carey, A.M., Norton, G.J., Deacon, C. et al. (2011). Phloem transport of As species from flag leaf to grain during grain filling. *New Phytol.* 192: 87–98.

Carey, A.M., Scheckel, K.G., Lombi, E. et al. (2010). Grain unloading of As species in rice. *Plant Physiol.* 152: 309–319.

Castrillo, G., Sanchez-Bermejo, E., de Lorenzo, L. et al. (2013). WRKY6 transcription factor restricts arsenate uptake and transposon activation in *Arabidopsis. The Plant Cell* 25: 2944–2957.

Chandrakar, V., Naithani, S.C., and Keshavkant, S. (2016). As-induced metabolic disturbances and their mitigation mechanisms in crop plants: a review. *Biologia* 71: 367–377.

Chen, Y., Hua, C.Y., Chen, J.X. et al. (2019). Expressing arsenite antiporter PvACR3;1 in rice (*Oryza sativa* L.) decreases inorganic arsenic content in rice grains. *Environ. Sci. Technol.* 53 (17): 10062–10069.

Chen, Y., Sun, S.K. et al. (2017). The nodulin 26-like intrinsic membrane protein OsNIP3;2 is involved in arsenite uptake by lateral roots in rice. *J. Exp. Bot.* 68 (11): 3007–3016.

Choudhury, B., Chowdhury, S., and Biswas, A.K. (2011). Regulation of growth and metabolism in rice (*Oryza sativa* L.) by As and its possible reversal by phosphate. *J. Plant Interact.* 6: 15–24.

Clemens, S. (2001). Molecular mechanisms of plant metal tolerance and homeostasis. *Planta 212*: 475–486.

Coman, V., Oprea, I., Leopold, L.F. et al. (2019). Soybean interaction with engineered nanomaterials: a literature review of recent data. *Nanomaterials* 9 (9): 1248.

Conway, J.R., Adeleye, A.S., Gardea-Torresdey, J., and Keller, A.A. (2015). Aggregation, dissolution, and transformation of copper nanoparticles in natural waters. *Environ. Sci. Technol.* 49 (5): 2749–2756.

Crohns, M. (2010). *Antioxidants, Cytokines and Markers of Oxidative Stress in Lung Cancer: Associations with Adverse Events, Response and Survival.* Finland: Tampere University Press, University of Tampere.

Cui, J., Li, Y., Jin, Q., and Li, F. (2020). Silica nanoparticles inhibit arsenic uptake into rice suspension cells via improving pectin synthesis and the mechanical force of the cell wall. *Environ. Sci. Nano.* 7 (1): 162–171.

Curie, C., Alonso, J.M., Le Jean, M. et al. (2000). Involvement of NRAMP1 from *Arabidopsis thaliana* in iron transport. *Biochem. J.* 347: 749–755.

Dar, T.A., Uddin, M., Khan, M.M.A. et al. (2015). Jasmonates counter plant stress: a review. *Environ. Exp. Bot.* 115: 49–57.

Das, I., Ghosh, K., Das, D.K., and Sanya, S.K. (2013). Assessment of arsenic toxicity in rice plants in areas of West Bengal. *Chem. Speciat. Bioavailab.* 25: 201–208.

De Oliveira, A.B., Alencar, N.L.M., and Gomes-Filho, E. (2013). Comparison between the water and salt stress effects on plant growth and development. In: *Responses of Organisms to Water Stress.* (ed. S. Akinci). Rijeka, Croatia: Intech.

De Vizcaya-Ruiz, A., Barbier, O., Ruiz-Ramos, R., and Cebrian, M.E. (2009). Biomarkers of oxidative stress and damage in human populations exposed to arsenic. *Mutat. Res.* 674: 85–92.

Debona, D., Rodrigues, F.A., and Datnoff, L.E. (2017). Silicon's role in abiotic and biotic plant stresses. *Annu. Rev. Phytopathol.* 55: 85–107.

Deng, S., Li, Z., Huang, J., and Yu, G. (2010). Preparation, characterization and application of a Ce—Ti oxide adsorbent for enhanced removal of arsenate from water. *J. Hazard. Mater.* 179 (1–3): 1014–1021.

Du, L., Xia, X., Lan, X. et al. (2017). Influence of arsenic stress on physiological, biochemical, and morphological characteristics in seedlings of two cultivars of Maize (*Zea mays* L.). *Water Air Soil Pollut.* 228: 55.

Ehlert, K., Mikutta, C., and Kretzschmar, R. (2014). Impact of birnessite on As and iron speciation during microbial reduction of As-bearing ferrihydrite. *Environ. Sci. Technol.* 48 (19): 11320–11329.

Fariduddin, Q., Yusuf, M., Ahmad, I., and Ahmad, A. (2014). Brassinosteroids and their role in response of plants to abiotic stresses. *Biol. Plant.* 58: 9–17.

Farmer, E.E. and Ryan, C.A. (1990). Interplant communication: airborne methyl jasmonate induces synthesis of proteinase inhibitors in plant leaves. *Proc. Natl Acad. Sci. USA* 87: 7713–7716.

Farooq, A.M., Hong, Z., Islam, F. et al. (2021). Comprehensive proteomic analysis of As induced toxicity reveals the mechanism of multilevel coordination of efficient defense and energy metabolism in two *Brassica napus* cultivars. *Ecotox. Environ. Safe.* 208: 111744.

Farooq, M.A., Islam, F., Yang, C. et al. (2018). Methyl jasmonate alleviates As-induced oxidative damage and modulates the ascorbate–glutathione cycle in oilseed rape roots. *Plant Growth Regul.* 84: 135–148.

Finnegan, P.M. and Chen, W. (2012). As toxicity: the effects on plant metabolism. *Front. Physiol.* 3: 182.

Flora, S.J. (2011). Arsenic-induced oxidative stress and its reversibility. *Free Radic. Biol. Med.* 51: 257–281.

Fu, S.F., Chen, P.Y., Nguyen, Q.T.T. et al. (2014). Transcriptome profiling of genes and pathways associated with arsenic toxicity and tolerance in *Arabidopsis*. *BMC Plant Biol.* 14: 94.

Gasic, K. and Korban, S.S. (2007). Transgenic Indian mustard (*Brassica juncea*) plants expressing an *Arabidopsis* phytochelatin synthase (AtPCS1) exhibit enhanced As and Cd tolerance. *Plant Mol. Biol.* 64: 361–369.

Gramss, G. (2012). Potential contributions of oxidoreductases from alfalfa plants to soil enzymology and biotechnology: a review. *J. Nat. Sci. Sustain. Technol.* 6: 169.

Grennan, A.K. (2007). Protein S-nitrosylation: potential targets and roles in signal transduction. *Plant Physiol.* 144: 1237–1239.

Groß, F., Durner, J., and Gaupels, F. (2013). Nitric oxide, antioxidants and prooxidants in plant defence responses. *Front. Plant Sci.* 4: 419.

Guo, J., Dai, X., Xu, W., and Ma, M. (2008). Overexpressing GSH1 and AsPCS1 simultaneously increases the tolerance and accumulation of cadmium and arsenic in *Arabidopsis thaliana*. *Chemosphere* 72: 1020–1026.

Hasanuzzaman, M., Bhuyan, M.H.M.B., Zulfiqar, F. et al. (2020). Reactive oxygen species and antioxidant defense in plants under abiotic stress: revisiting the crucial role of a universal defense regulator. *Antioxidants* 9: 681.

Hasanuzzaman, M. and Fujita, M. (2013). Exogenous sodium nitroprusside alleviates arsenic induced oxidative stress in wheat (*Triticum aestivum* L.) seedlings by enhancing antioxidant defense and glyoxalase system. *Ecotoxicology* 22: 584–596.

Hayashi, S., Kuramata, M., Abe, T. et al. (2017). Phytochelatin synthase OsPCS1 plays a crucial role in reducing arsenic levels in rice grains. *Plant J.* 91: 840–848.

Hayat, K., Menhas, S., Bundschuh, J., and Chaudhary, H.J. (2017). Microbial biotechnology as an emerging industrial wastewater treatment process for arsenic mitigation: a critical review. *J. Clean. Prod.* 151: 427–438.

Hollenbach, B., Schreiber, L., Hartung, W., and Dietz, K.J. (1997). Cadmium leads to stimulated expression of the lipid transfer protein genes in barley: implications for the involvement of lipid transfer proteins in wax assembly. *Planta* 203: 9–19.

Huang, J.H. (2014). Impact of microorganisms on arsenic biogeochemistry: a review. *Water Air Soil Pollut.* 225: 1848.

Huang, Q., Liu, Q., Lin, L. et al. (2018). Reduction of arsenic toxicity in two rice cultivar seedlings by different nanoparticles. *Ecotoxicol. Environ. Saf.* 159: 261–271.

Huang, T.L., Nguyen, Q.T.T., Fu, S.F. et al. (2012). Transcriptomic changes and signalling pathways induced by arsenic stress in rice roots. *Plant Mol. Biol.* 80: 587–608.

Hughes, M.F., Beck, B.D., Chen, Y. et al. (2011). Arsenic exposure and toxicology: a historical perspective. *Toxicol. Sci.* 123 (2): 305–332.

Ishimaru, Y., Takahashi, R., Bashir, K. et al. (2012). Characterizing the role of rice NRAMP5 in manganese, iron and cadmium transport. *Sci. Rep.* 2: 286.

Ismail, G.S.M. (2012). Protective role of nitric oxide against arsenic induced damages in germinating mung bean seeds. *Acta Physiol. Plant.* 34: 1303–1311.

Jha, A. and Dubey, R. (2004). Carbohydrate metabolism in growing rice seedlings under arsenic toxicity. *J. Plant Physiol.* 161: 867–872.

Jia, H.F., Ren, H.Y., Gu, M. et al. (2011). The phosphate transporter gene OsPht1;8 is involved in phosphate homeostasis in rice. *Plant Physiol.* 156: 1164–1175.

Jia, Y., Huang, H., Sun, G.X. et al. (2012). Pathways and relative contributions to arsenic volatilization from rice plants and paddy soil. *Environ. Sci. Technol.* 46: 8090–8096.

Job, C., Rajjou, L., Lovigny, Y. et al. (2005). Patterns of protein oxidation in *Arabidopsis* seeds and during germination. *Plant Physiol.* 138: 790–802.

Jobe, T.O., Yu, Q., Hauser, F. et al. (2021). The SLIM1 transcription factor is required for arsenic resistance in *Arabidopsis thaliana*. *FEBS Lett.* 595 (12): 1696–1707.

Kamiya, T., Islam, R., Duan, G. et al. (2013). Phosphate deficiency signaling pathway is a target of arsenate and phosphate transporter OsPT1 is involved in As accumulation in shoots of rice. *Soil Sci. Plant Nutr.* 59 (4): 580–590.

Kamiya, T., Tanaka, M., Mitani, N. et al. (2009). NIP1;1, an aquaporin homolog, determines the arsenite sensitivity of *Arabidopsis thaliana*. *J. Biol. Chem.* 284: 2114–2120.

Kaur, S., Singh, H.P., Batish, D.R. et al. (2012). Arsenic (As) inhibits radical emergence and elongation in phaseolus aureus by altering starch-metabolizing enzymes vis-à-vis disruption of oxidative metabolism. *Biol. Trace Elem. Res.* 146: 360–368.

Keramat, B., Kalantari, K.M., and Arvin, M.J. (2010). Effects of methyl jasmonate treatment on alleviation of cadmium damages in soybean. *J. Plant Nutr.* 33: 1016–1025.

Khalid, S., Shahid, M., Niazi, N.K. et al. (2017). Arsenic behaviour in soil-plant system: Biogeochemical reactions and chemical speciation influences. In: *Enhancing Cleanup of Environmental Pollutants*. (ed. N. Anjum, S. Gill and N. Tuteja), 97–140. Berlin, Germany: Springer.

Khan, M.I.R., Iqbal, N., Masood, A., and Mobin, M. (2016). Modulation and significance of nitrogen and sulfur metabolism in cadmium challenged plants. *Plant Growth Regul.* 78: 1–11.

Khan, M.R. and Rizvi, T.F. (2017). Application of nanofertilizer and nanopesticides for improvements in crop production and protection. In: *Nanoscience and Plant–Soil Systems*, vol. 48 (ed. M. Ghorbanpour, K. Manika and A. Varma), 405–427. Berlin, Germany: Springer.

Kidwai, M., Dhar, Y.V., Gautam, N. et al. (2019). *Oryza sativa* class III peroxidase (OsPRX38) overexpression in arabidopsis thaliana reduces arsenic accumulation due to apoplastic lignification. *J. Hazard. Mater.* 362: 383–393.

Kligerman, A.D., Malik, S.I., and Campbell, J.A. (2010). Cytogenetic insights into DNA damage and repair of lesions induced by a monomethylated trivalent arsenic. *Mutat. Res.* 695: 2–8.

Kobra, M., Mohammad, A., and Muhammad, A. (2019). Mitigation of arsenic toxicity in wheat by the exogenously applied salicylic acid, 24-Epi-Brassinolide and silicon. *J. Soil Sci. Plant Nutr.* 20: 577–588.

Konate, A., He, X., Zhang, Z. et al. (2017). Magnetic (Fe_3O_4) nanoparticles reduce heavy metals uptake and mitigate their toxicity in wheat seedling. *Sustainability* 9 (5): 790.

Krishnamurthy, A. and Rathinasabapathi, B. (2013). Auxin and its transport play a role in plant tolerance to arsenite-induced oxidative stress in *Arabidopsis thaliana*. *Plant Cell Environ.* 36: 1838–1849.

Kumar, N., Dubey, A.K., Upadhyay, A.K. et al. (2017). GABA accretion reduces Lsi-1 and Lsi-2 gene expressions and modulates physiological responses in *Oryza sativa* to provide tolerance towards arsenic. *Sci. Rep.* 7 (1): 1–11.

Lamattina, L., García-Mata, C., Graziano, M., and Pagnussat, G. (2003). Nitric oxide: the versatility of an extensive signal molecule. *Annu. Rev. Plant Biol.* 54: 109–136.

Lata, S. and Samadder, S.R. (2016). Removal of arsenic from water using nano adsorbents and challenges: a review. *J. Environ. Manag.* 166: 387–406.

LeBlanc, M.S., McKinney, E.C., Meagher, R.B., and Smith, A.P. (2013). Hijacking membrane transporters for arsenic phytoextraction. *J. Biotechnol.* 163: 1–9.

Li, G.W., Santoni, V., and Maurel, C. (2014). Plant aquaporins: roles in plant physiology. *Biochim. Biophs. Acta.* 1840: 1574–1582.

Li, R.Y., Ago, Y., Liu, W.J. et al. (2009). The rice aquaporin *Lsi1* mediates uptake of methylated arsenic species. *Plant Physiol.* 150 (4): 2071–2080.

Li, Y.J., Dhankher, O.P., Carreira, L. et al. (2004). Overexpression of phytochelatin synthase in *Arabidopsis* leads to enhanced arsenic tolerance and cadmium hypersensitivity. *Plant Cell Physiol.* 45: 1787–1797.

Li, Z., Zhang, D., Xiong, X. et al. (2017). A potent Cas9-derived gene activator for plant and mammalian cells. *Nat. Plant.* 3 (12): 930–936.

Li, Z.G., Ding, X.J., and Du, P.F. (2013). Hydrogen sulfide donor sodium hydrosulfide-improved heat tolerance in maize and involvement of proline. *J. Plant Physiol.* 170: 741–747.

Liang, Y., Chen, Q.I.N., Liu, Q. et al. (2003). Exogenous silicon (Si) increases antioxidant enzyme activity and reduces lipid peroxidation in roots of salt stressed barley (*Hordeum vulgare* L.). *J. Plant Physiol.* 160 (10): 1157–1164.

Limmer, M.A., Mann, J., Amaral, D.C. et al. (2018). Silicon-rich amendments in rice paddies: effects on arsenic uptake and biogeochemistry. *Sci. Total Environ.* 624: 1360–1368.

Liu, C., Wei, L., Zhang, S. et al. (2014). Effects of nanoscale silica sol foliar application on arsenic uptake, distribution, and oxidative damage defense in rice (*Oryza sativa* L.) under arsenic stress. *RSC Adv.* 4 (100): 57227–57234.

Liu, J., Simms, M., Song, S. et al. (2018). Physiological effects of copper oxide nanoparticles and arsenic on the growth and life cycle of rice (*Oryza sativa* L. japonica 'Koshihikari'). *Environ. Sci. Technol.* 52 (23): 13728–13737.

Liu, X., Wang, L., Liu, L. et al. (2011). Alleviating effect of exogenous nitric oxide in cucumber seedling against chilling stress. *Afr. J. Biotechnol.* 21: 4380–4386.

Liu, X., Zhang, S., Shan, X., and Zhu, Y.G. (2005). Toxicity of arsenate and arsenite on germination, seedling growth and amylolytic activity of wheat. *Chemosphere* 61: 293–301.

Ma, J.F. and Yamaji, N. (2006). Silicon uptake and accumulation in higher plants. *Trends Plant Sci.* 11 (522): 392–397.

Ma, J.F., Yamaji, N., Mitani, N. et al. (2008). Transporters of arsenite in rice and their role in arsenic accumulation in rice grain. *Proc. Nat. Acad. Sci. USA* 105: 9931–9935.

Ma, X., Sharifan, H., Dou, F., and Sun, W. (2020). Simultaneous reduction of arsenic (As) and cadmium (Cd) accumulation in rice by zinc oxide nanoparticles. *Chem. Eng. J.* 384: 123802.

Mahimairaja, S., Bolan, N.S., Hedley, M.J., and Macgregor, A.N. (1994). Losses and transformation of nitrogen during composting of poultry manure with different amendments: an incubation experiment. *Bioresour. Technol.* 47: 265–273.

Maity, J.P., Chen, C.Y., Bhattacharya, P. et al. (2020). Advanced application of nano-technological and biological processes as well as mitigation options for arsenic removal. *J. Hazard. Mater.* 405: 123885.

Malik, J.A., Goel, S., Sandhir, R., and Nayyar, H. (2011). Uptake and distribution of arsenic in chickpea: effects on seed yield and seed composition. *Commun. Soil Sci. Plant Anal.* 42: 1728–1738.

Mandal, A. (2015). Transgenic tobacco plants expressing ACR2 gene of *Arabidopsis thaliana* exhibit reduced accumulation of arsenic and increased tolerance to arsenate. *J. Plant Physiol. Pathol.* 2: 5.

Marin, A.R., Masscheleyn, P.H., and Patrick, W.H. Jr. (1992). The influence of chemical form and concentration of arsenic on rice growth and tissue arsenic concentration. *Plant Soil* 139 (2): 175–183.

Martinson, C.A. and Reddy, K.J. (2009). Adsorption of arsenic (III) and arsenic (V) by cupric oxide nanoparticles. *J. Colloid Interface Sci.* 336 (2): 406–411.

Mawia, A.M., Hui, S., Zhou, L., and Li, H. (2020). Inorganic arsenic toxicity and alleviation strategies in rice. *J. Hazard. Mater.* 408 (1): 124751.

Meadows, R. (2014). How plants control arsenic accumulation. *PLoS Biol.* 12 (12): e1002008.

Meharg, A.A. and Hartley-Whitaker, J. (2002). Arsenic uptake and metabolism in arsenic resistant and non-resistant plant species. *New Phytol.* 154: 29–43.

Meharg, A.A. and Macnair, M.R. (1990). An altered phosphate uptake system in arsenate tolerant *Holcus lanatus* L. *New Phytol.* 116: 29–35.

Memelink, J., Verpoorte, R., and Kijne, J.W. (2001). ORCAnization of jasmonate-responsive gene expression in alkaloid metabolism. *Trends Plant Sci.* 6 (5): 212–219.

Metwally, A., Finkemeier, I., Georgi, M., and Dietz, K.J. (2003). Salicylic acid alleviates the cadmium toxicity in barley seedlings. *Plant Physiol.* 132: 272–281.

Mousavi, S.R., Niknejad, Y., Fallah, H. et al. (2020). Methyl jasmonate alleviates arsenic toxicity in rice. *Plant Cell Rep.* 39: 1041–1060.

Mukhopadhyay, R., Bhattacharjee, H., and Rosen, B.P. (2014). Aquaglyceroporins: generalized metalloid channels. *Biochim. Biophys. Acta* 1840: 1583–1591.

Namdjoyan, S. and Kermanian, H. (2013). Exogenous nitric oxide (as sodium nitroprusside) ameliorates arsenic induced oxidative stress in watercress (*Nasturtium officinale* R. Br.) plants. *Sci. Hortic.* 161: 350–356.

Nath, S., Panda, P., Mishra, S. et al. (2014). Arsenic stress in rice: redox consequences and regulation by iron. *Plant Physiol. Biochem.* 80: 203–210.

Neill, S., Barros, R., Bright, J. et al. (2008). Nitric oxide, stomatal closure, and abiotic stress. *J. Exp. Bot.* 59: 165–176.

Okamoto, Y., Chou, P.H., Kim, S.Y. et al. (2008). Oxidative DNA damage in XPC-knockout and its wild mice treated with equine estrogen. *Chem. Res. Toxicol.* 21: 1120–1124.

Pandey, C. and Gupta, M. (2015). Selenium and auxin mitigates arsenic stress in rice (*Oryza sativa* L.) by combining the role of stress indicators, modulators and genotoxicity assay. *J. Hazard. Mater.* 287: 384–391.

Peña-Garcia, Y., Shinde, S., Natarajan, P. et al. (2021). Arsenic stress-related F-box (ASRF) gene regulates arsenic stress tolerance in *Arabidopsis thaliana*. *J. Hazard. Mater.* 407: 124831.

Pommerrenig, B., Diehn, T.A., and Bienert, G.P. (2015). Metalloido-porins: essentiality of nodulin 26-like intrinsic proteins in metalloid transport. *Plant Sci.* 238: 212–227.

Prasad, R., Bhattacharyya, A., and Nguyen, Q.D. (2017). Nanotechnology in sustainable agriculture: recent developments, challenges, and perspectives. *Front. Microbiol.* 8: 1014.

Raab, A., Williams, P., Meharg, A., and Feldmann, J. (2007). Uptake and translocation of inorganic and methylated arsenic species by plants. *Environ. Chem.* 4: 197–203.

Rady, M.M. (2011). Effects of 24-epibrassinolide on growth, yield, antioxidant system and cadmium content of bean (*Phaseolus vulgaris* L.) plants under salinity and cadmium stress. *Sci. Hortic.* 129: 232–237.

Rafiq, M., Shahid, M., Abbas, G. et al. (2017a). Comparative effect of calcium and EDTA on arsenic uptake and physiological attributes of *Pisum sativum*. *Int. J. Phytoremed. 19*: 662–669.

Rafiq, M., Shahid, M., Shamshad, S. et al. (2017b). A comparative study to evaluate efficiency of EDTA and calcium in alleviating arsenic toxicity to germinating and young *Vicia faba* L. seedlings. *J. Soils Sediments* 1–11. https://doi.org/10.1007/s11368-017-1693-5.

Raghu, K., Mahesh, K., Divya Sri, N., and Seeta, R.R.S. (2014). Effect of brassinosteroids on seed germination and seedling growth of radish (*Raphanus sativus)* under As toxicity stress. *Int. J. Dev. Res.* 4 (9): 1929–1933.

Rahman, A., Mostofa, G.M., Alam, M.M. et al. (2015). Calcium mitigates arsenic toxicity in rice seedlings by reducing arsenic uptake and modulating the antioxidant defense and glyoxalase systems and stress markers. *BioMed Res. Int.* 2015: 340812. UK: Hindawi Publishing Corporation.

Rahman, M.A., Kadohashi, K., Maki, T., and Hasegawa, H. (2011). Transport of DMAA and MMAA into rice (*Oryza sativa* L.) roots. *Environ. Exp. Bot.* 72: 41–46.

Rai, A., Tripathi, P., Dwivedi, S. et al. (2011). Arsenic tolerances in rice (*Oryza sativa* L) have a predominant role in transcriptional regulation of a set of genes including sulphur assimilation pathway and antioxidant system. *Chemosphere* 82: 986–995.

Ramakrishna, B. and Rao, S.S.R. (2013a). Preliminary studies on the involvement of glutathione metabolism and redox status against zinc toxicity in radish seedlings by 28-homobrassinolide. *Environ. Exp. Bot.* 96: 52–58.

Ramakrishna, B. and Rao, S.S.R. (2013b). 24-Epibrassinolide maintains elevated redox state of AsA and GSH in radish (*Raphanus sativus* L.) seedlings under zinc stress. *Acta Physiol. Plant.* 35: 1291–1302.

Rao, K.P., Vani, G., Kumar, K. et al. (2011). Arsenic stress activates MAP kinase in rice roots and leaves. *Arch. Biochem. Biophys.* 506 (1): 73–82.

Ratnaike, R.N. (2003). Acute and chronic arsenic toxicity. *Postgrad. Med. J.* 79: 391–396.

Regis, E. (1996). *Nano: The Emerging Science of Nanotechnology*. Boston, MA: Back Bay Books.

Roduner, E. (2006). Size matters: why nanomaterials are different. *Chem. Soc. Rev.* 35 (7): 583–592.

Rossi, L., Sharifan, H., Zhang, W. et al. (2018). Mutual effects and in planta accumulation of co-existing cerium oxide nanoparticles and cadmium in hydroponically grown soybean (*Glycine max* (L.) Merr.). *Environ. Sci. Nano* 5 (1): 150–157.

Roychowdhury, T., Tokunaga, H., Uchino, T., and Ando, M. (2005). Effect of arsenic contaminated irrigation water on agricultural land soil and plants in West Bengal, India. *Chemosphere* 58: 799–810.

Santino, A., Taurino, M., De Domenico, S. et al. (2013). Jasmonate signaling in plant development and defense response to multiple (a)biotic stresses. *Plant Cell Rep.* 32: 1085–1098.

Sarkar, A. and Paul, B. (2016). The global menace of arsenic and its conventional remediation - a critical review. *Chemosphere* 158: 37–49.

Sharma, I. (2012). Arsenic induced oxidative stress in plants. *Biologia* 67: 447–453.

Sharma, I., Pati, P.K., and Bhardwaj, R. (2011). Effect of 28- homobrassinolide on antioxidant defence system in *Raphanus sativus* L. under chromium toxicity. *Ecotoxicology* 20: 862–874.

Sharma, P., Bhardwaj, R., Arora, N. et al. (2008). Effects of 28-homobrassinolide on nickel uptake, protein content and antioxidative defence system in *Brassica juncea*. *Biol. Plant.* 52: 767–770.

Shi, S., Wang, T., Chen, Z. et al. (2016). OsHAC1;1 and OsHAC1;2 function as arsenate reductases and regulate arsenic accumulation. *Plant Physiol.* 172: 1708–1719.

Shin, H., Shin, H.-S., Dewbre, G.R., and Harrison, M.J. (2004). Phosphate transport in *Arabidopsis*: Pht1;1 and Pht1;4 play a major role in phosphate acquisition from both low- and high-phosphate environments. *Plant J.* 39: 629–642.

Singh, A.P., Dixit, G., Kumar, A. et al. (2017a). A protective role for nitric oxide and salicylic acid for arsenite phytotoxicity in rice (*Oryza sativa* L.). *Plant Physiol. Biochem.* 115: 163–173.

Singh, A.P., Dixit, G., Kumar, A. et al. (2016). Nitric oxide alleviated arsenic toxicity by modulation of antioxidants and thiol metabolism in rice (*Oryza sativa* L.). *Front. Plant Sci.* 6: 1272.

Singh, A.P., Dixit, G., Mishra, S. et al. (2015b). Salicylic acid modulates arsenic toxicity by reducing its root to shoot translocation in rice (*Oryza sativa* L.). *Front. Plant Sci.* **6**: 340.

Singh, H.P., Batish, D.R., Kohli, R.K., and Arora, K. (2007). Arsenic induced root growth inhibition in mung bean (*Phaseolus aureus* Roxb.) is due to oxidative stress resulting from enhanced lipid peroxidation. *Plant Growth Regul. 53*: 65–73.

Singh, N., Ma, L.Q., Srivastava, M., and Rathinasabapathi, B. (2006). Metabolic adaptations to arsenic induced oxidative stress in *Pteris vittata* L. and *Pteris ensiformis* L. *Plant Sci. 170*: 274–282.

Singh, P.K., Indoliya, Y., Chauhan, A.S. et al. (2017b). Nitric oxide mediated transcriptional modulation enhances plant adaptive responses to arsenic stress. *Sci. Rep.* 7 (1): 1–13.

Singh, V.P., Singh, S., Kumar, J., and Prasad, S.M. (2015a). Hydrogen sulfide alleviates toxic effects of arsenate in pea seedlings through up-regulation of the ascorbate glutathione cycle: possible involvement of nitric oxide. *J. Plant Physiol.* 181: 20–29.

Soares, C., Sousa, A.D., Pinto, A. et al. (2016). Effect of 24-epibrassinolide on ROS content, antioxidant system, lipid peroxidation and Ni uptake in *Solanum nigrum* L. under Ni stress. *Environ. Exp. Bot.* 122: 115–125.

Song, W.Y., Park, J., Mendoza-Cozatl, D.G. et al. (2010). Arsenic tolerance in *Arabidopsis* is mediated by two ABCC-type phytochelatin transporters. *Proc. Natl. Acad. Sci. USA* 107: 21187–21192.

Srivastava, S., Chiappetta, A., and Beatrice, M. (2013). Identification and profiling of arsenic stress-induced miRNAs in *Brassica juncea*. *J. Exp. Bot.* 64: 303–315.

Srivastava, S., Srivastava, A.K., Suprasanna, P., and D'Souza, S.F. (2009). Comparative biochemical and transcriptional profiling of two contrasting varieties of *Brassica juncea* L. in response to arsenic exposure reveals mechanisms of stress perception and tolerance. *J. Exp. Bot.* 60: 3419–3431.

Stoeva, N. and Bineva, T. (2003). Oxidative changes and photosynthesis in oat plants grown in As-contaminated soil. *Bulg. J. Plant Physiol. 29*: 87–95.

Sun, J., Wang, R., Zhang, X. et al. (2013). Hydrogen sulfide alleviates cadmium toxicity through regulations of cadmium transport across the plasma and vacuolar membranes in *Populus euphratica* cells. *Plant Physiol. Biochem.* 65: 67–74.

Takahashi, Y., Minamikawa, R., Hattori, K.H. et al. (2004). Arsenic behavior in paddy fields during the cycle of flooded and non-flooded periods. *Environ. Sci. Technol.* 38: 1038–1044.

Tang, Z., Chen, Y., Miller, A.J., and Zhao, F.J. (2019). The C-type ATP-binding cassette transporter OsABCC7 is involved in the root-to-shoot translocation of arsenic in rice. *Plant Cell Physiol.* 60 (7): 1525–1535.

Tanou, G., Job, C., Rajjou, L. et al. (2009). Proteomics reveals the overlapping roles of hydrogen peroxide and nitric oxide in the acclimation of citrus plants to salinity. *Plant J.* 60: 795–804.

Tian, S., Lu, L., Zhang, J. et al. (2011). Calcium protects roots of *Sedumal fredii* H. against cadmium-induced oxidative stress. *Chemosphere* 84 (1): 63–69.

Tripathi, D.K., Singh, S., Singh, V.P. et al. (2016). Silicon nanoparticles more efficiently alleviate arsenate toxicity than silicon in maize cultivar and hybrid differing in arsenate tolerance. *Front. Environ. Sci.* 4: 46.

Tripathi, R.D., Tripathi, P., Dwivedi, S. et al. (2012). Arsenomics: omics of arsenic metabolism in plants. *Front. Physiol.* 23 (3): 275.

Trotel, P., Bouchereau, A., Niogret, M.F., and Larher, F. (1989). The fate of osmoaccumulated proline in leaf discs of rape (*Brassica napus* L.) incubated in a medium of low osmolarity. *Plant Sci.* 118: 31–45.

Usman, M., Farooq, M., Wakeel, A. et al. (2020). Nanotechnology in agriculture: current status, challenges and future opportunities. *Sci. Total Environ.* 721: 137778.

Verma, G., Srivastava, D., Narayan, S. et al. (2020b). Exogenous application of methyl jasmonate alleviates arsenic toxicity by modulating its uptake and translocation in rice (*Oryza sativa* L.). In: *Ecotoxicology and Environmental Safety*. 201.110735.

Verma, P.K., Verma, S., Tripathi, R.D., and Chakrabarty, D. (2020a). A rice glutaredoxin regulates the expression of aquaporin genes and modulates root responses to provide arsenic tolerance. *Ecotoxicol. Environ. Saf.* 195: 110471.

Vromman, D., Lutts, S., Lefèvre, I. et al. (2013). Effects of simultaneous arsenic and iron toxicities on rice (*Oryza sativa* L.) development, yield-related parameters and As and Fe accumulation in relation to As speciation in the grains. *Plant Soil 371*: 199–217.

Wang, C., Na, G., Bermejo, E.S. et al. (2018b). Dissecting the components controlling root-to-shoot arsenic translocation in *Arabidopsis thaliana*. *New Phytol.* 217: 206.

Wang, C., Zheng, L., Tang, Z. et al. (2020). OASTL-A1 functions as a cytosolic cysteine synthase and affects arsenic tolerance in rice. *J. Exp. Bot.* 71: 3678–3689.

Wang, F.Z., Chen, M.X., Yu, L.J. et al. (2017). OsARM1, an R2R3 MYB transcription factor, is involved in regulation of the response to arsenic stress in rice. *Front. Plant Sci. 8*: 1868.

Wang, H., Xu, Q., Kong, Y.H. et al. (2014). *Arabidopsis* WRKY45 transcription factor activates phosphate transporter1;1 expression in response to phosphate starvation. *Plant Physiol.* 164: 2020.

Wang, P., Zhang, W., Mao, C. et al. (2016). The role of OsPT8 in arsenate uptake and varietal difference in arsenate tolerance in rice. *J. Exp. Bot.* 67 (21): 6051–6059.

Wang, S., Liu, H., Zhang, Y., and Xin, H. (2015). The effect of CuO NPs on reactive oxygen species and cell cycle gene expression in roots of rice. *Environ. Toxicol. Chem.* 34 (3): 554–561.

Wang, X., Sun, W., and Ma, X. (2019). Differential impacts of copper oxide nanoparticles and Copper(II) ions on the uptake and accumulation of arsenic in rice (*Oryza sativa*). *Environ. Pollut.* 252: 967–973.

Wang, X., Sun, W., Zhang, S. et al. (2018a). Elucidating the effects of cerium oxide nanoparticles and zinc oxide nanoparticles on arsenic uptake and speciation in rice (*Oryza sativa*) in a hydroponic system. *Environ. Sci. Technol.* 52 (17): 10040–10047.

Wang, Z., Von Dem Bussche, A., Kabadi, P.K. et al. (2013). Biological and environmental transformations of copper-based nanomaterials. *ACS Nano* 7 (10): 8715–8727.

Warren, L.A., Outridge, P.M., and Zimmerman, A.P. (1995). Geochemical partitioning and bioavailability of copper to aquatic plants in an artificial oxide-organic sediment. *Hydrobiologia* 304 (3): 197–207.

Wu, F., Fang, Q., Yan, S. et al. (2020a). Effects of zinc oxide nanoparticles on arsenic stress in rice (*Oryza sativa* L.): germination, early growth, and As uptake. *Environ. Sci. Pollut. Res. Int.* 27: 26974–26981.

Wu, F., Fu, Z., Liu, B. et al. (2011). Health risk associated with dietary co-exposure to high levels of antimony and arsenic in the world's largest antimony mine area. *Sci. Total Environ.* 409: 3344–3351.

Wu, X., Hu, J., Wu, F. et al. (2020b). Application of TiO_2 nanoparticles to reduce bioaccumulation of As in rice seedlings (*Oryza sativa* L.): a mechanistic study. *J. Hazard. Mater.* 124047.

Wysocki, R., Bobrowicz, P., and Ułaszewski, S. (1997). The *Saccharomyces cerevisiae* ACR3 gene encodes a putative membrane protein involved in arsenite transport. *J. Biol. Chem.* 272: 30061–30066.

Xu, J., Shi, S., Wang, L. et al. (2017). OsHAC4 is critical for arsenate tolerance and regulates arsenic accumulation in rice. *New Phytol.* 215: 1090–1101.

Xu, W., Dai, W., Yan, H. et al. (2015). *Arabidopsis* NIP3;1 plays an important role in uptake and root-to-shoot translocation under arsenite stress conditions. *Mol. Plant* 8: 722–733.

Xu, X.Y., Mcgrath, S.P., Meharg, A.A., and Zhao, F.J. (2008). Growing rice aerobically markedly decreases accumulation. *Environ. Sci. Technol.* 42: 5574–5579.

Yan, S., Wu, F., Zhou, S. et al. (2021). Zinc oxide nanoparticles alleviate the arsenic toxicity and decrease the accumulation of arsenic in rice (*Oryza sativa* L.). *BMC Plant Biol.* 21: 150.

Yan, Z., Chen, J., and Li, X. (2013). Methyl jasmonates modulator of Cd toxicity in *Capsicum frutescens* var. fasciculatum seedlings. *Ecotoxicol. Environ. Saf.* 98: 203–209.

Yang, J., Gao, M.X., Hu, H. et al. (2016). OsCLT1, a CRT-like transporter 1, is required for glutathione homeostasis and arsenic tolerance in rice. *New Phytol.* 211 (2): 658–670.

Ye, J., Rensing, C., Rosen, B.P., and Zhu, Y.G. (2012). Arsenic biomethylation by photosynthetic organisms. *Trends Plant Sci.* 17: 155–162.

Ye, Y., Li, P., Xu, T. et al. (2017). OsPT4 contributes to arsenate uptake and transport in rice. *Front. Plant Sci.* 8: 2197.

You, T., Liu, D., Chen, J. et al. (2018). Effects of metal oxide nanoparticles on soil enzyme activities and bacterial communities in two different soil types. *J. Soils Sediments* 18 (1): 211–221.

Yu, L.J., Luo, Y.F., Liao, B. et al. (2012). Comparative transcriptome analysis of transporters, phytohormone and lipid metabolism pathways in response to arsenic stress in rice (*Oryza sativa* L.). *New Phytol.* 195: 97–112.

Zanella, L., Fattorini, L., Brunetti, P. et al. (2016). Overexpression of AtPCS1 in tobacco increases arsenic and arsenic plus cadmium accumulation and detoxification. *Planta* 243: 605–622.

Zhang, X.W., Dong, Y.J., Qiu, X.K. et al. (2012). Exogenous nitric oxide alleviates iron-deficiency chlorosis in peanut growing on calcareous soil. *Plant Soil Environ.* 58 (3): 111–120.

Zhang, X.W., Zhang, M., Wang, Q.H. et al. (2011a). Effect of exogenous nitric oxide on physiological characteristic of peanut under iron-deficient stress. *Plant Nutr. Fert. Sci.* 17: 665–673.

Zhang, Z., Wang, H., Wang, X. et al. (2011b). Nitric oxide enhances aluminum tolerance by affecting cell wall polysaccharides in rice roots. *Plant Cell Rep.* 30: 1701.

Zhao, F.J., Ma, J.F., Meharg, A.A., and McGrath, S.P. (2009). Arsenic uptake and metabolism in plants. *New Phytol.* 181: 777–794.

Zhao, F.J., McGrath, S.P., and Meharg, A.A. (2010). Arsenic as a food chain contaminant: mechanisms of plant uptake and metabolism and mitigation strategies. *Annu. Rev. Plant Biol.* 61: 535–559.

Zhao, Y. (2020). A new era of metal organic framework nanomaterials and applications. *ACS Appl. Nano Mater.* 3 (6): 4917–4919.

Zhou, S., Peng, L., Lei, M. et al. (2015). Control of As soil to rice transfer (*Oryza sativa* L.) with nano-manganese dioxide. *Acta Sci. Circumst.* 35: 855–861.

Zhou, Z.S., Guo, K., Elbaz, A.A., and Yang, Z.M. (2009). Salicylic acid alleviates mercury toxicity by preventing oxidative stress in roots of *Medicago sativa*. *Environ. Exp. Bot.* 65: 27–34.

Zulfiqar, F., Navarro, M., Ashraf, M. et al. (2019). Nanofertilizer use for sustainable agriculture: advantages and limitations. *Plant Sci.* 289: 110270.

20

A Potential Phytoremedial Strategy for Arsenic from Contaminated Drinking Water Using *Hygrophilla spinosa* (Starthorn Leaves)

Nilanjana Roy Chowdhury, Debapriya Sinha, Antara Das, Madhurima Joardar, Anuja Joseph, Iravati Ray, Deepanjan Mridha, Ayan De, and Tarit Roychowdhury

School of Environmental Studies, Jadavpur University, Kolkata, West Bengal, India

20.1 Introduction

Arsenic contamination in the Ganga-Meghna-Brahmaputra plain is a long standing concern for the past 40–45 years, where West Bengal plays a major role. In rural area of Bengal, groundwater is available round the year at a shallow level in a sufficient quantity but after checking, the exact depth for palatable water along with the development of tube wells for water supply is found to be at stake (Chowdhury et al. 2018). At present 132 blocks are As affected among 341 blocks. Major health problem in this area included excess amount of As in drinking water above the permissible limit of $10\,\mu g\,l^{-1}$. People who are affected with As have no effective medicinal treatment. However, As-free water can help them to get rid of this situation (Joardar et al. 2021). So, the foremost priority is As-free water for human population, which is urgently needed to mitigate this toxic situation, and shield the population's physical condition and their safety. There are options available for water supply in As prone zones including avoidance from As and treatment of As-contaminated water. We can use the huge number of unused water sources (declared as a contaminated water supply) by treating them, and it can be instrumental in saving the human population. Different types of methodologies are available to eliminate As from drinking water, like oxidation, coprecipitation, coagulation-flocculation, lime treatment, adsorption on to sorptive media, ion-exchange resin, and membrane technologies. While searching, new methodologies involving the eradication of As-like toxicant from drinking water have been garnering a lot of attention to "biosorption," based on the theory of metal binding capacity. Adsorption is the contact based adherence or connection of ions or molecules on to the surface of other molecules. This process can be designated as the capability of biological materials to assimilate As from contaminated water by physicochemical processes or metabolically mediated pathways (Volesky 2001). Biosorbent materials have some good advantages, like very low costing, huge efficacy rate, lower amount of sludge, no extra

Arsenic in Plants: Uptake, Consequences and Remediation Techniques, First Edition.
Edited by Prabhat Kumar Srivastava, Rachana Singh, Parul Parihar, and Sheo Mohan Prasad.

nutrient necessity, regeneration, chance of metal recovery, broad range of target pollutants, huge sorption ability, and ecofriendly behavior (Kamsonlian 2011; Suresh et al. 2011). Biosorption is very difficult to define because of its overlapping mechanistic pathway with others throughout. Some factors involved here are adsorbate material, adsorbent characteristics, environmental factors, existence or nonexistence of metabolic pathways. *Bio*, this word mainly denotes any connection with biological entity like plant, living organism, or any derived product from a living creature (Gadd 2009). Adsorption is the most affordable among other cleanup technologies, and it is a general form of sorption method, where a sorbate material interacts with a sorbent material in a system resulting in accumulation or assimilation at the interface of sorbent–sorbate (Borda and Sparks 2008).

Adsorption study is mainly carried out in a simple closed system with the help of batch method. Industrial applications require mostly a flow-through or continuous methodology. An entire sorption isotherm is required to process and evaluate whole bio-adsorption study (Volesky 2001). Adsorption processes have been paying significant attention owing to their capacity to treat As-contaminated aqueous solutions. To assess the adsorption capacity of adsorbent materials, examination of different factors that rely on the system like adsorbent dose, pH, metal concentration, presence of cations, temperature, rpm, etc., is required. The sorbent material is allowed to assimilate/accumulate the sorbate material to equilibrium. The initial adsorption equation is $Q_e = V(C_i - C_f)/S$. Q_e is the adsorbate uptake as weight per unit dry weight of the adsorbent, V is the volume of the mentioned solution or adsorbate material, C_i and C_f are initial and final concentrations of the adsorbate, and S is the dry weight of adsorbent. But the Freundlich isotherm primarily defines the isotherm regarding adsorption to a heterogeneous surface. The Freundlich isotherm equation is $q_e = KC\beta$, where q_e is the stability value of adsorbate accumulation by the adsorbent, C is the adsorbate concentration, K is an affinity factor, and β is a dimensionless heterogeneity factor. The value of β is inversely proportional with heterogeneity (Kinniburgh 1985). $\beta = 1$, when the Freundlich equation reduces to a linear isotherm. It is rigorously suitable for metal adsorption at a low metal containing solution but used over a broad range of concentrations (Sposito 1989). Data are generally fitted to the logarithmic form of the equation, $\text{Log } q_e = \log K + \beta \log C$. Finally, it will give a straight line by plotting $\log q_e$ versus $\log C$ of slope β and an intercept of $\log K$ for $C = 1$ ($\log C = 0$).

Different materials have been examined for their adsorption capacity to eliminate As from drinking water (Table 20.1). But there is a limited number of work regarding materials derived from biological sources. Earlier findings also depict that agricultural waste can be a great option as an adsorbent in a developing country like India (Mohan et al. 2007) where humans are dependent on irrigation for their livelihood and agricultural waste plays a significant role. Such materials include almond shells, olive stones, peach stones, coconut shells, fertilizer slurry, pine saw dust, eucalyptus wood char, peanut shells, oat hulls, olive pulp, olive shells, etc.

So, adsorption has been known as an affordable method for eradication of As from drinking water in the third world countries. Researchers have put in hard work to make a new adsorbent, which is effortlessly reachable, accessible, handy, and locally obtainable along with cheap cost, easy preparation techniques and operational strategy, and managing policy. With the objective of preparing requisite characteristics of adsorbent to reduce As from

Table 20.1 Adsorbent materials prepared from biological sources to remove As from drinking water.

Sr. no.	Name of bio-adsorbents	References
1.	Waste tea, sugarcane bagasse, banana pseudo-stem, saw dust, dried jute, banana pith, mushroom	Brima and Haris (2014)
2.	Chitosan	Mcafee et al. (2001); Dambies et al. (2002)
3.	Alginate	Zouboulis and Katsoyiannis (2002)
4.	Orange waste	Ghimire et al. (2003)
5.	Fungal biomass	Loukidou et al. (2003)
6.	Methylated yeast biomass	Seki et al. (2005)
7.	Chicken feathers	Teixeira and Ciminelli (2005)

tainted drinking water, Starthorn (*H. spinosa*) leaf biomass may act as a cost-effective natural product in the near future. Its adsorption capacity has been evaluated for As eradication under balance and dynamic investigational circumstances.

20.2 Methodology

20.2.1 Adsorbent

The current work had thus utilized adsorbent materials prepared from iron (Fe)-rich plant leaves of Kulekhara (Starthorn plant) (*H. spinosa*) (Figure 20.1).

20.2.2 Sample Collection and Preparation of Adsorbent

20.2.2.1 Sampling Site

For this bio-adsorption study, the Starthorn leaves were collected from Heritage Institute of Technology, Anandapur, Kolkata (Figure 20.2).

20.2.2.2 Preparation of Material

For the bio-adsorption study, the collected plant materials (Kulekhara) were thoroughly washed under running tap water to remove all the dirt and silt present on them. The segregated leaves were again washed thoroughly by double distilled water and desiccated for three days in hot air oven at 50–55 °C until they were completely dry. The dried leaves were then ground into a rough powder using a mortar and pestle. It was further ground into a fine powder using Bajaj Rex-500-watt grinder. The fine powder was then sieved through a pore size of 0.15 mm and then treated with 1 : 2 HCl solution to bring about desorption and remove any impurity that might be present. The wet adsorbent was then filtered and again desiccated for three days in hot air oven at 50–55 °C. The dried materials were then stored into an air tight container for future use (Figure 20.3).

Figure 20.1 Leaves of Starthorn plant (*H. spinosa*).

Figure 20.2 Heritage Institute of Technology, Anandapur, Kolkata. *Source:* Google earth.

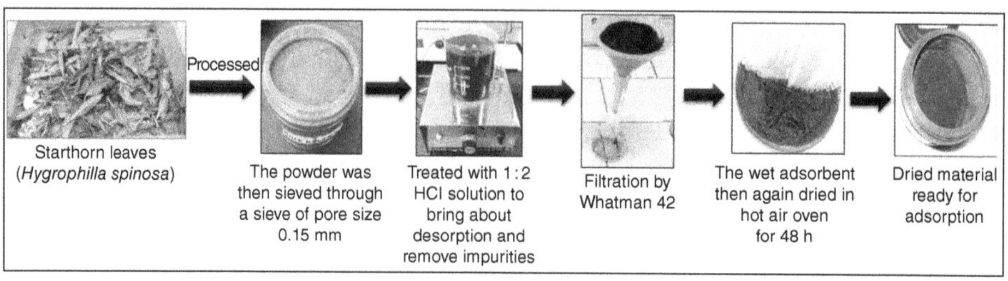

Figure 20.3 Preparation of adsorbent material.

20.2.3 Adsorbate

For As estimation, a standard stock solution of As (V) (1000 mg l^{-1} as, traceable to SRM from NIST H_3AsO_4 in HNO_3, 0.5 M) from Merck, Darmstact, Germany, was used (Das et al. 2020). The different concentrated As solutions like 100, 10, 1, and 0.1 mg l^{-1} were prepared by serial dilutions. The prepared 100 μg l^{-1} of As solution was used for the adsorption study.

20.2.4 The Batch Adsorption Study

The batch study was carried out to examine the result of optimizing parameters like contact time, adsorbent dosage/dose, rpm, and pH. For each study, the adsorbent material was weighed and 20 ml of respective As solution was added to each sample in individual plastic bottles. pH has been altered to the required value using 0.01 (N) HCl and/or 0.01 (N) NaOH. The plastic containers have been positioned on a temperature-controlled rotary shaker operated at 30 °C for respective number of hours at required rpm and pH using required adsorbent dosages. The required parameters have been clearly explained in Table 20.2.

The adsorbent materials have been detached from the solution (adsorbate) by filtration method with the help of Whatman ashless Grade Paper No: 42 (Sigma-Aldrich) and then desiccated for three days in the hot air oven at 50–55 °C for future use or reference. The remaining solution was further filtered by passing through Millipore filter paper of 0.45 μm. The reduction in the concentration of As was studied through the phenomena of adsorption too.

20.2.5 Estimation of As

Arsenic concentration was measured by hydride generation-atomic absorption spectrometry(HG-AAS) methodology coupled with VGA-77 with the software version 5.1 (Chowdhury et al. 2020).

20.2.6 Estimation of Fe

For colorimetric estimation of iron, acetic acid-sodium acetate buffer, hydroxylamine hydrochloride (10%), and 1,10 *o*-phenenthroline (0.25%) have been used. The information regarding the procedure and reagents has been mentioned previously (Fries et al. 1977).

Table 20.2 Controlling parameters during optimization under study.

Controlling parameters	pH	Adsorbent dose (g)	rpm	Time (h)	Temp. (°C)	Conc. of As (μg l^{-1})
pH	4–9	1	30	5	30	100
Adsorbent dose	7	0.1–1.0 g with an interval of 0.1	30	5	30	100
Time	7	1	30	1, 2, 3, 4, 5	30	100
rpm	7	1	30, 35, 40, 45, 50	5	30	100

20.2.7 Calculation

Arsenic uptake Q_e (mg g^{-1}) has been determined from the following equation:

$$Q_e = (C_0 - C_e) \times V/M$$

Where C_0 is initial and C_e is final As concentration of adsorbate material (mg l^{-1}), V is volume of the adsorbate in l, and M (g) is the adsorbent mass (Mondal et al. 2008).

20.2.8 Quality Control and Quality Assurance

The accuracy and meticulousness of the generated methodical data have been validated through standardization and analysis of blank and spiked samples. In some cases, inter-laboratory testing is also performed (Rahman et al. 2002). Quality control tests have been also done by estimating replicates and calculating recovery of the spiked samples.

20.2.9 Statistical Evaluation

Regression study and other numerical expressions including statistical associations among different parameters have been designed to comprehend their interdependence and importance through Microsoft Excel 2016 (Das et al. 2020).

20.3 Results and Discussion

During the study of As removal, the percentage of As adsorption on Starthorn leaves was observed at the initial concentration of 100 μg l^{-1} under optimized adsorbent dose, contact time, pH, and rpm. The comparison between As removal capacities of different adsorbents from various water sources like drinking water, waste water, and groundwater has been elaborately studied (Mohan et al. 2019). The batch adsorption study was carried out for leaves of *H. spinosa* (Kulekhara). The changes in concentration of As in each adsorbent material was measured against controlling the different parameters like adsorbent dosage, contact time, pH, and rpm.

Phenomenon in which a material is released from a surface or through a surface is known as desorption. Hence in a way it is opposite to sorption. For this batch adsorption study, the adsorbent materials mentioned earlier have undergone desorption so that the pores now become empty and can absorb the targeted contaminant, i.e. As. Analysis of the adsorbent materials were done before and after desorption (Table 20.3) to verify whether there had been any change in the concentration of iron and arsenic.

20.3.1 Effect of Adsorbent Dosage

The data from the batch experiment conducted for optimization of adsorbent dosage at its initial concentration of As with varying dosage of 0.1–1.0 g under optimized environmental conditions are presented in Table 20.4 and Figure 20.4.

Table 20.3 Overview of Kulekhara leaves before and after desorption.

	Pre-desorption		Post-desorption	
Sample name	**Iron (mg kg^{-1})**	**Arsenic (μg kg^{-1})**	**Iron (mg kg^{-1})**	**Arsenic (μg kg^{-1})**
Kulekhara leaf	1505	753	100	BDL*

BDL*: below detection limit; the detection limit of As in HG-AAS system is 3 μg l^{-1} with 95% confidence.

Table 20.4 Effect of adsorbent dosage on the reduction efficiency of As concentration using adsorbent ground leaves of *H. spinosa*.

Weight of the adsorbent (g)	Initial As conc. (μg l^{-1})	Final As conc. (μg l^{-1})	% of removal
0.1	100	98.7	1.2
0.2	100	96.5	3.4
0.3	100	94.2	5.7
0.4	100	87.6	12.9
0.5	100	81.2	18.6
0.6	100	71.3	28.2
0.7	100	55.8	44.9
0.8	100	46.0	53.3
0.9	100	40.6	59.7
1.0	100	37.4	62.6

The adsorbent dosage was increased from 0.1 to 1.0 g with a gap of 0.1 g to check the effect of increase of adsorbent dosage on the reduction of As concentration in the solution.

For this study, the following parameters were kept constant:

- pH – 7
- Temperature – 30 °C
- rpm – 30
- Contact Time – five hours

From the graph (Figure 20.4), it can be seen that maximum reduction has been for the adsorbent dosage of 1.0 g where 62.6% of As removal has taken place. It has been observed that 2.09 mg As was removed by 1 g of adsorbent. The removal percentage was increased from 1.2 to 62.6% according to the increasing amount of adsorbent dose. There is a direct relationship between adsorbent dose and removal percentage throughout. Higher rate of As removal from water has been seen due to the presence of active adsorbent sites for more adsorption at higher adsorbent dosage. Mass of adsorbent and its capacity of sorption are defined through Freundlich isotherm (Figure 20.5 and Table 20.5). This isotherm is used to test the adsorption intensity of As on the adsorbent surface.

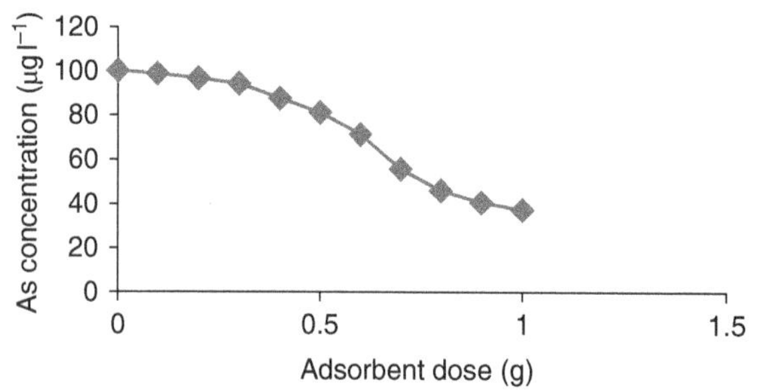

Figure 20.4 The effect of adsorbent dosage on reduction efficiency of As concentration using ground leaves of *H. spinosa*.

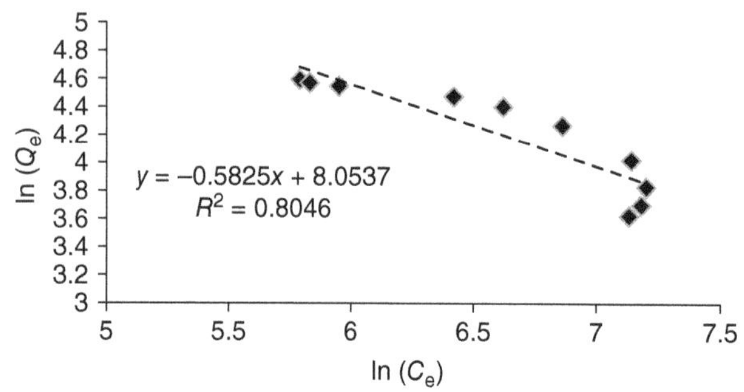

Figure 20.5 Adsorption isotherm showing the effect of adsorbent dosage on efficiency of As reduction by ground leaves of *H. spinosa*.

Table 20.5 Adsorption isotherm for the effect of adsorbent dosage using ground leaves of *H. spinosa*.

Adsorbent dose (g)	Q_e (μg g^{-1})	ln (Q_e)	C_e (μg l^{-1})	ln (C_e)
0.1	348	5.85	98.2	4.58
0.2	342	5.83	96.5	4.57
0.3	384	5.95	94.2	4.54
0.4	619	6.42	87.6	4.47
0.5	750	6.62	81.2	4.39
0.6	954	6.86	71.3	4.26
0.7	1262	7.14	55.8	4.02
0.8	1348	7.20	46.0	3.83
0.9	1319	7.18	40.6	3.70
1.0	1251	7.131	37.4	3.62

20.3.2 Effect of Contact Time

The mentioned data depicted in Figure 20.6 along with Table 20.6 revealed that the effect of contact time at optimum dosage of 1 g can be considered as value for As removal by adsorption on Starthorn leaves. Different stages of adsorption can be noticed, but in each case, these indicate high influence of contact time on removal capacity of As by adsorption.

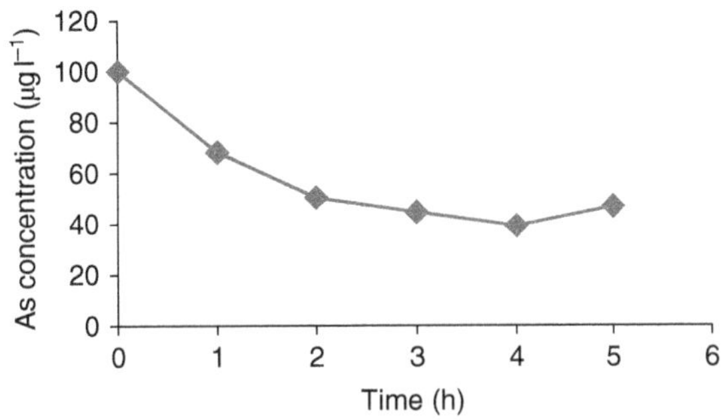

Figure 20.6 The effect of contact time on reduction efficiency of As concentration using adsorbent ground leaves of *H. spinosa*.

Table 20.6 The effect of contact time on reduction efficiency of As concentration using adsorbent ground leaves of *H. spinosa*.

Contact time (h)	Initial As conc. ($\mu g\, l^{-1}$)	Weight of adsorbent (g)	Final As conc. ($\mu g\, l^{-1}$)	% of removal
1	100	1.000	68.2	31.7
2	100	1.006	50.3	49.6
3	100	1.003	44.5	55.4
4	100	1.001	39.1	60.8
5	100	1.005	46.8	53.1

The increasing tendency of contact time initially had a positive impact on the increase of As uptake by the adsorbent but later it was observed that after the end of four hours, there was a decrease in the uptake of As by the same adsorbent.

In this study, the contact time was differed from one to five hours with the following controlling parameters being kept constant:

- Temperature – 30 °C
- rpm – 30
- pH – 7
- Adsorbent dose – 1 g

After one hour interval, it shows 31.7% As reduction but the removal percentage increased with time. Finally, at the time interval of four hours, the removal percentage reaches up to 61% but after this, adsorption decreased with increasing contact time. From the graph, it can be seen that maximum reduction has been for contact time of four hours where 60.9% of As removal has taken place. It has been observed (Figure 20.7 and Table 20.7) that 25.7 mg of As is removed by 1 g of adsorbent. So, it can be concluded that the contact time of four hours may act as an equilibrium time for adsorption of As (Mohan et al. 2019).

20.3.3 Effect of pH

The pH of the adsorbate material is a significant deciding factor in the adsorption mechanistic pathway. The impact of pH (4–9) on the reduction efficiency of As for a steady biosorbent dosage, contact time, and metal ion concentration was studied.

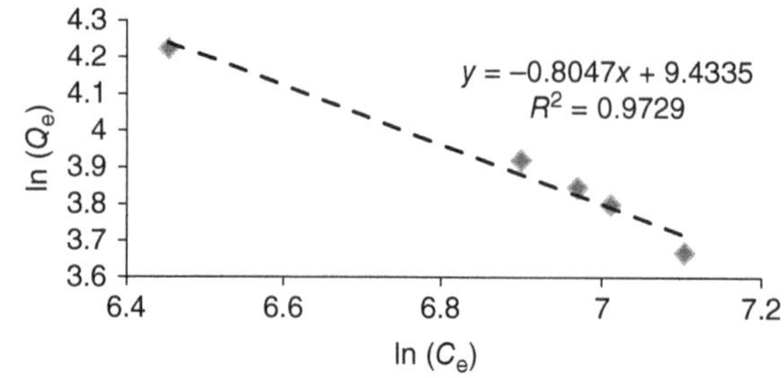

Figure 20.7 Adsorption isotherm to explain the effect of contact time on reduction efficiency of As concentration using adsorbent ground leaves of *H. spinosa*.

Table 20.7 The adsorption isotherm for the effect of contact time using ground leaves of *H. spinosa*.

Adsorbent dose (g)	Contact time (h)	Q_e (μg g^{-1})	ln (Q_e)	C_e (μg l^{-1})	ln (C_e)
1.012	1	635	6.45	68.8	4.22
1.029	2	992	6.90	50.3	3.91
1.013	3	1108	7.01	44.5	3.80
1.002	4	1216	7.10	39.1	3.66
1.015	5	1063	6.97	46.8	3.84

Thus, the role of hydrogen ion concentration was examined at pH values 4, 5, 6, 7, 8, and 9.

For this study, the following controlling parameters were kept constant:

- Temperature – 30 °C
- rpm – 30
- Contact time – four hours
- Adsorbent dose – 1 g

From this experiment, we inferred that pH had a noticeable impact on the metal accumulation and assimilation. The adsorbate concentration was decreased with increase of pH up to 5 and then it increased with an additional increase of pH. So there is a direct relationship between reduction efficiency and the pH up to 5. From Table 20.8 and Figure 20.8, it can be seen that maximum reduction has been observed at pH: 5 where 76.6% of As removal has taken place.

The optimal pH for the reduction of As was found to be 5. It can be said that the redox potential (Eh) of As gradually decreases in instances of increasing alkalinity (pH) (Pandey et al. 2009).

The quality and concentration of adsorbate material play a major function at the time of its uptake through the adsorbent material. Temperature remains constant throughout the experiment. Generally, the quantity of substance (adsorbate) adsorbed is evaluated as a function of the concentration at a stable temperature all through, and the consequent function is known as an adsorption isotherm. The Freundlich isotherm has a broad aspect, and it is used to represent impact on surfaces, which have a heterogeneous energy distribution during adsorption. It has been observed through Freundlich isotherm that 37.8 mg As is removed by 1 g of adsorbent (Figure 20.9 and Table 20.9).

Table 20.8 Effect of pH on the reduction efficiency of As concentration using adsorbent ground leaves of *H. spinosa*.

pH	Initial As conc. ($\mu g\,l^{-1}$)	Weight of adsorbent (g)	Final As conc. ($\mu g\,l^{-1}$)	% of removal
4	100	1.214	37.2	63.7
5	100	1.062	24.3	76.6
6	100	1.005	31.5	69.4
7	100	1.013	48.1	52.8
8	100	1.042	52.8	48.1
9	100	1.002	58.1	42.0

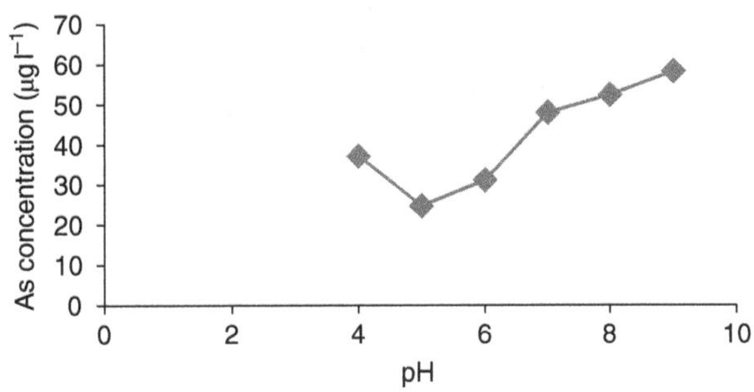

Figure 20.8 Effect of pH on the reduction efficiency of As concentration using adsorbent ground leaves of *H. spinosa*.

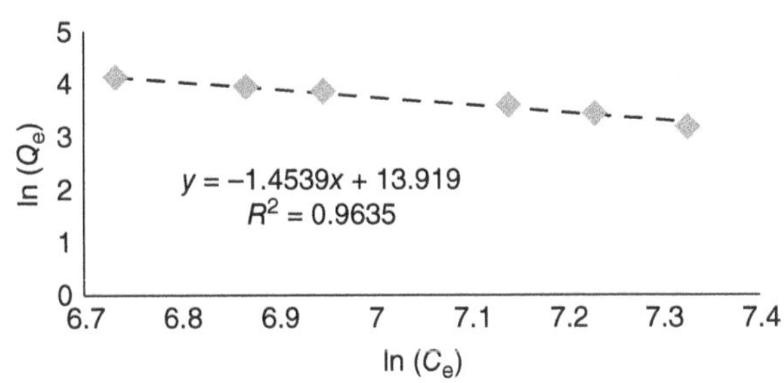

Figure 20.9 Adsorption isotherm to explain the effect of pH on the reduction efficiency of As concentration using adsorbent ground leaves of *H. spinosa*.

20.3.4 Effect of RPM

The agitation speed or rpm has a noteworthy impact on the sorption capacity of the adsorbent. In this study 30–50 rpm was taken into consideration for batch adsorption study with an interval of five.

For this study, the following controlling parameters were kept constant:

- Temperature – 30 °C
- pH – 5
- Contact time – four hours
- Adsorbent dose – 1 g

Table 20.9 To study the adsorption isotherm for the effect of pH using ground leaves of *H. spinosa*.

Adsorbent dose (g)	pH	Q_e (μg g^{-1})	ln (Q_e)	C_e (μg l^{-1})	ln (C_e)
1.001	4	1260	7.139	37.2	3.61
1.012	5	1520	7.326	24.4	3.18
1.003	6	1380	7.229	31.1	3.43
1.016	7	1040	6.947	48.2	3.87
1.024	8	960	6.867	52.7	3.95
1.008	9	840	6.733	58.3	4.06

Table 20.10 Effect of rpm on the reduction efficiency of As concentration using adsorbent ground leaves of *H. spinosa*.

rpm	Initial As conc. (μg l^{-1})	Weight of adsorbent (g)	Final As conc. (μg l^{-1})	% of removal
30	100	1.102	60.3	39.6
35	100	1.029	52.4	47.5
40	100	1.011	47.8	52.2
45	100	1.001	36.9	63.1
50	100	1.004	44.2	55.7

It was found that rpm of 45 was ideal to eradicate arsenic from contaminated water to the highest degree by this adsorption mechanism using Starthorn leaves. Table 20.10 and Figure 20.10 depicts that maximum reduction was observed for rpm of 45 where 63.1% of As removal took place. The concentration of adsorbate material is sharply decreased with the increasing rpm gradually up to an equilibrium value of 45 rpm. After that a further increase had a minor impact on adsorption methodology. Thus other optimized conditions were fixed at rpm 45 for the rest of the batch experiment to certify the ability of equilibrium (Shafique et al. 2012).

During adsorption this step deciphers that 45 rpm was an optimized condition or speed to encourage the occurrence of all the surface binding sites for the binding of As. At the time of lower agitation speed, the adsorbent materials were conglomerated in a specific zone instead of spreading that concealed all the active sites under the top layers. Concurrently, adsorption happened at top layer and other layers did not work, since it had no relation with adsorbate. On the other hand, higher speed causes random friction and collision between the solution and adsorbent, which did not give much time to the As ions to create a bond with surface of the adsorbent.

Figure 20.11 and Table 20.11 depict the slope and intercept of Freundlich model, which were used to compute the constant indicative of relative adsorption capacity and adsorption intensity. It has been observed that 30.1 mg As has been removed by 1 g of adsorbent.

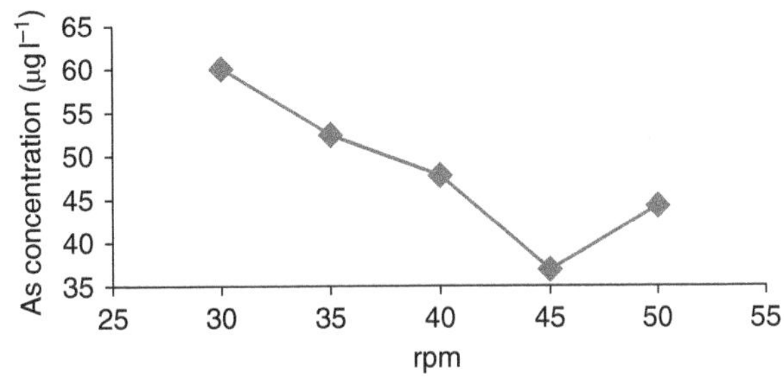

Figure 20.10 The effect of rpm on reduction efficiency of the As concentration using adsorbent ground leaves of *H. spinosa*.

Table 20.11 Adsorption isotherm for the effect of rpm using ground leaves of *H. spinosa*.

Adsorbent dose (g)	rpm	Q_e (μg g^{-1})	ln (Q_e)	C_e (μg l^{-1})	ln (C_e)
1.017	30	793	6.67	60.3	4.10
1.023	35	951	6.85	52.4	3.96
1.010	40	1044	6.95	47.7	3.86
1.016	45	1262	7.14	36.8	3.61
1.014	50	1115	7.01	44.2	3.79

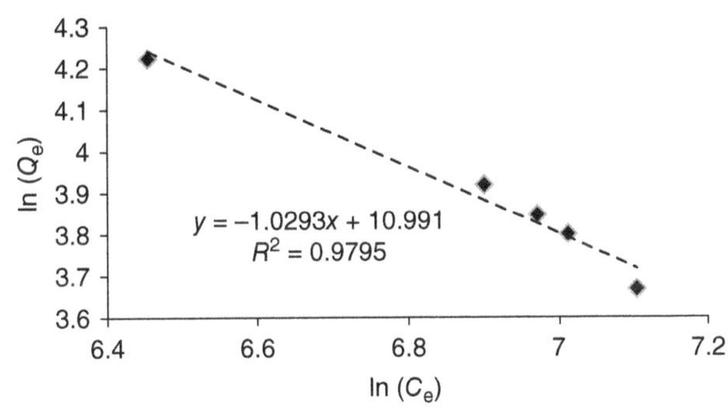

Figure 20.11 Adsorption isotherm to explain the effect of rpm on the reduction efficiency of arsenic concentration using adsorbent ground leaves of *H. spinosa*.

20.4 Conclusion

Different ranges of laboratory experiments were conducted to investigate and estimate the adsorptive capacity of low cost adsorbents prepared from plant materials for removal of As from drinking water. This study was carried out in order to understand the As removal capacity of the leaves of *H. spinosa* from drinking water by the application of bio-adsorption. The main reason to choose these plants lay in the fact that they are easily available and rich in tannins and iron. The results indicated that the used adsorbent in this study has the capacity to treat As-contaminated water. Adsorbent prepared from Starthorn leaf biomass can be effectively used as economical materials for removal of As from drinking water. Adsorbent dose, contact time, pH, and rpm were observed to be among the most controlling parameters for As adsorption (Table 20.12).

Table 20.12 Optimizing conditions for best removal of arsenic by Starthorn leaves.

Dose of adsorbent	Time	pH	rpm	% of As removal
Adsorbent dose: 1.0 g				62.6
Adsorbent dose: 1.0 g	Contact time: four hours			60.9
Adsorbent dose: 1.0 g	Contact time: four hours	pH: 5		76.6
Adsorbent dose: 1.0 g	Contact time: four hours	pH: 5	rpm: 45	63.1***

Drinking water As contamination and its rigorous health impacts have been a worldwide concern during the past few decades. A variety of chemical based treatments, i.e. coagulation, ion exchange, and reverse osmosis, are used in removing As from contaminated water. But most of these methods are not ecofriendly and have numerous limitations (high material cost, high energy needs, sludge generation, etc.) in large-scale practical applications. To overcome these limitations, some studies are considered for potential low-cost and ecofriendly treatments of As-contaminated water. It has been found that a wide range of biomass such as plant parts, fruit wastes, and agricultural wastes are low-cost, recyclable, not sludge generating, and have highly effective biosorption capacity in removing arsenic from water. Thus, biosorption has emerged as an ecofriendly and cost-effective technique in arsenic remediation. Starthorn leaves have the highest capability of successfully removing As from the contaminated water across all controlling parameters of the adsorption study and can thus be successfully used as an adsorbent for As removal. The adsorption of As on the adsorbents was confirmed by Freundlich isotherm in our study.

References

Borda, M.J. and Sparks, D.L. (2008). Kinetics and mechanisms of sorption–desorption in soils: a multiscale assessment. In: *Biophysico-Chemical Processes of Heavy Metals and Metalloids in Soil Environments* (ed. A. Violante, P.M. Huang and G.M. Gadd), 97–124. Wiley.

Brima, E.I. and Haris, P.I. (2014). Arsenic removal from drinking water using different biomaterials and evaluation of a phytotechnology based filter. *International Research Journal of Environmental Sciences-ISCA* 3 (7): 39–44.

Chowdhury, N.R., Ghosh, S., Joardar, M. et al. (2018). Impact of arsenic contaminated groundwater used during domestic scale post harvesting of paddy crop in West Bengal: arsenic partitioning in raw and parboiled whole grain. *Chemosphere* 211: 173–184.

Chowdhury, N.R., Das, A., Mukherjee, M. et al. (2020). Monsoonal paddy cultivation with phase-wise arsenic distribution in exposed and control sites of West Bengal, alongside its assimilation in rice grain. *Journal of Hazardous Materials* 400: 123206.

Dambies, L., Vincent, T., and Guibal, E. (2002). Treatment of arsenic-containing solutions using chitosan derivatives: uptake mechanism and sorption performances. *Water Research* 36 (15): 3699–3710.

Das, A., Das, S.S., Chowdhury, N.R. et al. (2020). Quality and health risk evaluation for groundwater in Nadia district, West Bengal: an approach on its suitability for drinking and domestic purpose. *Groundwater for Sustainable Development* 10: 100351.

Fries, J., Getrost, H., and Merck, D.E. (1977). *Organic Reagents Trace Analysis*, 975. Darmstadt, Germany: E. Merck.

Gadd, G.M. (2009). Biosorption: critical review of scientific rationale, environmental importance and significance for pollution treatment. *Journal of Chemical Technology & Biotechnology: International Research in Process, Environmental & Clean Technology* 84 (1): 13–28.

Ghimire, K.N., Inoue, K., Yamaguchi, H. et al. (2003). Adsorptive separation of arsenate and arsenite anions from aqueous medium by using orange waste. *Water Research* 37 (20): 4945–4953.

Joardar, M., Das, A., Mridha, D. et al. (2021). Evaluation of acute and chronic arsenic exposure on school children from exposed and apparently control areas of West Bengal, India. *Exposure and Health* 13: 33–50.

Kamsonlian, S., Balomajumder, C., Chand, S., and Suresh, S. (2011). Biosorption of Cd (II) and As (III) ions from aqueous solution by tea waste biomass. *African Journal of Environmental Science and Technology* 5 (1): 1–7.

Kinniburgh, D.G. (1985). *ISOTHERM. A Computer Program for Analyzing Adsorption Data.* Wallingford: British Geological Survey.

Loukidou, M.X., Matis, K.A., Zouboulis, A.I., and Liakopoulou-Kyriakidou, M. (2003). Removal of As (V) from wastewaters by chemically modified fungal biomass. *Water Research* 37 (18): 4544–4552.

Mcafee, B.J., Gould, W.D., Nadeau, J.C., and da Costa, A.C. (2001). Biosorption of metal ions using chitosan, chitin, and biomass of *Rhizopus oryzae*. *Separation Science and Technology* 36 (14): 3207–3222.

Mohan, D. and Pittman, C.U. Jr. (2007). Arsenic removal from water/wastewater using adsorbents – a critical review. *Journal of Hazardous Materials* 142 (1–2): 1–53.

Mohan, D., Dey, S., Dwivedi, S.B., and Shukla, S.P. (2019). Adsorption of arsenic using low cost adsorbents: guava leaf biomass, mango bark and bagasse. *Current Science* 117 (4): 649–661.

Mondal, P., Majumder, C.B., and Mohanty, B. (2008). Treatment of arsenic contaminated water in a batch reactor by using *Ralstonia eutropha* MTCC 2487 and granular activated carbon. *Journal of Hazardous Materials* 153 (1–2): 588–599.

Pandey, P.K., Choubey, S., Verma, Y. et al. (2009). Biosorptive removal of arsenic from drinking water. *Bioresource Technology* 100 (2): 634–637.

Rahman, M.M., Mukherjee, D., Sengupta, M.K. et al. (2002). Effectiveness and reliability of arsenic field testing kits: are the million dollar screening projects effective or not? *Environmental Science & Technology* 36 (24): 5385–5394.

Seki, H., Suzuki, A., and Maruyama, H. (2005). Biosorption of chromium (VI) and arsenic (V) onto methylated yeast biomass. *Journal of Colloid and Interface Science* 281 (2): 261–266.

Shafique, U., Ijaz, A., Salman, M. et al. (2012). Removal of arsenic from water using pine leaves. *Journal of the Taiwan Institute of Chemical Engineers* 43 (2): 256–263.

Sposito, G. (1989). *The Chemistry of Soils*. New York: Oxford University Press.

Suresh, S., Kamsonlian, S., Majumder, C.B., and Chand, S. (2011). Biosorption of As (v) from contaminated water onto tea waste biomass: sorption parameters optimization, equilibrium and thermodynamic studies. *Journal on Future Engineering and Technology* 7: 34–41.

Teixeira, M.C. and Ciminelli, V.S. (2005). Development of a biosorbent for arsenite: structural modeling based on X-ray spectroscopy. *Environmental Science & Technology* 39 (3): 895–900.

Volesky, B. (2001). Detoxification of metal-bearing effluents: biosorption for the next century. *Hydrometallurgy* 59 (2–3): 203–216.

Zouboulis, A.I. and Katsoyiannis, I.A. (2002). Arsenic removal using iron oxide loaded alginate beads. *Industrial & Engineering Chemistry Research* 41 (24): 6149–6155.

Index

Arsenic in Plants: Uptake, Consequences and Remediation Techniques, First Edition.
Edited by Prabhat Kumar Srivastava, Rachana Singh, Parul Parihar, and Sheo Mohan Prasad.

www.ingramcontent.com/pod-product-compliance
Lightning Source LLC
LaVergne TN
LVHW062204200926
848484LV00006B/1313